Push your Career Publish your Thesis

Science should be accessible to everybody. Share the knowledge, the ideas, and the passion about your research. Give your part of the infinite amount of scientific research possibilities a finite frame.

Publish your examination paper, diploma thesis, bachelor thesis, master thesis, dissertation, or habilitation treatises in form of a book.

A finite frame by infinite science.

Infinite Science Publishing

An Imprint of
Infinite Science GmbH
MFC 1 | Technikzentrum Lübeck
BioMedTec Wissenschaftscampus
Maria-Goeppert-Straße 1
23562 Lübeck
book@infinite-science.de
www.infinite-science.de

Student Conference Proceedings 2016

5th Conference on Medical Engineering Science
1st Conference on Medical Informatics

© 2016 Infinite Science Publishing
          University Press and
          Academic Printing

Imprint of Infinite Science GmbH,
MFC 1 | BioMedTec Wissenschaftscampus
Maria-Goeppert-Straße 1
23562 Lübeck, Germany

Cover Design and Illustration: Uli Schmidts, metonym
Editorial: Universität zu Lübeck

Publisher: Infinite Science GmbH, Lübeck, www.infinite-science.de
Printed in Germany, BoD, Norderstedt

ISBN Paperback: 978-3-945954-18-8

Bibliografische Information der Deutschen Nationalbibliothek:
Die Deutsche Nationalbibliothek verzeichnet diese Publikation in der Deutschen Nationalbibliografie; detaillierte bibliografische Daten sind im Internet über http://dnb.d-nb.de abrufbar.

Student Conference Proceedings 2016

5th Conference on Medical Engineering Science
1st Conference on Medical Informatics

Editors in Chief

T. M. Buzug and H. Handels

Associate Editors

C. Debbeler, C. Kaethner, J.-H. Wrage

Editorial Board

Barth, Erhardt; Birngruber, Reginald; Brinkmann, Ralf; Buzug, Thorsten M.;
Fischer, Stefan; Gehring, Hartmut; Handels, Heinz; Heinrich, Mattias;
Huber, Robert; Hübner, Christian; Hüttmann, Gereon; Ingenerf, Josef;
Klein, Stephan; Koch, Martin; Linz, Norbert; Madany Mamlouk, Amir;
Mertins, Alfred; Metzner, Christoph; Müller, Stefan; Paulsen, Hauke;
Rafecas, Magdalena; Rostalski, Philipp; Schmidt, Christian; Schweikard, Achim;
Vogel, Alfred; Wehrig, Stephan

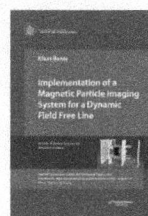

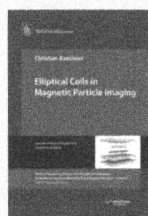

VisiConsult

X-ray Systems & Solutions

Brandenbrooker Weg 2-4 / D-23617 Stockelsdorf / - Germany -
Phone: 0049 (451) 290 286-0 / Fax: 0049 (451) 290 286-22
E-Mail: info@visiconsult.de / Internet: www.visiconsult.de

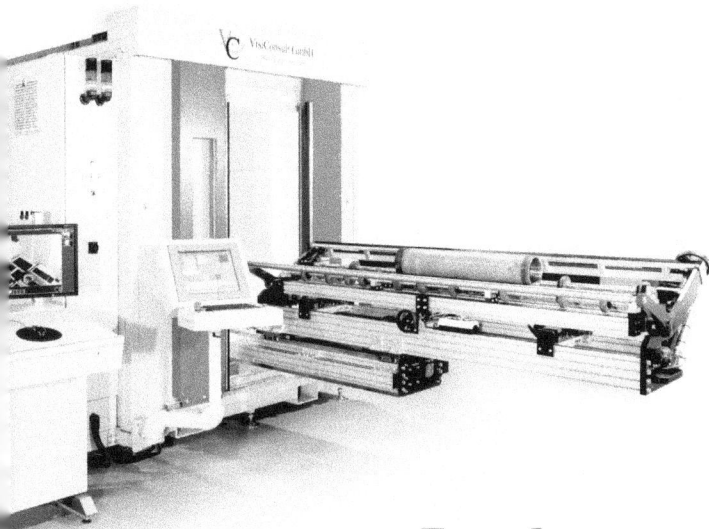

Unsere Werte - Unser Versprechen

Die VisiConsult GmbH ist ein familiengeführtes Unternehmen im Norden Deutschlands. Alle unsere Produkte werden vor Ort entwickelt, produziert und gebrauchsfertig geliefert. In Kombination mit lokalen Organisationen wie die HanseBelt e.V. stärken wir die lokale Industrie. Dies führt zu einer tiefen Verbindung mit geringen Kommunikationswegen gegenüber unseren Lieferanten und gewährleistet eine maximale Qualität. Das ist unser Verständnis des Prädikats "Made in Germany"!

Über uns

VisiConsult ist ein familiengeführtes Unternehmen in Norddeutschland und spezialisiert sich für kundenspezifische und standardisierte Röntgensysteme. Unsere Produkte werden in Premium-Qualität entwickelt und als schlüsselfertige Lösungen produziert. Wir sorgen für eine zerstörungsfreie Prüfung (ZfP) und öffentliche Sicherheit. Unser Ziel ist es, die Probleme unserer Kunden mit maßgeschneiderten Systemen zu lösen und garantieren einen Premium-Post-Sales-Service. Wir freuen uns immer auf neue Herausforderungen und sind stolz darauf ein zuverlässiger Partner mit nachhaltigen Produkten zu sein.

HanseBelt
region ahead

VisiConsult
X-ray Systems & Solutions

Jobs und Karriere

- Software
- Elektrotechnik
- Wirtschaftsingenieure
- Techniker
- Ausbildung zum Elektriker
- und andere technische Richtungen

Wir suchen Mitarbeiter mit Leidenschaft. Sie möchten sich verändern? Sie haben ein klares Ziel vor Augen? Sogar Karriere machen? Wir bieten ihnen die Möglichkeit, sich persönlich und beruflich zu entwickeln. Sie werden Teil eines vielfältigen Teams mit freundlichen Kollegen und anspruchsvollen Projekten. Wir wissen auch, wie wichtig es ist, ein Gleichgewicht zwischen Arbeit und Privatleben zu haben und unterstützen sie dabei. Unser junges und dynamisches Team freut sich auf sie!

Wir fördern Studenten und Young Professionals

YXLON für Young Professionals
- **Traineeprogramm**
- **Direkteinstieg**

Bewerbung an:

Job@hbg.yxlon.com

YXLON International GmbH
Frau Simten Yilmazalp
Essener Bogen 15
22419 Hamburg
Tel.: 040 527 29-225

Sie haben Ihr Studium oder Ihre Ausbildung abgeschlossen?
Sie möchten Ihr Wissen nun in die Praxis umsetzen?
Sie möchten Ihre Leidenschaft für Technologie mit uns teilen?

Dann starten Sie bei uns als Trainee oder mit einem Direkteinstieg!
- Sie übernehmen von Anfang an interessante Aufgaben und erarbeiten Lösungen im Team
- Sie können Verantwortung übernehmen und direkt zum Unternehmenserfolg beisteuern
- Sie arbeiten in einem internationalen Umfeld
- Als YXLON Mitarbeiter/in werden Ihre Fähigkeiten gezielt gefördert

YXLON für Studenten
- **Praktika von 3 bis 6 Monaten**
- **Betreuung von Abschlussarbeiten (Bachelor, Master, Doktorarbeiten)**
- **Werkstudenten**

Sie suchen während Ihres Studiums einen interessanten Praktikumsplatz oder ein spannendes Thema für die Abschlussarbeit?

Dann sind Sie bei YXLON genau richtig!

Wir bieten Ihnen Einblicke in folgende Abteilungen:

• **Entwicklung**	• **Vertrieb**
• **Produktmarketing**	• **Produktion**
• **Supply Chain**	• **Personal**

YXLON
Technology with Passion

VKK® Patentanwälte

PATENTANWÄLTE · EUROPEAN PATENT ATTORNEYS
HAMBURG · KEMPTEN · MÜNCHEN · FRANKFURT

Vkk Patentanwälte sind eine traditionsreiche, international tätige Patentanwaltskanzlei, spezialisiert auf die Erlangung, Verteidigung und Durchsetzung gewerblicher Schutzrechte weltweit mit Kompetenz im nationalen und internationalen Patent- und Gebrauchsmusterrecht sowie im nationalen und internationalen Marken- und Designrecht.

Ansprechpartner Dr. Philipp Knoop, Absolvent des Institutes für Medizintechnik, Universität zu Lübeck 1999, Patentanwalt, European Patent Attorney.

An der Alster 84
20099 Hamburg
Tel.: +49 - (0) 40 - 28 08 13 0
Fax: +49 - (0) 40 - 28 08 13 31
hamburg@vkkpatent.com

Edisonstraße 2
87437 Kempten
Tel.: +49 - (0) 831 - 232 91
Fax: +49 - (0) 831 - 177 15
kempten@vkkpatent.com

Sckellstraße 6
81667 München
 +49 - (0) 89 - 23 08 93 0
muenchen@vkkpatent.com

www.vkkpatent.com

Student Conference Proceedings 2016

Medical Engineering Science

Medical Informatics

Conference Chairs

Thorsten M. Buzug (Chair of Medical Engineering Science Conference), Institute of Medical Engineering, Universität zu Lübeck
Heinz Handels (Chair of Medical Informatics Conference), Institute of Medical Informatics, Universität zu Lübeck
Hartmut Gehring (Co-Chair), Department of Anesthesiology, University Medical Center Schleswig-Holstein (UKSH)
Stephan Klein (Co-Chair), Center for Biomedical Technology, Lübeck University of Applied Sciences

Local Coordination

Christina Debbeler, Institute of Medical Engineering, Universität zu Lübeck
Christian Kaethner, Institute of Medical Engineering, Universität zu Lübeck
Johanna Degen, Institute of Medical Informatics, Universität zu Lübeck
Jan-Hinrich Wrage, Institute of Medical Informatics, Universität zu Lübeck
Gisela Thaler, Institute of Medical Engineering, Universität zu Lübeck
Susanne Petersen, Institute of Medical Informatics, Universität zu Lübeck
Saskia Koch, Medisert, BioMedTec Science Campus

Scientific Program Committee / Editorial Board

Barth, Prof. Dr. Erhardt	Institute of Neuro- und Bioinformatics; Universität zu Lübeck
Birngruber, Prof. Dr. Reginald	Institute of Biomedical Optics; Universität zu Lübeck
Brinkmann, Dr. Ralf	Medizinisches Laserzentrum Lübeck
Buzug, Prof. Dr. Thorsten M.	Institute of Medical Engineering; Universität zu Lübeck
Fischer, Prof. Dr. Stefan	Institute of Telematics; Universität zu Lübeck
Gehring, Prof. Dr. Hartmut	Clinic for Anesthesiology; University Medical Center Schleswig-Holstein (UKSH)
Handels, Prof. Dr. Heinz	Institute of Medical Informatics; Universität zu Lübeck
Heinrich, Prof. Dr. Mattias	Institute of Medical Informatics; Universität zu Lübeck
Huber, Prof. Dr. Robert	Institute of Biomedical Optics; Universität zu Lübeck
Hübner, Prof. Dr. Christian	Institute of Physics; Universität zu Lübeck
Hüttmann, PD Dr. Gereon	Institute of Biomedical Optics; Universität zu Lübeck
Ingenerf, PD Dr. Josef	Institute of Medical Informatics; Universität zu Lübeck
Klein, Prof. Dr. Stephan	Lab for Medical Sensor and Device Technology, Lübeck University of Applied Sciences
Koch, Prof. Dr. Martin	Institute of Medical Engineering; Universität zu Lübeck
Linz, Dr. Norbert	Institute of Biomedical Optics; Universität zu Lübeck
Madany Mamlouk, PD Dr. Amir	Institute of Neuro- und Bioinformatics; Universität zu Lübeck
Mertins, Prof. Dr. Alfred	Institute of Signal Processing; Universität zu Lübeck
Metzner, Dr. Christoph	Institute of Robotics and Cognitive Systems; Universität zu Lübeck
Müller, Prof. Dr. Stefan	Lab for Medical Sensor and Device Technology, Lübeck University of Applied Science
Paulsen, PD Dr. Hauke	Institute of Physics; Universität zu Lübeck
Rafecas, Prof. Dr. Magdalena	Institute of Medical Engineering; Universität zu Lübeck
Rostalski, Prof. Dr. Philipp	Institute of Medical Electronics; Universität zu Lübeck
Schmidt, Prof. Dr. Christian	Isotope Laboratory; Universität zu Lübeck
Schweikard, Prof. Dr. Achim	Institute of Robotics and Cognitive Systems; Universität zu Lübeck
Vogel, Prof. Dr. Alfred	Institute of Biomedical Optics; Universität zu Lübeck
Wehrig, Prof. Dr. Stephan	Department of Construction Engineering, Lübeck University of Applied Sciences

Preface and Acknowledgements

After the great success of the previous meetings from 2012 to 2015, the Student Conference 2016 shows continuing growth both in quality and quantity of scientific contributions. In this year, the 5th Student Conference on Medical Engineering Science is hold together with the 1st Student Conference on Medical Informatics for the very first time.

The organization team of the BioMedTec Science Campus Lübeck, the Institutes of Medical Engineering and Medical Informatics in cooperation with Life Science North Management GmbH, the North German Life Science Cluster Agency, have spared no effort to provide an excellent conference, where master students of the campus present their recent research results to a broad public of academics and industry.

The contributions show how new approaches and methods in medical engineering and medical informatics can advance medicine, health and health care. Moreover, this conference offers a good opportunity for both students and companies to get in touch at the Recruiting Fair, a satellite meeting with industrial exhibition, and to get to know each other from a different point of view.

Students from the Life Sciences programs at the BioMedTec Science Campus present their results from projects carried out at the laboratories and institutes of Lübeck's Universities, in international research facilities or research-oriented industrial companies. The conference focus has been placed on topics from medical engineering and medical informatics. The interdisciplinary field of medical engineering has also been established at the Lübeck University of Applied Sciences for decades and Medical Engineering Science (Medizinische Ingenieurwissenschaft – MIW) is an important bachelor and master program at the Universität zu Lübeck as well. Both universities jointly offer the international master degree course Biomedical Engineering (BME). Furthermore, in the young master program Medical Informatics (Medizinische Informatik – MI) the 1st Student Conference on Medical Informatics is integrated as an important element where project results in the emerging field of digital medicine are presented by the students.

This is complemented with further life science oriented programs of the University (Computer Sciences, Mathematics in Medicine and Life Sciences, Molecular Life Science, Medicine etc.) which contribute to the success of the Medical Engineering Science and Biomedical Engineering and Informatics programs. The Student Conference on Medical Engineering Science and Medical Informatics 2016 has been complemented by concepts and designs of students of architecture of Lübeck University of Applied Sciences, which perfectly illustrates the dynamic development of the BioMedTec Science Campus – towards a broader variety of topics and an increasing identification with the campus itself.

As Conference Chairs we thank all the people who worked with enthusiasm and dedication to make the conference a successful event. We have to thank the companies and organisations who support the meeting by their contributions to the Recruiting Fair: Coherent GmbH Lübeck; Drägerwerk AG & Co. KGaA, Lübeck; Infinite Science GmbH, Lübeck; OLYMPUS Surgical Technologies Europe, Hamburg; Senspec GmbH, Rostock; VisiConsult X-ray Systems & Solutions GmbH, Lübeck; VKK Patentanwälte, Hamburg; MÖLLER-WEDEL GmbH & Co. KG, Wedel; Stryker GmbH & Co. KG, Kiel; YXLON International GmbH, Hamburg; Chamber of Commerce and Industry (IHK), Lübeck; Life Science Nord Management GmbH, Hamburg. Moreover, our thanks go to the BioMedTec Management GmbH. The professional management of Kanina Botterweck and her team has contributed substantially to the success of this conference.

Personally and on behalf of all colleagues of the BioMedTec Science Campus, we especially want to thank Christian Kaethner, Christina Debbeler from the Institute of Medical Engineering and Jan-Hinrich Wrage from the Institute of Medical Informatics. They have been the central contact points for all questions of students and the program committee members. Their in-depth overview of all details of this event is the key to the success of the Student Conference 2016 at the BioMedTec Science Campus.

Lübeck, March 9–11, 2016

Prof. Dr. Thorsten M. Buzug
Vice President of the Universität zu Lübeck
Chair of the 5th Student Conference on
Medical Engineering Science 2016

Prof. Dr. Heinz Handels
Chair of the Examination Board Medical Informatics
Chair of the 1st Student Conference on
Medical Informatics 2016

Sponsors of the Recruiting Fair

SENSPEC
SPECTRAL MONITORING SYSTEMS

Dräger

stryker

YXLON

OLYMPUS

COHERENT

V≪
VKK® Patentanwälte

VisiConsult
X-ray Systems & Solutions

HS MÖLLER-WEDEL INTERNATIONAL

Infinite Science

IHK Lübeck

LIFE SCIENCE NORD

Contents

Biomedical Optics

Biochemical Physics I

Biochemical Physics II

Biomedical Engineering

Safety and Quality

E-Health

Signal Processing

Medical Imaging

Image Processing I

Image Processing II

Image Processing III

1

Biomedical Optics

A graphical user interface-based automatized analysis routine for the intracellular luminescence nanothermometry

D. Weng [1], G. Hüttmann [2], M. Suzuki [3], and Y. Miura [2]

[1] Medizinische Ingenieurwissenschaft, Universität zu Lübeck, daniel.weng@student.uni-luebeck.de
[2] Institute of Biomedical Optics, Universität zu Lübeck, {miura,huettmann}@bmo.uni-luebeck.de
[3] Waseda Bioscience Research Institute in Singapore (WABIOS), Waseda University, suzu_mado@aoni.waseda.jp

Abstract

Measurement of the local temperature inside a single cell is of a great interest, and recently introduced luminescent ratiometric nanothermometer (RNT) may serve as a novel tool to realize it. One of our interests is temperature-resolved cellular responses during and after sublethal laser-induced temperature rise in retinal pigment epithelial (RPE) cells. Aim of the study was to implement an automatized analysis routine for the intracellular temperature measurement of cultured RPE cells with internalized RNT by measuring RNT's containing fluorophores (Eu-TTA and rhodamine 101) luminescence intensity ratio. Automatized analysis was performed using an own implemented graphical user interface-based ImageJ-Plugin, which includes the correction for photo bleaching of the RNT's luminescence, and tracking of moving RNT. Within a runtime of a few seconds the plugin allows fully automatized analysis of temperature measurement using RNT. First measurements showed promising results. Results of the ratio values may be better, if improved in signal-to-noise ratio.

1   Introduction

The retinal pigment epithelium (RPE) plays a crucial role in retinal function, thus its dysfunction is largely related to the pathogenesis of many retinal diseases. Therefore, seeking the way to the functional regeneration of RPE cells is of great interest among researchers and physicians. A previous study has shown that laser-induced sublethal hyperthermia of RPE cells may have a protective effect from oxidative stress [1], and thus might be a useful new treatment strategy for disease prophylaxis. By means of an optoacoustic technique it has become possible to measure and control the temperature increase during laser irradiation at the irradiated site [2]. In order to utilize this newly developed technology in the most beneficial way, it is necessary to elucidate the temperature-resolved RPE cell responses in more detail. The most detailed information on subcellular responses associated with temperature might be provided by the intracellular temperature measurement. A nanoparticle-based approach for the intracellular temperature measurement has been made with ratiometric nanothermometer (RNT), which was developed by a research group of M. Suzuki (WABIOS, Waseda University, Singapore). RNT consists of a hydrophilic, positively charged shell (hydrodynamic diameter: 143 ± 36 nm) and a hydrophobic core (diameter: 99.4 ± 30.7 nm). Two kinds of fluorophores rhodamine 101 (λ_{ex}=560 nm, λ_{em}=587 nm; fluorescence) and europium(III)-thenoyltrifluoroacetonate (Eu-TTA) (λ_{ex}=341 nm, λ_{em}=614 nm; phosphorescence) are embedded in the hydrophobic core of Poly(methyl methacrylate) (PMMA). The luminescence intensity of Eu-TTA is temperature-sensitive, while the fluorescence intensity of rhodamine 101 is temperature-insensitive. These luminescence properties allow the determination of the temperature by calculating the intensity ratio of Eu-TTA and rhodamine 101 luminescences [3]. Our own previous research has verified the applicability of RNT to cultured RPE cells [4]. Before conducting the temperature measurement of cultured RPE cells during laser irradiation, two matters have to be first addressed that may influence the measurement of RNT: 1) Photo-bleaching of Eu-TTA through the light exposure for measurement, 2) movement of the single RNT spots, especially during laser irradiation. Therefore, in this study experiments were conducted for the automatized analysis routines toward the optimal implementation of RNT-based intracellular temperature measurement with cell culture.

2   Material and Methods

In the following the RPE cell culture and RNT, which are used for fluorescence ratio imaging as well as the setup and the programmed analysis algorithms are explained in detail.

2.1   Cell Culture

All experiments were performed with primary porcine RPE cells prepared as previously described in [1]. The cultures of third generation were used for experiments.

2.2 RNT

The RNT were prepared as previously described in [3]. The temperature-dependent change in luminescence intensity was examined by measuring emission spectra of RNT suspension (RNT suspended in water), by measuring the spectrum of rhodamine 101 and Eu-TTA at $\lambda_{ex} = 560$ nm and 341 nm, respectively, with a fluorescence spectrophotometer (Hitachi F-2700, Tokyo, Japan). The RNT suspension was heated up to $50°$C, followed by the cooling to approximately $35°$C. The spectra were measured in the cooling time period ($50°$C $\rightarrow$ $35°$C). Directly afterwards, the RNT suspension was heated again to $50°$C, and the spectral measurement was repeated a second time again cooling.

2.3 Uptake of the RNT in RPE Cells

In order to verify the intracellular uptake of the RNT, fluorescence microscopy was conducted as following: cultured RPE cells were incubated with RNT-containing medium for 30 minutes letting the RNT intracellularly internalize. This was followed by replacing the RNT-containing medium with one without RNT, and live cell staining for mitochondria (MitoTracker Deep Red FM, invitrogen) was performed. Luminescence of Eu-TTA in the RNT and MitoTracker was detected at $\lambda_{ex/em}$=341/614 nm and 644/665 nm, respectively, with a fluorescence microscope (Olympus IX83, Tokyo, Japan).

2.4 Setup for Laser Irradiation and Luminescence Measurement of RNT

Fluorescence imaging of the RNT in cultured RPE cells during laser irradiation has been performed at a fluorescence microscope. A schematic representation of the setup is shown in Fig. 1. The setup consists of a fluorescence microscope (Zeiss Axiovert 200, Jena, Germany), coupled with a polychromatic illumination system (Visichrome, Visitron Systems GmbH, Puchheim, Germany). The image acquisition and processing were performed with the software VisiView®(Visitron System GmbH). The tip of the laser fiber of Thulium laser (Starmedtec, Starnberg, Germany) is fixed 12 cm above the cell culture dish, and the irradiation timing is controlled with a LabVIEW®(National Instruments) programm. The irradiation protocol is composed of 5 seconds waiting time (at room temperature), 10 seconds heating by irradiation (8 W) and 15 seconds cooling. While executing the irradiation protocol, a sequence of fluorescence microscopic images of rhodamine 101 and Eu-TTA were acquired. Images were acquired for rhodamine 101 and Eu-TTA at every 1004 ms, with an acquisition time of 500 ms for each fluorophore and a 2 ms change over time of the shutter for each excitation wavelength.

2.5 Analysis

Since the luminescence intensity of Eu-TTA decays through light exposure, the influence of bleaching on the luminescence intensity has to be corrected by the analysis routine.

Therefore, a reference sequence of 15 fluorescence microscopic images (totally 7,500 ms exposure time) of Eu-TTA was acquired before each experiment. This was done at constant room temperature. The image analysis was conducted with the software ImageJ (1.50c, Wayne Rasband, National Institute of Health, USA). For automatized and more intuitive analysis, the basic functions of ImageJ were supplemented by an self-implemented Plugin.

2.5.1 Influence of the age of RNT on Photo Bleaching of Eu-TTA

The influence of the age of the RNT preparations on photo bleaching of Eu-TTA was investigated. The reference fluorescence sequence (15 consecutive exposure with 500 ms exposure time) of the RNTs in 12, 22, 25 and 41 days after preparation was measured at $\lambda_{ex/em}$= 341 nm/614 nm, and correction factors for the 4 different ages were determined.

2.5.2 Programming of Plugins for RNT Luminescence Analysis

First a Plugin was programmed, which allows to determine a bleaching correction factor from the decreasing luminescence intensity. This factor was then used for all following analysis steps. For calculating the correction factor, the data from the reference sequence measurement were used. Assumingly the luminescence intensity should not change at constant temperature. Thus, the linear regression analysis was conducted between the frame number and the luminescence intensity in the reference sequence measurement, and the correction factor k_c was determined by calculating the

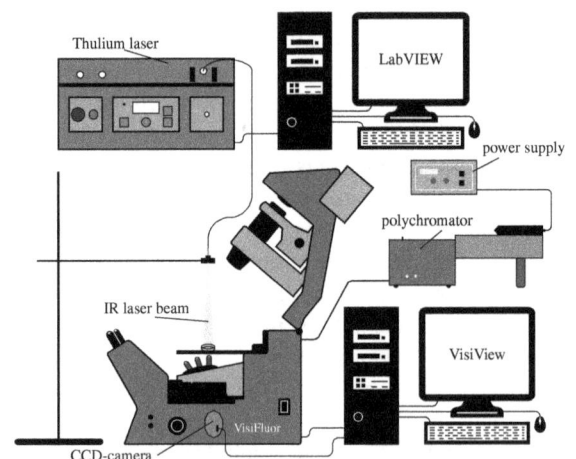

Figure 1: Setup for fluorescence imaging of RNT. Fluorescence microscopy was performed with a fluorescence microsope coupled with a polychromator as light source. The acquired image was processed with VisiView®. A temperature increase was obtained via IR laser irradiation from above the cell culture dish. The source of IR laser beam was a 1.94 μm Thulium laser system. The laser beam was coupled into a fiber, which ends 12 cm above the cell culture dish, such that the beam diameter at the height of cell culture dish is of the same size as the cell culture dish diameter.

slope of the regression line by

$$k_c = \frac{\sum_{i=1}^{n=stack\ size}(x_i - \bar{x}) \cdot (y_i - \bar{y})}{\sum_{i=1}^{n=stack\ size}(x_i - \bar{x})^2} \qquad (1)$$

with frame number x_i, average frame number $\bar{x}$, luminescence intensity y_i and average luminescence intensity $\bar{y}$. The corrected luminescence intensity $I_c(n)$ is given by:

$$I_c(n) = I_m(n) + k_c \cdot n \ , \quad n \in [0, \text{stack size} - 1] \qquad (2)$$

with the measured intensity $I_m(n)$, the correction factor k_c and the frame number n.

Moreover, one can observe a movement of intracellular RNT while image acquisition. This is due to the natural movement of the intracellular organelles to which the RNT are bound, and might also be due to the slight expansion of water, cell, or material of culture dish bottom during heating. That is why a simple automatized tracking algorithm was programmed to the analysis routine. For each frame of the acquired image stack, starting at the last known position, the implemented tracking algorithm finds the new position of moving RNT in the next frame as follows: assuming the maximum intensity originates from the center of the RNT, the algorithm is searching for moving RNT within a user defined field of view (FOV) by calculating the maximum intensity. The image pixel with the maximum intensity is set to the new center of moved RNT. Once the RNTs of interest are selected by a point region of interest (ROI) and thresholds for rhodamine 101 and Eu- TTA are set (to eliminate background), the analysis routine calculates the ratio values of each single selected RNT of interest for the complete image stack as shown in Fig. 2.

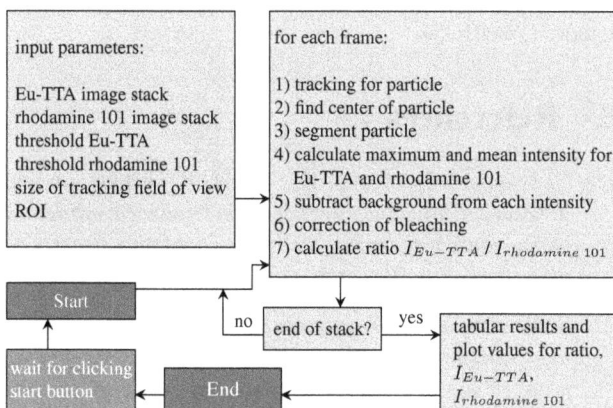

Figure 2: Scheme of programmed analysis routine.

3 Results and Discussion

Fig. 3 shows the temperature dependent luminescence intensity of Eu-TTA (emission spectrum at $\lambda_{ex} = 341$ nm) and rhodamine 101 (emission spectrum at $\lambda_{ex} = 560$ nm). Luminescence of Eu-TTA showed the maximum at 614 nm, which was increased with a temperature decrease. After

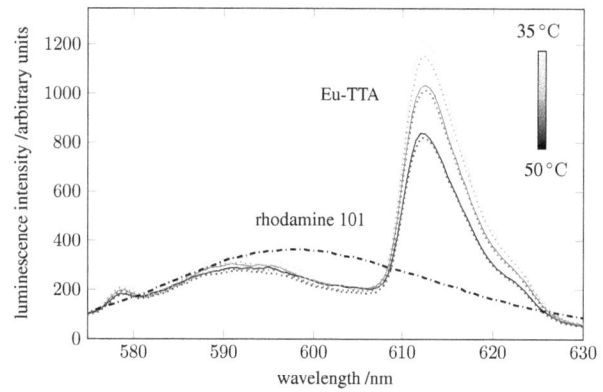

Figure 3: Temperature-dependent emission spectra of rhodamine 101 (dash-dotted, $\lambda_{ex} = 560$ nm) and Eu-TTA ($\lambda_{ex} = 341$ nm) in RNT. Spectra were measured while cooling two times in total: round 1 after heating up to 50 °C (solid line) and round 2 after re-heating (dotted).

re-heating (the second measurement) the luminescence intensity of Eu-TTA reached almost its initial value, but at the lowest temperature (35°C) the intensity was slightly lower than the first measurement. Different from Eu-TTA, the fluorescence of rhodamine 101 did not show a temperature-dependent intensity change. Intracellular uptake as well as localization of RNT in RPE cells was observed with fluorescence live cell imaging after minimum incubation time of 30 minutes with RNT. Exemplary results are shown in Fig. 4. After 30 minutes of incubation with RNT, several RNT particles were observed inside cells, many of which are located near the nucleus.

To assess the influence of particle age on photo-bleaching of RNT luminescence, the correction factors for 12, 22, 25 and 41 day-old RNT were determined. Their averaged values are plotted in Fig. 5. The average bleaching correction factor for 12 days was 0.18±0.04, for 22 days 0.26±0.09, for 25 days 0.31±0.15, and for 41 days 0.44±0.20. These results imply that the age of the RNT is linearly correlated with the increase rate of photo bleaching, and moreover, wider standard deviation in the old RNTs suggests an increase of nonuniform characteristics in the older RNTs. Finally, a RNT luminescence measurement during laser irra-

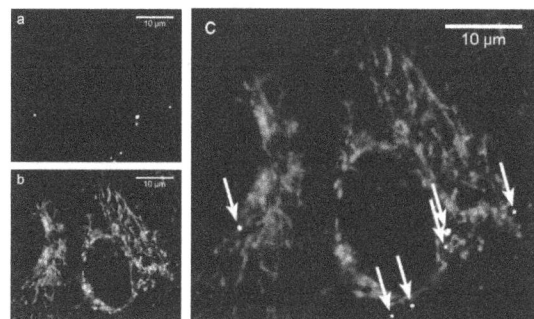

Figure 4: Intracellular uptake and localization of RNT in RPE cells. (a) RNT, (b) mitochondria and (c) merge of RNT (arrows) and mitochondria.

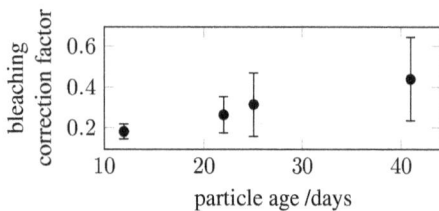

Figure 5: Influence of RNT age on the bleaching of Eu-TTA. Correction factors calculated from 15 consecutive reference measurement are correlated to the particle age.

diation was conducted using the programmed analysis routine. The exemplary measures are shown in Fig. 6. Results are calculated from two different RNT spots in the same cell culture dish. Laser irradiation was switched on at 5 s, and switched off at 15 s. The irradiation power was 8 W, and the temperature increase at the observed region was about 12°C (22°C → 34°C). Both the luminescence intensity of Eu-TTA and the ratio decreases during laser irradiation (=heating) and start to increase at the end of irradiation. The calculated ratio values ($\mathrm{Int}_{341\,nm}$ / $\mathrm{Int}_{560\,nm}$), corrected and uncorrected by bleaching, are plotted in Fig. 6. The ratio curves show a decrease until the end of the irradiation, followed by an increase after irradiation. One can observe, that the corrected values as well as the uncorrected ones dosn't return back to the initial level. However, this is expected because thermal relaxation takes longer than the observation time. Uncorrected ratio values are generally much lower than the corrected ones. The corrected ratio values of RNT spots 1 and 2 at the time point of 15 s were 10.73 and 5.82, respectively. Since temperature-ratio correspondence in this imaging setup has not been accomplished yet, we may so far interpret ratio value changes as a qualitative indicator of temperature alterations. Compared to the result on Eu-TTA intensity, the signal-to-noise ratio (SNR) of the ratio is very low, assumingly due to the very weak fluorescence intensity of rhodamine 101 at fluorescence microscopic images (not shown). Improving the SNR of rhodamine 101 may improve the SNR of resulting ratio values

and allows for quantitative measurements.

The implemented tracking algorithm resulted in reliable tracking of moving RNT. However, it may be disturbed if a particle of higher luminescence intensity than the particle of interest is moving into the same tracking FOV. Particles, which are affected by this disturbance, have to be rejected from the analysis. To improve the tracking algorithm, detection of movement using approaches of computer vision [5], like optical flow and block matching, will be implemented to the algorithm in the near future. This concepts, in addition to the RNT's center image pixel, includes all RNT's image pixel and the nearest neighborhood. Thus, the algorithm may be less affected by disturbances.

4 Conclusion

The presented results show that our first programmed automatized analysis routine for luminescence imaging of RNT may provide viable data. The analysis runtime takes only a few seconds. Thus, the analysis was greatly improved in time compared to manually analysis (with average time spent of minutes to hours). The implemented tracking algorithm as well as the correction of bleaching work well. However, in the near future the tracking algorithm needs to be improved by implementation of approaches of computer vision, for catching exceptions, like two RNTs are moving into the same FOV.

Acknowledgement

The work has been carried out at Institute of Biomedical Optics, Universität zu Lübeck. Authors thank the helpful cooperation Waseda Bioscience Research Institute in Singapore (WABIOS).

5 References

[1] H. Iwami, J. Pruessner, K. Shiraki, R. Brinkmann and Y. Miura, *Protective effect of a laser-induced sub-lethal temperature rise on RPE cells from oxidative stress.* Experimental Eye Research, vol. 128, pp. 37–47, 2014.

[2] S. Koinzer et al., *Temperature-controlled retinal photocoagulation reliably generates uniform subvisible, mild, or moderate lesions.* translational vision science & technology, vol. 4, no. 5:9, 2015.

[3] Y. Takei et al., *A nanoparticle-based ratiometric and self-calibrated fluorescent thermometer for single living cells.* ACS Nano, vol. 8, no. 1, pp. 198–206, 2014.

[4] D. Weng, *In vitro investigation for nanothermometry on retinal pigment epithelial cells.* Bachelor Thesis, Luebeck, Universität zu Lübeck, Institute of Biomedical Optics, 2014.

[5] R. Szeliski, *Computer Vision. Algorithms and Applications.* London: Springer London, 2011.

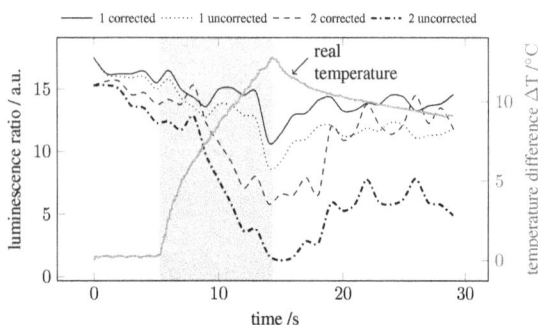

Figure 6: Exemplary results for luminescence ratio imaging of RNT and the correction for photo bleaching. The ratio values as well as the corresponding real temperature difference are plotted against time. Gray shading indicates the duration of the laser irradiation.

Multi-pump diode arrangement of a diode-pumped solid state laser

T. Schmidtke [1], R. von Elm [2], and C.L. Schmidt [3]

[1] Medizinische Ingenieurwissenschaft , Universität zu Lübeck, thomas.schmidtke@student.uni-luebeck.de
[2] Coherent LaserSystems GmbH & Co KG Lübeck, Ruediger.vonElm@coherent.com
[3] Isotopenlabor der Sektion Naturwissenschaft, Universität zu Lübeck, christian.schmidt@isolab.uni-luebeck.de

1 Abstract

In this study a solid state laser based on Pr^{3+}:$LiYF_4$ as the laser-active crystal is used. In medical technology the emitted light finds application in fluorescence spectroscopy, as well as in the disinfection of the devices. The focus is on constructing a pump mechanism with an arrangement of multiple laser diodes in continuous wave-mode (cw-mode). Two pumping units, each consisting of five laser diodes should be realised for a two-way pumping of the laser crystal. Thereby the geometry should be as compact and as stable as possible to achieve a high beam quality and a close focus in the laser crystal. Especially the emission of the crystal at 522nm is interesting. Deep ultraviolet light could be created by frequency doubling in cw-mode. A total pumping capacity in the multiwatt range is achieved. Through a simulation of the beam guidance and corresponding literature, the arrangement could be implemented successfully.

2 Introduction

Solid state lasers based on Pr^{3+}:$LiYF_4$ as active laser crystals are of great importance, due to the emission from Pr^{3+} at various lines in the visible spectrum from red (720nm) to green (522nm) [1]. Especially the emission at 522nm is interesting. Deep ultraviolet light at 261nm can be generated by internal resonator frequency doubling in cw-mode. This wavelength is of great importance in many sectors of the industry, for example, for producing and examing ever smaller microchips. In medical technology, the UV-light is used for disinfection of devices [1]. For this application, a stable laser-intensity is required to ensure a safe disinfection without damaging any equipment. Furthermore the UV-light is used in biology for spectroscopy, to stimulate fluorescent dye-marked cells [2]. The present method for generating a wavelength close to 261 nm uses a ND:YAG laser. Twofold doubling is required to generate a wavelength of 266nm. As a result, the production costs are significantly higher, and also the compactness of the laser is restricted. The current process also works with lower efficiency. In addition the ND:YAG laser is in most cases a pulse laser, which is not suitable for many applications. In contrast the Pr^{3+}:$LiYF_4$ laser can operate in cw–mode and thereby creates a steady intensity [3]. This paper presents a new approach to reach a wavelength of 261nm. It is also aspired to massively increase the input power to generate a high output power, despite high losses by the setup and various other optical components. Therefore an arrangement of multiple GaN-Laser diodes (444nm) is constructed to analyze the pumping in cw-mode. The upgrading to the multiwatt range, as well as the beam shaping are primary

objectives. Possible pump wavelengths of the crystal are 444 nm, 468 nm, and 479 nm (see Fig. 1) [1]. Two pumping units were used for the two-sided pumping of the laser crystal. First, one pump unit was implemented with 5 laser diodes; the second pump unit already exists [5]. A compact and stable geometry has to be found, which allows a high beam quality and also a close focus in the laser crystal with high focus depth (Rayleigh length). Another aspect is the manipulation of the emission wavelength of the diodes. Different approaches for the trimming of the diode laser wavelength have been tested to reach the absorption band from Pr^{3+}:$LiYF_4$ at 444nm.

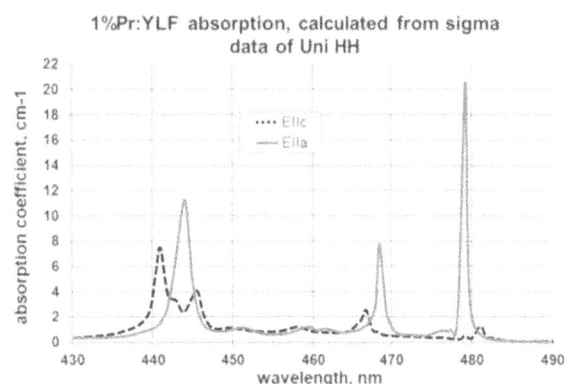

Figure 1: Absorption spectrum of the Pr^{3+}:$LiYF_4$ crystal while pumping with parallel polarised radiation (E∥a) and with vertical polarised radiation (E∥c) to the c-axis of the crystal [1].

3 Material and Methods

In the following the main components of the setup and criteria for their selection will be described. Especially the diodes and their beam guidance and -shaping will be discussed.

3.1 Diodes

A GaN-based laser diode consists of a semiconductor material, based on a p-n junction (see Fig. 2). Because of composition variation and physical processes, it is very difficult to produce diodes with a precise emmission wavelength. Instead, the manufactured diodes have to be selected for the desired wavelength, resulting in an increase of the price. The wavelength can be controlled slightly by changing the temperature, but also by the current [2].

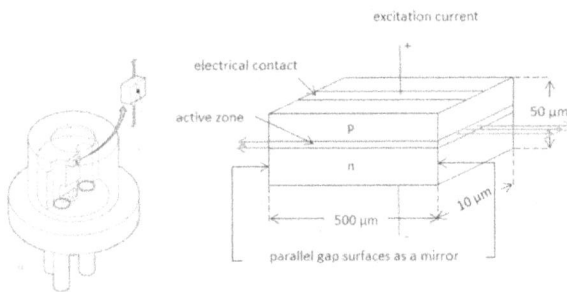

Figure 2: Setup of a complete laser diode (left). In the right picture, the emitter of the diode is shown enlarged. Because of its smallness it is difficult to produce a diode with an exact emission wavelength. Variations of the properties of the emitter also cause problems in beam shaping [4].

3.1.1 Power losses

Actually an output of 1 Watt per diode with a current of 1,2 A is desired, but this is very difficult to achieve. The output-measurement of each diode showed different variations. The output-losses are caused by the following factors.

- construction form
- optical components such as prisms or lenses
- positioning of laser diodes at the device
- heat development

Because individual laser diodes do not have the required power for pumping the laser medium, the performance has to be increased by a multi pump diode arrangement. The increase of the pump power opens the possibility for new laser applications.

3.2 Beam shaping

The laser diodes have a strongly diverging beam. The fast axis (high divergence) and the slow axis (low divergence) show a different divergence (see Fig. 3). These differences are caused by the geometrical shape of the p-n transition. The fast axis is 0.15 μm up to 1 μm long, in contrast to the slow axis that is 6 μm up to 100 μm long. The divergence is greater the smaller the opening angle is [4].

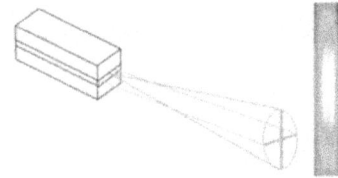

Figure 3: Emitter of a laser diode and the resulting beam propagation. The beam divergence can be divided in fast axis and slow axis. Without aspherical lenses the divergence in fast axis is many times higher than in slow axis [4].

3.2.1 Simulation of beam guidance and beam shaping

Because the divergence in the slow axis and the fast axis is not equal, they must be adjusted by means of a beam shaping optics. The beam shaping is simulated with a program named "Optic Tool", designed by Coherent. Fig. 4 shows a strongly diverging beam in fast axis and the slow diverging beam in slow axis". The goal is to focus diode laser light into the Pr:LiYF crystal over the total length (5.5 mm) with a beam diameter under 200 μm to achieve the best possible results for the pumping of the crystal. At first, the beam is collimated with an aspherical lens with a focal length f = 2.54 (see Fig.4). The focusing capability in the slow axis is much worse than in the fast axis, so the beam must first be expanded with a plano-concave cylindrical lens with a focal length of f = -75 mm. As a result, stronger focusing with a high power density can be achieved. Thereafter, the slow axis is focused with a plano-convex cylindrical lens having the focal length f = 75 mm. The fast axis is focused with a plano-convex cylindrical lens with the focal length f = 150 mm. It has to be noted that the beam radius should be smaller than half the diameter of the lens to avoid aberrations. With this simulation a general assesment can be made, which experimental setup may work. So the distance between the lenses can be determined not only experimental but also virtually. The program represents the beam path under perfect conditions. In Fig. 4 the focus ability of the crystal over the entire length is shown. In this case, the pumping beam has less than 200 μm diameter in the slow axis (about 160 μm) as well as in fast axis (about 120 μm).

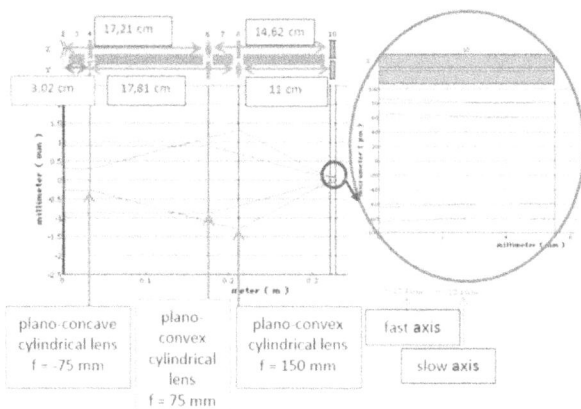

Figure 4: Simulation in "Optic Tool". Guidance and shaping of a laser diode beam in fast and slow axis with suitable lenses. In slow axis: After 3,02 cm, expansion of the beam with a plano-concave cylindrical lens. After a distance of 17,81 cm, focusing with a plano-convex lens. Then after 11 cm the beam is focused. In fast axis: After 17,31 cm the beam is expanded with a plano-concave cylindrical lens. Then after 14,62 cm the beam is focused on the crystal. In both cases, a beam radius less than 200 μm is achieved.

3.3 Setup

The entire pumping unit consists of aluminium components, because they show a good thermal conductivity and are easy to handle. Free mobility for the fine tuning of the laser diodes, lenses and prisms is elementary. A good beam shaping is only possible under these conditions. Because the laser diodes, as well as the lenses are very small and delicate, the smallest deviations and impurities lead to unwanted optical effects. This has an adverse effect on beam guidance and –quality. Each unit has its own bracket, whereby all brackets are interconnected. For the heat transmission of laser diodes, an Indium film is additionally required, which is attached to the rear side of the diode. The film conforms to the smallest irregularities and ensures that the heat is transmitted more efficiently. During extended operation, it is important to install additionally an air ventilation system at the rear side of the experimental setup.

For the selection of the diodes, a control unit with a series connection was installed. Every single diode has a toggle switch, so the diodes can be switched individually.

3.3.1 Pump arrangement

The prisms are arranged at the Brewster angle at which the beam can be transmitted as lossless as possible. The refractive index of the material depends on the wavelength. The Sellmeier equation relates the change of the refractive index of the medium to changes of the wavelength (1). For the calculation of Sellmeier equation the constants of the dispersion formula of the material is used. These constants can be taken from the data sheet or from the literature. The prisms are manufactured from S11-glass with an anti-reflective coating.

The refractive index of the Prism is

$$n = \sqrt{1 + \frac{A_1 \cdot \lambda^2}{\lambda^2 - B_1} + \frac{A_2 \cdot \lambda^2}{\lambda^2 - B_2} + \frac{A_3 \cdot \lambda^2}{\lambda^2 - B_3}} \quad (1)$$

$$= 1,821552105 \quad (2)$$

The Brewster angle is $61,23°$ and can be calculated by

$$\vartheta_{BR} = \arctan(\frac{\eta t}{\eta i}) = \arctan(\frac{1,8215}{1}) \quad (3)$$

$$= 61,23° \quad (4)$$

4 Results and Discussion

Fig. 5 shows a photograph of the pumping arrangement. The pump-lasers are arranged semicircular, which has proven to be most suitable for space-saving, as well as for the beam guidance (Fig. 6). In addition, four prisms are attached, that lead the beams in the desired direction to achieve a collimated beam.

Figure 5: pumping arrangement

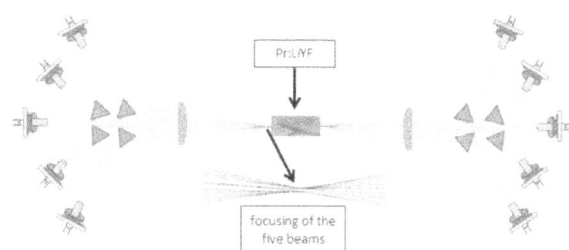

Figure 6: Semicircular arrangement of the pump-diodes and the last focusing step of the beam. The crystal is pumped from both sides.

Fig. 7 shows two gaussian profiles of the beam of a laser diode in slow and fast axis. To stimulate the crystal optimally, the collimated beam must have a diameter of less than 200 μm. With the optical design developed by simulations of the beam guidance in slow and fast axis a focus of less than 200μm could be achieved. The beam diameter is 155 μm (slow axis) and 139 μm (fast axis). This experiment was first performed with just one laser diode.

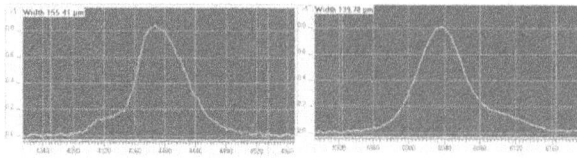

Figure 7: Two approximately gaussian profiles of a diode laser beam. The purpose was to achieve a beam radius of less than 200 μm. For the fast axis (right picture) a beam radius of 155 μm has been achieved, for the slow axis (left picture) 139 μm.

In Fig. 8 the setup is shown during laser operation. All laser diodes are positioned in a semicircular arrangement. A collimation of all beams was achieved through the beam guidance with prisms. The purpose of finding a space-saving geometry was also achieved. Finally, the collimated beam had the correct wavelength of 444nm to stimulate the Pr^{3+}:$LiYF_4$ crystal optimally. This way, a pump performance of approximately 4 watts has been achieved.

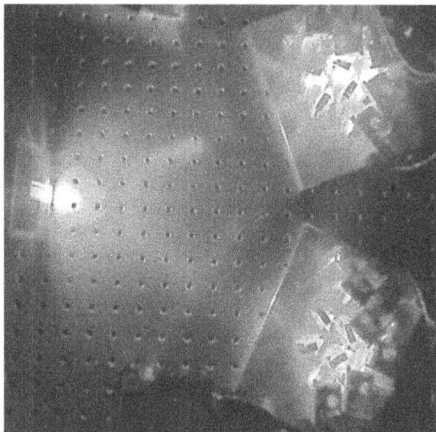

Figure 8: pump arrangement during operation

One of the biggest challenges was the precise attachment of the aspheric lens immediately in front of the laser diode. To achieve a perfect focus without losses, the following conditions have to be fulfilled:

- the distance between lens and diode has to be about 0,75 mm

- the emitter must be located on the optical axis of the aspherical lens

- the position of the prisms must be perfectly adjusted

Because of the size of the lens (ca. 4 mm) and the size of the emitter (ca. 10 μm) it is very difficult to fulfill these conditions. Also the brackets for the prisms need to be improved, because so far there is no possibility of movement in z-direction. Another point is the condition of the prisms. On the one hand it has to be supposed that the surface on

which the prisms lie is planar to 100%. On the other hand the bracket of the prisms causes a slight angular change, which is enough to broaden the beam. These tasks have to be solved in the future. Now that the pumping unit has been succesfully configured, the next step is frequency doubling of the Pr^{3+}:$LiYF_4$ laser emission to achieve an emission wavelength of 261 nm. Overall the experiment was successfully completed. A very important result is that it is possible to increase the pump power with an arrangement of multiple diodes.

5   Conclusion

A compact pumping unit with ten laser diodes could be sucessfully realised. A semicircular arrangement shows the most suitable conditions. The beams of all diodes were combined to one collimated beam with high power. The beam has the necessary wavelength to stimulate the Pr^{3+}:$LiYF_4$ crystal optimally. The experiment shows that an improved performance is possible in this way. In some areas, potential for optimization exists.

Acknowledgement

The work has been carried out at Coherent LaserSystems GmbH & Co KG, Lübeck. I thank the company Coherent for the possibility to complete my internship and to collaborate on interesting projects. Furthermore I thank my internal supervisor Rüdiger for all the time and the solutions, we worked out in the laboratory. Prof. Dr. C.L. Schmidt I want to thank for being my supervisor and to support me in many matters.

6   References

[1] A. Richter, V. Ostroumov, E. Heumann, W. Seelert, G. Huber, *Frequency doubling of visible Pr-laser radiation in continuous wave and pulsed mode.* Institute of Laser Physics, University of Hamburg , Coherent Lübeck GmbH, CLEO Europe 2007

[2] J.Eichler H. J.Eichler, *Laser Bauformen, Strahlführung, Anwendungen.* Springer 6. Edition, 2006, Berlin, Heidelberg, New York

[3] Roman Schwarz, *Entwicklung eines resonatorintern frequenzverdoppelten Pr:YLF CW-UV-Lasers.* diploma thesis, University of Applied Sciences Lübeck, 2011

[4] Helmbrecht Bauer, *Lasertechnik: Grundlagen und Anwendungen.* p. 121ff, Vogel Business Media/VM 1991

[5] P. von Brunn, *Construction of a 444nm laser diode pumped Pr3+:LiYF4 laser.* master thesis, University of Lübeck, 2014

Fiber based ytterbium amplified nanosecond pulses for 2-Photon- and stimulated Raman scattering microscopy

H. Hakert [1], M. Eibl [2], and R. Huber [2]

[1] Medizinische Ingenieurwissenschaft, Universität zu Lübeck, hubertus.hakert@student.uni-luebeck.de
[2] Institute of Biomedical Optics, Universität zu Lübeck, {eibl, robert.huber}@bmo.uni-luebeck.de

Abstract

Optical biomedical applications are often constructed around a pulsed laser light source whose characteristics highly influence the performance of the desired image modality. The constructed laser setup gives a hint for future light sources, especially in the field of non linear image modalities which requires high powers. A master osscilator power amplifier (MOPA) is presented for non linear imaging modalities like two-photon excited fluorescence (TPEF) microscopy and stimulated Raman scattering (SRS) microscopy. Nanosecond pulses at 1064 nm get amplified up to ~ 2 kW peak power in a fully ellectronically controllable setup. A Raman shifter allows to switch the wavelength to 1122 nm and 1186 nm. The shifting process was evaluated in terms of pump power in the last amplifier stage. The setup is easy controllable and provide nanosecond pulses with a spectral width below 0.1 nm for all achieved wavelengths that do not suffer from chromatic dispersion or self phase modulation.

1 Introduction

Two-photon excited fluorescence (TPEF) microscopy is a powerful imaging modality for high-resolution and deep tissue imaging. This non linear imaging modality has the advantage of intrinsic axial sectioning capability due to the small focus volume where TPEF is possible [1]. In addition the longer wavelengths used for excitation, compared to one photon excitement have a better tissue penetration, which results in imaging depths of ~ 100 μm-1 mm [1]. The required high intensities are usually applied by ultra short pulsed lasers (~ 100 fs-10 ps) working at a fixed repetition rate of ~ 80 MHz to reach the desired peak power. Therefore, longer pulses at the same peak power can be used if the repetition rate is lowered to achieve the same average power [2], limited by the damage threshold of the sample. In some cases lower repetition rates have a benefit because they avoid dark states of the fluorophores [3]. Longer pulses (0.5 ns-5 ns) have an important advantage, especially in fiber delivery. As the pulses have a narrow spectral width their shape and duration is not effected by chromatic dispersion (CD). Furthermore, the much smaller intensity slope prevents self phase modulation (SPM). For long interaction lengths, that occur in fiber based setups, the negligible influence of these effects are desired. As a consequence, the fiber guided pulses provide a stable, spatially pure beam quality and have the advantage to be applicable in fiber endoscopes to bypass the limited penetration depth of light (~ 100 μm-1 mm) in tissue. The presented TPEF system is fiber based and uses nanosecond pulses from a master oscillator power amplifier (MOPA). Pulse repetition rate as well as pulse length are freely programmable. This allows easy synchronization with a signal detection chain and a Fourier domain mode locked (FDML) laser which will be used in the future for stimulated Raman scattering (SRS) microscopy. The basic setup allows to amplify a 1064 nm seed pulse from 100 mW to kilowatt peak power, corresponding to an amplification factor of 40 dB. An additional Raman shifter allows to switch the wavelength from 1064 nm to 1122 nm and even further to 1186 nm, enabling to excite different fluorophores and making the setup more feasible for different situations. During the shifting process, spectral width as well as pulse duration is preserved, resulting in a cost effective laser setup constructed with usual off-the-shelf components. In this paper, the general setup of the laser system is discussed in detail. Further, the wavelength shifting process is evaluated, in terms of spectral width and peak power. This is done by various pump currents in the last amplifier stage to controll the shifting process.

2 Material and Methods

The general setup is based on the amplification of a spectral narrow pulse, emitted from a premodulated 1064 nm semiconductor laser diode. The premodulation is controlled by a waveform generator and results in pulses from 20 ns-40 ns to reduce the average power in the following puls picking process. For generating pulses in the nanosecond region (500 ps-5 ns) an electro-optical amplitude modulator (EOM) is used (Fig. 1). The EOM is based on a Mach-Zehnder interferometer using lithium niobate ($LiNbO_3$) crystals for phase changes in the reference arm, which allows extinction ratios up to 30 dB by a given offset voltage.

The EOM is driven by an electronic pulse board to change the offset voltage. Hence the EOM acts as a electronically controllable switch for light and enables to cut out the desired pulse length of the premodulated 1064 nm light. Leaving the EOM the pulses have about 20 mW of peak power. Typically, we choose 1 ns pulses with a repetition rate of 50 kHz. These pulses are amplified to about 100 W peak power in two ytterbium doped fiber amplifiers (YDFAs). Both YDFAs consist of 2 meters single-mode, core pumped ytterbium doped fiber. Pumplight is inserted through a wavelength division multiplexer (WDM) in counter propagating direction with two laser diodes at 976 nm delivering 450 mW power in continuious wave (CW) mode. Protection of the laser diodes is provided by optical Faraday isolators. After each amplification stage a laser line filter at 1064 nm is put in place to block the amplified spontaneous emission (ASE) and transmitting ~80 % of the 1064 nm light. The third YDFA is constructed of 5 meters double-clad ytterbium doped fiber (DC-YDF). In comparison to the core guided YDFAs the pump light is guided in the cladding of the fiber, which allows higher pump powers without destroying the fiber and the use of multi mode diodes. A high power multimode diode at 976 nm with up to 9 W optical power is coupled into in the cladding with an additional $(2\times 1+1)$ beam combiner. Usually about 3 W are used, resulting in 500 mW of CW power. The DC-YDF and further fibers have a core diameter of 10 μm. These fibers preserve single-mode operation at all achieved wavelengths. In the DC-YDF stage the 1064 nm pulses get amplified up to 1 kW of peak power. A significant TPEF signal is usually achieved at peak powers of 100 W [4], making the laser setup feasible for TPEF microscopy. With increasing power a significant portion of the pulse energy gets shifted by stimulated Raman scattering (SRS) [5]. This effect limits the maximal power of the 1064 nm pulse energy, leading to unwanted energy redistribution if higher 1064 nm power levels are desired. Stimulated Raman scattering is an inelastic scattering process of the light inside the silica based fiber. This results in a broad spectral gain plateau around 1122 nm. The Raman shifter mentioned before can use this process to achieve a wavelength shifting process, preserving the same pulse quality as the narrowband 1064 nm pulses.

Raman shifter

The high power 1064 nm pulses after the DC-YDFA stage can be used as a pump for stimulated Raman scattering in the fiber, resulting in shifting the 1064 nm light by 14.4 THz to a broad band gain plateau at 1122 nm. To achieve a narrow spectrum of the Raman shifted pulses, an additional seed diode with 2 mW optical power is inserted in the fiber. The high power 1064 nm light propagates together with the low power 1122 nm light in the fiber. Stimulated amplification of the 1122 nm occur when a 1064 nm photon interacts with an optical phonon, lifting the molecule into a higher virtual niveau. At the same time a 1122 nm photon depletes the higher virtual niveau, resulting in two 1122 nm photons. The instantaneous presence of the pump and seed photon

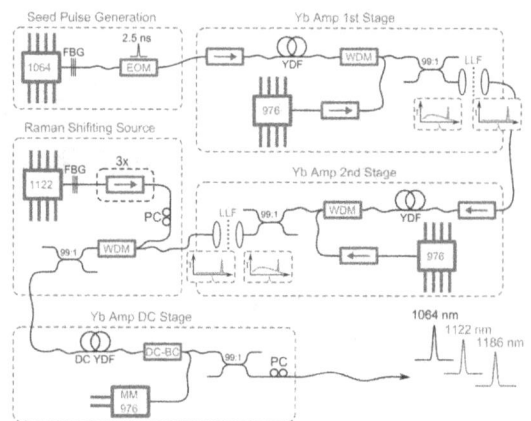

Figure 1: Setup of the fiber based MOPA light source. 1064 nm premodulated laser source ~30 ns pulses. Selectable, shorter pulse length (0.5-5 ns) created by an electro optical modulator (EOM) driven by an electronic pulse board (EPB). Pulses get amplified by two ytterbium doped fiber amplifiers (YDFA). Afterwards a laser line filter (LLF) is inserted to block most of the amplified spontaneous emission (ASE). For seeding the stimulated Raman scattering (SRS) in the glass fiber, a narrowband 1122 nm laser diode is coupled in by a wavelength division multiplexer (WDM). The shifting process is controlled by the pump power of the double clad YDFA (DC-YDFA). High power pulses at different wavelengths are obtained, which are controllable in timing and power. FBG: Fiber back grating

results in stimulated emission, as it takes place in conventional laser systems. The 1122 nm seed light is coupled into the fiber before the DC-YDFA with a WDM. In contrast to combine the light after the DC-YDFA the broadband Raman gain of the fiber is supressed and spectral narrowband pulses are achievable. Additionally, the power presented at the WDM is reduced, because optical components are usually sensitive to high powers. With increasing power of the 1122 nm pulse, a cascaded Raman shift arise which limits the power of the 1122 nm Raman band. The next Raman band is located at 1186 nm, resulting in an additional shift of 14.4 THz of the 1122 nm light. Against the assumption of a broad band gain plateau at 1186 nm, the same spectral narrowband characteristics of the 1064 nm and 1122 nm light can be observed without seeding the stimulated Raman scattering with a laser diode at 1186 nm. We estimate the origin of this process in an 1186 nm photon that is generated in a four wave mixing (FWM) process [6]. This phase critical mechanism arises by interaction of one 1064 nm photon and two 1122 nm photons. The equidistant shift of 14.4 THz from peak to peak supports this assumption. Phase matching of the waves only takes place in short segments of the fiber, resulting in an oscillating phase match and mismatch of the two waves. Only in the period of phase matching a FWM related 1186 nm photons experience a high stimulated Raman gain leading to an amplified narrowband pulse at 1186 nm. Stimulated Raman scattering, either by seeding with a diode at 1122 nm or the FWM related seeding, highly depends on the power level in the system, making a modulation attempt to regulate the power in the system a feasible solution to controll the shifting process.

3 Results and Discussion

In order to evaluate the peformance of the laser setup different aspects were tested. First of all the spectral characteristics of the amplified 1064 nm pulses and the shifted pulses at 1122 nm and 1186 nm are observed. Second, the shifting process is quantified in terms of pump currents in DC-YDFA stage, resulting in a maximum gain of the shifted pulses.

3.1 Spectral characteristics

The presented spectrum in Fig. 2 is recorded with an optical spectrum analyzer (OSA). In the first place the high pedestal under the narrowband peaks can be observed. The socket under the 1064 nm light is ASE emitted by the YD-FAs, transmitted by the laser line filters (Thorlabs, FL1064-10) with a transmission window of 10 nm around 1064 nm. Considering the CW nature of ASE, this leads to an error in its height in the spectrum, since the OSA records an integrated spectrum, which is longer than the repetition rate of the pulsed 1064 nm light. As a consequence the ASE light has to be accounted for the duty cycle:

$$D_c = \frac{1}{1ns \cdot 50kHz} = 20000 \qquad (1)$$

i.e lowered by $\sim$ 40 dB. The sockets below the Raman shifted peaks at 1122 nm and 1186 nm correlate to a spontaneous Raman scattering process. The portion of shifted energy to a narrowband spectral band during the stimulated Raman scattering is associated with the power of the seed light. Therefore, the pedestal under the 1122 nm pulses is smaller in comparison to the pedestal under 1186 nm pulses, because the CW power ($\sim$ 2 mW) of the 1122 nm laser diode is much higher than the power generated in the oscillating FWM process for the 1186 nm pulses [7]. The bottom part of Fig. 2 shows a zoom in of the narrowband pulses for all three wavelengths. A spectral width below 0.1 nm at full width half maximum (FWHM $\sim$ -3 dB) is clearly identified, proofing that the narrow spectral width of the 1064 nm pump light is preserved during the shifting process. The before mentioned negligible effect of chromatic dispersion and self phase modulation of nanosecond pulses in fiber delivery get supported by the recorded spectrum.

3.2 Raman shifting process

To evaluate the shifting process in detail, the required power for Raman scattering has to be observed. Without seeding, the Raman scattering, a spectral broad gain plateau arround 1122 nm can be observed (Fig. 3). In comparison to Fig. 2, it is clearly observable that seed light is needed to preserve the narrowband spectral characteristics of the 1064 nm pump light. Further, the portion of shifted energy of the 1064 nm pump light scales with the pump power in

the DC-YDFA stage. For currents below 2.9 A no sufficient Raman gain around 1122 nm arise, which results in a threshold of $\sim$ 2.7 A (1.8 W optical power) in terms of pump power in the DC-YDFA. Hence, the power of the 1064 nm light is limited to $\sim$ 2 kW peak power. Further increasing leads to a large portion of shifted energy of the 1064 nm light to the broad gain plateau arround to 1122 nm, due to stimulated Raman scattering (Fig. 3). When the 1122 nm seed diode is used, most of the energy is shifted to the narrowband 1122 nm seed light due seeded stimulated Raman scattering. The maximum peak power of the 1122 nm light is 2.8 kW at a forward current of 3.1 A for the DC-YDFA. Higher powers of the seeded light at 1122 nm result in shifting the energy to the next Raman band gain plateau around 1186 nm. The maximum peak power for the 1186 nm pulses are achieved at 4.5 A pump pwer in DC-YDFA, leading to the same peak power as the 1122 nm light of 2.8 kW by a difference of $\sim$ 20 dB to the pedestal (Fig. 4). As the Raman scattering occurs in fiber, its length has a crucial influence on the shifting process. However, first experiments show that a longer fiber in the light delivery path results in a spectral broadening of the pulses, which stands in contrast to the desired pulse qualities for all wavelengths [7]. Future work will show if the cascaded Raman shifting can achieve narrowband pulses at longer wavelengths. First results show that some light is visible at 1257 nm, which corresponds to an additional shift of 14.4 THz from the 1186 nm light. We believe that this light has its origin in an additional FWM process of one 1186 nm photon and two 1122 nm photons. Amplification of the 1257 nm seed light was not achieved and will be researched in future works.

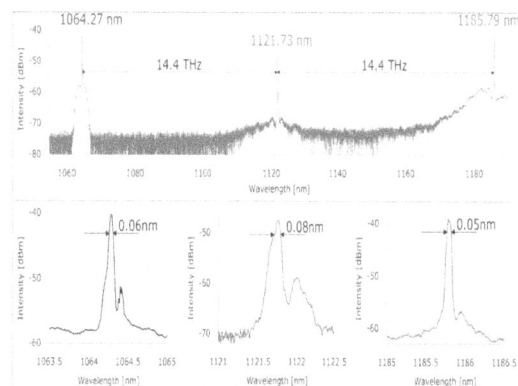

Figure 2: Top: Raman spectrum of the laser. Most of the energy is shifted to 1186 nm. ASE pedestal under the 1064 nm pulses due to the laser line filters is observeable. The socket under the 1122 nm and 1186 nm light has its origin in spontaneous Raman scattering. Portion of shifted energy to the narrowband pulses scales with the seed power for the shifting process. FWM seed light has less power than the light emitted from the 1122 nm laser diode → Socket under the narrowband 1186 nm light is higher. Bottom: Spectral characteristics of the narrowband 1064 nm light gets preserved in the shifting process. Spectral width at full width half maximum (FWHM, -3 dB) < 0.1 nm for all wavelengths. Signal to noise ratio (SNR) $\sim$ 20 dB.

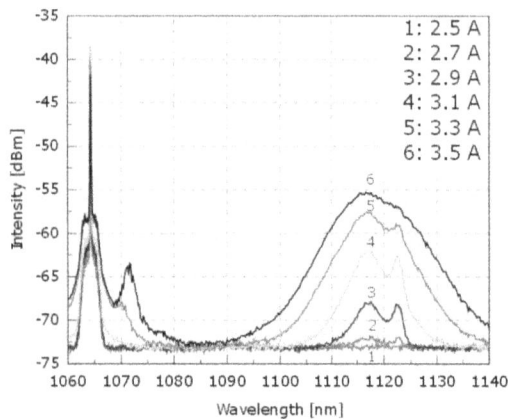

Figure 3: Spontaneous Raman gain arround 1122 nm for different pump powers in the double clad ytterbium doped fiber amplifier (DC-YDFA) is illustrated. With increasing pump power the portion of shifted energy from the 1064 nm pump light to the broad gain plateau around 1122 nm arises, resulting in a Raman threshold of $\sim$ 3 A (2 W optical power). Without seed light from a narrowband 1122 nm laser diode the spectral characteristics of the 1064 nm pump light get lost.

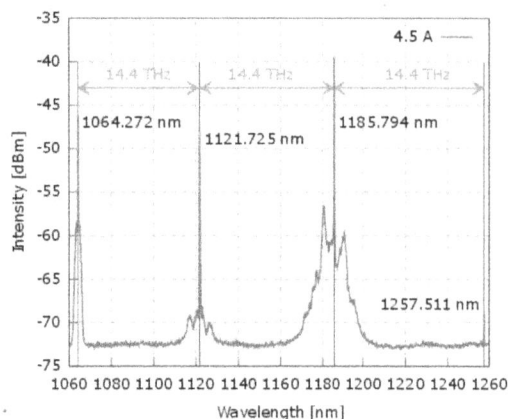

Figure 4: Spectrum of all achieved wavelengths. A pump power of 4.5 A in the DC-YDFA leads to a high energy shift to 1186 nm. Additional four wave mixing of one 1186 nm and two 1122 nm photons results in light at 1257 nm. The equidistant shift of 14.4 THz proofs this.

4 Conclusion

The presented light source is a feasible high power solution for different non linear imaging modalities like TPEF- and SRS microscopy and gives a hint for future light sources that stand in competition to common pulsed high power lasers. The narrowband spectral characteristics of the pulses and the ability to work at different wavelengths and repetition rates enables this cost effective setup to work in various situations in the broad field of biomedical optics. The performance of the setup was investigated in terms of spectral characteristics and a controllable wavelength switching process, preserving the narrowband and single mode proper-

ties of the pulses. Experiments show that these features are preserved during amplification of the 1064 nm light as well as in the shifting process if the seed light is narrowband. Current modulation of the last amplifier stage provides a feasible solution for the shifting process, whereas the controllable pulse length through an EOM enables to create ns pulses that do not suffer from chromatic dispersion or self phase modulation in a fiber based setup. Future work will investigate the shifting process in more detail, especially the amplification of the 1257 nm light. Besides the implementation of the setup in a TPEF microscope, synchronization of the laser setup with an additional Fourier domain mode locked laser will lead to a multimodal imaging platform that combines TPEF microscopy and SRS spectroscopy and microscopy.

Acknowledgement

The work has been carried out at the Institue of Biomedical Optic, Universität zu Lübeck, Luebeck, Germanny.

5 References

[1] E. E. Hoover and J. A. Squier, "*Advances in multiphoton microscopy technology*," Nature photonics, vol. 7, no. 2, pp. 93–101, 2013.

[2] Y. Chen, H. Guo, W. Gong, L. Qin, H. Aleyasin, R. R. Ratan, S. Cho, J. Chen, and S. Xie, "*Recent advances in two-photon imaging: technology developments and biomedical applications*," Chinese Optics Letters, vol. 11, no. 1, p. 011703, 2013.

[3] G. Donnert, C. Eggeling, and S. W. Hell, "*Major signal increase in fluorescence microscopy through dark-state relaxation*," Nature methods, vol. 4, no. 1, pp. 81–86, 2007.

[4] S. Karpf, M. Eibl, B. Sauer, F. Reinholz, G. Hüttmann, and R. Huber, "*Fiber-based TPEF Microscopy with ns-Pulses*," in Non-Linear Raman Spectroscopy and Its Chemical Application, 2016.

[5] S. Karpf, M. Eibl, W. Wieser, T. Klein, and R. Huber, "*Multi-color fiber MOPA with electronically switchable wavelengths shifter for non-linear laser spectroscopy*," in Non-Linear Raman Spectroscopy and Its Chemical Application, 2016.

[6] M. Baumgartl, T. Gottschall, J. Abreu-Afonso, A. Díez, T. Meyer, B. Dietzek, M. Rothhardt, J. Popp, J. Limpert, and A. Tünnermann, "*Alignment-free, all-spliced fiber laser source for CARS microscopy based on four-wave-mixing*," Optics express, vol. 20, no. 19, pp. 21010–21018, 2012.

[7] M. Eibl, "*Stimulated Raman spectroscopy with a Rapidly Wavelength Swept Laser*," Master's thesis, Ludwig-Maximilian-University Munich, 2012.

Ex-vivo validation of a new speckle-based retinal photocoagulation control

L. Effe [1], L. Stockmann [1], K. Bliedtner [2], E. Seifert [2], and R. Brinkmann [2,3]

[1] Medizinische Ingenieurwissenschaft, Universität zu Lübeck, {lisa.effe,leonie.stockmann}@student.uni-luebeck.de

[2] Medizinisches Laserzentrum Lübeck GmbH, Lübeck, Germany, {bliedtner,eric.seifert}@mll.uni-luebeck.de

[3] Institut für Biomedizinische Optik, Universität zu Lübeck, brinkmann@bmo.uni-luebeck.de

Abstract

Photocoagulation is a well-established laser treatment for several retinal diseases. The effects inflicted to the target tissue are not uniform. A feedback technique able to evaluate the photocoagulation induced temperature in real-time would be a first step towards a control mechanism. This can be realized by optical or acoustical approaches. In this work a new set-up of a laser speckle based optical approach is further investigated. A HeNe laser with a wavelength of 633 nm is used for illumination and a frequency-doubled Nd:YAG laser (532 nm) is used for heating. The backscattered light is detected by a CMOS camera to monitor the speckle movement (movement proportional to temperature). The results are compared with data using a 855 nm diode laser for illumination [1]. It can be demonstrated that the 633 nm illumination laser leads to an information gain regarding the resulting backscattering in comparison with the 855 nm diode laser.

1 Introduction

Photocoagulation of the retina has become a standard treatment for a variety of retinal diseases, such as diabetic retinopathy, branch vein occlusion and central vein occlusion [2]-[5].

In this treatment light in the green, yellow or near IR spectrum is applied to the fundus. It gets absorbed in the Retinal Pigment Epithelium (RPE) and Choroid. Due to thermal conduction the temperature in surrounding tissue increases. In photocoagulation the laser parameter were chosen to induce thermal denaturation along with cell apoptosis or necrosis. The therapeutic effect which is a biological response to cell death differs according to the disease and consequently different laser parameter are used. Depending on the mode of treatment the laser power varies between 50 and 500 mW, the irradiation time of the laser between 50 and 200 ms and the spot diameter between 50 and 500 μm [3]-[5].

Strong intra- and inter-individual variations of pigmentation density of the RPE and choroid prohibit constant heat development [6], [7]. This effect is intensified by light scattering properties of the anterior eye media varying with age and disease. Consequently the damage to the RPE cells is not predictable and inconstant. After denaturation of the retina a white lesion is visible. Whitening is caused by increased backscattering, induced either by coagulation (if whitening appears immediately after the irradiation) or by edema (if whitening appears after seconds or later) [8]. Nowadays the subjective evaluation of whitening is the only treatment control. During the treatment the laser power can be adjusted for each spot to achieve a white lesion, but due to several hundreds of spots this method is time-consuming. A feedback technique able to evaluate the photocoagulation induced damage would be a first step towards a control mechanism for this treatment. Recent approaches to achieve this goal by optoacoustic or OCT (optical coherence tomography) techniques show promising results [8], [9]. While the drawback of the optoacoustic technique is the decreasing performance in the characterization of intense lesions the disadvantage of the recent OCT techniques is the costs. The actual approach to quantify retinal coagulation may overcome both disadvantages if a comparable performance measurement can be achieved. This can be realized by optical or acoustical approaches. In this work a new set-up of a laser speckle based optical approach is investigated.

The new setup detects the temperature induced tissue changes caused by thermal expansion and denaturation. Backscattered light of the laser produces a speckle pattern which is detected by a camera and analysed by two different algorithms to evaluate the tissue damage. Previous in-vivo experiments in rabbits showed a large speckle noise, most likely caused by the blood flow in the choroid. To reduce this noise a shorter illumination wavelength is used, because of the shorter penetration depth. Since a change of wavelength can lead to a change in a variety of phenomena, previously performed in vitro-measurements [1] have been repeated with 633 nm illumination.

2 Material and Methods

In this section the fundamentals of the speckle evaluation and the optical set-up for the measurements are described.

2.1 Fundamentals

Since a speckle pattern is an interference phenomena, changes of the medium interacting with the light (phase or amplitude changes) lead to changes in the speckle pattern. Lateral movements of the object lead to lateral speckle movements ("translational speckle") and axial movements of the object (e.g. induced by thermal expansion) or structural changes (e.g. induced by denaturation) lead to random changes of the speckle pattern (boiling speckle). The shape of translational speckles remain unchanged even after displacements and the speckle grain moves as a whole when the diffuser moves. Whereas boiling speckles do not move even when the diffuser moves, they just change in size, disappear and reappear [10].

Since the speckle pattern is temporally stable in an in-vitro situation, most of the observed changes can be expected to be induced by the laser irradiation. Thus in this early stage it is expected to be sufficient to analyze the difference from one image acquired at the time t ($I_t(x,y)$) to the next image ($I_{t+1}(x,y)$) pixel wise. The sum of the absolute values of this difference image represents the amount of changes from one acquisition to the next. To reduce the influence of external influences (vibrations) on this value, it gets normalized to the difference value (calculated in the same way) of two consecutive images (I_1 and I_2) of the same record set, acquired before the treatment laser is activated. This sum is called "Generalized Differences value" (GD value):

$$GD_t = \frac{\sum_x \sum_y |I_t(x,y) - I_{t+1}(x,y)|}{\sum_x \sum_y |I_1(x,y) - I_2(x,y)|} \qquad (1)$$

To set the starting point of GD values to zero, the offset of 1 is subtracted from the GD_t value. Since every change in the speckle pattern leads to an increase of the corresponding GD_t value, a sum of the GD values (GD_t) over the heating- (and cooling-) time represents the amount of laser induced changes:

$$S_{GD} = \sum_{exposuretime} GD_t \qquad (2)$$

The temperature increase can lead to denaturation and permanent damage of the overheated tissue. Temperature induced changes in the optical properties will occur especially in the retina. It will change from mostly transparent over opaque/grayish to white depending on the degree of coagulation. Such visible changes in the neural retina originate from changes of the backscattering properties. To have a measure scaling with the backscattered light, the sum of the intensity of all pixels in each frame ($I_t(x,y)$) is calculated (3). The IB value IB_t indicates the amount of light reaching the sensor. The increased backscattering S_{IB} is the sum of the IB value IB_t during exposure time:

$$IB_t = \sum_x \sum_y I_t(x,y) \qquad (3)$$

$$\implies S_{IB} = \sum_{exposuretime} IB_t \qquad (4)$$

2.2 Optical Set-up

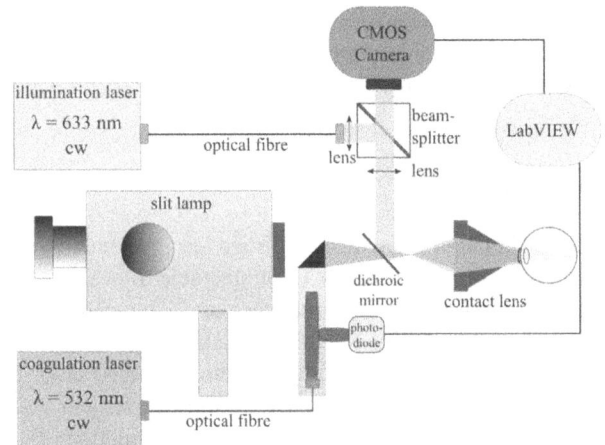

Figure 1: Optical set-up: coagulation and illumination laser are superimposed and focused on the retina. The backscattered light is detected by the CMOS camera.

For photocoagulation a frequency-doubled Nd:YAG laser with a wavelength of 532 nm (Novus Spectra, LUMENIS) and as illumination laser a HeNe laser with a wavelength of 633 nm is used, see Fig. 1. The measured data are compared with measurements obtained with an 855 nm diode laser [1]. The illumination and the coagulation laser are integrated into a slit lamp. The coagulation laser is coupled into a slit lamp via multimode fibre (NA = 0.11; core diameter = 50 µm; length = 2 m). After exiting the fibre the laser light passes a dichroic mirror and a contact lens before being focused on the retina. The illumination laser exits the optical fibre (NA = 0.11; core diameter = 50 µm; length = 20 m), passes a beamsplitter and the dichroic mirror and is superimposed with the coagulation laser beam before being focused on the retina. The backscattered light of the illumination laser passes the dichroic mirror, the beamsplitter and an aperture with a diameter of 12 mm. The light is afterwards detected by a CMOS camera (CL600x2, Optronis). The collected information is analysed and calculated by a LabVIEW program, written by the photocoagulation workgroup. Due to losses at each optical element 50 % of the laser power is lost in the system.

2.3 Experiments on porcine eyes

The experiments were done with a pulse duration of 200 and 500 ms, a minimum of five porcine eyes each, a coagulation laser spot size of 100 µm and a illumination laser spot size of 500 µm. The maximum frame rate for the 855 nm laser is 1000 fps and for the 633 nm laser between 300 and

500 fps. Enucleated porcine eyes with a strong pigmented retina were used as test objects.

During treatment the opthalmoscopically visible damage (visible or not visible) is noted. After treatment the treated retina is removed from the RPE cells using Ringer's solution. The RPE cells are stained with Calcein AM (acetoxylmethyl), a non-fluorescent and hydrophobic substance, which converts into a fluorescent substance inside the cell. The sample is analysed by a fluorescence microscope. Intact RPE cells appear green and dead cells dark. Each spot is analysed and classified as either unknown (cell damage cannot be determined), invisible (intact RPE cells), RPE damage (damage surrounded by intact RPE cells) or whitening (damage visible via slit lamp).

3 Results and Discussion

In this section the results of the porcine eye treatment using a 633 nm and 855 nm [1] illumination laser are compared.

3.1 Time-dependent Measurements

Temporal changes of the observed retinal tissue are induced by temperature. Before the already explained coagulation effects from section 2.1 take place thermal expansion of cellular liquid as well as contraction of cell components (e.g. collagen) can induce changes in the backscattered light (in the observed speckle pattern).

These movements are represented by the first increase of the GD graph (see arrow No. 1 in Fig. 2) and the second increase originates from tissue relaxation, see arrow No. 2 in Fig. 2. The tissue temperature decreases exponentially after laser exposure because of water expansion regression and following collagen relaxation [12]. The IB graph shows this increase only with the treatment of the 633 nm illumination laser and not with 855 nm, see arrows 3 and 4 in Fig. 2 and Fig. 3. The amount of backscattered light increases during photocoagulation using the 633 nm laser and not in a measurable amount for the 855 nm laser.

Thus, regarding the investigated wavelength for irradiation, the IB method is only applicable for 633 nm.

3.2 S_GD depending on S_IB

Fig. 4 shows the amount of speckle change of each spot (S_{GD}) plotted over the increased backscattering (S_{IB}). Due to the strict exclusion criteria and the low number of eyes the number of not visible spots (17 spots) is much smaller than spots where whitening could be observed (133 spots). The lower graph of Fig. 4 shows a distribution of values in each category (whitening and invisible). Using $\tau = 200$ ms leads to similar results. As a first performance measure the system should show that it is able to distinguish between ophthalmoscopically visible and non-visible spots. The estimated statistical probability to correctly identify an ophthalmoscopically visible or non-visible spot is called sensitivity or specificity respectively. These are the performance measures for the feedback techniques. As it can be seen in

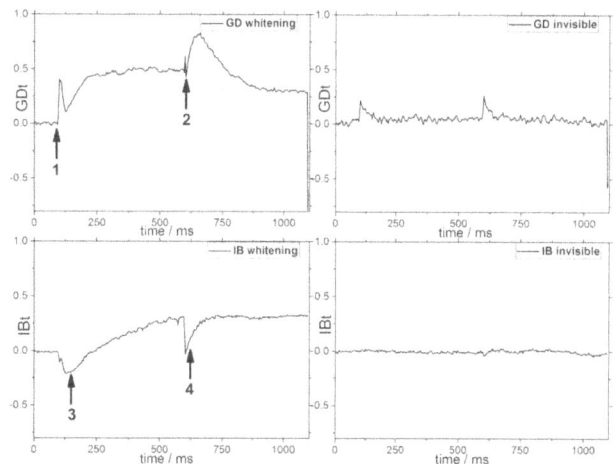

Figure 2: GD_t and IB_t over time at 633 nm and $\tau = 500$ ms

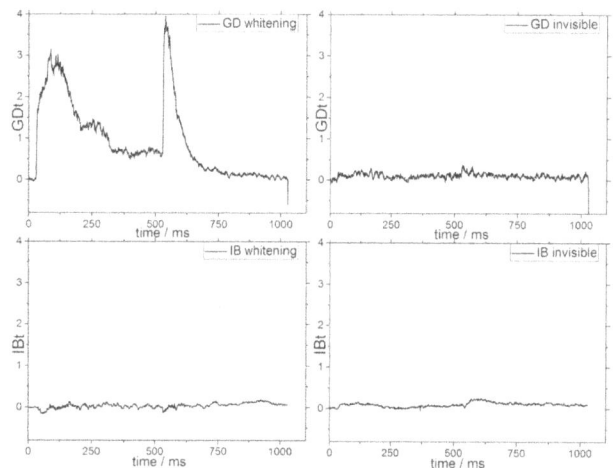

Figure 3: GD_t and IB_t over time at 855 nm and $\tau = 500$ ms

Table 1 the performance measure pairs achieve higher values if 633 nm illumination is used.

The lower diagram of Fig. 4 shows that severe lesions lead to higher IB values using the 633 nm laser. The linear relationship between the IB and GD algorithm at the ophthalmoscopically visible spots indicates a strong influence of the increased backscattering on the GD value. The GD algorithm is desired to indicate speckle movement. Under this aspect the influence of increased backscattering on the GD values can be seen as noise source for the GD algorithm. Additional algorithm elements as the normalization on the overall intensity can eliminate this noise source. On the other hand speckle modulations induced by coagulation processes can be so fast that the applied cameras may not be fast enough to acquire these processes adequately. In this scenario an evaluation of the backscattered light can deliver valuable information about the actual state of coagulation. Hence it is reasonable to apply both approaches parallel in the evaluation routine.

Table 1: Sensitivity and Specifity values for $\tau = 500$ ms

Wavelength	633 nm	855 nm
GD-sensitivity	0.91	0.69
GD-specifity	1.00	0.86
IB-sensitivity	0.84	0.83
IB-specifity	0.94	0.43

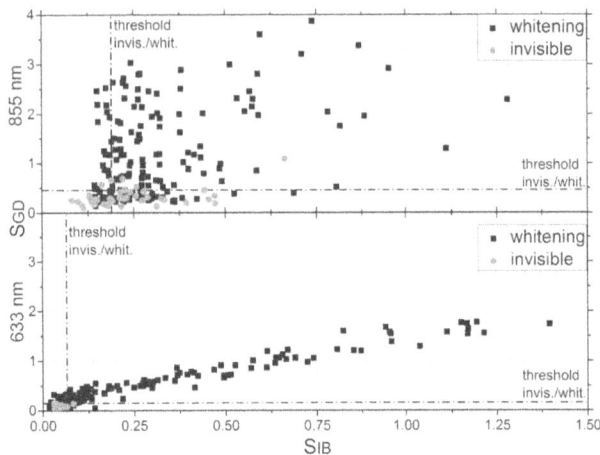

Figure 4: S_{GD} plotted over S_{IB} at 855 nm (top) and 633 nm (bottom) and $\tau = 500$ ms

4   Conclusion

Laser-heating induced changes in a speckle pattern as a measure of induced cell damage have been studied in this project while performing photocoagulation on porcine eyes. Two different illumination lasers (wavelength 633 and 855 nm) have been used for illumination of the fundus. It has been found that the GD respectively IB value can be used to differentiate between invisible and white lesions. The 633 nm illumination laser leads to an information gain regarding the resulting backscattering in comparison with the 855 nm laser. This leads to an increase in the dependent statistical probability to correctly identify visible and non-visible lesions.

For future in-vivo measurements it is advised to use the 633 nm illumination laser due to the smaller penetration depth compared to the 855 nm laser. This will decrease the influence of the blood flow located in the choroid and enables the characterization of strong lesions by the analysis of increased backscattering. The GD algorithm should be redesigned in a way to be independent from an increase in the backscattering properties. Therefore another algorithm should be applied parallel to evaluate the changing of the backscattering properties only. The exact way of combining these algorithms will be one aspect of the next steps of this project

Acknowledgement

The work has been carried out at Medizinisches Laserzentrum Lübeck GmbH, Lübeck, Germany.

5   References

[1] L. Stockmann, *Speckle-basierte Überwachung der retinalen Photokoagulation*. Bachelor thesis, Universität zu Lübeck, 2015.

[2] A. M. Shah, N. M. Bressler and L. M. Jampol, *Does Laser still have a role in the management of retinal vascular and neovascular diseases*. American Journal of opthalmology, vol. 152, no. 3, pp. 332–339, 2011.

[3] Early Treatment Diabetic Retinopathy Study Research Group, *Photocoagulation for diabetic macular edema. Early Treatment Diabetic Retinopathy Study report number 1*. Arch. Opthalmol., vol. 103, no. 12, pp. 1796–1806, 1985.

[4] Branch Vein Occlusion Study Group, *Argon laser scatter photocoagulation for prevention of neovascularization and vitreous hemorrhage in branch vein occlusion. A randomized clinical trial*. Arch. Opthalmol., vol. 104, no. 1, pp. 34–41, 1986.

[5] The Central Vein Occlusion Study Group, *Natural history and clinical management of central retinal vein occlusion*. Arch. Opthalmol., vol. 115, no. 4, pp. 486–491, 1997.

[6] S. Y. Schmidt and R. D. Peisch, *Melanin concentration in normal human retinal pigment epithelium. Regional variation and age-related reduction*. Investigative ophthalmology & visual science, vol. 27, no. 7, pp. 1063–1067, 1986.

[7] W. J. Geeraets, et al., *The relative absorption of thermal energy in retina and choroid*. Investigative ophthalmology & visual science, vol. 1, no. 3, pp. 340–347, 1962.

[8] R. Brinkmann, et al., *Real-time temperature determination during retinal photocoagulation on patients*. Journal of Biomedical Optics, vol. 17, no. 6, pp. 061219—1, 2012.

[9] H. H. Müller, et al., *Imaging thermal expansion and retinal tissue changes during photocoagulation by high speed OCT*. Biomedical Optics Express, vol. 3, no. 5, 2012.

[10] H. J. Rabal and R. A. Braga Jr., *Dynamic laser speckle and applications*. Vol. 139, CRC Press, 2009.

[11] E. Seifert, K. Bliedtner and R. Brinkmann, *Laser speckle tracking for monitoring and analysis of retinal photocoagulation*. In SPIE BiOS, pp. 89460F–89460F, International Society for Optics and Photonics, 2014.

[12] R. Birngruber, W. Weinberg, G. Chan, V.-P. Gabel and F. Hillenkamp, *Fundusreflektometrie, thermische Modellrechnungen und Temperaturmessungen als Hilfsmittel zur Optimierung der Bestrahlungsparameter bei der Photokoagulation der Netzhaut*. Ionisierende Strahlen in der Augenheilkunde, JAEGER Publishing, vol. 76, JF Bergmann-Verlag, pp. 563–567, 1962.

Imaging cold-induced vasodynamic behaviour using optical coherence tomography for microangiography

M. Casper [1,2,3], C. Nguyen [2], M. Evers [2,3], H. Schulz-Hildebrandt [3], R. Birngruber [3], D. Manstein [2] and G. Hüttmann [3]

[1] Medizinische Ingenieurwissenschaft, Universität zu Lübeck, malte.casper@student.uni-luebeck.de
[2] Department of Dermatology, Massachusetts General Hospital/Harvard Medical School, MJCasper@mgh.harvard.edu
[3] Institut für Biomedizinische Optik, Universität zu Lübeck, huettmann@bmo.uni-luebeck.de

Abstract

The ability to image the physiology of microvasculature with high spatial resolution in three dimensions while reacting to changes in temperature is crucial for understanding the complex processes of vasodynamics, which lead to constriction and dilation of vessels. However, previous studies using Laser-Doppler flowmetry and imaging could not provide reliable angiographic images which allow to quantify changes in blood vessel diameter. Here, we report the first angiographic imaging of microvasculature in a mouse ear pinna using speckle variance optical coherence tomography (svOCT) during and after localized cooling. In contrast to the general opinion in literature, we were able to observe that the majority of vessels with a diameter larger than $20\,\mu$m maintain perfused with a constant diameter when the tissue is cooled from baseline to subzero temperatures. During this procedure vasoconstriction only appeared in a few venous branches. The results of this preliminary study lead us to reconsider essential aspects of previous understanding of temperature-induced vasodynamics in cutaneous microvasculature.

1 Introduction

Treatments based on external application of cooling are well established in modern medicine. They mainly address the reflex of constriction of blood vessels in order to selectively reduce blood perfusion in cooler areas of the body. Localized cooling is most commonly used to reduce swelling, pain, inflammation and bleeding e.g. after orthopedic surgery [1]. In dermatology the reflexes of vasoconstriction and vasodilation are known as important mechanisms of thermoregulation of the inner body [2]. Dilation of cutaneous blood vessels is supposed to provide a loss of heat, while vasoconstriction leads to less perfusion which prevents the body from cooling down. These mechanisms of interaction of the skin with the environment can easily be noticed at outdoor activities. However, it has been reported that the response to localized cooling is characterized by an initial decrease in cutaneous blood flow, followed by a transient vasodilation, and a secondary progressive vasoconstriction [3]. Several techniques have been applied to detect and measure such adaptations in blood perfusion. Laser-Doppler flowmetry (LDF) uses monochromatic light which is reflected at moving particles, like red blood cells, unveiling their average velocity by analyzing the shift of frequency [1], [4], [5]. Another approach to this phenomena is used in Laser-Doppler perfusion imaging (LDPI) as a scanning equivalent of LDF in order to obtain images of perfusion areas. Unfortunately, flow measuring techniques like LDF, LDPI and other imaging techniques based on the Doppler effect are dependent on angle to the direction of

flow and lack of depth resolution [6]. Furthermore changes in flow velocity might not correlate with physiological dynamics like vasoconstriction.

Optical coherence tomography (OCT) is based on the interferometric analysis of backscattered light providing cross-sectional, high-resolution volumetric images of in-vivo tissue. Over the past decade OCT has gotten numerous extensions for angiographic imaging following two different approaches of signal analysis. On the one hand information about flow velocity is obtained by detecting frequency shifts caused by the Doppler effect in order to construct volumetric maps of flux with high-resolution. These approaches either determine the Doppler frequency shift by an average shift of phase between sequential A-Scans (phase-resolved Doppler OCT) or intensity information (intensity-based Doppler variance) [7]. However, these Doppler based techniques still generate low-quality information since, like in LDPI, the angle of measurement to flow direction remains unknown while the information is angle-dependent. Additionally the angle also limits its sensitivity [7]. Other algorithms like speckle variance (svOCT) [8] and optical micro-angiography (OMAG) [9] operate on OCT intensity images analyzing the temporal variation of each pixel's intensity to obtain information about movement. In OCT intensity images, pixel in areas of perfusion, e.g. in blood vessels, appear to have high fluctuations in intensity values over sequential scans while pixel in static tissue maintain constant intensity. By computing the average difference or variance of each pixel for consecutive scans at the same position, a map of perfusion can be created indicating movement of scattering particles like blood cells, but without information of direction or velocity. Recently Doppler analysis has been coupled with OMAG in order to map information of flow velocity onto the corresponding angiogram [10]. Nevertheless the velocity is still only measured in axis of the OCT beam. In this article, a commercial OCT system is applied to perform high-resolution in-vivo angiography using a speckle variance algorithm at a mouse ear pinna in order to image the dynamics of vasoconstriction. To the best of our knowledge, this is the first simultaneous angiographic OCT imaging of the response of the cutaneous microvascular system to induced local temperature changes.

2 Material and Methods

Non-invasive in-vivo imaging was performed at the top of the mouse ear pinna of healthy approximately 18 weeks old female mice while simultaneously cooling from the bottom (Fig. 1). During the experiments, the mice were anesthetized using 1.5% isoflurane. To maintain the body temperature, the animal was placed in supine position on a heated blanked at a constant temperature of $37°$ C. The animal's right ear was depilated using Nair hair removal lotion (Church & Dwight Co., NJ, USA) and taped onto the surface of the handheld cooling device in order to provide thermal conduction.

The custom built cooling device consists of a 5240 TEC-Source temperature controller (Arroyo Instruments, CA) regulating a thermoelectric cooler module embedded in a hand-piece using Peltier element (TE Technology Inc., MI). Additionally a chiller Oasis 160 (Solid State Cooling Systems, NY) was attached to the backside of this element to remove heat. An interface in LabVIEW (National Instruments, 2012, TX) was implemented to control and monitor temperature of the cooler's surface.

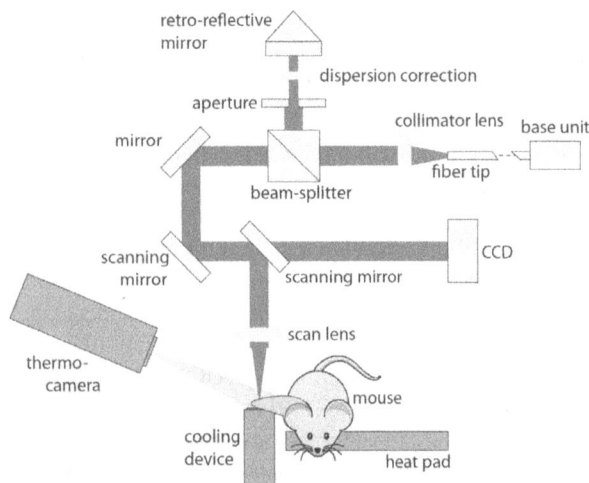

Figure 1: OCT imaging of a mouse ear from the top while cooling from the bottom. Thermo camera is measuring temperature at OCT's field of view.

In order to monitor the ear's top surface temperature with

a precision of $\pm 1°$ C, the infrared thermal camera A600-Series (FLIR, OR) was set up. The arrangement of the experiment is outlined in Fig 1. OCT scanning was performed using a commercial spectral-domain OCT scanner (TELESTO II, Thorlabs Inc., NJ) with a center wavelength of $1.3\,\mu$m and a manufacturer-specified axial resolution of $5.5\,\mu$m (in air). Using two different lenses (LSM02 & LSM03, Thorlabs Inc.) enables imaging with lateral resolutions of $7\,\mu$m and $13\,\mu$m, respectively. In order to avoid pressure on the observed tissue, no spacer was attached onto the imaging probe as it is common for many applications of optical coherence angiography (OCA). Using the LSM03 lens scanning was performed over a FOV of 5×4 mm and 2.5×2.5 mm using LSM02.

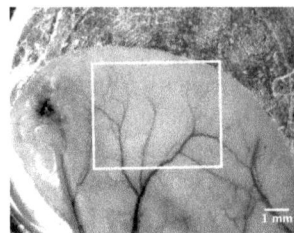 

Figure 2: Field of view (4.8×4 mm) which was imaged with lens LSM03

Figure 3: Field of view (2.5×2.5 mm) which was imaged with lens LSM02

Pairs of co-located B-scans (615×735 pixels in x and z) were acquired at a pixel size of $6.5 \times 3.5\,\mu$m (z scan depth of 3.6 mm in air) using the lens LSM03 and assessed for speckle decorrelation. Each C-scan consisted of 738 B-scan pairs with a spacing of $6.5\,\mu$m. In order to obtain images with higher resolution, the LSM02 lens was used performing a similar scanning protocol with 713×735 pixel sized B-Scans at a pixel size of $3.5 \times 3.5\,\mu$m and 713 pairs of B-Scans acquired at a spacing of $3.5\,\mu$m. Both scanning protocols were performed at an A-Scan rate of 76 kHz leading to a data acquisition time of 18.3 s and 19.6 s for each C-Scan. For each set consisting of N sequential B-scans acquired at the same location, one speckle variance image was calculated based on

$$svOCT_{ij} = \sigma_{ij}^2 = \frac{1}{N}\sum_{k=1}^{N}\left(I_{ijk} - \frac{1}{N}\sum_{k=1}^{N}I_{ijk}\right)^2 \quad (1)$$

with location of pixel in a B-scan $i \in x$ and $j \in z$. The number of repeated B-scans N is a factor with an averaging influence on the algorithm while $k \in 1, ..., N$. In our case $N = 2$ in order to to obtain raw data of decent size. To provide an angiographic impression of the volumetric speckle-variance data, maximum intensity projections (MIP) of the *enface*-view have been computed over the depth of focus for either lens with $310\,\mu$m and $140\,\mu$m, respectively, covering the area of cutaneous blood vessel.

Cooling was applied in steps of $2 - 5°$ C starting at the baseline temperature of $27°$ C down to $-10°$ C. Images were acquired right after reaching the temperature and with a delay of 90 s maintaining the temperature. Once $-10°$ C was reached, two paths of the ear's rewarming were imaged,

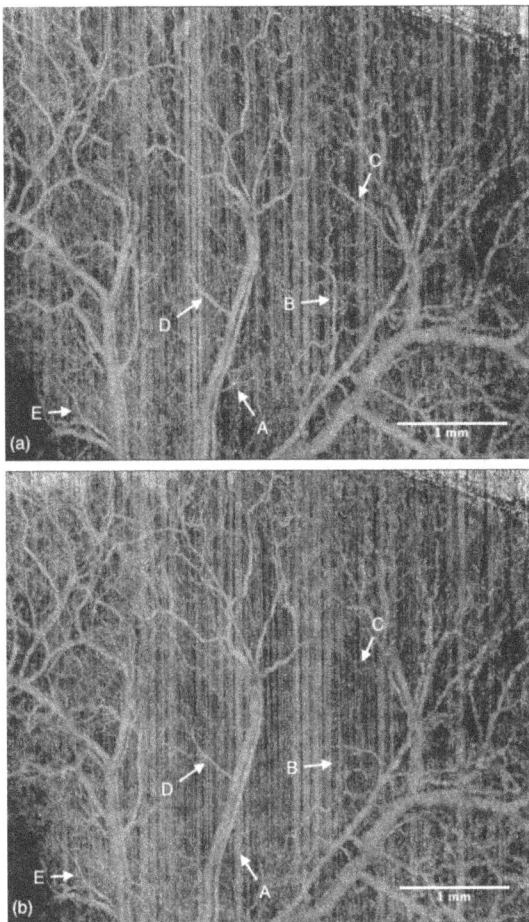

Figure 4: *Enface* MIPs in the top $310\,\mu$m of svOCT data showing the mouse ear's vasculature at temperatures of (a) $25°$ C and (b) $5°$ C while using the LSM03 lens. Arrows indicate branches of veins which constricted at a certain temperature (A,B,C) and similar venous branches which maintained perfused until $-5°$ C was reached (D,E)

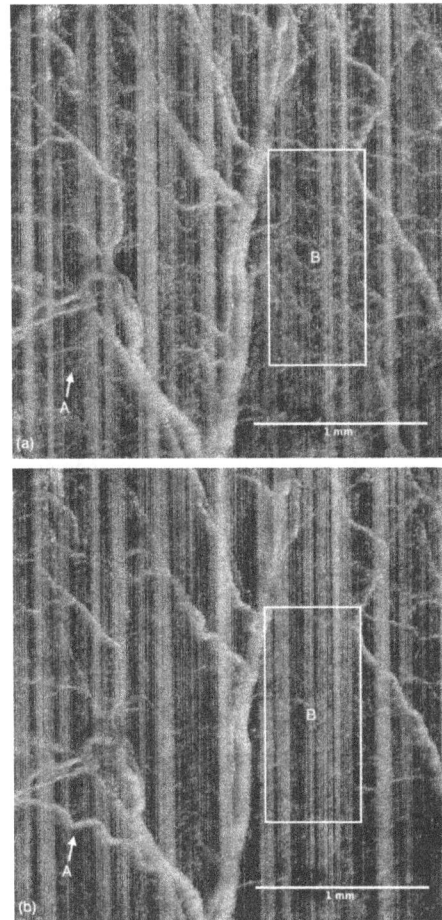

Figure 5: *Enface* MIPs in the top $140\,\mu$m of svOCT data showing the mouse ear's vasculature at temperatures of (a) $25°$ C and (b) $20°$ C while using the LSM02 lens. Arrow A points on a major arterial branch which appeared at the temperature of $20°$ C. In area B constriction of microvasculature was observed as a stop of perfusion and its reappearances during further cooling.

self-rewarming by switching off the TEC and stepwise actively rewarming using the temperature controller.

3   Results and Discussion

The angiograms at the location on the mouse ear shown in Fig. 2 unveil the perfused vascular network and are shown in Fig. 4 at temperatures of $25°$ C and $5°$ C (Fig. 4 (a) and (b)). The three major arteries $\varnothing : 50-150\,\mu$m are paired with veins lying underneath. To our surprise, these arteries maintained a constant diameter at different temperatures between $+27°$ C and $-5°$ C. Vasoconstriction was only observed at three branches of two veins indicated as A,B,C in Fig. 4. Branch A ($\varnothing \sim 30\,\mu$m) started to constrict between $15°$ C to $10°$ C and was closed after the 90 s delay. Branches B and C (both $\varnothing \sim 40\,\mu$m) slightly started to constrict between $10°$ C to $5°$ C and perfusion stopped after the 90 s delay. Other venous branches of the same veins and comparable size maintained perfused with constant diameter, e.g. branches D and E, until $-5°$ C. Interestingly,

the only constricting vessels are venous branches with a diameter $\varnothing < 40\,\mu$m reaching in-between the paired main arteries and veins. The perfusion of all vessels stopped between $-5°$ C and $-10°$ C while increasing ice formation in the skinlayer was visible. Unfortunately the ice formation affects the properties of scattering in tissue leading to image blur. During the process of rewarming condensate appeared on the mouse ear's surface causing a similar image blur. Additionally, movement between the specimen and OCT probe caused a variance of speckle in supposedly static areas of the image. This leads to artifacts of vertical lines in the *enface*-projections.

Since the perfusion in areas of microvasculature with $\varnothing < 15\,\mu$m cannot be imaged properly using the LSM03 lens, additional experiments using the LSM02 lens were performed. In Fig. 5 the angiograms of the location indicated in Fig. 3 show a pair of a major artery and vein ($\varnothing \sim 80\,\mu$m) with a bifurcation while temperatures of (a) $25°$ C and (b) $20°$ C were present. As observed before, these vessels maintained their diameter of perfusion over different steps of temperature between $+27°$ C and $-8°$ C. During cooling from $27°$ C to $20°$ C both, arterial and venous microvasculature constricted slowly. By reaching $20°$ C their perfusion stopped and on the bottom left of the image, another artery indicated as A appeared ($\varnothing \sim 40\,\mu$m) and remained perfused with constant diameter until it started to constrict when subzero temperatures were reached. Microvasculature with $\varnothing \leq 20\,\mu$m in the indicated area B reappeared in the angiogram during a 90 s delay maintaining $10°$ C. Minor reappearances of microvasculature during the cooling process were observed but no certain order was found.

The general opinion in literature is that vasodynamics like constriction and dilation are mechanisms induced by autonomic neural physiology for thermoregulation of the body [11]. However, in this preliminary study we could not observe that exposure to cold stress either elicits a generalized cutaneous vasoconstriction as proposed in [2] or a clear sequence of vasoconstrictions and -dilations as suggested in [3], [12]. The hypothesis of Johnson that mechanisms of vasoconstrictor response might vary between arteries and veins or among vessel size [13] could be supported, since we noticed vasoconstriction only in veins of certain size. Furthermore, constriction only occurred in veins branching off perpendicularly to their origin into the space between the main vessels while other branches of the same origin and size remained perfused.

4   Conclusion

In this article, we have demonstrated the feasibility of using OCT with a speckle variance algorithm for monitoring the changes of cutaneous vasculature in response to different temperatures of localized cooling. The size of perfused vasculature can be well displayed with sufficient resolution, while, surprisingly, vasoconstriction was observed very rarely and only in veins. Unfortunately, svOCT is unable to provide functional information such as flow velocity which is essential for further understanding of vasodynamics. Moreover, the conclusions made from experiments with the mouse ear might not be transferable to behavior of human's cutaneous microvasculature. In further experiments, the role of the mouse ear for thermoregulation needs to be evaluated. Afterwards this approach will be applied to the human body.

5   References

[1] S. Khoshnevis, N. K. Craik, R. M. Brothers, and K. R. Diller, "Cryotherapy-induced persistent vasoconstriction after cutaneous cooling: Hysteresis between skin temperature and blood perfusion," *Journal of Biomedical Engineering*, 2016.

[2] N. Sawasaki, S. Iwase, and T. Mano, "Effect of skin sympathetic response to local or systemic cold exposure on thermoregulatory functions in humans," *Autonomic Neuroscience: Basic and Clinical*, vol. 87, no. 2-3, pp. 274–281, 2001.

[3] C. T. Minson, "Thermal provocation to evaluate microvascular reactivity in human skin," *Journal of Applied Physiology*, vol. 109, no. 4, pp. 1239–1246, 2010.

[4] F. Yamazaki, R. Sone, K. Zhao, G. E. Alvarez, W. A. Kosiba, and J. M. Johnson, "Rate dependency and role of nitric oxide in the vascular response to direct cooling in human skin," *Journal of Applied Physiology*, vol. 100, no. 1, pp. 42–50, 2006.

[5] G. E. Alvarez, K. Zhao, W. A. Kosiba, and J. M. Johnson, "Relative roles of local and reflex components in cutaneous vasoconstriction during skin cooling in humans," *Journal of Applied Physiology*, vol. 100, no. 6, pp. 2083–2088, 2006.

[6] J. Allen and K. Howell, "Microvascular imaging: techniques and opportunities for clinical physiological measurements," *Physiological measurement*, vol. 35, no. 7, p. R91, 2014.

[7] A. Zhang, Q. Zhang, C.-L. Chen, and R. K. Wang, "Methods and algorithms for optical coherence tomography-based angiography: a review and comparison," *Journal of Biomedical Optics*, vol. 20, no. 10, pp. 100901–100901, 2015.

[8] A. Mariampillai, M. K. Leung, M. Jarvi, B. A. Standish, K. Lee, B. C. Wilson, A. Vitkin, and V. X. Yang, "Optimized speckle variance oct imaging of microvasculature," *Optics Letters*, vol. 35, no. 8, pp. 1257–1259, 2010.

[9] L. An, J. Qin, and R. K. Wang, "Ultrahigh sensitive optical microangiography for in vivo imaging of microcirculations within human skin tissue beds," *Optics Express*, vol. 18, no. 8, pp. 8220–8228, 2010.

[10] R. K. Wang and L. An, "Doppler optical microangiography for volumetric imaging of vascular perfusion in vivo," *Optics express*, vol. 17, no. 11, pp. 8926–8940, 2009.

[11] S. F. Morrison, "Chapter 50 - central thermoregulation," in *Primer on the Autonomic Nervous System (Third Edition)*, pp. 243 – 247, San Diego: Academic Press, third edition ed., 2012.

[12] J. T. Shepherd, N. J. Rusch, and P. M. Vanhoutte, "Effect of cold on the blood vessel wall.," *General pharmacology*, vol. 14, no. 1, pp. 61–4, 1983.

[13] J. M. Johnson, "Mechanisms of vasoconstriction with direct skin cooling in humans.," *American journal of physiology. Heart and circulatory physiology*, vol. 292, no. 4, pp. H1690–H1691, 2007.

Development And Simulation Of An Optical Coherence Tomography Catheter For Early Screening Of Colorectal Cancer

S. Lohmann [1] M. Gora [2,3,4], J. Dong [2], T. Ford [2], L. Quénéhervé [2], G. Tearney [2,4] and R. Birngruber [5]

[1] Medizinische Ingenieurwissenschaft, Universität zu Lübeck, sabrina.lohmann@student.uni-luebeck.de

[2] Wellman Center for Photomedicine, Massachusetts General Hospital, {Gora.Michalina,jdong,lqueneherve}@mgh.harvard.edu, gterney@helix.mgh.harvard.edu

[3] French National Center for Scientific Research, ICube Laboratory, gora@unistra.fr

[4] Department of Pathology, Massachusetts General Hospital and Harvard Medical School, Boston, Massachusetts, gterney@helix.mgh.harvard.edu

[5] Institut für Biomedizinische Optik, Universität zu Lübeck, bgb@bmo.uni-luebeck.de

Abstract

Colorectal Cancer is one of the most common cancers in the United States. Since the current standard of care does not enable a high detection rate and accurate characterization of colonic polyps, there is a need for a new screening device to be introduced to a clinical practice. This paper presents development and simulation of a trans rectal catheter utilizing optical coherence tomography to provide microscopic cross sectional images of the entire human colon without need of excising tissues. The device is composed of an optical probe that is enclosed in a protective sheath and introduced into the colon through a working channel of the standard colonoscope. The development process towards the first clinical prototype including simulation of the optical performance and mechanical design is described in details.

1 Introduction

The human colon is a tubular organ that is divided into sigmoid, ascending, transverse and descending segments as well as caecum and rectum (Fig. 1). Its main function is to absorb water, electrolytes, bile acids and vitamins [1]. The wall of a healthy human colon is composed of four concentric layers: mucosa, submucosa, muscularis propria and serosa [1]. The most superficial mucosa layer is further divided into epithelium, lamina propria and muscularis mucosa (Fig. 2(a)). It is common in adults to develop colon polyps, which are growth that form in the normal lining of the colon. Even though only a small percentage of polyps have cancerous potential, colorectal cancer (CRC) is one of the most common cancers in the United States with 136,830 new cases per year (2014 [2]). Among various polyp types present in the human colon, adenomas (Fig. 2(b)) with a size larger than 1 cm have the highest risk of transformation into a colonic adenocarcinoma[3]. Different serrated forms of polyps are also common: hyperplastic polyps have a low risk of transformation and serrated adenomas and sessile serrated adenomas (Fig. 2 (b)) increased risk of transformation was recently highlighted [4]. Depending on the type of polyp an appropriate treatment and follow up is recommended. The analysis of the colon mucosa in targeted patients is necessary to detect and remove the pre-neoplastic lesions in order to avoid the onset of cancer. Currently white light colonoscopy with biopsies or removal of the polyps for pathological analysis is recommended regularly for evalua-

tion. The interval between two colonoscopies depends on the type, size and number of identified polyps. White light endoscopy can detect most nonflat polyps in the field of view of the colonoscope, but cannot easily distinguish between adenomas and benign polyps without biopsy. It also has a limited detection rate because of the complex geometry of the colon with folds blocking the field of view [10].

New technologies are needed to improve polyps detection rate and ability for real-time characteristics of polyps. Among others, narrow band imaging [6] and confocal endomicroscopy [7] has gained increased attention as having potential to enhance *in situ* imaging and diagnostic capabilities of the current standard of care.

Optical coherence tomography (OCT) is another optical imaging technique that allows the acquisition of tomographic images similar to ultrasound imaging but with much higher resolution. As a medical imaging technique OCT is most often used in ophthalmology, dermatology and cardiology. In the digestive tract it has been mostly used for imaging of the esophagus [9] but preliminary data showing potential of OCT for differentiation of colon polyps has been presented [10].

In order to address current limitations of the standard of care endoscopic examination a device was developed that aims on improving detection of polyps and their *in situ* characterization using OCT to enable a complete, minimally invasive, early stage screening of the mucosa throughout the whole length of the colon. This paper describes in details the design process from the design inputs, through optimization of the optical design to a proposed mechanical design of the catheter that is compatible with the standard of care colonoscope.

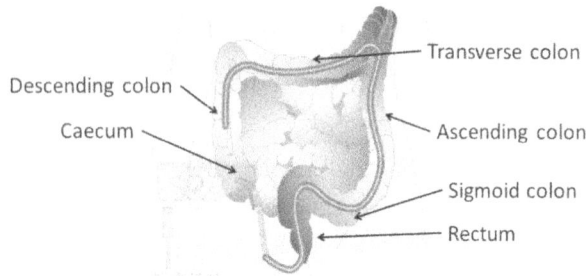

Figure 1: The human colon [5]

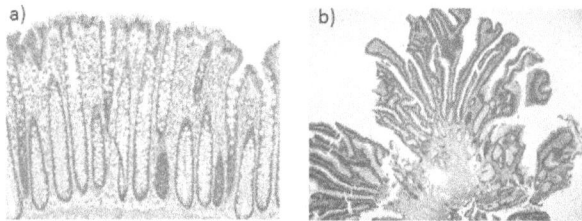

Figure 2: (a) Healthy Mucosa [1], (b) Hyperplastic adenoma [8]

2 Material and Methods

The device comprises a swept source OCT (SS-OCT) console and an optical catheter guided through the working channel of an endoscope. In a single measurement the SS-OCT system provides depth-resolved reflectance information from the tissue, called an A-line. For volumetric imaging scanning of the area of interest (AOI) is required. Three dimensional imaging of the luminal colon is achieved by radial scanning of the optical beam and longitudinal translation of the optical probe to achieve a helical scan pattern. In order to fulfill the Nyquist sampling criteria the radial scanning frequency has to be chosen based on the lateral resolution of the probe and the expected circumference of the lumen.

OCT Imaging System

The SS-OCT system called optical frequency domain imaging system, is based on a swept laser and a Mach Zehnder interferometer, where the light reflected from the mirror in the reference arm interferes with the scattered light from the tissue in the sample arm. The interference pattern is detected by a photo diode and after discrete Fourier transform one depth-resolved reflectivity array (A-line) is achieved. The laser emits near infrared light at 1300 nm with a bandwidth of 0.06 nm which provides an axial resolution of 9 μm and a penetration of depth up to 2 mm. The output laser power of this system is less than 50 mW [11].

Probe Design

The trans-rectal OCT catheter is composed of an outer sheath that protects an inner optical core, which transmits the light from the OCT imaging console towards the tissue. The inner optical core is further composed of an optical fiber centered with a ferrule in front of a 1 mm spacer, GRIN lens and prism assembly (Fig. 5). The optical fiber (Fig. 3) is enclosed in a driveshaft that delivers a torque from the proximal end to the distal end of the catheter (Fig. 3). The outer sheath is composed of a clear polyamide tubing with a tapered most distal end of 20 cm and a proximal extension that connects to a telescopic segment for reconnecting the catheter to the imaging system (Fig. 3).

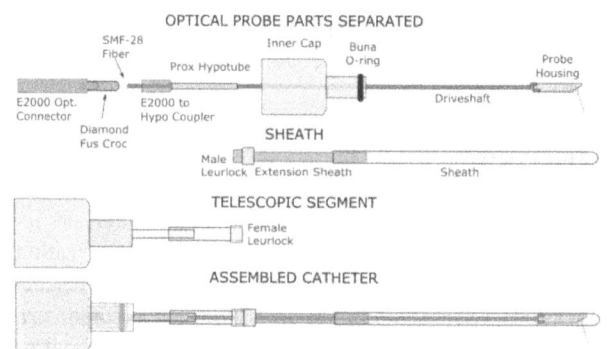

Figure 3: Schematic representation of OCT Catheter

Colonoscope

The probe design is compatible with colonoscopes available at the Gastrointestinal Unit at Massachusetts General Hospital (MGH) that are used for standard of care examination. The exemplary scope is EC-2990Li (Pentax) that has an outer diameter of 10.2 mm, a high definition white light endoscopic visualization, and one insertion channel with an inner diameter of 2.8 mm. The length of the insertion channel is 1.700 m and the total length of the scope is 2.023 m (Fig. 6).

Design Inputs

The key requirement for this design where high patients safety, minimal alteration of the standard of care procedure, and circumferential imaging of the colonic wall. In order to find a design, which meets the requirements for imaging colon tissue and compatibility with the current standard of care, a few limitations are faced. The key requirement for the optical design (focal and Rayleigh lengths) of the probe is based on the dimensions of the human colon (Fig. 4 (a)). The radius of the human colon lumen varies from 20 to 30 mm. For the mechanical design of the catheter the most important limitations come from the diameter and the rigidity of the working channel of the colonoscope and typical curvatures in the human colon that can reach up to 90°. This will determine the acceptable dimensions and flexibility of the optical core and the outer sheath.

Another key input for the design was to choose the most appropriate scanning method. There are different ways to effectuate radial scanning. It is possible to use a motorized rotating joint at the proximal end outside of the patient. In this case a driveshaft is used to rotate the whole optical core inside the static sheath. Another considered option is to use an angle polished micro-motor placed at the most distal end of the catheter and in front of the optical setup without a prism. This way electrical current needs to be led through the catheter to the tip of the probe where the motor is mounted. Both scanning options require designated optical designs that take into consideration different distances for the light to travel to the focal region. The details of the optical and mechanical design of the final design are presented below.

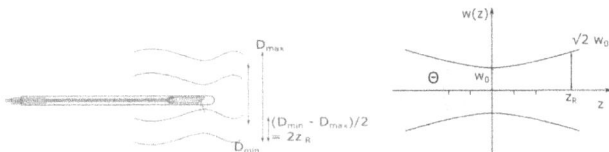

Figure 4: (a) Schematic drawing of human colon varying diameter, (b) Gaussian focus: beam waist ω_0, Rayleigh length = depth of focus z_R

Geometric vs. Gaussian propagation

Zemax is a specialized software that simulates light propagation including aberrations, through optical components (e.g. lenses) defined. However Zemax provides only geometrically simulations of the optical setup.

For geometrical ray tracing only paraxial rays can be considered. Since the focal spot is very small, results of Gaussian ray tracing are very similar to the geometrical Ray tracing. An optical lens makes a Fourier transform between the two focal planes. Gaussian functions stay perfectly Gaussian after the Fourier transform. A Gaussian function at the beam waist stays a Gaussian function at the beam waist behind the lens. Only its size changes. The larger the Gaussian focal spot is the higher is the difference between the Gaussian beam waist and the geometrical focus and the more important it is to calculate the Gaussian beam propagation additionally.

Gaussian calculation

These calculations of z_R are made at first to find out the required Rayleigh length to adjust focusing to the thickness of the colon wall and its expected misplacement due to peristalsis for example (Fig. 4 where $2z_R$ equals the focal length and D_{max}/D_{min} are the average diameter of the human colon). For a given focal range calculated by

$$\frac{D_{max} - D_{min}}{2} = 2z_R, \qquad (1)$$

the numerical aperture

$$NA = n \sin \frac{\Theta}{2} \qquad (2)$$

as well as the beam waist

$$\omega_0 = \sqrt{\frac{z_R \cdot \lambda}{\pi}} \qquad (3)$$

can be calculated (compare Fig. 4).

In order to get the highest possible lateral resolution the whole surface of the refractive lens, in this case the GRIN lens, has to be used. A certain length of the spacer realizes this requirement. Depending on the length of the GRIN lens different focal lengths with different NA can be realized. A trade-off has to be found between the required focal length and the lateral resolution.

3　Results and Discussion

Gaussian Approach

Tab. 1 shows the Rayleigh length and lateral resolution [12]. A perfect beam without any aberrations would lead to a beam waist of 65 μm at a Rayleigh length of 10 mm, for a shorter Rayleigh length a higher lateral resolution is possible. The following results are concluded from this calculation and the accounting for aberrations. As shown in Fig. 4 the diameter of the area to be imaged varies over a large range. Thus a long focal length needs to be achieved, which is $10 - 30$ mm to image the radius. This leads to a solution of a focal length of around 20 mm, where the focal spot is put into the middle of the diameter range of the human colon.

Table 1: Optical parameter Gaussian optics

imaging diameter D_{min} to D_{max}	Rayleigh length	NA	ω resolution
$20 - 60$ mm	10 mm	0.32	65 μm
$20 - 40$ mm	5 mm	0.37	45 μm

Optical Core Design

Due to the limited diameter of the accessory channel of the colonoscope, an external scanning method using a rotary junction and a driveshaft was chosen, as opposed to an integrated motor. Reasons include that the motor would have to be very small, which would increase the cost of the device and it would be challenging to provide double insulation for the electric current as required by the guidelines for medical devices. For the proximal scanning solution, the final optical design with a long focal length of 25 mm and a Rayleigh length of 10 mm is shown in Fig. 5 a. A small astigmatism is introduced by the outer sheath. Since this astigmatism is negligible, no additional optical correction is considered as necessary (Fig. 5 b).

Complete Trans Rectal OCT imaging system

The final imaging system is a combination of an OCT catheter guided through the working channel of the colonoscope as shown in Fig. 6. The inner optical probe is housed

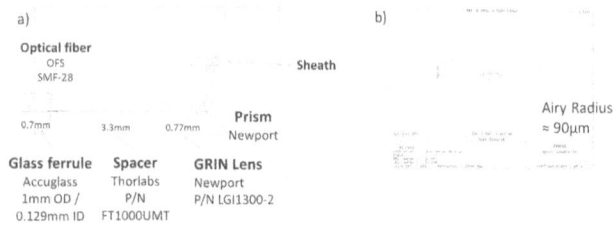

Figure 5: (a) Schematic drawing of the optical core used in combination with driveshaft (b) Airy disc of proposed optics

by a combination of a clear sheath used for imaging and a telescopic sheath and segment for shielding purposes. The catheter will be connected to the SS-OCT system that has 6.4 mm imaging range in the air. It is expected that this distance may not be long enough and a different light source will have to be used with a longer imaging range or an active adjustment of the reference arm will have to be introduced.

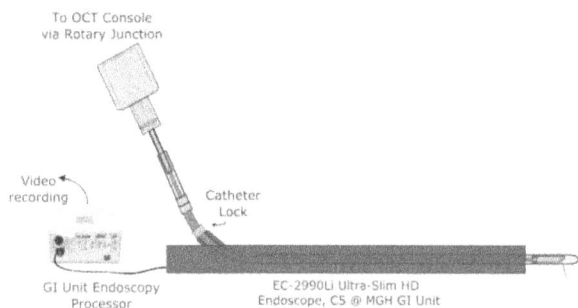

Figure 6: Schematic representation of complete catheter

4 Conclusion

The design of the OCT catheter for imaging of the whole circumference of the colon is presented here. Because the diameter of the colon varies over a range diameter, based on a Zemax simulation the transverse resolution needs to be decreased to image the entire range of possible diameters. The final simulated optical design meets the requirements to image a human colon. The mechanical design of the catheter is compatible with the standard of care colonoscopes and is compatible with the intended clinical use and sterilization. The final combination of the FDA approved endoscope and an optimized complete setup enables to begin the process of device development, towards first-in-human use.

Acknowledgement

The work has been carried out at the Wellmann Center for Photomedicine in Boston, MA. We would like to thank Catriona Grant, Amna Soomro and Thomas Cerruto for helpful feedback on clinical material and translation.

5 References

[1] J. F. Reinus and D. Simon, *Gastrointestinal Anatomy and Physiology: The Essentials*, Wiley-Blackwell, 2014.

[2] R. Siegel, et. al., *Cancer Statistics, 2014*, CA Cancer Journal Clinicians, 2014.

[3] M. J. O'Brien, S. J. Winawer, A. G. Zauber, et al., *The National Polyp Study. Patient and polyp characteristics associated with high-grade dysplasia in colorectal adenomas*, in: Gastroenterology, vol. 95, pp. 3053–63, 1990.

[4] J. H. Bond, *Colon Polyps and Cancer*, Endoscopy, 2003.

[5] Société Française d'Endoscopie Digestive, *Coloscopie*, Available: http://www.sfed.org/patients/examens/... coloscopie/coloscopie [last accessed on Jan 10th 2016].

[6] A. Ignjatovic et. al., *Opdical diagnosis of small colorectal polyps at routine colonoscopy (Detect Inspect Chsracterise Resect and Discard; DISCARD trial): a prospective cohort study*, Lancet Oncology, vol. 10, pp. 1171 -1178, 2009.

[7] A. M. Buchner et. al., *Comparison of Probe-Based Confocal Laser Endomicroscopy With Virual Chromoendoscopy for Classification of Colon Polyps*, Gastroenterology, vol. 138, pp. 834 – 842, 2010.

[8] D. Moussata, et. al., *Endoscopic and histologic characteristics of serrated lesions*, World Journal of Gastroenterology, vol. 21, pp. 2896 – 2904, 2015.

[9] M. J. Gora, et. al, *Tethered capsule endomicroscopy enables less invasive imaging of gastrointestinal tract microstructure*, nature medicine, vol. 19, pp. 238 – 241, 2013.

[10] S. Jaeckle, et.al, *In Vivo Optical Coherence Tomography of the Human Gastrointestinal Tract - Toward Optical Biopsy*, Endoscopy, vol. 32, no. 10, pp. 743 - 749, 2000.

[11] S. H. Yun, et. al. *High-speed optical frequency-domain imaging*, Optical society of America, vol. 11, no. 22, pp 2953–2963, 2003.

[12] E. Hecht, *Optics*, vol 2, Pearson education (Addison-Wesley, 2002), 2002.

Sensitivity Characterization of an Epifluorescent Microscope for Tracking of Photoluminescent Nanoparticles in Chick Embryo Tumour Models

I. Kuschnerus [1], A. Nadort [2], Y. Lu [2], A. Guller [2] and A.V. Zvyagin [2]

[1] Medizinische Ingenieurwissenschaft, Universität zu Lübeck, inga.kuschnerus@miw.uni-luebeck.de

[2] ARC Centre of Excellence for Nanoscale BioPhotonics, MQ Photonics, Macquarie University, NSW 2109, Australia, {annemarie.nadort, yiqing.lu, anna.guller, andrei.zvyagin}@mq.edu.au

Abstract

The visualization of tumours in cancer therapy research is an important application of photoluminescent nanoparticles. The chick embryo chorioallantoic membrane (CAM) is a suitable model for tumour visualization because it is responsible for the gas and nutrient exchange during embryonic development and human tumours can be grafted on it. Epifluorescent stereomicroscopy can be used to observe nanoparticles but characterization of the microscope sensitivity is essential to ensure reliable visualization. In this project, the sensitivity was estimated using an optical phantom, i.e. a thin microcapillary tube filled with photoluminescent silica nanoparticles suspended in water. The results show the potential to image 1000 nanoparticles/pixel in capillary systems. This sensitivity is appropriate for tumour visualization and could form the basis for further studies of nanoparticle-tumour interactions such as the accumulation of nanoparticles in tumours, toxicity effects of nanoparticles in vivo and the therapeutic effects of drug-loaded nanoparticles.

1 Introduction

Nanomaterials hold promise in cancer research and clinical oncology as contrasting and therapeutic agents or as drug delivery vehicles. In particular, photoluminescent (fluorescent) nanoparticles are applicable for diagnostic imaging owing to their contrast-rendering properties, and therapy owing to their ability to carry drug cargo to pathology lesions. There is a great demand in studies of biocompatibility and behaviour of the photoluminescent nanoparticles in vivo, hence suitable models are needed [1]. Following the ethical limitations and challenging conditions for real-time imaging of photoluminescent nanoparticles in conventional animal models (need in special high-end equipment for imaging in mice, rats, etc.), significant attention was attracted to tumour models on the chick embryo chorioallantoic membrane (CAM) [2], [3]. CAM is an extraembryonic gas-exchanging membrane. It is well vascularized and relatively easily accessible for tumour grafting and observation [2]. Fluorescent silica nanoparticles represent a promising high-contrast drug delivery nano-vehicle in CAM due to their relatively easy and cheap production and their biocompatibility [4]. Stereomicroscopy is an affordable method for imaging of tumour progression on CAM for example to analyse the tumour angiogenesis [5]. An example of grafting human colon cancer cells on CAM imaged with the Olympus stereomicroscope (Tokyo, Japan) is shown in Fig. 1. There are publications discussing stereomicroscopy applications for the analysis of the distribution of various dyes and luminescent nanomaterials, e.g. [6]. However, almost nothing is known about the sensitivity of stereomicroscopy systems in detection of (certain types of) photoluminescent nanoparticles, while this is critically needed for the generation of quantitative assays of the nanoparticle's blood circulation times and accumulation in tissues. In this study we therefore structurally assessed the sensitivity limits of stereomicroscopy for vascular fluorescence imaging.

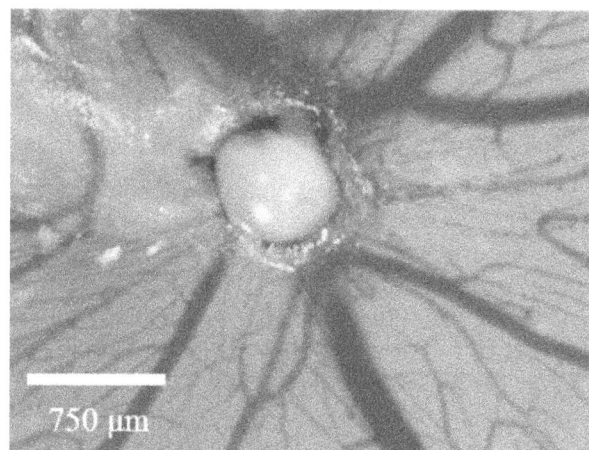

Figure 1: Human colon cancer cells grafted on chick embryo chorioallantoic membrane (CAM), imaged with the Olympus stereomicroscope (image by *ARC Centre of Excellence for Nanoscale BioPhotonics*).

2 Material and Methods

To provide an understanding of the experimental setup, the materials will be listed and the experimental procedure will be described.

Epifluorescent Stereomicroscope In this project, the sensitivity of the epifluorescent stereomicroscope Olympus MVX10 (Tokyo, Japan) is characterized. The stereomicroscope features an eyepiece magnification range of 0.63x to 6.3x and provides a 3D view of the sample. This is achieved by having two optical pathways through two eyepieces which are focused on the same spot. Due to this special optical setup, two separate images at slightly different angles are produced which allows stereoscopic depth perception [7]. The MVX10 operates with a fixed focus and two objective lenses, NA 2x/0.5 and 1x/0.25. The samples are illuminated from above using a 100 W mercury lamp. To image the photoluminescent nanoparticles, a filter cube containing one single-edge short pass dicroic beamsplitter (Semrock, New York, USA, transmits wavelengths greater than 750 nm) and a single-band band pass filter as the emission filter (Semrock, New York, USA, transmits wavelengths between 510 nm and 570 nm) are used. An electron-multiplication CCD camera (Andor EMCCD iXon 885 DU, Belfast, UK) is mounted on the microscope port. Table 1 shows the camera features in detail. The EMCCD camera is controlled by the Andor SOLIS software, version 4.22.

Table 1: Camera features in detail.

Camera	Andor EMCCD
A/D resolution in bit	14
Readout speed in MHz	27
Pixels	1004 x 1002
CCD sensitivity in electrons/A/D count	2.53
Quantum efficiency (QE) in %	60

Photoluminescent Nanoparticles The sensitivity of the stereomicroscope is characterized with the use of green photoluminescent silica nanoparticles *sicastar®-F*, purchased from Micromod (Rostock, Germany, product no. 42-00-102). They are produced by hydrolysis of orthosilicates, have a hydrophilic surface with terminal Si-OH (silanol) moieties and contain covalently bound FITC [4]. The main characteristics of the aqueous colloid of spherical nanoparticles are as follows: the concentration is 50 mg/ml, the mean diameter of the monomodal size distribution is 100 nm and the nanomaterial density is 2 g/cm^3. This amounts to $9.5 \cdot 10^{11}$ nanoparticles/mg. The excitation and emission spectra peaks lie at 485 nm and 510 nm, respectively. They are stable in water and organic solvents [4].

Chick Embryo Chorioallantoic Membrane (CAM) CAM functions similar to the placenta in humans. It is responsible for the gas and nutrient exchange in the embryo. It is widely used as a cancer research tool because of relative ease of observation. Due to the lack of a fully developed immune system in a chick embryo of certain age, several types of human tumours can be grafted on CAM. Rats and mice have been used as an in vivo model. However, tumours grafted on CAM are easily accessible for microscopic observation. The production is also less complicated and less expensive than using rats or mice [2]. The procedure used for previous experiments was the *ex ovo culturing*. After three days of incubation, the fertilized eggs are opened and transferred to a petri dish, covered with sterile plastic wrapping and further incubated [2]. Tumours are usually grafted on day 8 to 10 of embryonic development.

Colloidal Sample Preparation The first sample type is prepared using the following procedure: The nanoparticles are further diluted in distilled water and 10 colloids with silica concentrations from 5 to $2.5 \cdot 10^{-4}$ µg/µl are produced. The nanoparticles are added to a centrifuge tube with a volume of 0.5 ml and filled with distilled water. Next, the samples are mixed by vortexer for approximately 30 seconds. For each colloid, five slides are prepared: 1 µl of the sample is placed on each slide and dried. After that, the imaging is performed by the Olympus MVX10. The second sample type is prepared by diluting the nanoparticles in distilled water according to the previous concentration range. Then 2 µl colloid is filled in a microcapillary glass tube by Sigma-Aldrich (Castle Hill, Australia), with an outer diameter of 1 mm and an inner diameter of 0.94 mm and a total capacity of 20 µl. These tubes are used to simulate the blood vessels of CAM. They are sealed to prevent drying. Fig. 2 shows images of the tested samples acquired by the epifluorescent microscope.

Optical Microscopy During the measurements, the eyepiece magnification is always set to 0.8x. With the Andor EMCCD camera, it is possible to set the electron-multiplication gain (EM gain). With this EM gain the CCD sensor of the camera is able to detect single photons [8]. The EM gain is adjustable in real time. The exposure time and the EM gain while performing the measurements are continuously adjusted by histogram in order to get the same (the highest) signal level in all images. Dark images are also recorded for accurate calibration. For these measurements, several images of the microscope slide are acquired and processed with the excitation light switched off.

Characterization of the Sensitivity In detail, the sensitivity is represented by the lowest particle concentration visible. To characterize the sensitivity, the signal amplitude is determined for each picture taken. The signal is expressed in terms of number of photons/second/pixel. The signal average for each picture is determined using the software ImageJ version 2.0.0. Prior to the estimation of the sensitivity, the average signal is normalized according to the individual

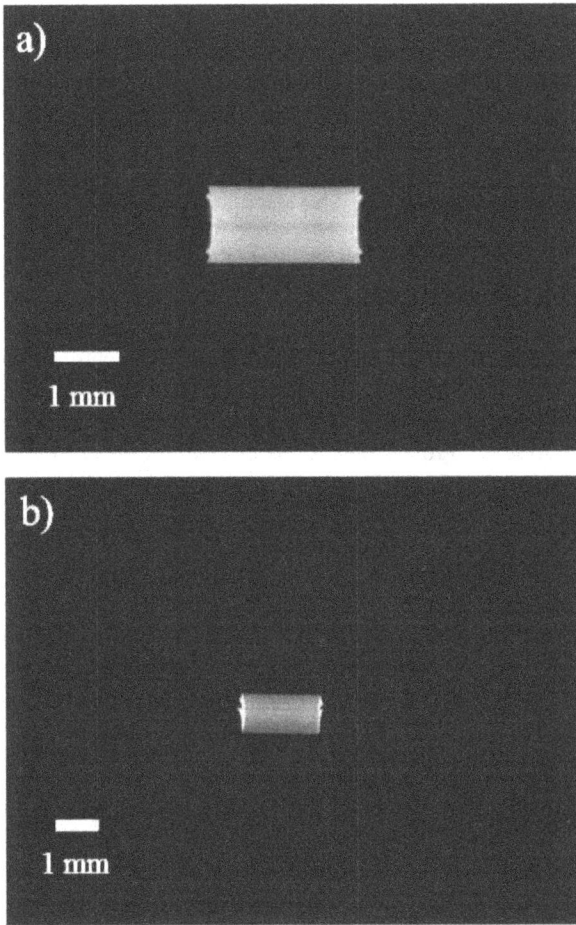

Figure 2: Images of the filled microcapillary tubes. a) Image taken with the 0.5 NA lens and b) with the 0.25 NA lens.

exposure time (ET) and the EM gain. Next, the sensitivity is determined using the following equations:

$$I_{cor} = I_{avg} - I_{dark}, \qquad (1)$$

$$N_{photon} = \frac{I_{cor}}{S_{CCD}}, \qquad (2)$$

$$S_{photon} = \frac{N_{photon}}{QE}. \qquad (3)$$

I_{avg} describes the average signal of the image and I_{dark} the signal of the dark image. I_{cor} is the corrected image signal. N_{photon} represents the number of electrons/photon, S_{CCD} the CCD sensitivity (electrons/A/D count). QE is the quantum efficiency and S_{photon} the sensitivity in terms of photons/second/pixel. The concentration is estimated in terms of the number of particles/pixel by dividing the total amount of particles in the sample by the total number of pixels representing the sample. For the tubes the depth of focus (DOF) has to be taken into account since the particles outside the focal plane contribute less to the image signal. The DOF is estimated using the Berek formula [9]:

$$DOF = \frac{\omega \cdot 250000}{NA \cdot M} + \frac{\lambda}{2(NA)^2}. \qquad (4)$$

ω represents the resolving power of the eye (0.0014, given by Olympus), NA the numerical aperture, λ the wavelength (emission wavelength of 510 nm) and M the total magnification. M is estimated by multiplying the eyepiece magnification with the objective lens magnification. Using the 0.5 NA objective the estimated DOF is smaller than the inner tube diameter. To calculate an appropriate amount of particles the sample volume is adjusted according to the DOF.

3 Results and Discussion

The results of the sensitivity characterization are plotted using a common logarithmic scale. Since the EMCCD camera has a linear response to optical signals, the results should show a linear increase of the signal with increasing concentration. The following graphs show the results of comparing the sensitivity of the stereomicroscope with the EMCCD camera using two objective lenses. As shown in Fig. 3, for the 0.25 NA lens and concentrations over $5 \cdot 10^{-4}$ µg/µl with around $1.3 \cdot 10^2$ particles/pixel the sensitivity increases linearly with a linear correlation coefficient of $R^2 > 0.97$. The 0.5 NA lens shows a greater linear increase of the signal with $R^2 > 0.98$ for the same concentration. At lower concentrations than the last-mentioned, significant scatter is dominating. This is mainly caused by reflection of excitation light by the glass slide which leaks through the emission filter. The graph in Fig. 4 shows the

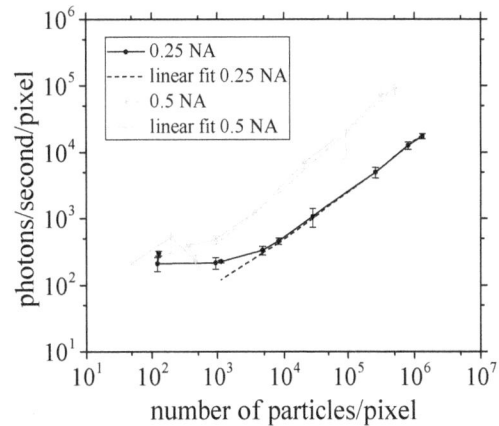

Figure 3: Sensitivity of the imaged samples on glass slides.

sensitivity of the microscope imaging the filled microcapillary tubes. Again, the glass of the tubes reflects the excitation light which causes the noise for lower concentrations. The 0.25 NA lens shows a linear correlation coefficient of $R^2 > 0.98$ for concentrations over $1 \cdot 10^4$ particles/pixel with $2.5 \cdot 10^{-2}$ µg/µl. Imaging the tubes, we see significant noise using the 0.5 NA lens at greater concentrations (around $3.5 \cdot 10^4$ particles/pixel and $2 \cdot 10^3$ particles/pixel) compared to the microscope slides. The reason for this error might be staining, e.g. dust on the tube. Ignoring this error, there would be a linear increase for concentration over

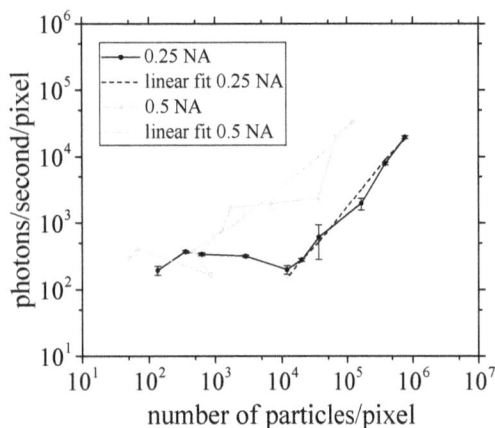

Figure 4: Sensitivity of imaged samples in capillary tubes.

$5 \cdot 10^{-3}$ µg/µl ($5 \cdot 10^2$ particles/pixel) with $R^2 > 0.94$. Overall, the microcapillary tubes show a slightly lower signal with $3 \cdot 10^4$ photons/second/pixel as the greatest value. In comparison, the highest signal which can be measured for the samples on slides is approximately $9 \cdot 10^4$ photons/second/pixel. On the slides, all particles are in the focal plane, which means it is a highly idealised model since it does not represent the shape of blood vessels in CAM. Regarding the application of nanoparticles injected in CAM, this means that we probably cannot expect detection of $5 \cdot 10^{-4}$ µg/µl, which are around $1.3 \cdot 10^2$ particles/pixel (as for the samples on slides using the 0.25 NA lens), in this in vivo model. This is due to the fact that in CAM interfering factors such as auto-fluorescence and further reflection by blood and tissue has to be taken into account. However, we expect that the concentration of $5 \cdot 10^{-3}$ µg/µl with $1 \cdot 10^3$ particles/pixel can be detected easily in blood vessels of CAM in real time by epifluorescence microscopy. Previous tumour-on-CAM studies used concentrations of $1 \cdot 10^{-1}$ µg/µl of cowpea mosaic virus (CPMV) particles with a size of 30 nm labelled with fluorescent dyes [3].

4   Conclusion

The results of this project show that epifluorescence stereomicroscopy can be used as a reliable and affordable alternative method for the detection of nanoparticles in blood vessels in CAM. In detail they show the capability of the Olympus stereomicroscope with the Andor EMCCD camera to detect $1 \cdot 10^3$ particles/pixel. The sensitivity should be high enough to detect these nanoparticles in blood vessels of in vivo models such as the chick embryo. An important aspect is that the sensitivity should change with different types of nanoparticles (due to different size etc.). For CAM, the sensitivity is obviously lower due to auto-fluorescence and scattering of the tissue. The next step would be the study of the sensitivity for nanoparticles in biocompatible solvents such as phosphate-buffered saline (PBS) or blood. For this purpose the nanoparticles might need further polymer coating such as polyethylene glycol (PEG) to increase their circulation lifetime. However, solvent as well as coatings can influence the fluorescence properties of the nanoparticles [10]. After these experiments, further studies would include the injection in CAM vessels and the final characterization of the imaging sensitivity to detect silica nanoparticles in this in vivo model.

Acknowledgement

The work has been carried out at the ARC Centre of Excellence for Nanoscale BioPhotonics, MQ Photonics, Macquarie University, NSW 2109, Australia. The author would like to thank Prof. Dr. C. G. Hübner, Institute of Physics, Universität zu Lübeck.

5   References

[1] K. Liu, X. Liu, Q. Zeng, Y. Zhang, L. Tu, T. Liu et. al, *Covalently Assembled NIR Nanoplatform for Simultaneous Fluorescence Imaging and Photodynamic Therapy of Cancer Cells*. in: ACS Nano, vol. 6, no. 5, pp. 4054–4062, 2012.

[2] A. Nadort, L. Liang, E. Grebenik, A. Guller, Y. Lu, Y. Qian et. al, *Systematic assessment of blood circulation time of functionalized upconversion nanoparticles in the chick embryo*. Proc. SPIE, vol. 9668, pp. 96683Y–96683Y-7, 2015.

[3] H. S. Leong, N. F. Steinmetz, A. Ablack, G. Destito, A. Zijlstra, H. Stuhlmann et. al, *Intravital imaging of embryonic and tumor neovasculature using viral nanoparticles*. in: Nature Protocols, vol. 5, no. 8, pp. 1406–1417, 2010.

[4] Micromod Partikeltechnologie GmbH, *Products 2016*. 2016.

[5] D. Ribatti, *History of Research on Tumor Angiogenesis*. Springer Netherlands, pp. 8–10, 2008.

[6] J. S. Yoo, H.-M. Johng, T.-J. Yoon, H.-S. Shin, B. C. Lee, C. Lee et. al, *In vivo fluorescence imaging of threadlike tissues (Bonghan ducts) inside lymphatic vessels with nanoparticles*. in: Current Applied Physics, vol. 7, no. 4, pp. 342–348, 2007.

[7] K. Scoffin, *Stereomicroscope Technology*. in: American Laboratory, vol. 45, no. 6, pp. 28–30, 2013.

[8] Andor, *Andor System Performance Ixon plus*. Andor Technolody Limited, ed. 1, 2002.

[9] Olympus, *MVX10 MacroView*. 2008.

[10] G. Prencipe, S. M. Tabakman, K. Welsher, Z. Liu, A. P. Goodwin, L. Zhang et. al, *PEG Branched Polymer for Functionalization of Nanomaterials with Ultralong Blood Circulation*. in: Journal of the American Chemical Society, vol. 131, no. 13, pp. 4783–4787, 2009.

2

Biochemical Physics I

Simulation of UV Absorption of the Toxic Gas Hydrogen Sulfide

P. Olenik [1], O. Wille [2], and P. Rostalski [3]
[1] Medizinische Ingenieurwissenschaft , Universität zu Lübeck, philipp.olenik@student.uni-luebeck.de
[2] Dräger Safety AG & Co. KGaA,
[3] Institut für Medizinische Elektrotechnik, Universität zu Lübeck, philipp.rostalski@uni-luebeck.de

Abstract

One way of detecting toxic gases is to measure the absorption of light with a characteristic wavelength in the gas. We simulate the absorption characteristics of several toxic gases in the UV range. The spectral simulations base on numerical convolutions. This numerical convolutions include the emission characteristics of the light source, the molecular spectra depending on the environmental conditions, the filter curves and the sensitivities of different detector materials. The simulations identify suitable filter parameters and detector materials for different applications. The dependence of measurement results on the absorption length and the temperature is analyzed and discussed. We identify that a wavelength about 215 nm and a spectral width about 10 nm promises the best results. Furthermore a silicon detector has the best sensitivity. On the basis of this information a proposal for an experimental setup is made.

1 Introduction

Toxic gases like hydrogen sulfide, sulfur dioxide or benzene are extremely hazardous and jeopardize people in industrial environments. Fixed or portable sensors can help to protect people from being poisoned. The present work aims at the development of a fixed optical sensor for toxic gases. A typical optical gas sensor for industrial applications comprises of a light source, an absorption pathway through the surrounding atmosphere, one or more optical filters and several detectors like shown in Fig. 1.

Figure 1: Typical setup of an optical gas sensor.

The light source sends out an electromagnetic spectrum. Depending on the type of the lamp it consists of one spectral line (e.g. laser or LED), several lines (e.g. mercury vapor lamp) a continuous thermal distribution similar to a blackbody spectrum (e.g. halogen lamp) or several lines on an underlying thermal continuum (e.g. Xenon flash lamp). This light passes the absorption area and reaches the filter and the detectors behind. The optical filter narrows the detected spectral regions and an occurred absorption can be detected [1].
Infrared optical gas sensors (spectral range 3 μm to 4,5 μm) are widely-used for the detection of explosive hydrocarbons or carbon dioxide environments. Toxic gases like hydrogen sulfide, sulfur dioxide, nitrogen oxides or aromatic hydrocarbons show typical absorption bands in the UV range from 200 nm to 300 nm. They can also be detected with electrochemical sensors, but these sensors have a limited lifetime due to contamination. With a UV based sensor the detection of nitrogen oxide with a sensitivity of less than 1 ppm is demonstrated in [2]. There are also commercial UV based open path detectors for toxic gases, [3].

This work aims for a theoretical model of an optical gas sensor for toxic gases like hydrogen sulfide (H_2S). With this model the relevant parameters for the system (path length, filter parameters) and different materials for light source and detectors will be discussed. Furthermore a proposal for an experimental setup will be made.

2 Material and Methods

The spectra of the source, the detector and the filter are provided by the distributors [4], [5]. Additional to the spectra of these parts we must include the spectra of the gas in our model. The gas spectra are provided in a database by the Satellite-Group of the Max-Planck-Institute for chemistry in Mainz [6]. These spectra are a compilation of various scientific work. Fig. 2 shows the absorption spectrum of H_2S [7].
To simulate the absorption measurement we use Matlab [8].

2.1 Numerical Integration

The signal, measured by the detector can be written as a integral described in (1),

$$S(\lambda) = \int_0^\infty Q(\lambda) \cdot G(\lambda) \cdot F(\lambda) \cdot D(\lambda) \, d\lambda \qquad (1)$$

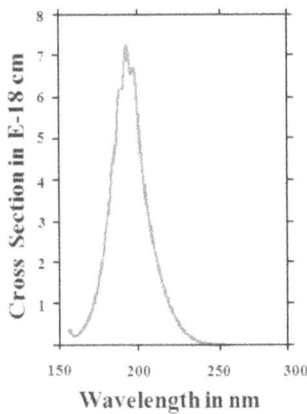

Figure 2: Absorption Spectrum of H$_2$S.

where S is the signal intensity, Q the emission spectrum of the source, G the transmission spectrum of the gas, F the transmission spectrum of the filter and D the spectral sensitivity of the detector. The spectra are not described by an analytical function, but by a finite number of sampling points. The numerical integration is approximated by a sum over the products at all sampling points using the trapezoidal rule as shown in (2)

$$S(\lambda) = \sum_{n=1}^{N} \frac{s(\lambda_{n+1}) + s(\lambda_n)}{2} \cdot (\lambda_{n+1} - \lambda_n). \quad (2)$$

In order to make sure that the sums give feasible results the spectral functions have to be approximated by a linear interpolation to have identical sets of sampling points.
Thus all spectra are filled with the necessary sampling points and each spectrum contains the same number and position of sampling points.
The absolute values of light intensities, which are allocated in the spectra, strongly depend on the experimental settings of the simulated experimental setup. So we normalized our simulations to one with respect to a configuration with no absorbing gas.

2.2   Filter Spectra

Commercial filters are specified by their center wavelength λ_0, their spectral width σ_{fwhm}, which is specified by the full width at half maximum (FWHM) and the peak transmission T$_{max}$. For simulation purposes they are approximated by a gaussian profile by (3):

$$T(\lambda) = T_{max} \cdot e^{-\frac{1}{2}(\frac{\lambda - \lambda_0}{\sigma^*})^2}. \quad (3)$$

T$_{max}$ is the position where the function of the filter has a maximum and σ^* is calculated by (4)

$$\sigma^* = \frac{\sigma_{fwhm}}{2 \cdot \sqrt{2 \cdot log(2)}}. \quad (4)$$

2.3   Signal intensity

Using the Beer-Lambert law (5), the transmitting signal intensity $I(\lambda)$ at a given wavelength can be calculated as a function of the specific absorption coefficient of a gas α, the concentration of the gas c and the experimental path length d. The coefficient is characteristic for a substance and represents its absorption characteristics [10]. In our simulated setup the experimental path length is constant so we can convert the empirical gas spectra [7] to any designated path length and concentration as follows:

$$I(\lambda) = I_0(\lambda) \cdot e^{-\alpha \cdot c \cdot d}. \quad (5)$$

I_0 describes the intensity without any gas.
Using (5) we can compute the specific absorption coefficient α for the gas of our interest. In this work we focus on hydrogen sulfide.
By normalizing I_0 to one, we get (6) as an expression for α where I_λ is the given measured value:

$$\alpha = \frac{ln(I(\lambda))}{c \cdot d}. \quad (6)$$

To get the intensity in relation to the concentration or distance, the converted intensity $I(\lambda, c^*, d^*)$ for each new concentration c^* or the new distance d^* can be calculated by (7)

$$I(\lambda, c^*, d^*) = e^{-\frac{ln(I(\lambda))}{c \cdot d} \cdot c^* \cdot d^*}. \quad (7)$$

2.4   Procedure

As detector material we simulate Silicon (Si) and Silicon-Carbide (SiC). The spectral sensitivities are available from the distributors [5].
All spectra are interpolated from 190 nm to 400 nm with 1000 sampling points per nm. To compare the absorption signals for different filter parameters the normalized signals in relation the center wavelength are calculated. We use various combinations of the center wavelength and the spectral width. Furthermore we vary the center wavelength from 200 nm to 300 nm in 2 nm steps and the width from 10 nm to 50 nm FWHM in 2 nm steps. For each filter configuration the absorption characteristics are calculated for concentrations from 0 ppm to 200 ppm H$_2$S. We consider absorption path length from 0.2 m, 0.4 m and 1 m. Additionally we simulate the influence of the temperature on the normalized signals from -40°C to 60°C. We consider typical spectral shifts for the filters [9]. The temperature effect on the gas spectrum can be neglected [7].
As a light source a Xenon flash lamp will be considered. We need a continuous spectrum with enough intensity. So lasers (too narrow) and LEDs (too short lifetime, too low intensity) cannot be used in this application. Halogen lamps do not have enough intensity in the UV region due to the absorption of the glass and are thus also not considered here.

3   Results and Discussion

In Fig. 3 the spectra of the two detectors are shown. The sensitivity for the Si-detector is much better in the spectral region below 270 nm and over 300 nm. Especially from 200 nm to 240 nm, where H_2S has its strongest absorption SiC is not the appropriate material for the detector (see Fig. 3). So we focus on Si for our simulations.

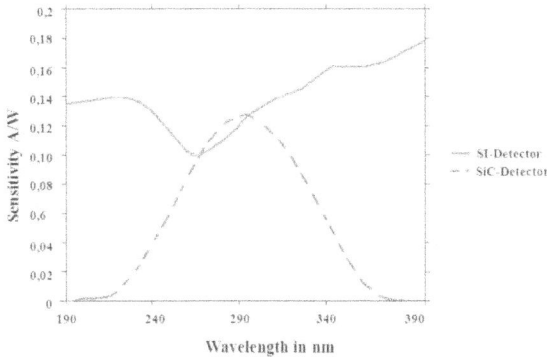

Figure 3: Spectral sensitivity for the Si- and SiC-Detector.

To find the best experimental parameter set for the filters we first compare the standardized absorption as a function of the concentration. Different center wavelengths with a constant spectral width are simulated. In Fig. 4 the normalized signals in relation to the concentration at different center wavelengths with a spectral width of 10 nm FWHM and an absorbing path length of 1 m are plotted. Center wavelength below 215 nm have not been considered due to oxygen absorption.

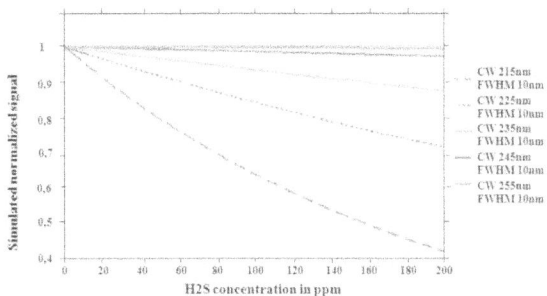

Figure 4: Normalized signal in relation to the concentration, plotted for different center wavelength with a spectral width of 10 nm FWHM and an absorbing path length of 1 m.

A center wavelength of 215 nm promises the best experimental resolution. The higher the center wavelength the less sensitive the measurement will be.
As the next step we compare the normalized absorption signals in relation to the center wavelength of the filter. Different spectral widths at a concentration of 100 ppm with an absorption path length of 1 m are considered.
Fig. 5 shows the dependence of the signal intensity in relation to the center wavelength at different spectral width. For a wavelength less than 230 nm a smaller spectral width of the filter gives a much better signal sensitivity. At a wavelength of 230 nm the simulated signals are identical and above they converge to one.
In Fig. 6 the normalized signals in relation to the concentration are plotted with different absorbing distances. A cen-

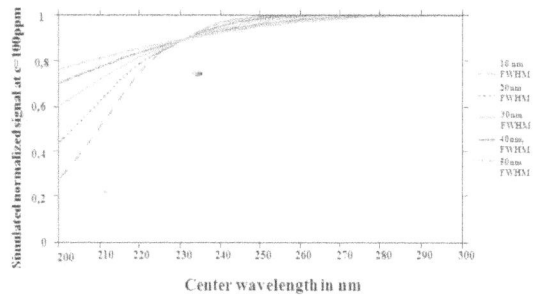

Figure 5: Normalized signals in relation to the center wavelength with various spectral width at a concentration of 100 ppm and an absorption path length of 1 m.

ter wavelength of 215 nm and a spectral width of 10 nm FWHM are chosen.

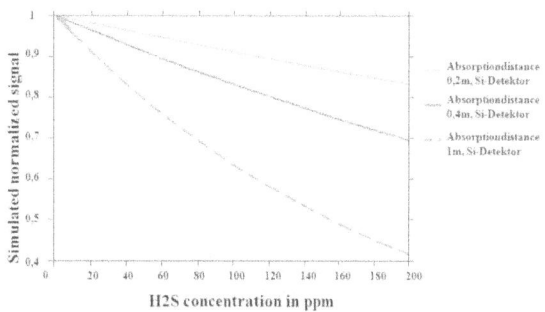

Figure 6: Normalized signals in relation to the concentration, plotted for different absorbing length at a spectral width of 10 nm and a wavelength of 215 nm.

The absorbed light and thus the resolution of the measurement increases with the absorption distance.
To evaluate the temperature dependence the normalized signals in relation to the center wavelength with different temperatures are simulated. A spectral width of $\sigma_{FWHM}=10$ nm and a concentration of 100 ppm are taken as simulated experimental parameters for the absorption measurement. The results are shown in Fig. 7. There are only minor deviations of 5% of the signal between -40°C and 60°C which can be numerical compensated in a future sensor.

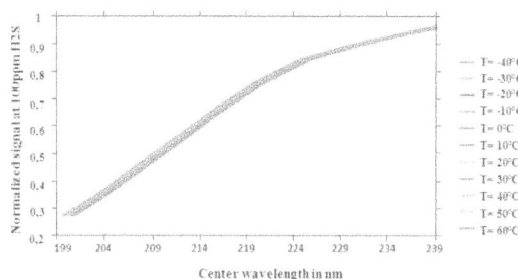

Figure 7: Normalized signals in relation to the center wavelength at different temperatures. The concentration is 100 ppm and the spectral width is 10 nm.

4   Conclusion

The normalized signals for different sets of filter parameters were simulated. In the spectral region from 200 nm to 220 nm the experimental resolution promises to be the best (see Fig. 4). Taking the atmospheric absorption of oxygen below 210 nm into account, a center wavelength of the filter in the region between 210 nm and 220 nm should be chosen for the absorption measurements of H_2S. A smaller spectral width also gives the best results so we recommend a width of 10 nm FWHM what is a common value for a distributed filter (see Fig. 5). A narrower filter would presumably reduce the measured signal intensity too much.

As an absorption path length we recommend a value of noticeably lager than 0.5 m. Otherwise the minimal resolution of the sensor will be too high. The larger the absorption distance is, the better the resolution will be, but also the larger the sensor will be. Therefore a compromise for the path length must to be found.

Compared to the temperature dependence of an infrared sensor our simulations only show a minor temperature dependence (see Fig. 7). In the considered concentration region at atmospheric pressure there are no further relevant temperature effects. However this should be explored experimentally for a compensation temperature model of a future sensor.

As a light source we choose the Xenon flash lamp. This lamp has the advantage of a broad band spectrum so it is more flexible than e.g. a laser source. Furthermore the lifetime of a sensor will be much longer than with a LED.

We should consider that the used spectra were measured under different experimental conditions. Hence our simulations are only an approximation. Furthermore the conditions like for example temperature will be different in a future sensor. It is expected that the simulations give us a good starting point for the parameters of an experimental setup and a future sensor design. A real system will always consist of more absorbing gases or other parts like glas plates. This is not included in our model yet.

We will verify our simulations with an experimental setup similar to Fig. 1. As hydrogen sulfide is hazardous we will have to use a closed cuvette. To test the influence of the path length on the minimal measurement resolution we will use linear cuvettes with different length. To realize a compact setup we also could use a x-shaped setup with reflected beams to reach a path length of about 1 m.

The experimental setup could look like shown in Fig. 8. In this setup we use two detectors, one measurement detector and an additional reference detector. The filter in front of the reference detector narrows the light distribution to a wavelength of 250 nm. The measurement detector D_1 and the reference detector D_2 detect the signals S_1 and S_2 which are proportional to the light intensity I_0 of the source, given by the Beer-Lambert law (5). If the optical pathway is contaminated or if the lamp intensity is drifting, both signals S_1 and S_2 should be lowered by the same amount, so a ratio $S_r = \frac{S_1}{S_2}$ of both signals will be independent from those influences.

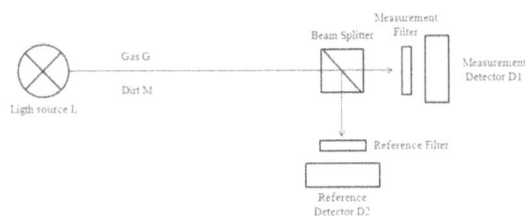

Figure 8: Experimental setup including a measurement and reference detector.

Acknowledgement

The work has been carried out at Dräger Safety AG & Co. KGaA, Gas Detection Sensors.

5   References

[1] W. Jessel, *Gase-Daempfe-Gasmesstechnik*. Dräger Safety AG & Co. KGaA, Lübeck, 2001.

[2] G. Wiegleb, *UV-Gassensor zum Nachweis von Stickoxiden*. Fachhochschule Dortmund, 2004.

[3] *Spectrex inc.* Available: http://www3.spectrex-inc.com [last accessed on 12.01.2015].

[4] *Excelitas   Technologies.*   Available: http://www.excelitas.com   [last   accessed   on 13.01.2015].

[5] *Hamamatsu.*   Available: http://www.hamamtsu.com [last accessed on 13.01.2015].

[6] Satellite Group Mainz, *The MPI-Mainz UV/VIS Spectral Atlas of Gaseous Molecules of Atmospheric Interest*. Available: http://satellite.mpic.de/spectral_atlas [last accessed on 07.01.2015].

[7] C. Y. Wu, *Temperature-Dependent photoabsorption cross sections of H_2S in the 1600-2600 ÅRegion*. University of Southern California, Los Angeles, 1997.

[8] *Mathworks.* Available: http://de.mathworks.com [last accessed on 13.01.2015].

[9] *Andover   Corpoiration.*   Available: http://www.andovercorp.com   [last   accessed   on 13.01.2015].

[10] M. Kaschke, K. Donnerhacke, M. Rill, *Optical Devices in Ophthalmology and Optometry*. Wiley-VCH, 2014.

The UV-visible absorption spectrum of all-trans retinal chromophore in the gas phase

A. Kluge [1]

[1] Medizinische Ingenieurwissenschaft, Universität zu Lübeck, anika.kluge@student.uni-luebeck.de

Abstract

The retinal chromophore with a protonated Schiff base is the light absorbing part in the rod and cone cells of the retina and is for this reason very important for the vision of mammals. In order to understand the intrinsic properties of this chromophore, the absorption profile in the UV and visible range of the synthesized all-trans retinal with a protonated Schiff base in the gas phase is measured in the present work. The electrostatic storage ring ELISA combined with an electrospray ion source and a laser is used to study the action absorption spectroscopy in the gas phase. We obtain absorption maxima for the $S_0 \rightarrow S_1$ (620 nm), $S_0 \rightarrow S_2$ (400 nm), $S_0 \rightarrow S_3$ (340 nm), $S_0 \rightarrow S_4$ (294 nm) transitions correlating well with calculated vertical excitation energies. In addition, another absorption maximum at around 250 nm is measured which may correspond to the $S_0 \rightarrow S_5$ transition of the all-trans retinal chromophore.

1 Introduction

The retinal chromophore is one of the most important parts in mammalian vision and is found in the photoreceptors of the eye. The photoreceptors in the retina can be divided into two different types. The cone cells which are responsible for the color vision and the rod cells for the night vision. The cone cells contain three different light-sensitive proteins (opsins) and another type of opsin is located in the rod cells (rhodopsin) which together gene-rate a broad absorption spectrum. However, in all types of opsins the same chromophore, the 11-cis-retinal chromophore with a protonated Schiff base, is the light absorbing part whereas the protein environment tunes its absorption maximum [1], [2]. The investigation of the absorption spectrum of the isolated retinal chromophore in the gas phase from 240 nm to 660 nm is therefore important to learn more about the intrinsic properties of the chromophore. Studying the chromophore in the gas phase has the advantage that the results are not influenced by the environment of the chromophore as it is the case in solution-phase experiments [4]. To measure the absorption spectrum of the all-trans retinal with a protonated Schiff base in the gas phase, the storage ring ELISA (ELectrostatic Ion Storage ring, Aarhus) is used in the present work. ELISA is able to store heavy ions in vacuum with the aid of electrostatic fields. The ions can be stored with a storage energy of 22 keV and the coupling of a laser beam in the beamline of the stored ions is possible [3]. By absorbing for instance a photon in the wavelength range around 600 nm, the retinal chromophore transitions from the ground state S_0 into the first excited singlet state S_1. Internal conversion brings the chromophore back into the ground state S_0 and transfers the electronic excitation

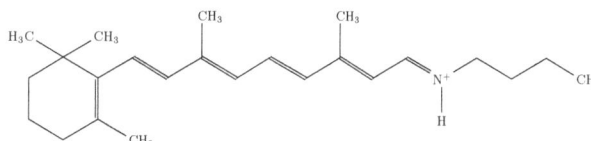

Figure 1: Chemical structure of the all-trans retinal chromophore with a protonated Schiff base (mass: 340 amu).

energy into vibrational energy. Due to the increased internal energy, the molecule is now in a hot ground state and dissociation may take place where a neutral fragment is formed, which can be detected in ELISA [4],[5].

2 Material and Methods

In this work the absorption spectrum of the all-trans retinal chromophore with a protonated Schiff base in the gas phase is studied. The chemical structure of the chromophore ion is shown in Fig. 1.

2.1 Electrospray Ionization and ELISA setup

The production of ions in the gas phase by electrospray ionization is based on the method of Fenn et. al [6] and was optimized for the ELISA setup by J. U. Andersen et al. [7]. The methanol-dissolved chromophore is filled in a syringe pump with a flow rate of about 150 μl/hour and reaches a steel needle. A heated capillary with a temperature of 80 °C is located in a distance of 5-10 mm to the needle. The potential difference between the needle and the heated

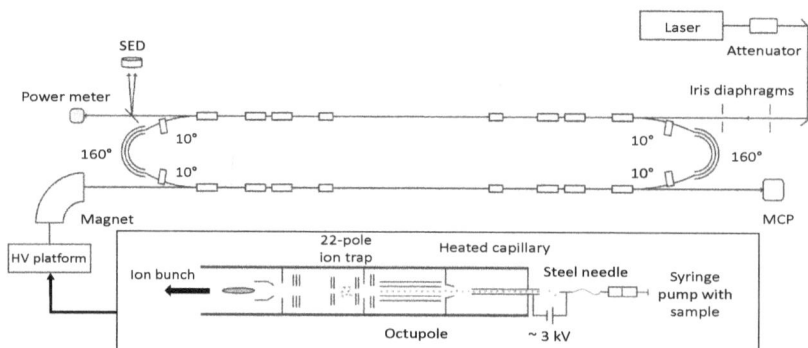

Figure 2: The ELISA setup with an electrospray ion source [4], [8].

capillary is on the order of 3 kV. Small charged droplets are formed and accelerated to the heated capillary if the coulomb force is larger than the surface tension [6], [7]. A first vacuum pump ensures a 1 mbar region around the capillary. Within the heated capillary, ions in the gas phase are created from the charged droplets and reach a second vacuum region (pressure about 10^{-3} mbar). Gas flow due to the pressure gradient transfers the ions to the octupole chamber [7]. The octupole bias is set to around 4 V so that the ions gain a kinetic energy of 4 eV. The pressure in the third vacuum region is about 10^{-6} mbar. The ions are accumulated in this region by a 22-pole ion trap for 40 ms. The pressure inside the ion trap is on the order of 10^{-4} mbar. For the accumulation, helium is used as a buffer gas to slow down the accelerated ions by low energy collisions to remove the kinetic energy of the ions. Finally, the gas-phase ions leave the trap as a bunch by switching the trap voltages. Since the ion source is located on a 22 kV high voltage platform, the ions gain a kinetic energy of 22 keV after their release from the ion trap [7],[8]. A mass separation has to be done since the storage condition of an electrostatic storage ring only depends on the kinetic energy over charge of the ions but not on the masses. In order to control the mass over charge ratio of the ions which are injected ELISA, a 90° bended magnet is used [3]. ELISA is able to store the entering ions in the gas phase for thousands revolutions with the aid of electrostatic devices. Basically ELISA consists of two 160° deflectors, four 10° deflectors and eight quadrupoles to guide the ion beam (see Fig. 2). The circumference of ELISA is about 8 m and the revolution time of an ion bunch is approximately 70 μs. The pressure inside ELISA is about 10^{-11} mbar [7]. A tunable diode pumped laser system (EKSPLA NT232-50-SH-SFG) is situated at the end of the straight section of ELISA and is able to emit light in the wavelength range from 210 nm to 2600 nm. The laser is running at 50 Hz and has a pulse duration of 4 ns. The laser is fired 13 ms after each injection, so that each ion bunch is affected by a single laser pulse. To detect the neutrals which are formed either by collisions with residual gas in ELISA or by the interaction with the laser beam, two detectors are installed at the end of both straight sections (see Fig. 2). The neutrals can be detected since they are not

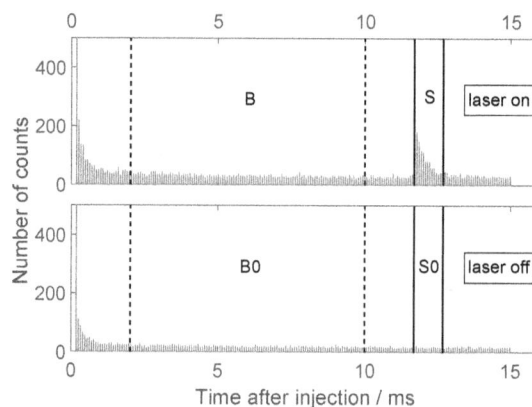

Figure 3: Typical measurement with a background time window (B, B0) and a signal time window (S, S0) after 1000 injections of the sample. The lower panel shows the number of detected neutrals with the two windows when the laser is off and the upper panel when the laser in on to calculate laser induced counts.

deflected by the electric fields in ELISA and hence leave the storage ring. Since the secondary electron detector (SED) can absorb light in the UV region, only the microchannel plate detector (MCP) is used in our measurements. Therefore, it is very important to have delayed fragments which are detectable for the MCP.

2.2 Spectroscopy and Data Analysis

The action absorption spectroscopy uses the number of neutrals, which are formed after photon absorption due to fragmentation of the molecule, to calculate the absorption cross section of the chromophore. In order to separate the number of neutrals, which are formed through laser absorption (signal counts) from the number of neutrals arising from collisions with residual gas in ELISA (background counts), two time windows are used. The first one, the background window, is set several μs after the injection of the ion bunch into ELISA and before the laser is fired. The second window, the signal window, is set a few μs after the laser is fired (see Fig. 3) [8]. The absorption cross section is then given by

$$\sigma(\lambda) \propto \frac{S(\lambda) - \frac{S_0}{B_0}B}{B\lambda E_{pulse}(\lambda)}, \qquad (1)$$

where λ represents the wavelength of the laser, S the number of counts in the signal window when the laser is on and S_0 the number of counts in the signal window when the laser is off. B_0 is the number of counts in the background window when the laser is off, B the number of counts in the background window when the laser is on and E_{pulse} is the laser pulse energy for each wavelength. The equation is based on the assumption that only one photon is absorbed by an ion bunch [8]. The control and data acquisition of the measurements is done by a LabView program and a MATLAB program enables the calculation of the absorption cross section.

3 Results and Discussion

The absorption cross section of the all-trans retinal in the gas phase is measured in the wavelength range from 240 nm to 390 nm at 2 nm steps and from 405 nm to 660 nm at 5 nm steps. For the UV region, a bandpass filter (240–395 nm) is added in the beam path to make sure that no visible light is coming through. The background window is set from 2000 μs to 10000 μs after the injection and the signal window from 11680 μs to 12680 μs (marked in Fig. 3).

3.1 Absorption Spectrum

Fig. 4 shows the measured absorption spectrum. In addition, calculated vertical excitation energies for the $S_0 \rightarrow S_1$, $S_0 \rightarrow S_2$, $S_0 \rightarrow S_3$ and $S_0 \rightarrow S_4$ transitions of the 6s-cis state (dashed vertical lines) as well as the 6s-trans state (solid vertical lines) of the retinal chromophore are plotted. These two states are the two minimum structures of the all-trans retinal and are very important since the beta-ionone ring can easily rotate between these conformations due to the low energy barrier [9], [10]. Table 1 summarizes the calculated values. The height of the vertical lines represents the oscillator strength of each transition. The data show a broad absorption band for the $S_0 \rightarrow S_1$ transition from 500 nm to 630 nm with a maximum at around 620 nm, which correlates with the calculated absorption wavelength (see Table 1) and with measurements that were done before [4], [8], [11]. It is striking that the $S_0 \rightarrow S_1$ absorption band is very broad. This can be explained by isomers of the all-trans retinal, which can be formed easily [11] and by the enormous vibrational broadening for instance due to the rotation between the two minimum structures of the molecule. The $S_0 \rightarrow S_2$ absorption band seems to be in the wavelength range from 370 nm to 440 nm (maximum at 400 ± 5 nm). However, the laser is performing very poor in this region and hence the $S_0 \rightarrow S_2$ transition can not be described completely. The wavelength range below 300 nm for the all-trans retinal in the gas phase has not been measured before. As can be seen in Fig. 4, the $S_0 \rightarrow S_3$ excitation is

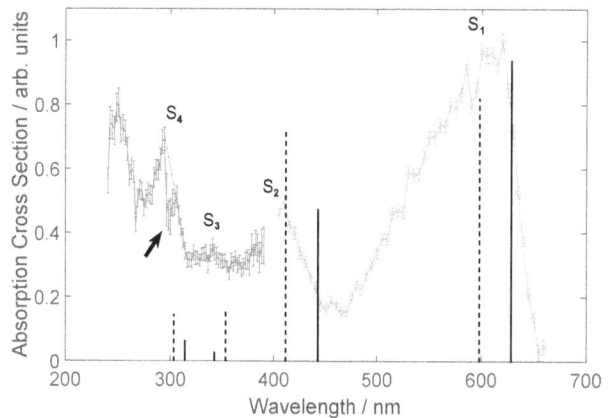

Figure 4: Absorption cross section of all-trans retinal with a protonated Schiff base in the gas phase. The UV region (240 nm–390 nm at 2 nm steps) and the visible region (405 nm–660 nm at 5 nm steps) were measured separately. The vertical lines represent the oscillator strength at the calculated excitation wavelength for each transition (solid: 6s-trans state, dashed: 6s-cis state). The oscillator strengths were all divided by two and the six smallest were additionally multiplied by a factor of ten.

Table 1: Vertical excitation energies and corresponding oscillator strengths. Calculated by A. V. Bochenkova.

	transition	$\lambda_{AbsCalc}$	oscillator strength
6s-trans	$S_0 \rightarrow S_1$	628 nm	1.812
6s-cis	$S_0 \rightarrow S_1$	598 nm	1.643
6s-trans	$S_0 \rightarrow S_2$	442 nm	0.095
6s-cis	$S_0 \rightarrow S_2$	411 nm	0.145
6s-trans	$S_0 \rightarrow S_3$	342 nm	0.006
6s-cis	$S_0 \rightarrow S_3$	353 nm	0.031
6s-trans	$S_0 \rightarrow S_4$	314 nm	0.013
6s-cis	$S_0 \rightarrow S_4$	303 nm	0.029

located from 320 nm to 355 nm with a maximum around 340 ± 2 nm. The calculated excitation wavelengths for this transition are 342 nm (6s-cis state) and 353 nm (6s-trans state). Directly after the $S_0 \rightarrow S_3$ absorption band, another absorption peak from 275 nm to 315 nm (maximum at 294 ± 2 nm) can be seen. This absorption band corresponds to the $S_0 \rightarrow S_4$ excitation. Laser instabilities are responsible for the local minimum at 296 nm (arrow in Fig. 4) and the expected shape of this absorption band is outlined with a dash-dot line. Moreover, the measured absorption profile shows another absorption band from 245 nm to 265 nm (maximum at 250 ± 2 nm). This band may correspond to the $S_0 \rightarrow S_5$ transition. It should be noted that the distance between the excitation energies in the UV is smaller compared to the visible. The excitation energies for the $S_0 \rightarrow S_1$ and $S_0 \rightarrow S_2$ transitions are 2 eV (620 nm) and 3.1 eV (400 nm) whereas the energies for the $S_0 \rightarrow S_3$ and $S_0 \rightarrow S_4$ transitions are 3.65 eV (340 nm) and 4.22 eV

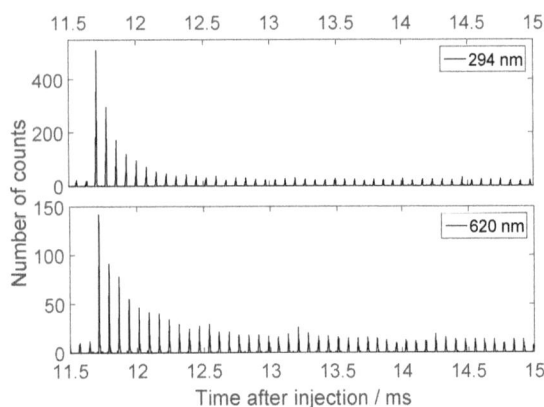

Figure 5: Decay of the detected neutrals after laser excitation with 294 nm (top) and 620 nm (bottom).

(294 nm). The distance between the energy levels decreases analogous to the rydberg states in atomic systems.

3.2 Decay Comparison

By comparing the decay of the number of detected neutrals after the laser excitation with 294 nm and 620 nm (Fig. 5), it can be seen that the decay after the excitation with 294 nm (Fig. 5 top) is faster than the decay after the excitation with 620 nm (Fig. 5 bottom). One reason for this is the higher energy of the photons with a wavelength of 294 nm (4.22 eV) compared to 620 nm (2 eV) resulting in a higher internal energy of the molecule after absorption. As a consequence, the dissociation of the molecule in the hot ground state is happening faster.

4   Conclusion

In this study, the UV and visible absorption of the all-trans retinal chromophore in the gas phase has been measured. The UV region has been measured for the first time which is an important step to understand the intrinsic properties of this chromophore. The molecule showed a relatively high absorption in the UV region. The measured absorption bands in the UV correlated with the calculated excitation energies for the $S_0 \rightarrow S_3$ and $S_0 \rightarrow S_4$ excitations. Moreover, it was possible to measure another local absorption maximum which may correspond to the $S_0 \rightarrow S_5$ transition. In order to obtain a continuous spectrum, further measurements should be performed where the laser instabilities can be reduced.

Acknowledgement

The work has been carried out at the Department of Physics and Astronomy, Aarhus University, Denmark. I thank Dr. H. V. Kiefer and Prof. Dr. L. H. Andersen for their support in the lab and their ideas and discussions on the subject. Dr. A. V. Bochenkova is acknowledged for the calculation of the theoretical values and for the helpful discussions. I would also like to thank Prof. Dr. C. G. Hübner, Institute of Physics, Universität zu Lübeck, Germany.

5   References

[1] G. G. Kochendoerfer, S. W. Lin, T. P. Sakmar and R. A. Mathies, *How color visual pigments are tuned.* Trends in biochemical sciences, vol. 24, no. 8, pp. 300–305, 1999.

[2] W. Wang, J. H. Geiger and B. Babak, *The photochemical determinants of color vision.* Bioessays, vol. 36, no. 1, pp. 65–74, 2014.

[3] L. H. Andersen, O. Heber and D. Zajfman, *Physics with electrostatic rings and traps.* Journal of Physics B: Atomic, Molecular and Optical Physics, vol. 37, no. 11, pp. R57, 2004.

[4] L. H. Andersen, et al., *Absorption of Schiff-base retinal chromophores in vacuo.* Journal of the American Chemical Society, vol. 127, no. 35, pp. 12347–12350, 2005.

[5] L. H. Andersen, et al., *Experimental studies of the photophysics of gas-phase fluorescent protein chromophores.* Physical Chemistry Chemical Physics, vol. 6, no. 10, pp. 2617–2627, 2004.

[6] J. B. Fenn, M. Mann, C. K. Meng, S. F. Wong and C. M. Whitehouse *Electrospray ionization for mass spectrometry of large biomolecules.* Science, vol. 246, no. 4926, pp. 64–71, 1989.

[7] J. U. Andersen, et al., *The combination of an electrospray ion source and an electrostatic storage ring for lifetime and spectroscopy experiments on biomolecules.* Review of Scientific instruments, vol. 73, no. 3, pp. 1284–1287, 2002.

[8] H. Kiefer, *Photophysics of gas-phase protein chromophore ions.* Ph.D. thesis, 2015.

[9] J. Rajput, et al., *Probing and modeling the absorption of retinal protein chromophores in vacuo.* Angewandte Chemie International Edition, vol. 49, no. 10, pp. 1790–1793, 2010.

[10] Y. Toker, A. Svendsen, A. V. Bochenkova and L. H. Andersen, *Probing the Barrier for Internal Rotation of the Retinal Chromophore.* Angewandte Chemie, vol. 124, no. 35, pp. 8887–8891, 2012.

[11] N. J. A. Coughlan, B. D. Adamson, L. Gamon, K. Catani and E. J. Bieske, *Retinal shows its true colours: photoisomerization action spectra of mobility-selected isomers of the retinal protonated Schiff base.* Physical Chemistry Chemical Physics, vol. 17, no. 35, pp. 22623–22631, 2015.

Lactate separation employing electrophoresis
– Separation of blood components in an electric field –

R. Gänger [1], S. Fiedler [2], and S. Müller [2]

[1] Technical Biochemistry, University of Applied Science Lübeck, rene.gaenger@stud.fh-luebeck.de
[2] Medical sensors and devices laboratory,University of Applied Science Lübeck,{felix.fiedler, stefan.mueller}@fh-luebeck.de

Abstract

Lactate is an important parameter in the human body and becomes important in the sports- and intensive care-diagnostics as well. Today, measuring the concentration of lactate requires biosensors.A big disadvantage of biosensors is the need of steady environment concerning e.g. temperature and humidity. This research aims in finding a new method, independent from the environmental conditions. For that purpose, e.g. spectroscopic methods can be applied, which are only precise for concentration higher than the physiological range so far. Using electrophoresis it is possible to move electrically charged molecules in a solution to increase the concentration. Here it is shown that lactate can be moved applying an electric field in PBS-buffer. The next step will be to create a impressiv measurement in blood plasma, to proof the separation of lactate in this media.

1 Introduction

Lactate is the salt of lactic acid which is transported via the blood circuit. As shown in figure 1, the concentration of lactate rises under anaerobic conditions inside the muscles. In this case glucose is reduced to two molecules pyruvate, in the glycolysis, and those are reduced to two molecules lactate to ensure the provision of energy in the form of adenosine triphosphate. Via the blood circulation the lactate is transported to the liver and oxidized again to glucose by the consumption of six adenosine triphosphates.

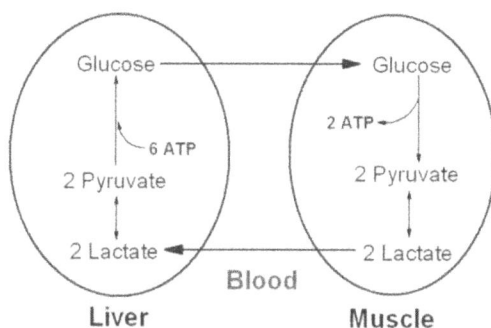

Figure 1: Cori-Circle, to show the energy production in the muscle and the energy consumption in the liver[1].

Lactate is not only an indicator for the oxygen saturation of muscle cells [2]. It is also involved in the sore healing and as well in the vascular reforming [3]. The analysis of the lactate concentration is important in intensive care diagnostics and sports.

To analyze lactate concentrations, biosensors using an enzymatic-amperometric measurement method are mostly applied [3]. These biosensors are used to detect the concentration of lactate. It is necessary to use lactate oxidase enzymes. The enzymes are immobilized on the sensor's surface to split lactate in the blood serve. A disadvantage of biosensors is the need of controlled environmental conditions. Particularly a steady temperature during transportation and storage is required. New possibilities to detect lactate and glucose with spectral methods and calculate their concentrations from the absorption spectra's are part of current research [1], [2], [3]. The low physiological concentration and the scattering of the blood cells affect the spectral analysis. To simplify the analysis of blood or blood plasma the smple should be unpicked and lactate be separated by the other plasma components. The dissolution should allow the measurement of concentrations with normal spectroscopic procedures.

2 Material and Methods

During the approach, theoretical calculations and simulations were made to describe the behavior of the molecules in a fluid. Conclusively for the movement of a molecule in an electric field is its charge, its size and the viscosity of the medium. The gel electrophoresis uses the same properties to separate fragments of the deoxyribonucleic acid. The direction of the molecular motion is important. Negatively charged molecules move into the direction of the anode and positively charged molecules towards the cathode. The following calcula- tions and experiments are made to get an overview of the movement of lactate in a solution. Lactate is dissociated in blood and has a charge of „-1". This results in slowing effects for the movement of all molecules in a solution, due to the molecule atmosphere and the presence of other macromolecules. Including to the theoretical calculations initial measurements were carried out by analyzing the movement of lactate in the electrical field and to make the theoretical calculations visible. Here differently to known electrophoretic procedures

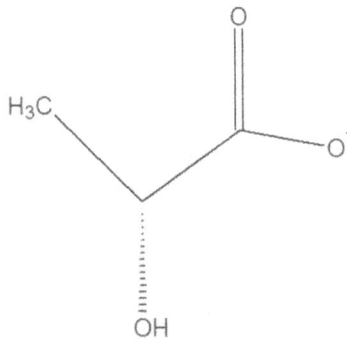

Figure 2: Lactate molecule in his dissociated form.

The proteins, which are contained in the blood plasma, are differently charged depending on the isoelectric point causing the specific pH value of the media. They may be charged positively or negatively. Compared to the smaller Lactate in size and has a very low mass.

Table 1: Masses of selected biopolymers [4-6]

Molecul	mass / kDa
Albumine	69
Fibrinogen	335
Globuline	36-1300
Glucose	0,18
Lactate	0.09

The masses in table 1 are averaged masses from different sources. The charge of the proteins can be calculated on the basis of the pH value. At a pH of 7.4 albumin has a molecular charge of -137. The velocity of molecules in a medium can be calculated via a power balance: The energy F which sets the molecules to motion, consists of the charge q [J] and the electric field strength E [V/m].

$$F = q \cdot E. \qquad (1)$$

F [N] is opposed to the friction effect of the molecule with its surrounding. With the stokes radius r [m] of the molecules, the dynamic viscosity η [Pa*s] of the medium and the speed v [m/s] of movement of the species it is possible to calculate the fric- tion effect.

$$F = 6 \cdot \pi \cdot r \cdot \eta \cdot \nu. \qquad (2)$$

From (1) and (2) results:

$$\nu = \frac{E \cdot q}{6 * \pi \, cdotr \cdot \eta}. \qquad (3)$$

The theoretical analysis of molecular speed must be checked in a real system. Factors that affect the entire system are the size of the hydrated molecules, the dissociation degree and the ion atmosphere [6]. This results to slowing effects, due to the periphery and the presence of other macromolecules, for the movement of all molecules in a solution. the separation of the molecules should take place in a low viscosity media, as far as possible directly in full blood samples ($\eta = 0,004$ Ps) Currently, it is assumed that this reaction in the blood plasma ($\eta = 0.0021$ Ps) takes place. According to theoretical calculations, the results are virtually validated. So it is possible to correct the calculations by empirically determined values. In this case The lactate concentration is detected by enzymatical biosensors

3 Results and Discussion

The calculations have been made varying three parameters. In the first the velocities of the molecules at different potentials have to be considered. Next, the distances of the electrodes changes field strength. The calculations have been set up, so it is assumed that a molecule which is negatively charged at the cathode needs to move all the way to the anode.

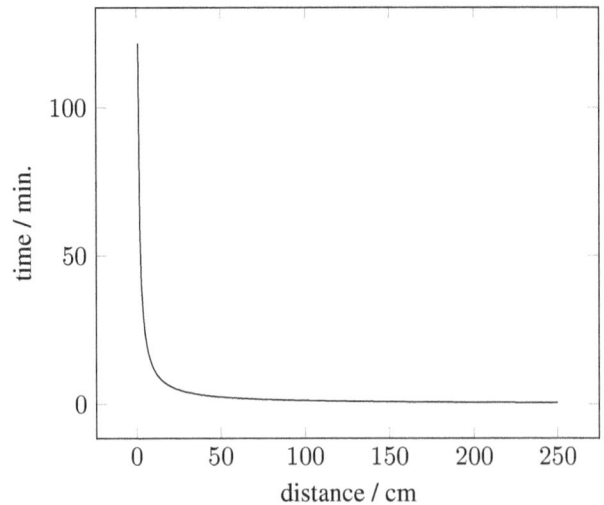

Figure 3: Required Time of albumin molecule for a distance of 10 cm in an electric field at a viscosity according to the blood plasma

As it can be seen in figure 33, with the applied voltage which increases, the speed of the molecules increases greatly, and in the high viscosity compared to a medium, the albumin is 10,000 times faster in a medium having a viscosity such as blood plasma.

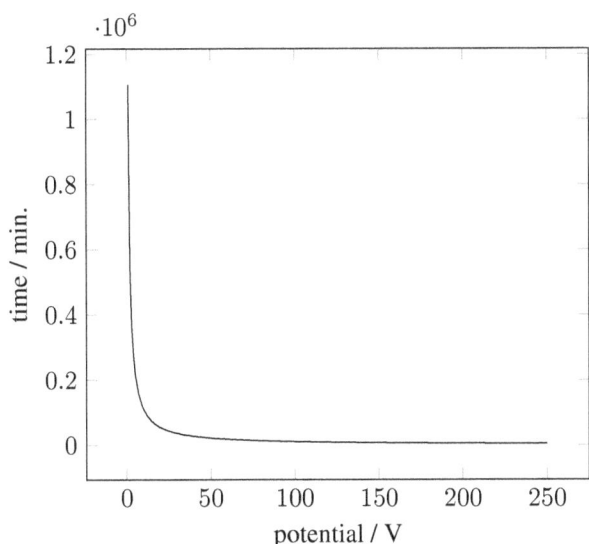

Figure 4: Required Time of albumin molecule with the distance of 10 cm in an SDS gel (Sodium dodecyl siloxane 50 % water and 50 % polymer $\eta = 19.1 Ps$) in an electric field)

At a constant voltage and a variable distance of the electrodes to each other, the migration velocity of the molecules on the distance prolong significantly as shown in figure 5.

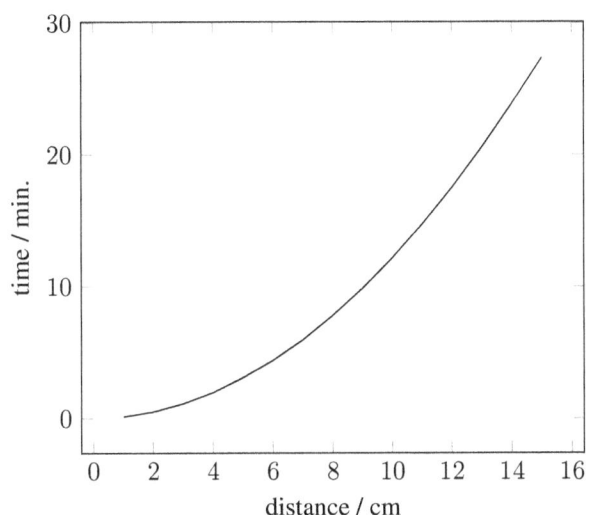

Figure 5: Time required for an albumin molecule, a voltage of 1 volt and different distances to the anode.

Lactate is in contrast to albumin with a negative charge of 1, uncharged. Compared to the masses is lactate in the ratio of 1:767 very small.
As seen in figure 7 lactate is faster then the albumine molecules over the distance and the potential.
The practical experiments over time show that lactate moves to the anode.

4   Conclusion

The results of the practical test series show that lactate moves in the electric field and the motion can be detected

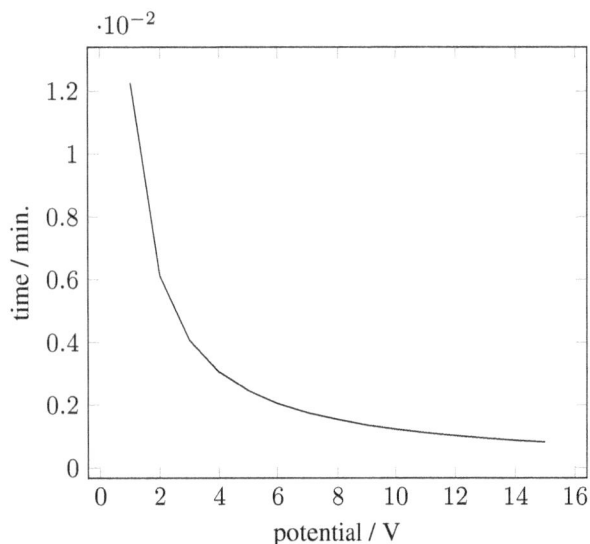

Figure 6: Time required for a lactate molecule, a voltage of 1 volt and different distances to the anode.

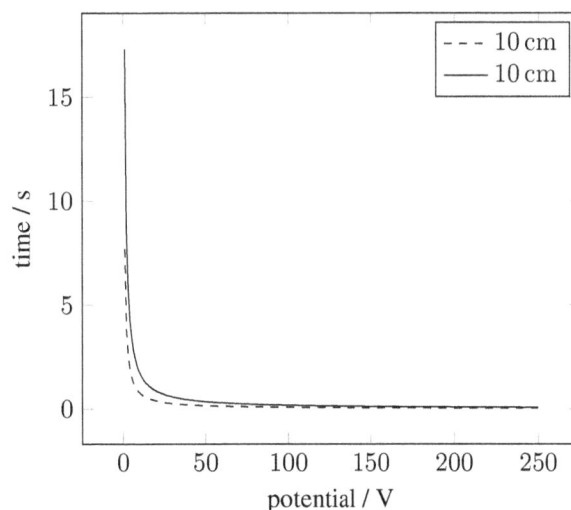

Figure 7: Time required for an albumin molecule, a voltage of 1 volt and different distances to the anode

Table 2: Compared to the time required from the cathode to the anode of lactate and albumin at a distance of electrodes of 10 cm in various voltages and a viscosity of PBS-buffer (n = 0.1 Ps)

potential	Albumin / min	Lactate /min
1	121	0,128
2	60	0,064
3	40	0,043
4	30	0,032
5	24	0,026
6	20	0,021
7	17	0,018

Table 3: Change of the lactate concentration at the anode over time at a voltage of 1 volt and a distance of 8.5 cm

time / s	conc. / $mmol \cdot L^{-1}$
0	5,4
1	5,4
5	5,6
10	5,9
15	6

by sampling at the anode. The theoretical calculations and practical evidence do not coincide in time. This may be due to the experimental set-up or a non-specific sampling.

5   Outlook

In the next steps specially prepared electrophoresis cell will be used, which has a smaller volume with specific geometrically dimensions. Furthermore, this cell may also be used with separation media such as membranes and gels are, for example, a " cut-off " insert which stays the negatively charged proteins. Characterized the liquid is free from impurities at the anode and a concentration of lactate on the permeate side is made possible to simplify the determination of the concentration.

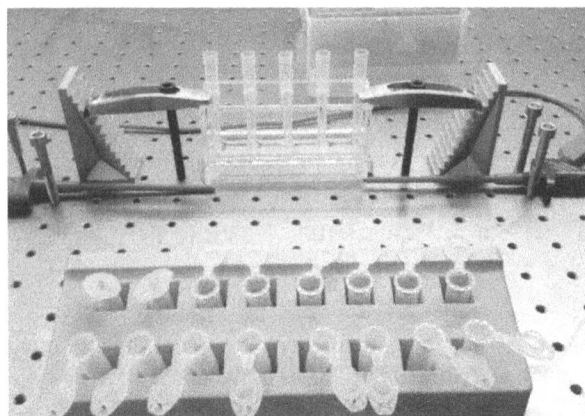

Figure 8: The new designed chamber with two elektrodes out of platinum and 5 points for getting samples.

With this new electrophoresis cell we will try to show the change of concentration over time and voltage.

6   Acknowledgments

The work has been carried out at a laboratory of the university of applied sciences luebeck for medical technics and devices. Special thanks for help and cooperations to Stephan Klein.

The work has been carried out at Example GmbH, Examplecity

7   References

[1] C. Bröll, and T. Kurth , *Book of Biochemistry.* Thiele, Berlin 2011

[2] A. Pjilip ,A. Macdonald and P. W. Watt , *Lactate - a Signal coordinating cell and systemic function.* Exp. Biol., 2005.

[3] P. Wahl, T. Bloch, W. Mester and J. Mester, *Modern approaches to the lactate : lactate an overemphasized , yet underappreciated molecule.* Schweizer Sportzeitschrift, Köln, 2009.

[4] Anderson NL and Anderson NG, *High Resolution Two-Dimensional Electrophoresis of Human Plasma Proteins.* Proceeding of the National Academy of Sciences, 1977.

[5] J.N. Adkins et al., *Toward a human blood serum proteome: analysis by multidimensional separation coupled with mass spectrometry.* Molecular and Cellular Proteomics, 2002.

[6] J. Berg, M.Tymoczko and L. Lubert Stryer, *Biochemistry*, Spektrum, Heidelberg, 2007.

[7] V. Knecht, H. Risselada, A. Mark, S. Marrink, *Electrophoretic mobility does not always reflect the charge on an oil droplet.* in: Journal of Colloid and Interface Science 2008, pp 318-319, 2008.

Method Optimisation and Preparation for Nanoparticle Induced Hyperthermia as a Novel Therapy to Treat Cutaneous Leishmaniasis

C. Grill [1], S. Oates [2], N. Telling [3], C. Hoskins [3] and H. Price [2]

[1] Medizinische Ingenieurwissenschaft, Universität zu Lübeck, christin.grill@student.uni-luebeck.de
[2] School of Life Sciences, Keele University, {s.l.oates,h.price}@keele.ac.uk
[3] Science and Technology in Medicine, Keele University, {n.d.telling,c.hoskins}@keele.ac.uk

Abstract

Leishmaniasis is a widely spread tropical disease. No vaccines exist and the few existing drugs are potentially toxic. Moreover, resistances occurred in the past years. A new idea for treatment is to kill the parasites by nanoparticle induced hyperthermia. Therefore, nanoparticles were made and their size, surface charge, concentration and heating potential were analysed and hyperthermia experiments were performed. A viability assay for investigation of the percentage of live and dead cells was optimised and a temperature gradient experiment was performed. The experiments resulted in nearly 100% dead parasites at 42 °C while 80% of the macrophages survived. This temperature can be reached with the tested nanoparticles. The concept of this technology is promising but further work is required to validate this.

1 Introduction

Leishmaniasis is a widely spread tropical disease. About 350 Million people live in risk of infection, which are mainly poor people in developing countries in Africa, Asia and Latin America [1]. Some 2 million new cases appear and an estimated number of 20,000 to 40,000 leishmaniasis deaths occur every year [1]. Hence, it is the second most lethal tropical parasitic infection in the world after malaria. Only a few drugs are available, some of them are potentially toxic and an increasing amount of resistances occurred over the last two decades, and no vaccines exist to prevent the disease [2]. Therefore, it is important to investigate possible new treatments. The use of nanoparticles to induce hyperthermia is a new approach being investigated to treat leishmaniasis and is the subject of this project.

1.1 Leishmaniasis

The term leishmaniasis comprises several different diseases caused by over 20 known different species of the protozoan parasite *Leishmania spp.*. Humans and some animals, e.g. dogs, are infected by the bite of female phlebotomine sandflies, which carry the parasite. There are three main types of the disease, which are visceral, cutaneous and mucocutaneous [2]. The visceral form is the most serious form and approximately 0.2 to 0.4 million new cases occur every year [1] with an estimated mortality of 10% mainly in poor regions including India, Sudan and Brazil [2]. The visceral form damages vital organs and in the process leads to irregular attacks of fever, loss in weight, abdominal disten-

sions, anaemia and enlargement of liver and spleen. If not treated visceral leishmaniasis is lethal [2]. However, the most prevalent form of the disease is cutaneous leishmaniasis with 0.7 to 1.2 million cases annually [1]. Cutaneous leishmaniasis leads to skin lesions, which result in permanent scars, disfigurement, alteration in skin pigmentation and sometimes to disabilities [2]. Mucocutaneous leishmaniasis is a complication of the cutaneous form and results in mutilation of mucous membranes of nose, mouth and throat in total or in parts [2]. Approximately 35,000 cases occur mostly in Bolivia, Brazil an Peru every year [2].

In this project the focus is on the treatment of cutaneous leishmaniasis, because the superficial lesions are easily accessible and the body would only be partly exposed to nanoparticles and an alternating magnetic field. But if it works a treatment of the other forms of leishmaniasis may be possible.

1.2 The Parasite

The *Leishmania* parasite goes through different stages during its life cycle. In the blood-feeding female sand fly the parasite lives as an extracellular form, the flagellated procyclic promastigote and has a size of 15 to 20 μm. This needs to differentiate into the infective metacyclic promastigote before it is transferred into the mammalian hosts. Once inside the host, the parasites invade macrophages and differentiate into the intracellular amastigotes with a diameter of 3 to 5 μm. Amastigotes evolve and duplicate in the macrophages until they are released via cell lyses to infect other macrophages [2]. The parasites transform back into promastigotes when they are ingested by a female sandfly sucking blood from an infected host [2].

1.3 Macrophages

Macrophages are part of the immune system. Approximately 3 to 8% of peripheral blood leukocytes are monocytes, which circulate for 1 to 3 days in the blood before entering the tissue to become macrophages. Macrophages are phagocytes, which eliminate debris, e.g. invading microorganisms like *Leishmania* parasites via phagocytosis [3].

1.4 Nanoparticles and Hyperthermia

Magnetic nanoparticles (MNP) are particles consisting of magnetic material like iron, nickel and cobalt with a size between 1 nm and 1 μm. MNPs are becoming more commonly used in medical and scientific applications. For medical applications non-toxicity is an important requirement. Ferrous or ferric oxide are therefore the primary component in magnetic particles for the utilisation in living beings. Possible applications are drug targeting, separation and isolation of cells and molecules and magnetic resonance imaging with particles as contrast agents. There is also a lot of research about cancer diagnosis, targeting and treatment via MNPs [4].

Magnetic nanoparticles have special physical characteristics due to their small size. They are able to produce a localized heating effect when exposed to an alternating magnetic field. The entire heat generated by magnetic nanoparticles is defined by the specific absorption rate (SAR). A high SAR value is essential for a heating effect that allows killing the parasites. The efficiency of the hyperthermia depends on magnetic characteristics of the particles, their core and hydrodynamic size (size of the core, coating and the due to surface charge attached solvent layer) distribution in the nanoparticle solution and the relaxation process [5]. Heating is generated either through magnetic hysteresis for particles with multiple domains or through relaxation for single domain particles [6]. During Brownian relaxation a particle rotates physically and energy is converted into heat by friction between the particle and the surrounding media [6]. Therefore it is dependent on the viscosity of the surrounding fluid [5]. During Néel relaxation the particle itself does not move, only the magnetic moment changes its orientation due to thermal randomization of magnetic moments [5], which is as well a lossy process converting electromagnetical in thermal energy. The prevalent process is the one with the shorter relaxation time [6].

2 Material and Methods

The following methods were used as preparatory work for hyperthermia experiments.

2.1 Growing Macrophages

U937 cells (differentiated human monocytes) were cultured in complete RPMI 1640 medium (22409-015, Thermo Fisher Scientific) with Fetal Calf Serum for Leishmania and stored in vented flasks at 37 °C with 5% CO_2. They were diluted 1:10 every two to three days.

2.2 Growing and Differentiating *Leishmania*

The promastigotes were cultured in 10 ml Schneider's complete medium (21720-001, Thermo Fisher Scientific) in plugged flasks with a pH of 7.0, incubated at 26 °C and diluted 1:200 every two to three days. The number of promastigotes increases exponentially for approximately 7 days. After this so-called log-phase they differentiate into metacyclic promastigotes. At this point the metacyclic promastigotes can be differentiated further into amastigotes by changing the medium to Schneider's complete medium with a pH of 5.5, phosphoric acid was used to reduce the pH. Amastigotes were incubated at 32 °C for 24 hours.

2.3 Counting Cells

All cells were counted with the aid of a Neubauer haemocytometer, where 10 μl of cell solution were pipetted in. Both promastigotes and amastigotes were diluted with paraformaldehyde prior to counting. Monocytes were stained with tryptan blue to make live and dead cells distinguishable. The cells appearing in the grid of the counting chamber were counted under a microscope with a 20x magnification objective lens and gave the cell number times 10^4 per ml.

2.4 Viability Assay Optimisation on Untreated Cells

In order to verify cell viability an MTT (3-(4,5-dimethylthiazol-2-yl)-2,5-diphenyltetrazolium bromide) assay was optimised for the amastigotes. Between 1 x 10^7 and 1.5 x 10^5 cells were seeded in a 96-well plate via titration. Each well contained thereby half as much cells as the previous one. Three replicants of each cell number were seeded. Each well contained 100 μl media including the cells, or without cells as a control. Between 20 and 40 μl MTT (M2003, Sigma-Aldrich) were added and the plate was incubated at 32 °C in the dark because of the MTT's light sensitivity. Incubation times between one and four hours were tested. 100 μl DMSO were added to stop the reaction and the absorbance at 560 nm was measured using a plate reader (GloMax®-Multi Detection System, Promega). The absorbance was normalized against the absorbance of the wells without cells. As a positive control 2.5 x 10^6 cells were incubated with amounts of 100 μM, 10 μM, 1 μM, 0.1 μM and without Amphotericin B, which kills the amastigotes.

2.5 Temperature Gradient Experiment for Parasites and Macrophages

In order to determine which temperature is required during hyperthermia treatment to kill the amastigotes a temperature gradient experiment was performed. Either 5×10^6 amastigotes or 10^5 macrophages in 100 μl were seeded per well of an optical 96-well reaction plate (4316813, Thermo Fisher Scientific). 50 μl media were added to each well, because it was expected to evaporate during heating. The plate was occluded with an optical adhesive cover (4311971, Thermo Fisher Scientific). In each case three replicants were incubated with one of twelve different temperatures between 32 and 56 °C for one hour in a thermal cycler (MJ Research PTC-200 Thermal Cycler, GMI). As a control the same amount of cells were incubated for an hour under the best conditions as they were cultured. Three wells only had media and three wells had cells, but they were killed with SDS as a control as well. 100 μl of each sample from the optical 96-well plate were pipetted to the 96-well plate with the controls. 40 μl MTT were added to each well. After three hours of incubation at 32 °C for amastigotes or 37 °for macrophages 100 μl DMSO were added and the absorbance was measured using a plate reader.

2.6 Nanoparticle Preparation and Analysis

In order to kill *Leishmania* parasites by hyperthermia, MNPs that accumulate on or in the protozoa and generate sufficient heat in an alternating magnetic field, are necessary. So hybrid nanoparticles, containing a metallic core and a shell of a different metal [7], were produced as described in [7]. The iron cores (Magnetite, Fe_3O_4) were coated with a long polymer, gold seeds, a short polymer and again with gold as described in [7], resulting in biocompatible nanoparticles, which do not release free radicals [7]. These particles were imaged via transmission electron microscopy (JEOL 1230 TEM). The zeta potential (surface charge) was measured (Zetasizer Nano ZS, Malvern) and the nanoparticle concentration was determined by inductively coupled plasma (700 Series ICP-OES, Agilent Technologies).
Citric acid coated Maghemite (Fe_2O_3) particles were produced like described in [8] and characterized. Particle size and surface charge were measured via Zetasizer (Zetasizer 3000 HSA, Malvern Instruments). The nanoparticle concentration was determined via Ferrozine assay, which compares the absorbance of samples with a known concentration (standard curve) with the absorbance of the sample. The type of relaxation process were determined with an AC Susceptometer. First hyperthermia experiments with a frequency of 362 kHz and a magnetic field of 29.4 mT were performed.

3 Results and Discussion

The following results were obtained.

3.1 Viability Assay Optimisation on Untreated Cells

Fig. 1 shows the results of an optimised MTT assay on amastigotes with cell numbers between 1×10^7 and 3.125×10^5. The best parameters were an amount of 40 μl MTT and an incubation time of three hours for investigation of the viability. It is observable that the signal saturates. We can assume a curve linearity approximately until 5×10^6 cells. For macrophages the linearity is given until 10^5 cells. Therefore, these are good numbers to start a hyperthermia or temperature gradient experiments with and verify the viability afterwards. In the positive control all Amphotericin B concentrations killed the cells. The absorbance had circa the same value as the absorbance of the wells without cells.

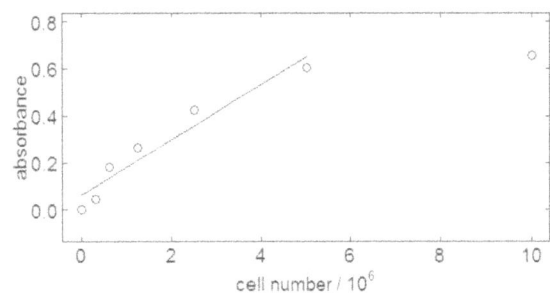

Figure 1: Results of the optimised MTT assay with an incubation time of 3 hours. Absorbance plotted over cell number with linear regression from 0 to 5×10^6 cells.

3.2 Temperature Gradient Experiment for Parasites and Macrophages

The temperature gradient experiment (data not shown) showed that at 38.5 °C approximately two third of amastigotes are alive. At 41.9 °C all parasites are dead as measured by absorbance. In contrast at this temperature circa 80% of macrophages are alive. The viability of macrophages is 0% at 54 °C. So, there is a clear temperature gap in surviving of parasites and macrophages. Therefore, there is a possibility to kill the parasites even if the nanoparticles are not in or attached to the parasites and the complete fluid in and around infected macrophages is heated to 42 °C.

3.3 Nanoparticle Preparation and Analysis

The pure magnetite particles are not chemically stable. In the human body they could build free radicals, which could interact with molecules and kill cells. They have a surface charge of 4.52 mV. The four coating steps generate chemically stable and thus biocompatible particles and they decrease the zeta potential to -28.9 mV. Fig. 2 shows the MNPs with visible gold seeds on the surface. It is observable that the particles agglomerate. They are not physically stable after these coating steps. To ensure physically stable particles which do not cluster and therefore would not lose their closely size linked characteristics, another coating like

PEG is necessary. Because of the negatively charged cell membrane MNPs with a positive charged surface could improve cellular targeting. The next step would be investigating their heating potential.

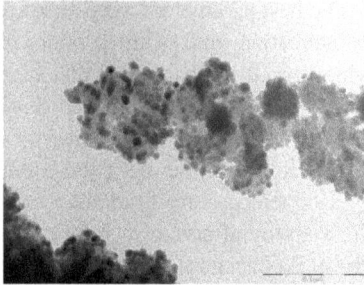

Figure 2: Magnetite Particle with PEI coating and distinctly and visibly gold seeds on the surface, seen via an electron microscope with magnification of x400,000. Bar, 0.1 μm.

The citric acid coated Maghemite particles had a size between 12 and 18 nm and a surface charge around -35 to -45 mV. An AC susceptibility measurement has shown that the prevalent relaxation is the Brownian relaxation. The concentration of the nanoparticle solution was 16.2 mg/ml. Fig. 3 shows the results of the first hyperthermia experiments. The achievable temperature is higher with a higher nanoparticle concentration as expected (curve a, b, c).

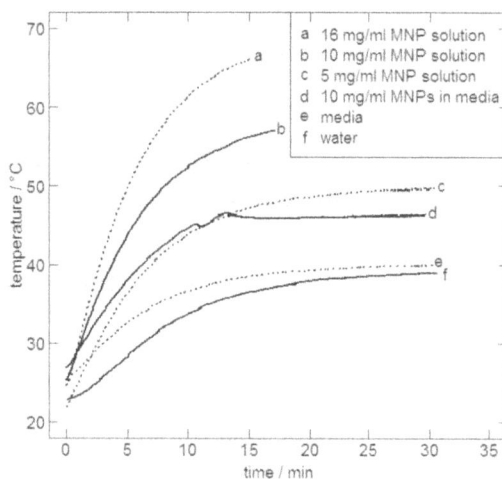

Figure 3: Heating effect of different concentrations of citric acid coated Maghemite in water and medium under a 29.4 mT magnetic field reached with 362 kHz alternating magnetic field.

Curve d shows the heating effect of particles in medium. It is observable that the temperature of particles with the same concentration in water (curve b) could not be reached. After the hyperthermia experiment with this sample the particles were not stable and sedimented on the bottom of the vial. The bend in the curve d probably shows the moment, when the particles started to agglomerate. The water and media sample were heated as well because of the coil generating heat. The temperature saturated at circa 40 °C. The necessary temperature of 42 °C to kill the amastigotes can easily

be reached with this particles, but the problem of unstable particles in media must be solved for an application.

4 Conclusion

At a temperature of 42 °C the parasites are dead while most of the macrophages are alive. This temperature can easily be reached with nanoparticles in an alternating magnetic field. The next steps of investigation are finding a solution to keep MNPs stable in media and performing more hyperthermia experiments with parasites as well as macrophages. Moreover different coatings and their ability to bind or enter the parasites can be tested.

Acknowledgements

The work has been carried out at School of Life Science and Institute of Science and Technology in Medicine, Keele University, UK. The internship was funded by Erasmus.

5 References

[1] J. Alvar, I. D. Vélez, C. Bern, M. Herrero, P. Desjeux, J. Cano, J. Jannin, M. d. Boer, and the WHO Leishmaniasis Control Team, "Leishmaniasis worldwide and global estimates of its incidence," *PLoS ONE*, vol. 7, p. e35671, 05 2012.

[2] D. Pace, "Leishmaniasis," *J. Infect.*, vol. 69 Suppl 1, pp. S10–18, Nov 2014.

[3] B. Hannigan, C. Moore, and D. Quinn, *Immunology*. [Scion Handbooks], Scion Publishing Limited, 2009.

[4] A. Ito, M. Shinkai, H. Honda, and T. Kobayashi, "Medical application of functionalized magnetic nanoparticles," *J. Biosci. Bioeng.*, vol. 100, pp. 1–11, Jul 2005.

[5] E. Cespedes, J. M. Byrne, N. Farrow, S. Moise, V. S. Coker, M. Bencsik, J. R. Lloyd, and N. D. Telling, "Bacterially synthesized ferrite nanoparticles for magnetic hyperthermia applications," *Nanoscale*, vol. 6, pp. 12958–12970, Nov 2014.

[6] R. V. Stigliano, F. Shubitidze, K. Kekalo, I. Baker, A. J. Giustini, and P. J. Hoopes, "Understanding mNP Hyperthermia for cancer treatment at the cellular scale," *Proc SPIE Int Soc Opt Eng*, vol. 8584, p. 85840E, Feb 2013.

[7] C. M. Barnett, M. Gueorguieva, M. R. Lees, D. J. McGarvey, R. J. Darton, and C. Hoskins, "Effect of the hybrid composition on the physicochemical properties and morphology of iron oxide–gold nanoparticles," *Journal of Nanoparticle Research*, vol. 14, no. 10, pp. 1–12, 2012.

[8] S. Campelj, D. Makovec, and M. Drofenik, "Preparation and properties of water-based magnetic fluids," *J Phys Condens Matter*, vol. 20, p. 204101, May 2008.

3

Biochemical Physics II

Construction of a Cryo Microscope for Single-molecule Measurements

K. Duda [1], V. Hirschfeld [2], J. Pavlita [2] and C. Hübner [2]

[1] Medizinische Ingenieurwissenschaft, Universität zu Lübeck, katharina.duda@student.uni-luebeck.de
[2] Institut für Physik, Universität zu Lübeck, {hirschfeld, pavlita, huebner}@physik.uni-luebeck.de

Abstract

Single-molecule fluorescence measurements of dye-labeled biomolecules at room temperature are a widespread and commonly performed method to obtain information about the individual behavior of molecules in complex environments. However, often the observation time of intrinsic dynamics and thus the information content is limited by photobleaching and diffusion of the molecules. To apply single molecule fluorescence spectroscopy methods at low temperature, an adequate sample preparation and an integrated cooling unit was developed. To ensure amorphous freezing, the sample is spin coated under liquid-nitrogen cooling. An integrated sample carrier was developed for a cryostat in order to measure immobilized fluorescent labeled molecules. The sample carrier allows a simple sample transfer and the adaptation to a confocal laser scanning microscope. This system meets the requirements of an effective and easy to handle sample cooling in a microscope equipped with a high NA short working distance microscope objective.

1 Introduction

In order to investigate the structure and folding of proteins, single-molecule measurements are increasingly gainingin popularity. For this purpose, the protein of interest is labeled with fluorescent dyes and analyzed by confocal microscopy. Proteins posses different folding states, conformations and show different dynamics and enzymatic processes. In contrast to fluorescence measurements of an ensemble providing only an average view on protein folding and structure, Single-molecule measurements give access to the sequence of processes in folding and function. The most powerful single molecule fluorescence technique for structural information is fluorescence resonance transfer (FRET). With FRET, conformational changes, dynamics, and protein-protein interactions can be investigated. Single molecule FRET measurements at room-temperature have two main advantages. On the one hand, it is possible to get unaveraged information about individual molecules under close-to physiological conditions. On the other hand, dynamic processes and interactions are accessible in real time. However, the exact determination of distributions of intramolecular distances using confocal microscopy at room temperature is restricted for processes like e.g. photobleaching of the dyes and diffusion motion. Here, cryo fixation of the sample can be beneficial. Temperature-cycle experiments still enable dynamics to be investigated. The dynamics step is occurring at elevated temperatures, while the single-molecule fluorescence measurement is performed at low temperature, preventing both, diffusion and premature photo bleaching

A crucial process, which is responsible for the quality of the measurement results, is the sample preparation. In order to maintain the structure of the biomolecules, vitrification of the water solvent is mandatory. The term vitrification refers to any process resulting in glass formation, the transformation from a liquid to an amorphous without crystallization. In order to avoid the crystallization processes it is crucial to cool very rapidly, i.e. with a cooling rate of $\geq 10^7 \mathrm{K \cdot s^{-1}}$ [8]. Several methods are applied to achieve the necessary high cooling rates, e.g. plunge-freezing or high pressure freezing [9]. A novel method developed in the Institute of Physics is spin-coating of the sample solution on a cooled cover slip [1]. So far, this cover slip was immersed in the low-temperature gas above liquid nitrogen. However, this set up does not provide well-defined temperatures. The quality of the sample is endangered, because glass possess a low thermal conductivity. We set out to optimize the process of preparation to improve the reproducibility and reliability of sample preparation. Therefore, a new sample holder for spin-coating is constructed allowing for contact cooling of the cover slip. It will also simplify the sample preparation and the cooling process and improve the quality of the sample rendering the measurements more reliable [1] [2] [5] [7].

2 Material and Methods

the following sections, first the cryo-microscope used for single-molecule measurements at low temperature be introduced. The set-up is based on a home-built laser scanning confocal microscope. The fluorescence measurement of single-molecules will not be explained in detail here. A detailed description can be found e.g. in [6]. The main focus of the present work is on the cryostat and the sample preparation process detailed in the second section.

2.1 Cryo Microscope

Single-molecule measurements were performed on a laser scanning confocal microscope. It contains a home-built cryostat system. The cryostat allows sample support and cooling the sample down to the temperature of liquid nitrogen (77 K). The set-up is sketched in Fig. 1. Two continuous wave lasers (488 nm and 594 nm) are fed through a single-mode optical fiber, collimated by a lens and then reflected by a dichroic mirror. The laser beams are focused into the sample by an objective (CFI Achromat 60 X, Nikon, NA 0.8). The emitted light, collected by the same objective, passes the scanning unit and the dichroic mirror. Finally the emitted light is detected by an avalanche photodiode [2] [3].

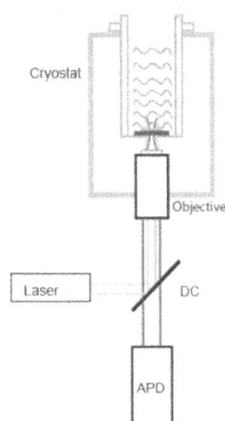

Figure 1: Simplified illustration of the laser scanning confocal microscope with cryostat. A laser beam is reflected via a dichroic mirror (DC) and collimated by an objective. The objective focuses the beam into the sample on the cover slip. The emitted light is collected by the same objective and passes the dichroic mirror. The light is detected by an avalanche photodiode (APD). For simplicity the scanning unit is omitted [1] [2].

2.2 Cryostat and Sample Preparation

Single-molecule experiments at the temperature of liquid nitrogen require a sophisticated preparation of the sample. Furthermore, a sample holder with effective sample cooling and a short working distance for the microscope is needed. To comply with the requirements, a custom built cryostat, consisting of two dewar vessels (outer and inner), is directly adapted to the microscope and to the sample preparation unit including a spin-coater (Fig. 1).

In our previous procedure [1], for the preparation of a thin sample layer, an inner dewar vessel is placed on a home-built spin-coating device made from stainless steel.

The space formed by the spin-coater and the inner cryostat vessel is filled with liquid nitrogen. The cover slip is placed on a Teflon-cap immersed in the cold nitrogen gas (Fig. 2). A drop of a few microliters of sample solution is spin cast onto the cooled cover slip rotating at 9000 rpm. The solution spreads very quickly and forms a thin, frozen layer. The cooled inner dewar with the cold cover slip is immediately transferred to the outer cryostat and refilled with liquid nitrogen To avoid fogging of the bottom surface of the cover slip the outer dewar vessel is evacuated immediately after attachment. The inner dewar vessel is immediately refilled with liquid nitrogen.

This sample preparation is a delicate task. The application of the drop presently is done manually and therefore prone to reproducibility errors. Furthermore, air humidity is detrimental for the preparation of a high quality sample. Another factor is that the sample on the cover slip is only cooled by the nitrogen gas. The cover slip is placed on a pierced Teflon-cap to ensure suction. Teflon has a low thermal conductivity; therefore the cover slip may not be cooled sufficiently and the temperature of the sample may exceed the critical glass transition temperature of 135 K. Additionally, ice condensation may take place and destroy the intramolecular structure.

After spin casting, the inner dewar vessel is lifted, picking up the cover slip, which as a result is closing the opening of the inner dewar. Centering of the cover slip on the opening is difficult to achieve this way. After successful positioning of the cover slip the inner dewar vessel is attached to the outer dewar vessel. During this process the sample is not continuously cooled by liquid nitrogen [2] [3] [4] [5].

In order to improve the sample preparation and thus the measurements, we developed a new cryo-sample holder. For the design of the holder and rotation base, the software 3D-CAD Inventor® was used. Both components were built in the machine shop of the University of Luebeck.

2.3 Sample holder and Rotation Base

The new sample holder (Fig. 3) for the cryostat combines the sample preparation and measurement chamber in one component.

The new sample holder features a split design and provides a positioning support surface for the cover slip and at the same time serves as the bottom of the cryostat, with an opening for the microscope objective and a sealing surface to enable evacuation of the space below the cryostat. In order to ensure efficient cooling of the cover slip, aluminum was chosen for its excellent thermal conductance. Aluminum has a significantly higher thermal conductance as compared to stainless steel (see Table 1).

The sample holder of the cryostat has a quadratic depression with the dimensions of the microscope cover slip (22 mm x 22 mm) and a depth of 1 mm. During the rotation, the cover slip fits in the depression so that it does not slip. The circular grooves in the cover slip seating improve gas tightness

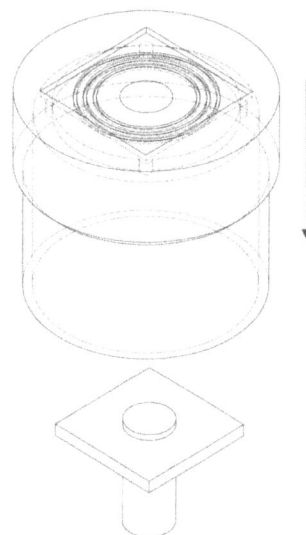

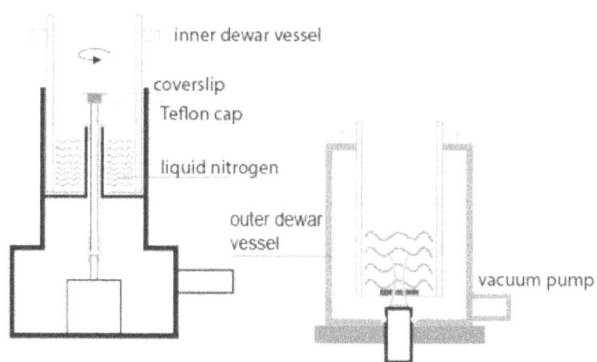

Figure 2: Left: spin-coating device. The spin-coater (black) contains the motor and is evacuated. The rotating hollow driveshaft extending into the dewar vessel carries the cover slip held by suction. Liquid nitrogen at the bottom of the dewar cools the cover slip. After spreading and freezing of a small drop of sample solution the dewar is transferred to the outer dewar vessel. Right: the outer dewar vessel with the attached objective. After preparation of the sample with the spin-coating device the inner dewar vessel gets attached to the outer vessel. The cover slip is closing the opening in the inner dewar vessel. The objective focuses the beam into the sample on the cover slip. The emitted light is collected by the same objective. After installation of the inner vessel the outer dewar vessel gets evacuated by a vacuum pump.

Figure 3: Technical drawing of the sample holder and the rotation base. Both components consist of aluminum. The rotation base fits into the sample holder driving it via contact edges of the quadratic top . The rotation base itself is attached to the spin-coater motor.

Table 1: Thermal conductivity of materials relevant for the cryo microscope [5].

material	thermal conductivity at 20 ° $W \cdot m^{-1} \cdot K^{-1}$
Teflon	0.2
Glass	0.9
Stainless Steel	15 – 45
Aluminum	220

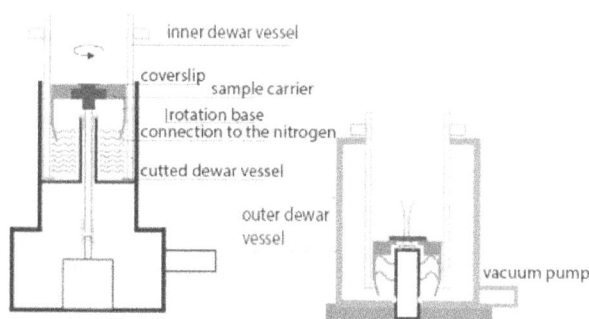

Figure 4: Left: Sample holder with the spin-coating device. After preparation, lifting off the inner dewar vessel the sample holder fits exactly into the opening at the bottom of this vessel. Right: Illustration of the cryostat with sample holder and microscope objective.

of the cover slip/sample holder interface. For spin-casting, the sample holder ist placed on the rotation base, and the rotation base is driving the sample holder via contacts of the edges of the quadratic top of it. The rotation base itself is attached to the motor of the spin-coater via a shaft.

While rotating, the bottom of the sample holder is immersed in liquid nitrogen ensuring continuous cooling. A drop of the sample solution is applied to the rotating microscope coverslip and thus spreading rapidly ensuring a fast cooling rate.

After sample preparation, the whole sample holder can be lifted off the spin-coater and placed inside the cryostat while maintaining the temperature well below the glass transition temperature of the vitrified sample. The sample holder has an indention on its circumference serving as the sealing surface in the cryostat. Evacuation of the outer cryostat results in a force pushing the sample holder onto the sealing surface ensuring a tight seal. Because all components are continuously cooled by the liquid nitrogen in this process, reproducible sample preparation will be achieved.

Figure 5: Sample holder (left) and rotation base (right).

3 Results and Discussion

In Fig. 5 the sample holder is shown together with the rotation base attached to the spin-coating device. The circular grooves are well visible. In order to test the improved sample preparation method, first experiments with the new sample holder were performed. First, it was tested if the cover slip will remain seated while rotating. It should be stabilized by the rotation, so that no further attachment should be necessary. The tests showed that rotation up to 95 revolutions per second did not cause a detachment of the cover slip from the sample holder.

Next, the spin-coating procedure was tested. For this, a pipette was fixed on a lab stand and filled with a fluorescent dye (10 μL, Atto 463). Instead of the inner dewar, which will be modified in the next step, a plastic tube with a rubber seal was used. For the experiment, liquid nitrogen was filled in the gap between the tube and the sample holder. After filling an appropriate amount of liquid nitrogen, the spin-coating device was turned on. While rotating at approximately $95\,\mathrm{r \cdot s^{-1}}$ in liquid nitrogen, a 10 μL drop of the dye solution was applied to the cover slip. In Fig. 6, the set-up for spin-casting and a photograph of the resulting film are shown. In the center of the cover slip, a thin frozen layer of the solution can be seen.

As can be clearly seen, frost from the air humidity as formed on top of the sample holder. This is will avoided in the future by covering the spin-coating device. It has to be emphasized that frost on top of the sample is not an issue as imaging is performed through the bottom of the cover glass.

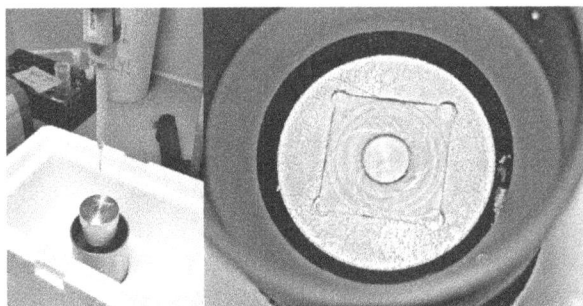

Figure 6: Left: sample holder on top of the spin coater with the pipette in a styrofoam box. Right: sample holder in a plastic cylinder. In the seating the cover slip with the spin-cast layer of the dye solution is visible.

4   Conclusion and Outlook

We have introduced a method to investigate single-molecule measurements at liquid nitrogen temperature. Therefore, a cryostat with a high thermal stability is required. A sample holder for the existing cryostat system fitting to a conventional laser scanning confocal microscope was developed. The preparation of a thin amorphous layer of fluorescently labeled molecules was performed on the new sample holder. The resulting films now have to be characterized. We plan to implement an additional infrared laser system in order to enable temperature cycle experiments. Further optimizations of the procedure will include jet-coating in order to obtain even thinner sample layers.

5   Acknowledgment

The work has been carried out at the Institute of Physics at the University of Luebeck. I thank Reinhard Schulz for producing the sample holder and rotation base.

6   References

[1] V. Hirschfeld, *Förster-Energietransfer an einzelnen Molekülen bei tiefen Temperaturen*. Inauguraldissertation, Universität zu Lübeck, 2011.

[2] V. Hirschfeld and C.G. Huebner, *A sensitive and versatile laser scnanning confocal optical microscope for single-molecule fluorescence at 77 K*. Review of Scientific Instruments 81, 1137051 – 1137057, 2010.

[3] V. Hirschfeld, H. Paulsen and C.G. Hübner, *The spectroscopic ruler revisited at 77 K*. Phys. Chem. Chem. Phys. 15, 17664 – 17671, 2013.

[4] W. Li, S. C. Stein, I. Gregor and J. Enderlein, *Ultrastable and versatile widefield cryo-fluorescence microscope for single-molecule localization with subnanometer accuracy*. Opt. Express 23, pp 3770 – 3783, 2015.

[5] H. Kuchling, *Taschenbuch der Physik*. Carl Hanser Verlag GmbH & Co. KG; 16. Auflage, pp 635 – 636, 1999.

[6] W. J. Greenleaf, M. T. Woodside and S. M. Block. *High-Resolution, Single-Molecule Measurements of Biomolecular Motion*. Annual review of biophysics and biomolecular structure 36, pp 171 -- 190, 2007.

[7] R. Zondervan, F. Kulzer, M. A. Kol'chenko and M. Orrit, *Photobleaching of Rhodamine 6G in Poly(vinyl alcohol) at the Ensemble and Single-Molecule Levels*. The Journal of Physical Chemistry A 108, pp 1657 – 1665, 2004.

[8] T. Loerting, V. Fuentes-Landete, P. H. Handle, M. Seidl, K. Amann-Winkel, C. Gainaru and R. Böhmer, *The glass transition in high-density amorphous ice*. Journal of Non-Crystalline Solids 407, pp 423 – 430, 2015.

[9] D. W. Galbraith and H. J. Bohnert, *Methods in Plant Cell Biology, PART A*. Academic Press, pp 4 – 6, 1995.

Evaluation of molecular dynamics calculations for the determination of diffusion coefficients

H. Quardokus [1] and H.Paulsen [2]

[1]Medizinische Ingenieurwissenschaft , Universität zu Lübeck, hanna.quardokus@student.uni-luebeck.de
[2] Institut für Physik , Universität zu Lübeck, paulsen@physik.uni-luebeck.de

Abstract

Diffusion of small molecules is an important process for any living organism. Better knowledge about the diffusion rates of different molecules may contribute to a deeper understanding of cell metabolism and may support the interpretation of spectroscopies. In this way the investigation of diffusion rates contributes to the knowledge, that forms a basic for any medical technology on the molecular level. In this project the focus was put on the theoretical determination of diffusion coefficients by means of molecular dynamics calculations, which were performed for small polyprolines in water. The slope of mean square displacement (MSD) of the polyprolines as a function of time can be used as an estimate for the diffusion coefficient D. The results obtained with this method raise the question whether the reciprocal relation between diffusion rate and molecular size as predicted for large molecules (Stokes-Einstein equation) is valid also for small molecules like 1-proline and 2-proline.

1 Introduction

The diffusion coefficients of molecules in solutions are important parameters from the point of fundamental research as well as for pharmacology and medical technology as far as the application of drugs or of tracers for medical imaging are considered. Diffusion coefficients can be determined experimentally by a variety of different methods or estimated theoretically, for instance by molecular dynamics (MD) calculations. On point of interest in this project is to verify whether the Stokes-Einstein equation also applies to small molecules. For large molecules this equation predicts a reciprocal relation between the diffusion coefficient D and the molecular radius r:

$$D = \frac{k_B T}{6\pi\eta r} \qquad (1)$$

(k_B- is the Boltzmann constant, T- the temperatur and η the viscosity). The latter approach is evaluated in this project on the example of three different polyprolines. This paper is organized as follows: in section 2 first the theoretical background and the calculational methods are explained and afterwards the molecules are presented for which the diffusion coefficients are calculated. In section 3 the obtained results are discussed and in the final section some conclusions are drawn from these data.

2 Materials and Methods

To theoretically calculate the rate of diffusion, it is of importance to have knowledge about particle motion in a solvent like water. For this purpose MD calculations are method of choice. The MD calculations of polyprolines in water have been done in previous scientific work [1] [4]. To achieve this, the polyprolines together with water molecules were inserted into a simulation box. All simulation boxes were cubic with a side length of 2.5 nm, except the large boxes where the side length was 3 nm. Afterwards the software package GROMACS (Groningen Machine for Chemical Simulations) [3] was used to create sufficiently long trajectories for the polyprolines. GROMACS is running under the linux operating system and provides a comprehensive and fast way for analysis and calculation of chemical and physical properties of molecules and their environment. The molecular dynamics simulations can be carried out with various different force fields. The output files of GROMACS can be edited a large number of programs, a fact that provides the opportunity for freely choosing how to further analyze the results.

2.1 Diffusion and the mean squared displacement (MSD) as a function of the time

Diffusion can be explained by the collision of particles in a solvent due to thermal noise which is considered as Brownian motion, which can be understood with the help of 'random-walk' model. This is a mathematical model, which describes random motion. Fig.1 shows this motion in x- and y-directions consisting of a large number of randomly chosen steps.

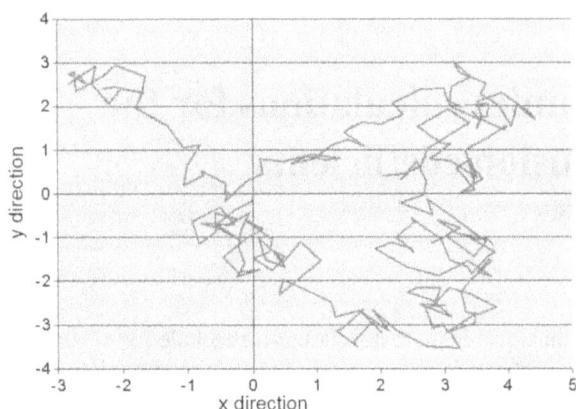

Figure 1: Graphical representation of a 2D random walk in x- and y-directions.

The mean square displacement (MSD) is defined as:

$$[x(t + \Delta t) - x(t)]^2 = (\Delta s)^2. \tag{2}$$

This provides the motion of a particle in every time interval Δt from its originating position, whilst Δs is the length of the step. After n steps of time Δt one gets for the time t=nΔt the MSD:

$$[x(t) - x(0)]^2 = \frac{t}{\Delta t}(\Delta s)^2. \tag{3}$$

2.2 Derivation of the autocorrelation function of the velocity

According to Fick's 1st law:

$$j = -D\frac{\partial c}{\partial x} \tag{4}$$

one may derive the diffusion constant D. Therefore the particle-current density j(x) and the particle concentration c(x) are to be defined. Examining the particle concentration alongside the x-axis so the particle-current density is vertical to the y-z-plane the average particle concentration in c_-(left)- und in c_+(right)-direction of the surface A can be defined as (the surface A is parallel to y-z-plane and cuts the x-axis on the point x):

$$c_{\pm} = c(x) \pm \frac{1}{2}\frac{\Delta c}{\Delta x}\Delta s. \tag{5}$$

To derive the particle current density, the difference in the particle concentration of right c_+- and left c_-- direction of motion is determined:

$$j = \frac{c_- A\Delta s - c_+ A\Delta s}{2A\Delta t} = -\frac{(\Delta s)^2}{2\Delta t}\frac{\Delta c}{\Delta x}. \tag{6}$$

According to Fick's 1st law results the rate of diffusion:

$$D = (\frac{(\Delta s)^2}{2\Delta t}). \tag{7}$$

And the MSD is:

$$[x(t) - x(0)]^2 = 2Dt. \tag{8}$$

The presented calculations imply a one-dimensional space. To define the particle movement in three-dimensional space the vector $\vec{r}$ is used.

$$[\vec{r}(t) - \vec{r}(0)]^2 = 6Dt. \tag{9}$$

The MSD after N steps is proportional to the time τ= NΔt and is defined as:

$$\langle \vec{r}^2 \rangle \sim \tau. \tag{10}$$

The angular brackets, wich denote an ensemble average, can be replaced by executing a time average, which leads to an equivalent result.

$$\frac{1}{t_1 - t_0}\int_{t_0}^{t_1}[\vec{r}(t + \tau) - \vec{r}(t)]^2 dt = 6D\tau \tag{11}$$

where t_0 and t_1 are start- and end-time of the calculations. As a result the diffusion rate is defined as [4]:

$$D = \frac{1}{6\tau}\langle[\vec{r}(t + \tau) - \vec{r}(t)]^2\rangle_{\tau}. \tag{12}$$

After numerous transformations and a diminutive τ the rate of diffusion is described with the following formula :

$$D = \frac{1}{3}\int_0^{\infty}\langle v(t + \tau) \cdot v(t)\rangle_t d\tau. \tag{13}$$

As a result it is theoretically possible to calculate with the acquired formula, the rate of diffusion not only with the average quadratic distance but also with the autocorrelation functions of the velocity (13). [2]

2.3 Materials

The calculation of the rate of diffusion was performed using the MSD as a function of the time as well as the autocorrelation function of the velocity for some α-amino acids and peptides, 1-proline, 2-proline and 3-proline in the water models TIP3P (Transferable Intermolecular Potential 3-Point) and TIP4P (Transferable Intermolecular Potential 4-Point) with various potentials and simulation boxes. [5]

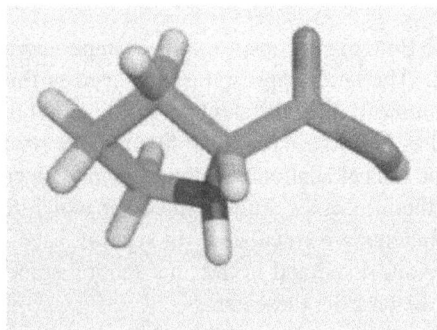

Figure 2: 3D representation of 1-proline.

For this project solutions of α-amino acid prolin (Fig.2, Fig.3 [4]) were used in the following forms:

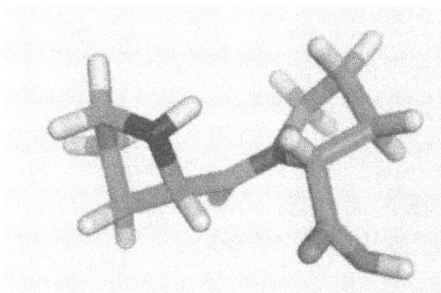

Figure 3: 3D representation of 2-proline. 2- proline is doublel of 1-proline

- 1-proline in water model TIP3P (proline 1 tip3p)

- 1-proline in water model TIP4P (proline 1)

- 2-proline in water model TIP4P (proline 2)

- 3-proline in water model TIP4P (proline 3)

- 3-proline in water model TIP4P in a large simulation box of size $(3nm)^3$ TIP4P (proline 3 box3)

- 3 -proline in the water model in a large simulation box of size $(3nm)^3$ TIP4P with potential 1 (proline 3 box3 ang res 1)

- 3-proline in the water model in a large simulation box of size $(3nm)^3$ TIP4P with potential 2 (proline 3 box3 ang res 2)

- 3-proline in the water model in a large simulation box of size $(3nm)^3$ TIP4P with potential 3 (proline 3 box3 ang res 1)

3 Results and Discussion

In this paper the rate of diffusion for various forms of α-amino acids 1-proline, 2-proline, 3-proline.
Following, the results of the calculation for the MSD will be described.
GORMACS uses the time interval for the calculation of the rate of diffusion in the command option –trestart. The ideal value for the time steps is the one, which was used for the protein simulation. Whereby the calculation with those values can take some time to compute. For that reason, additionally greater steps on the time axis were used to possibly reduce the time of calculation whilst the value of the rate of diffusion was not affected.
With –beginfit and –endfit a trajectory section for the rate of diffusion is chosen with consideration to the calculated trajectory for the particular MSD to the particular time interval (the numbers given stand for the time in ps).
The correct selection of the boundaries (–beginfit and –endfit) is to pick the most linear gain of the curve. The selection of the curve, where a loss of linearity occurs, leads to wrong values for of the rate of diffusion. These selected values are presented in the Table 1.

Table 1: The selected -beginfit and -endfit.

protein	-beginfit	-endfit
proline 1 tip3p	0	8950000
proline 1	0	47000000
proline 2	0	9950000
proline 3	0	9400000
proline 3 box3	0	3000000
proline 3 box3 ang res 1	0	4000000
proline 3 box3 ang res 2	0	1400000
proline 3 box3 ang res 3	0	700000

Table 2: The results for the calculation of the rate of diffusion of protein with the various –trestart values.

protein	diffusion constant $[10^{-5}cm^2/s]$
proline 1 tip3p	0,3712 ($\pm$0,2855)
proline 1	0,1360 ($\pm$ 0,0052)
proline 2	0,2261 ($\pm$ 0,1090)
proline 3	0,1291 ($\pm$ 0,0694)
proline 3 box3	0,3111 ($\pm$ 0,0523)
proline 3 box3 ang res 1	0,4072 ($\pm$0,5068)
proline 3 box3 ang res 2	0,2021 ($\pm$ 0,0040)
proline 3 box3 ang res 3	0,4632 ($\pm$ 0,1066)

The results of the average rate of diffusion with the alterations –trestart values (1000, 10000 and 100000) are shown in the Table 2. The resulsts of the rate of diffusion are shown average, because with the increase of –trestart by the power of 10, the change of the rate of diffusion is minimal (around $0,001 \cdot 10^{-5}cm^2/s$). Due to this observation it is possible to use greater steps on the time axis. This leads to shorter calculation times, whilst the calculation of the rate of diffusion is not falsified.
The calculations of 1-proline, 2-proline, 3-proline in the water model TIP4P gave evidence to expect the rate of diffusion decreases with the increasing size of the protein. However the practical work has shown, that the rate of diffusion of 2-proline is greater than the one of 1-proline and 2-proline, hence 2-proline diffuses at a faster. However, the standard deviation for 2-proline is quite large ($\pm$ 0,1090 refer to Table 2), meaning that a better statistic has to be gained in order to get significant results for 2-proline.
Fig. 4, Fig. 5 and Fig.6 show the MSD for prolines. The representation of curves of the proteins does not occur in an image. This happens for the reason, that the simulation time was not the same in each protein. In the figures can be seen, that MSD increases linearly with time. For the calculations of diffusion coefficients the time limits (-beginfit and -endfit) were chosen, where the curve is linear.
The last step of our work was the calculation of the rate of diffusion with the autocorrelation functions of the velocity.The calculated results which were created with the autocorrelation function were not analysable. This was due to the fact that the size of the time interval for the MD-simulation was set to high. The time interval between 0 and about 1 ns (Fig.7) shows the drop of the curve. These calculations have to be repeated with a better time resolution.

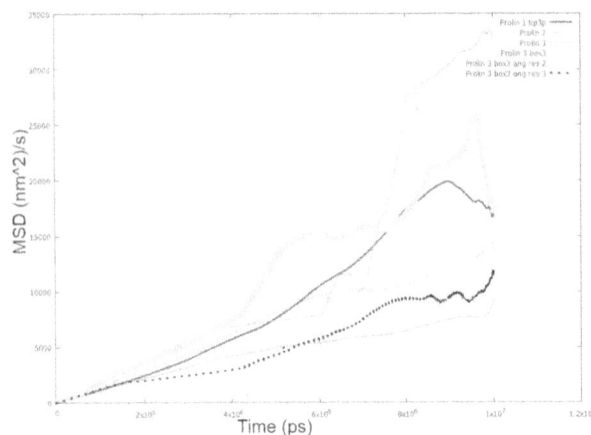

Figure 4: Representation of prolines.

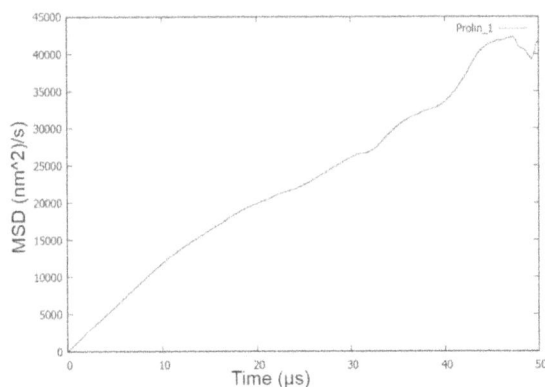

Figure 5: Representation of 1-proline.

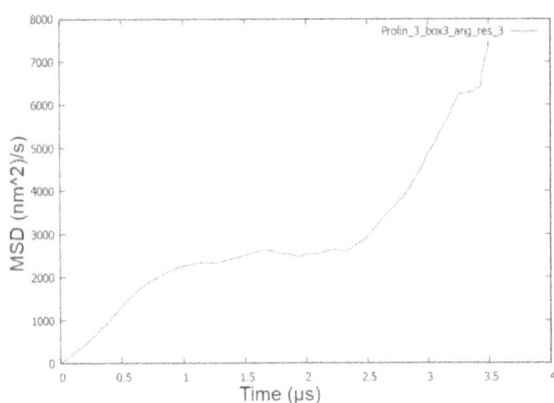

Figure 6: Representation of 3-proline TIP4P box3 ang res 3.

4   Conclusion

From the calculated data no unambiguous relation between the diffusion coefficient D and the size of the molecule could be extracted. For large molecules the Stokes-Einstein equation predicts a reciprocal relation between D and the molecular diameter, which is in accordance with the values obtained for 1-proline and 3-proline but in contradiction to the finding for 2-proline. Future investigations will

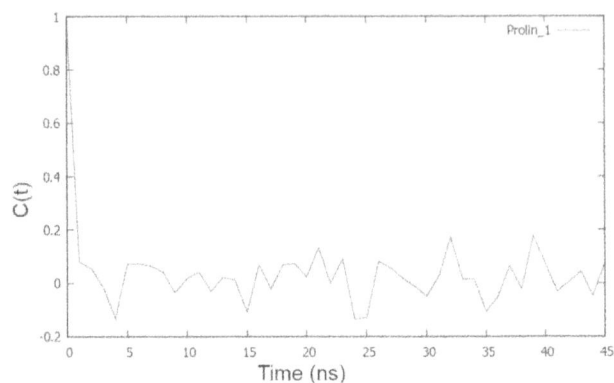

Figure 7: Trend of the scaled-up autocorrelation function for the velocity of 1-proline TIP4P.

have to show whether the results for 2-proline are an artifact, or whether the Einstein-Stokes-equation is not valid for very small molecules. The drastic influence of the size of the simulation box that has been observed for 3-proline suggests that also the calculations for 1-proline should be repeated with a large simulation box.

Acknowledgement

The work has been carried out at Institute of Physics , Universität zu Lübeck.

5   References

[1] F. Alotaibi, *Translation von Prolin und Polyprolin in Wasser unter den Einfluss externer Potentiale in Moleküldynamiksimulationen*, Bachelor Thesis, Lübeck, 2013

[2] B. J. Adler, D. M. Gass and T. E. Wainwright, *Studies in Molecular Dynamics. VIII. The Transport Coefficients for a Hard-Sphere Fluid*, in: The Journal of Chemical Physics 53, (3813), 1970

[3] Gromacs-Hompage.   Available: http://www.gromacs.org/ [last accessed on 22.09.15].

[4] J. Beer, *Moleküldynamik-Simulationen zur Diffusion von Prolinen in Wasser*, Bachelor Thesis, Lübeck, 2014

[5] M. Hömberger, *Untersuchung des Einflusses von Kraftfeld und Wassermodell in Moleküldynamiksimulationen von Proteinen anhand von spezifischen Parametern*, Master Thesis, Lübeck, 2013

Improving the optical properties of a fused silica capillary

Kim C. Reiter [1], Sven Schneider [2] and Christian G. Hübner [2]

[1] Medizinische Ingenieurwissenschaft, Universität zu Lübeck, kim.reiter@student.uni-luebeck.de

[2] Institute of Physics, Universität zu Lübeck, {schneider, huebner}@physik.uni-luebeck.de

Abstract

Fused silica capillaries are well suited for fluorescence measurements under pressure and perform much better than formerly used pressure cells since they have a high pressure stability in combination with a small wall thickness. These capillaries allow the use of high numerical aperture (NA) objectives. High NA-objectives and a square cross section can realize very small focal volumes, the basic requirement for single molecule (sm) fluorescence measurements. Customized square fused silica capillaries are usually very expensive with costs exceeding 1.000 $ per meter. Most suitable commercially available capillaries for the proposed application have inner and outer diameters of 50 μm and 300 μm respectively, which is not optimal for standard water immersion objectives and therefore not sufficient for sm-fluorescence measurements. Here, the optics are improved by putting the capillary on a 90 μm fused silica coverslip and an index matching gel in between. Results of fluorescence correlation benchmarks and fluorescence resonance energy transfer (FRET) histograms are presented, which are strongly improved when compared to the standard optical setup.

1 Introduction

Single molecule FRET (smFRET) has proven a powerful method for the study of the unfolded state of model folding proteins [1]. Unfolding by temperature and chemical denaturants has been elucidated by smFRET. However unfolding experiments using high pressure has not been reported to date. For high pressure applications one can take fused silica capillaries as a pressure cell [2]. Their advantage is a small wall thickness in combination with a high pressure stabililty, which allows the use of high NA-objectives for measurements with sm sensitivity. Unfortunately standard objectives are optimized to borosilicate coverslips, which do not match the refractive properties of fused silica layers. Here, an approach is presented, how to improve the optics of a commercially available fused silica capillary by putting it on a 90 μm fused silica coverslip and surrounding the capillary with a refractive index matching gel. In order to evaluate the optics of the capillary, fluorescence correlation benchmarks, FRET experiments and focus scans from the new arrangement and from bare capillary measurements are compared with those measured with a standard coverslip made of borosilicate, which serve as a reference for the experimental setup.

2 Material and Methods

This section provides a short overview of the experimental setup and materials used. Measurements and data analysis are also described.

2.1 Capillaries

The capillaries used (Z-FSF-050365, Postnova) are square and made of fused silica with a refractive index n of 1.46. They have an outer and inner diameter of 300 μm and 50 μm respectively and a 30 μm thick protection layer on the outside (see Fig. 1), which is removed before measurements. The cost of the capillaries is only about 54 $ for 1 m.

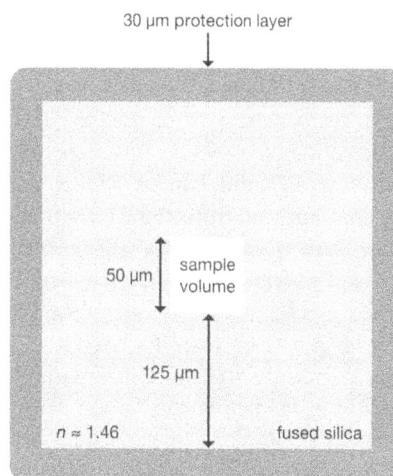

Figure 1: Scheme of the used fused silica capillary. The protection layer is removed before measurements.

2.2 Experimental Setup

The experimental setup consists of a custom built confocal microscope with a 60x water immersion (WI) objective (CFI Plan Apochromat 60x 1.2 WI, Nikon) with a NA of 1.2. Two continuous wave lasers with wavelengths of 488 nm (Spectra Physics Cyan, Newport Corp.) and 640 nm (OBIS 640LX, Coherent) can be used to excitate the fluorophores. With two piezo actuators (P-733.2CL, P-721.CLQ, Physik-Instrumente (PI) GmbH & Co. KG) the sample can be scanned in three dimensions within a range of $100 \times 100 \times 100 \, \mu m^3$. The fluorescence photons passing through a dichroic or 50/50 beam splitter and different filters depending on the type of measurement. They are then focussed on two avalanche photo diodes (APD) (SPCM-AQR 14, PerkinElmer Optoelectronics) with a quantum efficiency of 65%, which serve as detectors. To record the detected photons, a single-photon counting electronic (Timeharp 200, Picoquant GmbH) is used, which writes the detected photons in the tttr-mode (time-tagged-time-resolved) to a file.

2.3 Measurements and Samples

The measurements are performed with three different arrangements. One with a standard borosilicate coverslip ($n \approx 1.52$, Roth) with a thickness of 150 μm for which the objective is abberation corrected. Next with the bare capillary and finally with the new arrangement consisting of the capillary on top of a 90 μm fused silica coverslip and an index matching gel ($n \approx 1.46$, Cargille Laboratories) in between (see Fig. 2). To keep the gel in place a ring of glue (Fixogum, Marabu) is attached to the coverslip. FRET experiments of 30 minutes duration are performed with a laser power of 40 μW and a wavelength of 488 nm. The sample is a double stranded DNA (dsDNA) labeled with Alexa Fluor 488 as the donor dye and Atto 647N as the acceptor dye. The labeled dsDNA is dissolved in a solution of Mops-buffer and 0.1% Tween. The correlation benchmarks consist of several fluorescence correlation spectroscopy (FCS) measurements with different laser powers and a record time of 10 minutes. The benchmarks are used to quantify the excitation and detection efficiencies of the experimental setup for comparison of all arrangements. Since the optics in the capillary are improved for smFRET experiments under pressure, benchmark measurements are made with Alexa Fluor 488 as well as with Atto 647N. For Atto 647N measurements, the same sample is used as for the FRET experiments, but with a 640 nm laser for excitation. To avoid FRET effects during the 488 nm benchmark, pure Alexa 488 dissolved in Mops-buffer and Tween is used. To determine the focus dimensions of the setup, small fluorescent beads (FluoSpheres, Invitrogen) are immobilized in an agarose gel and then scanned with a laser power of 1 nW and a wavelength of 488 nm.

2.4 Theory and Data Analysis

Since standard WI objectives are optimized for borosilicate coverslips, the use of fused silica covers of the same thickness leads to aberrations. However, it is possible to reduce them by combining the lower refractive index of fused sil-

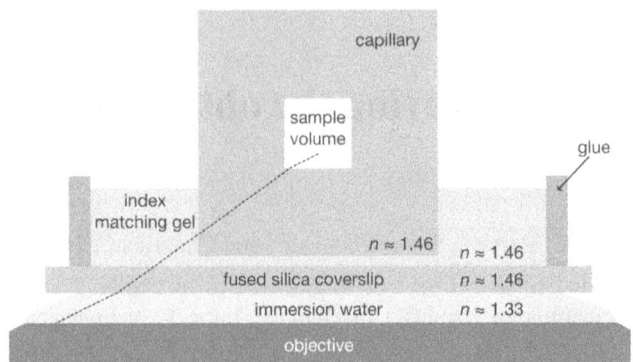

Figure 2: Scheme of the new arrangement. A combination of the capillary, a 90 μm fused silica coverslip and an index matching gel is used to improve the optics.

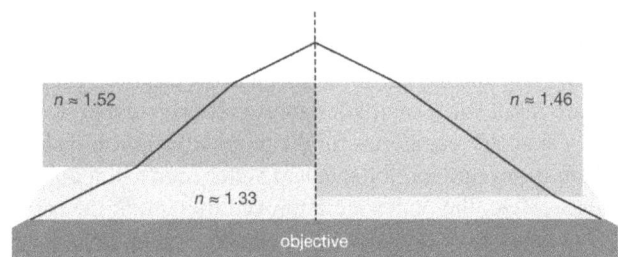

Figure 3: Schematic beam paths for the standard setup (left) and a fused silica coverslip with smaller refractive index (right). Combination of a smaller refractive index with a greater wall thickness reduces aberrations.

ica with a greater wall thickness (see Fig. 3). Therefore the capillary is placed on top of a 90 μm fused silica coverslip to extend the wall thickness. The gel is intended to prevent a transition of the laser beam from glass to air and to enable peripheral beams entering the capillary through the side walls without being refracted (dotted line in Fig. 2). All experimental data is analyzed with *IGOR Pro* from Wavemetrics. The detected photons are divided into time bins of 1 ms. For the FCS benchmarks, the binned data is pseudo autocorrelated by correlating the signals of the two detectors. In order to obtain the triplet fraction T and triplet time tt, the autocorrelation function is fitted with a theoretical model: Assuming a three-dimensional gaussian beam profile, the autocorrelation function can be expressed as [3]

$$G(\tau) = 1 + \frac{1}{\langle N \rangle} \left[1 + \frac{\tau}{\tau_D} \right]^{-1} \cdots \qquad (1)$$

$$\cdots \left[1 + S \frac{\tau}{\tau_D} \right]^{-1/2} \left[1 + \frac{T}{1-T} \exp\left(-T/tt\right) \right],$$

where $\langle N \rangle$ is the mean number of dye labeled molecules in the focal volume, τ the lag time and τ_D the diffusion time, which describes the mean time the molecules stay in the focal volume . $S = w_{xy}^2/w_z^2$ is the ratio between the lateral (w_{xy}) and axial (w_z) radii of the focal volume. T and tt are used to compute the transition rate to the triplet state $k_{isc} = T/tt$ for the excitation benchmark for each used laser power P. Then the saturation power P_{sat} is determined by fitting the transition rates with the relation

$$f(P) = k_{\text{isc}}/(1 + P_{\text{sat}}/P), \qquad (2)$$

where a lower P_{sat} indicates a smaller focus since the laser power is concentrated to a smaller volume and therefore the dyes are excited more efficiently. The molecular brightness ϵ for detection benchmarks can be calculated using the mean intensity $\langle I \rangle$ and $\langle N \rangle$ [4]:

$$\epsilon = \langle I \rangle / \langle N \rangle, \qquad (3)$$

whereby a higher ϵ at the respective P_{sat} implies a better detection. $\langle I \rangle$ is directly determined from the time trace of the measurements and $\langle N \rangle$ is obtained from the fit of the autocorrelation function. To analyze FRET experiments, the transfer efficiency E of each time bin is calculated with [5]

$$E = I_{\text{A}}/(I_{\text{A}} + I_{\text{D}}), \qquad (4)$$

where I_{A}, I_{D} are the acceptor and donor intensities. They are combined into a transfer efficiency histogram with a threshold level of 30 counts/ms for the sum of donor and acceptor intensities. Since a FRET histogram is always broadened due to shot noise, it is fitted with a gaussian function to obtain E. Bead scans were analyzed with a two dimensional gaussian fit to determine the dimensions of the focus.

3 Results and Discussion

Results of the excitation benchmarks are shown in Fig. 4. As one can see, the determined saturation powers from the capillary measurements differ significantly from the reference measurements made with the borosilicate coverslip, indicating poorer optics in the capillary. This holds for both dyes, Alexa Fluor 488 and Atto 647N. Compared to this, the saturation powers of the measurements with the new arrangement are in a similar range as those of the coverslip. Detection benchmarks in Fig. 5 show a similar behavior. The brightness of the two dyes is similarly high for the reference measurements and the new arrangement, whereas it is much worse for the capillary in both cases. As in the excitation benchmarks, this indicates an improvement of the optics for the new arrangement compared to the use of the fused silica capillary.

The results of the FRET experiments are shown in Fig. 6. Since the optics in the capillary are improved for experiments under pressure, the main interest is the number of events during the experiment and not the value of the FRET efficiency itself. More events indicate better optics because a smaller focus has a higher power density and is therefore more efficient to excite the dyes. Considering the number of events, there are distinct differences between the three arrangements. The capillary show worst results with about 4.5 times less events as compared to the coverslip. Although the result of the new arrangement is not as good as for the coverslip, it is about a factor of 3 better than with the capillary, indicating a significant improvement of the optics. Bead scans show the cause of the results of the benchmarks and FRET experiments (see Fig. 7). While the lateral dimensions of the focal volume are comparable for

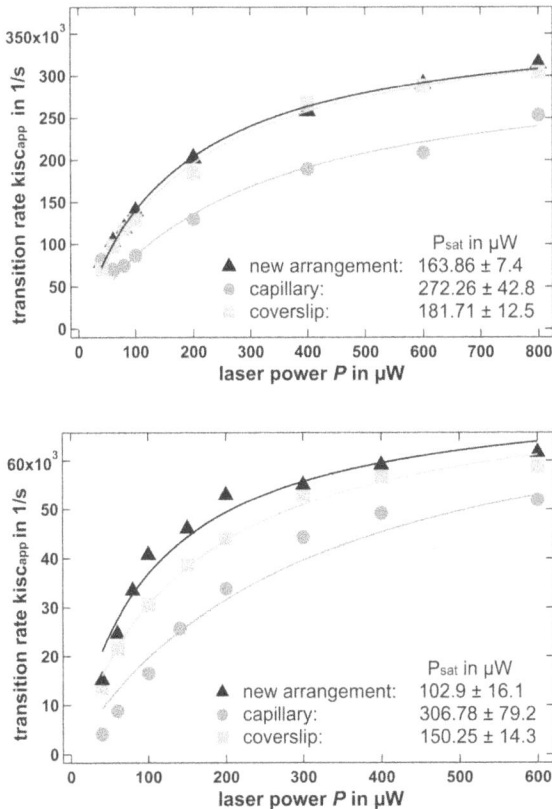

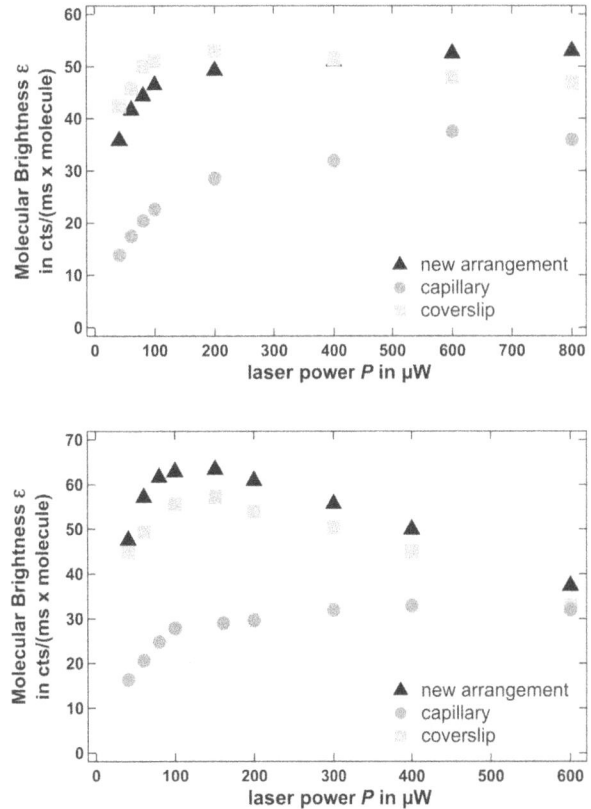

Figure 4: Excitation benchmarks from the measurements with Alexa Fluor 488 (above) and Atto 647N (below).

Figure 5: Detection benchmarks from Alexa Fluor 488 (above) and Atto 647N (below).

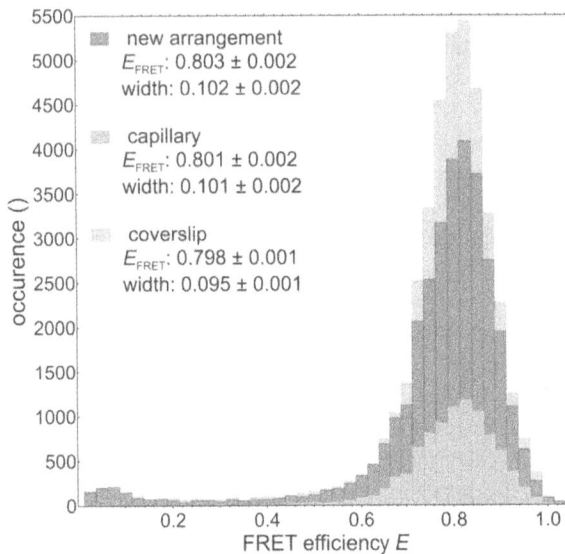

Figure 6: FRET histograms of all arrangements. Poor optics in the capillary are indicated by low number of events compared to the borosilicate coverslip. The new arrangement is not as good as the coverslip too, but greatly improved in comparison to the capillary.

all three arrangements, there are significant differences in the axial direction. The new arrangement and the coverslip show similar results, whereas the axial dimension of the focal volume in the bare capillary is almost twice as large. The unmatched refractive index in combination with a too thin wall thickness of the capillary leads to aberrations and a greatly elongated focal volume in the axial direction. Whereas results of the FRET experiments show differences for the coverslip and the new arrangement, bead scans show almost equal results for both arrangements. This may be due to chromatic aberrations, since in FRET experiments also acceptor photons are detected. This is not the case for the bead scans where such effects are not observed.

4   Conclusion

In this paper a cost effective way is presented, to improve the optics of commercially available fused silica capillaries for high pressure applications with single molecule sensitivity. In order to evaluate the performance of the optics in the capillary, fluorescence correlation benchmarks, FRET experiments and scans of the focal volume are made with three different configurations for comparison. One with a standard borosilicate coverslip serving as reference for the experiment. Next with the bare capillary and finally the proposed arrangement consisting of the capillary, a 90 μm fused silica coverslip and an index matching gel. As expected, all experiments show worst results for the bare capillary due to the unmatched refractive index and wall thickness of the capillary walls, which leads to aberrations. With the new arrangement, strongly improved results can be achieved with values in the range of those determined with the standard coverslip. Only FRET experiments show slightly worse results for the new arrangement, which are nevertheless still much better than the bare capillary FRET experiments. In summary, the new arrangement with the capillary on top of a thin fused silica coverslip and an index matching gel in between reveals greatly enhanced results for fluorescence correlation benchmarks and FRET experiments on the basis of a smaller focal volume of the experimental setup compared to bare capillary measurements due to reduced aberrations. This allows single molecule sensitivity in a fused silica capillary for high pressure experiments.

Acknowledgement

This work has been carried out at the Institute of Physics of the Universität zu Lübeck.

5   References

[1] Schuler, Lipman and Eaton, *Probing the free-energy surface for protein folding with single-molecule fluorescence spectroscopy*, Nature, Vol. 419, 743-747, 2002

[2] Tekmen and Müller, *High-pressure cell for fluorescence fluctuation spectroscopy*, Review of scientific instruments, Vol. 75, 5143-5148, 2004

[3] Widengren, Mets and Rigler, *Fluorescence Correlation Spectroscopy of Triplet States in Solution: A Theoretical and Experimental Study*, J. Phys. Chem., Vol. 99, 13368-13379, 1995

[4] Elson, *Fluorescence Correlation Spectroscopy: Past, Present, Future*, Biophysical J., Vol. 101, 2855-2870, 2011

[5] Chung, Louis and Eaton, *Distinguishing between Protein Dynamics and Dye Photophysics in Single-Molecule FRET Experiments*, Biophysical J., Vol. 96, 696-706, 2010

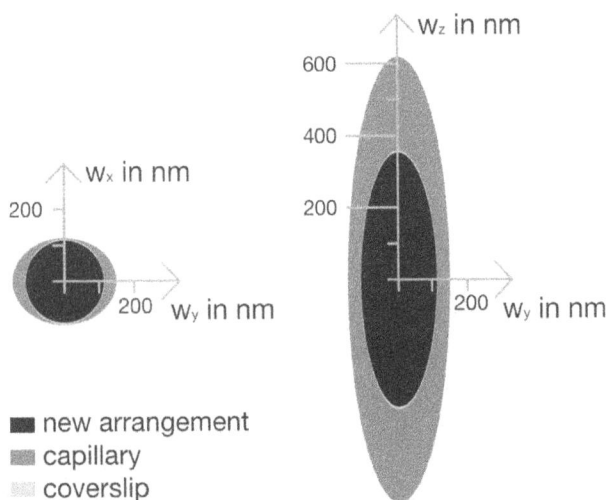

Figure 7: Schematic representation of the lateral (left) and axial (right) dimensions of the focal volume.

Inkjet printing of surfactants, proteins and enzymes for biomedical applications

K. Hering [1], S. Björklund [2], S. Klein [3], V. Kocherbitov [2] and T. Ruzgas [2]

[1] Biomedical Engineering, University of Applied Science Lübeck, kathrin.hering@stud.fh-luebeck.de
[2] Biofilms – Research Center for Biointerfaces, Malmö University, Sweden, {sebastian.bjorklund,vitaly.kocherbitov, tautgirdas.ruzgas}@mah.se
[3] Medizinische Sensor- und Gerätetechnik, University of Applied Science Lübeck, Germany, stephan.klein@fh-luebeck.de

Abstract

Ink solutions relevant for biomedical applications have been printed using a commercial inkjet printer. Specifically, water-based inks containing surfactant, protein and enzyme have been evaluated. Printability of surfactant ink was theoretically estimated from practically determined surface tension and viscosity of the solution. Quartz crystal microbalance with dissipation monitoring (QCM-D) was used to estimate the mass of inkjet printed surfactant. The effect of printing patterns and hydration on the QCM-D data was evaluated. Finally, horseradish peroxidase ink was printed on skin and an enzymatic reaction on skin was observed. Taken together, the results from this study provide a promising starting point from which inkjet printing of protein-enzyme mixtures on skin can be evaluated.

1   Introduction

Inkjet printing has not only been used for printing documents or pictures. This dispensing method has also been applied to print small structures for sensors or other applications relevant to medicine. Among the benefits, it is stated that printing of biosensors enhance their quality, decreases the amount of waste, etc. [1]. To date the most used drop-on-demand printing technique for printing biological inks is based on a piezoelectric actuation. Commercial "Epson" inkjet printer utilizes this principle of dispensing, which easily can be adopted for printing solutions containing biomolecules. The composition of biological ink should, preferably, possess the suitable physical and chemical properties, similar to commercial ink that ensures optimal printability.

The physical and chemical parameters that characterize the ink are the density, ρ, viscosity, η, and surface tension, γ. By knowing the radius of dispenser orifice, a, the printability can be estimated by calculating the inverse (Z) of Ohnesorge number (Oh). The number is a dimensionless number expressing the ratio of the Reynolds number (N_{Re}) and the Weber number (N_{We})[2]:

$$Z = \frac{N_{Re}}{(N_{We})^{\frac{1}{2}}} = \frac{\frac{v \cdot a \cdot \rho}{\eta}}{\frac{v^2 \cdot a \cdot \rho^2}{\gamma}} = \frac{(a\rho\gamma)^{\frac{1}{2}}}{\eta} \quad (1)$$

The Z value is used to define the optimal printability of a fluid. That means that a droplet can be formed without the presence of long ligaments or satellite drops [2][3]. As determined by Jang et al (2009), for good printability of an ink, Z should be in the range of $4 \leq Z \leq 14$ [2].

All these parameters should be kept in mind when designing a printing ink of interest. However, the possibility to print proteins, drug formulations and living cells has been demonstrated even without assessing Z number of the biological inks. E.g., printing of enzymes (glucose oxidase, horseradish peroxidase, etc.) has been used for making different biosensors [2]. Printing drug formulations provide an easy way to regulate the dose and, thus, help to improve safety and efficiency of medical treatment [4]. In addition to these options, living cells have also been used for printing living tissues [5].

The aim of the project was to assess R280 printer from Epson for printing micrometer size structures onto relevant surfaces. For this we used refillable cartridges with none modification of the printer. To evaluate the printing, quartz crystal microbalance with dissipation (QCMD) was used to measure the mass of different microstructures printed on QCM-D sensors. The printed patterns were also observed by optical microscopy. Additionally, we evaluated the possibility to print a defined layer on the surface of a skin membrane to enable colorimetric detection of hydrogen peroxidase penetration through skin membrane.

2   Material and Methods

2.1   Materials

Dimethyldodecylamine N-oxide (DDAO), horseradish peroxidase and bovine serum albumin (HRP-BSA), and diaminobenzidine (DAB), all purchased from Sigma-Aldrich, were dissolved in water and used to print onto glass slides, QCM-D sensors and skin membranes. The concentrations of the surfactant (DDAO) were chosen in proximity of the critical micelle concentration (cmc, 0.05 wt%). DDAO was chosen due to well known phase behavior of the DDAO-water system.

HRP catalysed DAB oxidation was utilized to probe function of an inkjet printed enzyme layer on skin membrane. This reaction is frequently used in immunostaining exploiting HRP-conjugated antibodies. The aim was to use this reaction (2) for monitoring H_2O_2 penetration through skin membrane.

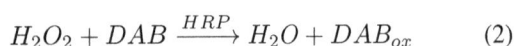

$$H_2O_2 + DAB \xrightarrow{HRP} H_2O + DAB_{ox} \qquad (2)$$

2.2   Methods

2.2.1   Procedure

Before starting with the printer, the physical properties of surfactant solution were assessed with respect to equation (1). Therefore, aqueous mixtures with different concentrations were produced. Subsequently, the surface tension and viscosity were measured.

To evaluate the homogeneity and mass of the printed structures QCM-D experiments were performed. In these experiments QCM-D sensors were placed on the surface of compact disk (CD) and the printing software from Epson for printing on CD was programmed to create a desired pattern. A full circle, half cycle, rings and half rings with different diameters and line thicknesses (minimum 0.1 mm, due to software limits) were printed. The amount of printed layers was increased systematically by printing 1, 5, 10, or 15 times after which the mass of printed surfactant was determined with the QCM-D method.

2.2.2   Surface Tension

Any solution possesses cohesiveness and when dispersed tends to create a droplet with minimal surface. The molecules of a liquid create small spheres due to the surface tension acting at the surface. The surface tension is defined as the force acting on a surface and is measured in mN/m [6]. The Lauda Drop Volume Tensiometer TVT 2 was utilized for measuring the surface tension of DDAO solution. This instrument slowly pumps solution through a syringe needle and determines the exact volume of a droplet at the moment of it detachment (dropping). The volume is used then to calculate the surface tension of the solution [7].

2.2.3   Viscosity

The viscosity was measured with capillaries and is based on the principle of Hagen-Pouseuille, which basically measures the pressure difference. As there were constant conditions during the measurement, the following relation could be used for calculating the viscosity of different solutions related to the viscosity of water ($\eta = 1\,\mathrm{mPa \cdot s}$) [8].

$$\eta = \frac{\Delta t}{const}[mPa \cdot s] \text{ with } const = \frac{8 \cdot L \cdot \Delta V}{\pi \cdot r^4 \cdot \Delta P} \qquad (3)$$

2.2.4   Measurements of printed mass

After successful printing surfactant structures on quartz sensors, QCM-D (Q-Sense, Biolin Scientific) was used for characterizing the areal mass of the printed films. The instrument measures the change of the resonance frequency, Δf, of the QCM-D quartz sensor. The simplest mass calculation is based on the Sauerbrey equation, which describes the proportionality between the decrease in resonance frequency and the mass of the film per area [9][10].

3   Results and Discussion

Before starting the inkjet printing protein and surfactant solutions, the Z parameter of the commercial ink was assessed. The ink had a surface tension of 25.36 mN/m and a viscosity of 2.84 mPa $\cdot$ s . According to eq. (3), this gives Z value of 12.53. Following the same methodology Z value of solutions with different concentrations of DDAO were evaluated.

3.1   Inkjet printing of DDAO

To assess printability parameter Z the surface tension of DDAO solutions was measured. As shown in Table 1, the surface tension decreases by increasing DDAO concentration in water and stabilizes at cmc, 0.05 wt%. To keep surface tension of DDAO ink constant, concentrations of 0.05 wt%, 0.1 wt% and 10 wt% were chosen for testing inkjet printing. The concentration of 10 wt% was chosen due to the fact, that a single layer printing with lower concentration never created a closed and continuous film. The viscosity of DDAO solutions was measured, where the nozzle radius and the density were assumed as $a = 50\,\mu m$ and $\rho = 0.983\,\mathrm{g/ml}$ respectively. It should be noted that the measured viscosity values are unexpectedly low at increasing DDAO concentration and should be treated with certain reservation. These might cause incorrect Z values. As can be seen from Table 1 the Z values of DDAO ink are outside of the printable range ($4 \leq Z \leq 14$). However, test prints showed that printing is anyway possible and thus these DDAO solutions were used for inkjet printing.

As described in the procedure (2.2 Methods), different patterns were printed. After several test prints, a closed film was obtained after 15 printed layers of a ring pattern with a line thickness of 0.1 mm (Fig. 1 and 2). Whereas the quality of the printout is shown magnified in Fig. 2.

Printing full circles resulted in single droplets of ink visible on the surface with free space in the center of the pattern. For this reason two semicircles were printed 15 times on top of one another and the result from this printed sensor is shown in Figs. 3 and 4. Magnified printouts showed overlapping profile after printing 15 layers (Fig. 4, microscopic), which might be due to printing several layers.

Table 1: Properties of DDAO solutions (c-concentration, ρ - density, γ-surface tension, η-viscosity, Z-value)

c[wt%]	ρ [kg/m^3]	γ [mN/m]	η [mPa · s]	Z
0	998	71.15	1	59.59
0.01	983	47.26	1.033	46.66
0.02	983	41.57	0.966	46.79
0.04	983	35.31	0.945	44.08
0.05	983	33.97	0.918	44.51
0.06	983	33.11	0.936	43.09
0.085	983	33.21	0.919	43.96
0.1	983	33.58	0.924	43.97

Figure 1: Rings printed using DDAO (10 wt%) solution, line thickness 0.1 mm

Figure 2: The part of the printed circle using DDAO (10 wt%) solution, 15 layers (10x magnification).

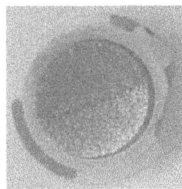

Figure 3: Two semicircles printed on a sensor.

Figure 4: Center of two semicircles, 15 layers.

DDAO printed with different amount of layers on QCM-D sensor have been characterized. First, simple calculation of mass was done by measuring the frequency decrease with each printed layer. The dependence of mass on number of printed layers is shown in Fig. 5. The printing was done using 0.01 wt% DDAO solution. Linear mass growth with an increasing number of layers was found (Fig. 5). From Fig. 5 the amount of dispensed ink per printed layer was calculated to be $8.5\,\mu l/cm^2$.

Additionally, the effect of hydration of the printed DDAO-structures were investigated by scanning the relative humidity (RH) from dry to humid conditions. During this scan, the water activity a_w, which is defined as the partial vapor pressure in a substance divided by the partial vapor pressure of pure water, was controlled. With this relation the RH could be calculated as: $RH = a_w \cdot 100$. The aim was to assess if the effect of hydration, measured by QCM-D, depends on the symmetry of the printed pattern and the total amount of printed DDAO. For a totally dry DDAO film, no splitting of overtones was observed (Figs. 6 and 7), which means that DDAO layers are solid and well coupled to the surface of QCM-D; Sauerbrey equation can be used to calculate the mass on QCM-D crystal.

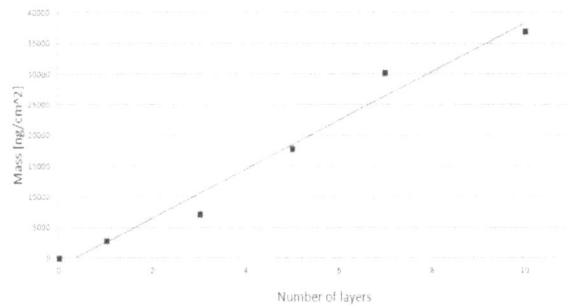

Figure 5: The dependence of mean mass of printed DDAO on number of printed layers. DDAO concentration was 0.01 wt%.

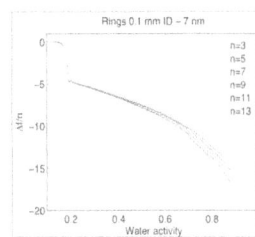

Figure 6: QCM-D frequencies for printed rings with increasing RH. Line thickness of 0.1 mm. (n - number of overtone).

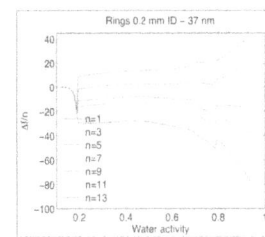

Figure 7: QCM-D frequencies for printed rings with increasing RH. Line thickness 0.2 mm.

With increase of DDAO layer hydration the QCM-D response becomes different for different overtones. It was found that a low printed amount of DDAO resulted into less deviation of the overtones (compare Fig. 6 and 7). Furthermore, the symmetry on the surfaces had an influence on the sensor behavior as well; a lower symmetry causes a higher deviation (compare Figs. 7 and 8).

It should be noted that phase transitions of the surfactant at higher humidity could be observed, i.e., when the layer becomes a liquid-like film, which induces abrupt changes of the frequency (see for example Fig. 8 and 9). The phase transitions are caused by water uptake by DDAO film at higher relative humidity. The transitions follow the order from dry solid, to lamellar, to cubic, and finally hexagonal phase [11]. As can be seen from comparison of Figs. 6-9, the phase transitions can be more easily observed for thicker films.

3.2 Inkjet printing of enzymes

HRP enzyme was chosen for printing to realize one of the most common reactions often used in immunostaining. Initially, HRP and DAB were printed on paper and onto skin. Finally, the reaction was activated with penetration of peroxide through paper or the skin membrane. After some test prints it was found, that one layer of HRP and 15 layers of DAB gave the best reaction on the paper, with a clear pattern border (Fig. 10a and 10b).

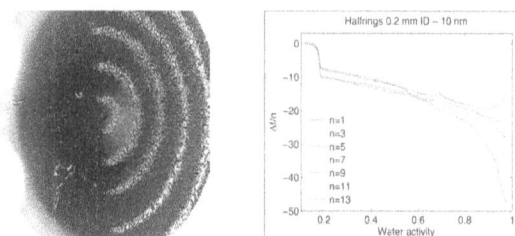

Figure 8: Printed halve rings with different diameter, a line thickness of 0.2 mm and frequency change with increasing RH.

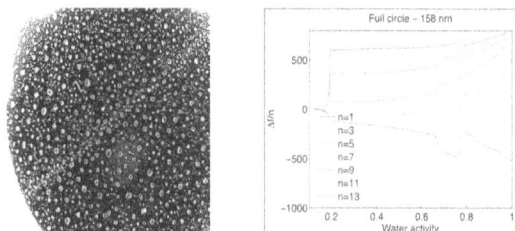

Figure 9: Printed halve circles and frequency change with increasing RH.

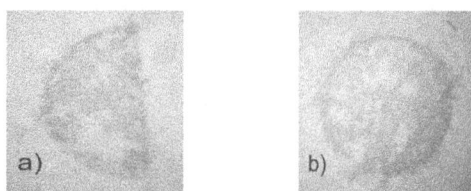

Figure 10: Printed HRP/BSA and DAB pattern on paper after initiation of staining reaction by hydrogen peroxide.

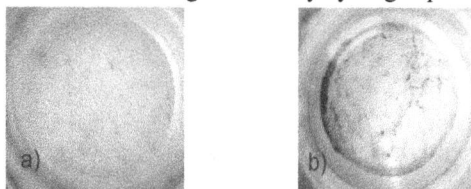

Figure 11: Printed pattern HRP/BSA and DAB on skin after initiation of staining reaction by hydrogen peroxide.

Thus, the same amount of layers was printed also on the skin (Fig. 11a). However, after the reaction initiation with peroxide, the pattern on the skin (Fig. 11b) was not as sharp as it was on the paper (Fig. 10a). This difference might be due to the difference in drying of HRP and DAB after printing on paper and skin. Drying on skin is much slower and thus probably results that multiple layer printing produce high amount of ink which can move as a big droplet over the surface of the skin.

4   Conclusion

This project shows that a simple commercial printer can be used to print surfactant and protein containing water based inks. Additionally, well-defined structures were successfully printed on QCM-D sensors, which enabled

characterization of the amount of printed ink and the effect of hydration of the surfactant system. These results illustrate that the presented printing method can be utilized to study other biologically relevant systems. Additionally, printing with enzymes onto skin is possible, even though the pattern was not as sharp as it was on the paper. Therefore, the printing and reaction conditions for this application need to be optimized.

Acknowledgement

The work has been carried out at Biofilms – Research Center for Biointerfaces, Malmö University, Sweden.

5   References

[1] J. T. Delaney, P. J. Smith, and U. S. Schubert *Inkjet printing of proteins*. Soft Matter, 5, 4866–4877, 2009.

[2] D. Jang, D. Kim, and J. Moon *Influence of fluid physical properties on ink-jet printability*. Langmuir, 25(5), 2629–2635, 2009.

[3] B. Derby, *Bioprinting: inkjet printing proteins and hybrid cell-containing materials and structures*. J. Mater. Chem., 18, 5717–5721, 2008.

[4] D. Raijada, et al. *A step toward development of printable dosage forms for poorly soluble drugs*. J. Pharm. Sci., 102(10), 3694-3704, 2013.

[5] C. Xu, et al. *Study of droplet formation process during drop-on-demand inkjetting of living cell-laden bioink*. Langmuir, 30(30), 9130–9138, 2014.

[6] J. Schatz and R. Tammer (ed), *Erste Hilfe-Chemie und Physik für Mediziner*. Springer, Heidelberg, 2015.

[7] Lauda Dr. R. Wobser GmbH & Co. KG, *Instruction Manual of Drop Volume Tensiometer TVT2*. 2. Edition, 1993.

[8] J. Sotres, *Viscosity of CMC solutions - Lab instructions*. Master Programme in Biomedical Surface Science. 2013.

[9] Q-Sense Biolin Scientific, *QCM-D Technology*. Available: http://www.biolinscientific.com/q-sense/technologies/. [last excess on 11.11.2015].

[10] S. Björklund and V. Kocherbitov, *Humidity scanning quartz crystal microbalance with dissipation monitoring setup for determination of sorption-desorption isotherms and rheological changes*. Rev. Sci. Instrum., 86(5), 2015.

[11] V. Kocherbitov and O. Söderman, *Hydration of dimethyldodecylamine-n-oxide: enthalpy and entropy driven processes*. J. Phys. Chem. B, 110(27), 13649–13655, 2006.

4

Biomedical Engineering

Micro alignment of optical components
– A comparative study about adhesives –

Nora Tobies [1]

[1] Medizinische Ingenieurwissenschaft, Universität zu Lübeck, nora.tobies@student.uni-luebeck.de

Abstract

This paper focuses on micro alignment of optical components used for medical devices. The project team tried to find out what is the best way to join optical components together through adherence. This pilot study, on behalf of a client, compares two different types of adhesive of which one should later on be used to adhere a delayer in front of a prism with the help of supporting blocks. This prism assembly is part of a medical measuring camera. The measurements were performed with test glass blocks that simulated the prism assemble. The main focus of the tests lay on the bond strength of each adhesive, tested with pressure shear tests under different temperature conditions, and on possible translation and rotational movement during the curing process, tested with an autocollimator. The project team was able to present the results to the client and create sufficient trust to go ahead with the project together.

1 Introduction

Technical joining is according to DIN 8580 a main group of manufacturing techniques and describes a process that creates a permanent conjunction of at least two components [1]. A subgroup of technical joining is adherence or adhesive bonding. An adhesive bond consists of at least two adherends and the adhesive layer in-between. The holding force (adhesion) arises from interaction and the mechanical form fit. To achieve an optimal wetting of the component, the adhesive needs to be liquid. It gains its inner force (cohesion) afterwards by curing it with physical methods or through chemical reactions [2]. This work is a pilot study for an optical component used for a new measuring camera. The object in focus was a prism assembly consisting of a beam splitter, a delayer and two supporting blocks, as seen in Fig. 1. In this work two previously chosen adhesives will

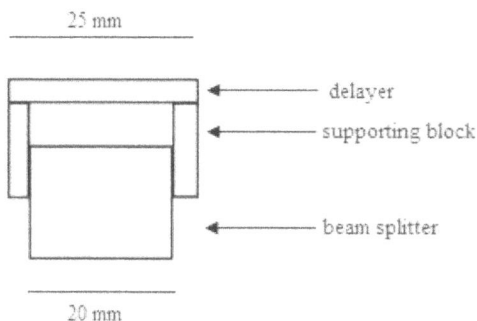

Figure 1: Illustration of the prism assembly.

be characterized. They will from now on be called adhesive A and adhesive A. Both are radiation curing adhesives and react to UV-light. One of the two main goals of the pilot

study was to find out which adhesive has the better bond strength under specific conditions. The second one was to approve that, if there are any translation or rotation movements of the delayer around the three axes XYZ (during curing or afterwards), these do not pass over the specification alignments.

2 Material and Methods

2.1 Rheometer

To get a first overview about the characteristics of the two adhesives A and B rheometer measurements were performed. A rheometer is a device which applies a shear deformation and thus determinates deformation and flow behavior of material. Through that one can find out about the needed curing time and radiation intensity. The measurements were done on a machine from the company Anton Paar, on which it was possible to install the UV-light source bluepoint LED of Hönle. One after another a small amount of adhesive A and adhesive B was put on a resting glass plate. After that a pin with a circular surface was driven into the adhesive until a predefined gap was reached. The measurements were performed with adhesive gaps of $0.1\,mm$ and $0.5\,mm$. After the start of the rheometer and the UV-light source the measurement begins, the pin performs oscillating movements until the adhesive is fully cured and no more movement is possible. The torque moment that is necessary to keep the movement alive is measured and used to determine the shear stress and thus the viscosity [3]. The results of these measurements are discussed in section 3.1.

(a) Glass samples with adhesive line on it; this amount is enough to bewet the desired area of 5 x 20 mm^2.

(b) Setting the desired adhesive gap between the two sample glass plates (a & b) with help of a gauge (c).

(c) Two adhered glass plates.

(d) Illustration of pressure shear test [4].

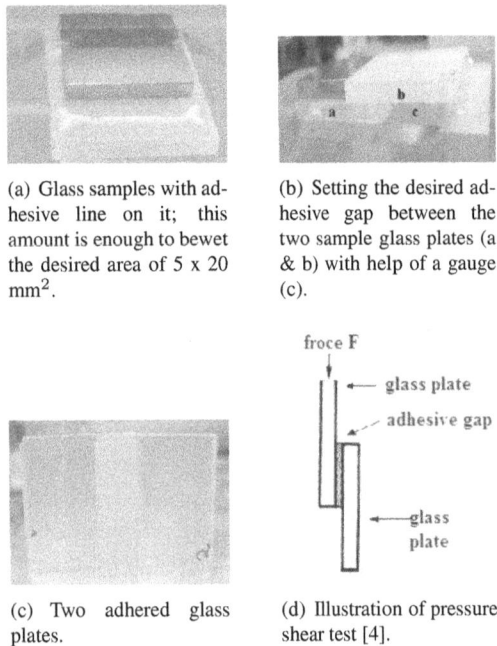

Figure 2: Glass samples for the pressure shear tests.

2.2 Pressure shear test

2.2.1 Samples at room temperature

Strength investigations can be done with a pressure shear test. To do so, samples out of two small glass plates with each a ground and a polished side were adhered overlapping with an adhesive thickness of $50\,\mu m$. The adhesive was applied in a thin line with a dispenser (see Fig. 2a), the gap width was set with a gauge (see Fig. 2b). The adhered glasses, as seen in Fig. 2c, were put into a pull tester machine and were loaded by a testing pin from above until the bond failed, as seen in Fig. 2d. The machine measures the maximum force required to cause the failure. Not only the needed force, but also the ways how the glass plates get separated from each other determinate the strength of the joint. An adhesive failure is a fracture between the surface of the adherend and the adhesive. It indicates an insufficient adhesiveness on the adherend's surface, possibly occurred through missing or wrong surface preparation. As opposed to this a coherent failure is a fracture in the adhesive and indicates poor stability, e.g. caused by too short curing time.[4] Ten samples with adhesive A were cured for $60\,s$, whereas another five samples with adhesive B were cured for $5\,min$ and five more samples were cured for $10\,min$. For comparison adhesive joints out of glass and metal were made. For this Invar (36% nickel and 64% iron) was chosen. The adhesive gap was once again $50\,\mu m$. The adhesive joining was performed as before. The results of the pressure shear tests can be seen in Fig. 6 and Fig. 7.

2.2.2 Samples from clime cabinet

The previous measurements were done at room temperature, but it should also be tested if the adhesive joint can withstand temperature variations within $-25\,°C - 80\,°C$.

Since the glass-Invar-joints didn't show good results, the tests were continued only with glass-glass-joints. New samples were adhered together with the already known procedure and after that they were put into a climate cabinet at $-25\,°C$ for $12\,h$ and finally measured in the pressure shear machine. The results can be seen in Fig. 8.

2.3 Autocollimator

The other import aspect of the pilot study was to find out if the delayer does any rotation around the three axes XYZ during the curing process or afterwards. Therefore a new test setup was created and new samples were produced. The new samples consisted of glass blocks with two stages on the top where the adhesive was applied. There were three different forms: The first version (Fig. 3a), called compensating adhesive gab, had two stages of the same size but two different amounts of adhesive were applied. The second version (Fig. 3b) had a step, so the stages were of different sizes, but an equal amount of adhesive was applied. In the third version the stages had a small groove in which the adhesive was added. One glass plate was put on each glass block. In the third version (Fig. 3c) the glass plate was in direct contact with the glass block, in the other two versions on the adhesive. The glass blocks were jointed through optical contact bonding on a zerodur block, which was used as a reference for the measurement setup. To observe if an ob-

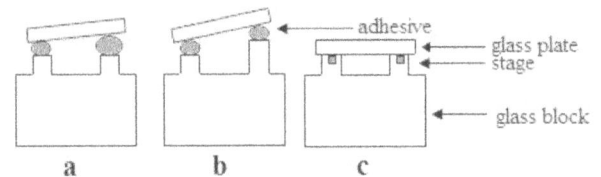

Figure 3: Glass blocks and glass plates for the autocollimator measurements in three versions: compensating adhesive gab (a), step (b) and direct contact (c).

ject moves an autocollimator was used. This is an optical instrument for non-contact measurements of angles and gives results in minutes and seconds of arc. An autocollimator directs optical signals at deflective surfaces and derives angle information from comparing the signal from the glass plates with the signal from the zerodur block [5]. Once again adhesive A (exposure time: $1\,min$) and adhesive B (exposure time: $5\,min$) were used. The autocollimator measurement setup can be seen in Fig. 4. With this setup it was possible to check if there are translation movements around Y and rotational movements around Z happening, as illustrated in Fig. 5. The results can be seen in Table 1.

3 Results and Discussion

3.1 Rheometer

The rheometer tests showed that adhesive A has a really fast curing time. The $0.1\,mm$ and the $0.5\,mm$ gap both were

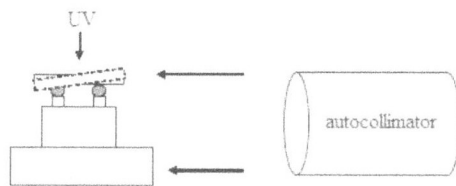

Figure 4: Illustration of the autocollimator measurement, the glass plates are adhered to the glass blocks which sit on a zerodur block. The UV light comes from above and cures the adhesive, the autocollimator measures from the side.

(a) Translation around axis Y. (b) Rotation around axis Z.

Figure 5: Movements of the glass plates that can be measured with the autocollimation setup. Glass block (dark grey), stages (black) and plate (light grey) seen from above.

fully cured in less than 5 s. On the contrary adhesive B showed a much longer curing time. The 0.1 mm gap took 5 min to be fully cured and the 0.5 mm gap another 5 min, a total of 10 min.

3.2 Pressure shear test

In order to assure reliable results about the strength of the two adhesives A and B pressure shear tests on the glass-glass-joints and the glass-Invar-joints were performed.

3.2.1 Samples at room temperature

The graphs in Fig. 6 and Fig. 7 show the results of the glass-glass-joints and the glass-Invar-joints that have been stored at room temperature. For adhesive A ten samples were cured for 1 min and tested, for adhesive B five samples were cured for 5 min and five more samples were cured for 10 min and tested. For the glass-glass-joints the average bond strength of adhesive A was $8.04 \, \mathrm{N\,mm^{-2}}$ with a standard deviation of $3.30 \, \mathrm{N\,mm^{-2}}$. The bond strength of adhesive B was almost just half as much as of adhesive A with a mean value of $4.29 \, \mathrm{N\,mm^{-2}}$. Therefore it had less variance ($s = 1.53 \, \mathrm{N\,mm^{-2}}$). The glass-Invar-joints showed similar results: Adhasive A had an average bond strength of $7.49 \, \mathrm{N\,mm^{-2}}$ ($s = 3.82 \, \mathrm{N\,mm^{-2}}$) and adhesive B only an average bond strength of $2.69 \, \mathrm{N\,mm^{-2}}$ ($s = 0.71 \, \mathrm{N\,mm^{-2}}$). By doubling the exposure time only the glass-glass-joints had a higher bond strength, whereas there are no differences for the glass-Invar-joints. At the glass-glass-joints adhesive

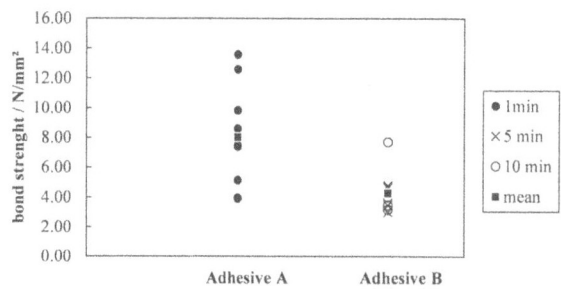

Figure 6: Results of the pressure shear test of the glass-glass-joints. The graph shows the bond strength in $\mathrm{N\,mm^{-2}}$ of the two adhesives A and B depending on the exposure times of 1 min, 5 min and 10 min. The squares represent the average values.

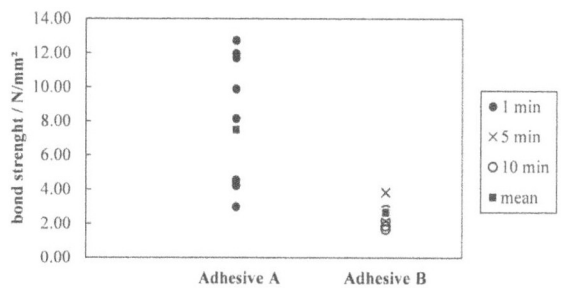

Figure 7: Results of the pressure shear test of the glass-Invar-joints. The graph shows the bond strength in $\mathrm{N\,mm^{-2}}$ of the two adhesives A and B depending on the exposure times of 1 min, 5 min and 10 min. The squares represent the average values.

A failed mostly on the polished glass side. On the other hand the glass-Invar-joints failed twice as often on the Invar side than on the glass side. These many failures on the Invar side indicated that the adherend is not capable for these adhesives. So the next tests were continued only with glass-glass-bonds.

3.2.2 Samples from clime cabinet

The pressure shear tests were repeated with glass-glass-joints that had been stored at $-25 \, \mathrm{°C}$ for 12 h. The results can be seen in Fig. 8, compared with the test results at room temperature from Fig. 6. The samples adhered with adhesive A that have been in the clime cabinet show an average bond strength of $6.78 \, \mathrm{N\,mm^{-2}}$ and a standard deviation of $2.98 \, \mathrm{N\,mm^{-2}}$. Comparing this to the samples at room temperature, it seem like adhesive A looses some of it's bond strength at cold temperatures. Adhesive B shows a mean value of $4.15 \, \mathrm{N\,mm^{-2}}$ ($s = 1.78 \, \mathrm{N\,mm^{-2}}$) at $-25 \, \mathrm{°C}$. This is about the same as at room temperature, so adhesive B seems to be not affected by the change of the storage temperature.

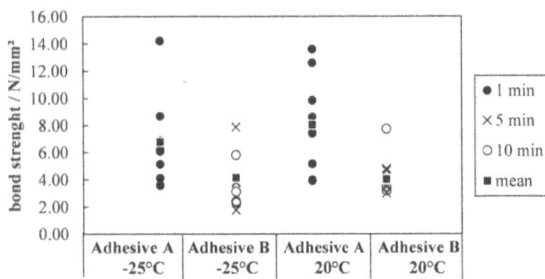

Figure 8: Results of the pressure shear test of the glass-glass-joints at different temperatures. The graph shows the bond strength in $N\,mm^{-2}$ of the two adhesives A and B depending on the exposure times of 1 min, 5 min and 10 min. The squares represent the average values.

Table 1: Results of autocollimator measurement: Translation around Y and rotation around Z of the glass plates in second of arc.

Measurement	Y	Z
Adhesive A	2.75	0.54
Adhesive B	0.45	0.26
Compensating gap	1.35	0.74
Step	1.03	0.37
Direct contact	2.57	2.61

3.3 Autocollimator

Table 1 shows the rotation around Z and the translation around Y of the adhered glass plates with the adhesives A and B. It shows that with adhesive B the movement around both axes is less than with adhesive A. For both adhesives the rotation around Z is less than the translation around Y. The table also shows the movement of the glass plates relative to the different adhesive geometries (compensating adhesive gab, direct contact and step) for both adhesives. Here, the rotation around Z is smaller than the translation around Y for all geometries once more. The geometries have more or less the same values except the direct contact setup. Those high values were not as expected and occurred probably because the adhesive was not only in the small groove of the stage but also on the stage due to improper work and too much adhesive.

4 Conclusion

Until this point of time the pilot study was successful. The team was able to find out about the characteristics of the two adhesives A and B and present the intermediate results satisfactorily to the client. The pressure shear tests showed that adhesive A has the greater overall strength at room temperature. In comparison to this adhesive B is not reactive to climate change at negative temperatures. The autocollimator measurements indicated that adhesive B is also steadier during curing and does less movements than adhesive A. The measured values are way below the highest possible

standard. One of the next steps is to extend the temperature tests: the time in the climate cabinet should be doubled to 24 h at $-25\,°C$. This way the changes of the bond strength can be monitored closely. Furthermore new samples should be produced and stored at $80\,°C$ for 12 h and 24 h to see if there is a reaction to warm temperatures. Also it should be tested if there is any translation around X happening during curing, this can e.g. be tested with an interferometer measurement setup. When all tests are finished a final decision about the manufacturing technique of the measurement camera can be made.

Acknowledgement

This work has been carried out at Berliner Glas KGaA Herbert Kubatz GmbH & Co., BU Medical Applications from October 12th 2015 till January 11th 2016. It has been supervised by Dr. Stefan Beyer, unit manager development. This work was made in collaboration with Dr. Gereon Hüttmann, Institut für Biomedizinische Optik, Universität zu Lübeck.

5 References

[1] G. Salvendy, *Handbook of Industrial Engineering: Technology and Operations Management.* John Wiley & Sons, New York , 2001.

[2] B. Müller, W. Rath, *Formulierung von Kleb- und Dichtstoffen.* Vincentz Network, Hannover, 2004.

[3] C.W. Macosko, *Rheology: Principles, Measurements and Application.* John Wiley & Sons, New York , 1994.

[4] G. Habenicht, *Kleben. Grundlagen, Technologien, Anwendungen.* Springer, Berlin/Heidelberg, 2009.

[5] Vermont Photonics, *Small Angles and Autocollimatores.* Available: http://www.vermontphotonics.com/small-angles-autocollimators/ [last accessed on 07.02.2016].

Setup for Estimation of Absolute Subcutaneous Water Content by Noninvasive Spatial Resolved Diffuse Reflectance Measurement

H. Köhler [1], O. Fugger [2], A. Vogel [3]

[1] Medizinische Ingenieurwissenschaft , Universität zu Lübeck, hannes.koehler@student.uni-luebeck.de
[2] Senspec GmbH, Rostock, oliver.fugger@senspec.com
[3] Institut für Biomedizinische Optik, Universität zu Lübeck, vogel@bmo.uni-luebeck.de

Abstract

A method for absolute subcutaneous water content estimation by noninvasive spatial resolved diffuse reflectance measurement is presented. The method provides local and continuous information on the skin water content via estimation of absorption coefficient μ_a in the NIR wavelength range from 900-990 nm. Sensor calibration is done by use of calculated forward data for epoxy phantoms with known optical properties. Estimation of optical properties on scattering media from measurements is performed by applying neural networks trained by Monte Carlo simulated forward data. Scattering coefficients in the range of 0.6 to 1.7 mm^{-1} of three different phantoms are measured with an error of 7 %. Absorption coefficients in the range of 0.03 to 0.13 cm^{-1} were estimated with an error of 15 % in the spectral range of 920-975 nm. The introduced method enables further investigations on the relation between state of hydration in humans and subcutaneous water content estimated by μ_a.

1 Introduction

Knowledge of body water content plays an important role in optimizing dialysis sessions and in preventing cardiovascular disease. Furthermore the state of hydration of older people is an interesting parameter in the field of assistant systems for monitoring vital parameters. To get information about the state of hydration indirectly by determining the absolute subcutaneous water content, knowledge of the optical properties of biological tissue is essential. To describe the interaction between light and biological tissue, the absorption coefficient (μ_a), anisotropy factor (g) and scattering coefficient (μ_s) are used. There are different methods to estimate these optical properties, for example measuring a) total transmittance and reflectance with an integrating sphere, b) time resolved reflectance and transmittance [3],[4] and c) spatial resolved diffuse reflectance. In Method a) using an integrating sphere [1],[2], a tissue sample is cut into a thin slab and placed at the entry- or exit port of the sphere. Total transmittance and reflectance can be measured while illuminating the probe. Optical properties of the tissue sample are estimated by fitting Monte Carlo (MC) simulated data of the setup to the measured data. With this method continuous measurements in vivo are not feasible. In the second case, the measurement of time resolved reflectance and transmittance, a short light pulse is directed on the sample surface. At some distance from the point of incidence a detector is placed to measure the remitted light by time-correlated single photon counting. The obtained time-of-flight histogram contains information about the optical properties. Source and

detector have to be capable of high time resolution. In vivo measurements are possible, but the equipment is expensive and complex. The third commonly used method is the spatial resolved diffuse reflectance as described in [5]. This method enables a noninvasive or non-contact estimation of living tissue optical properties. Spatial resolved absolute diffuse reflectance can be measured by use of a laser beam or spectrally-broadband light source in combination with a spectrograph. For spatial resolution a camera or fiber bundles at different distances from the source can be used.

In this paper, a method for measuring noninvasive the spatial resolved diffuse reflectance based on principles shown by [6] is presented. We use five near-infrared LEDs placed at different distances to a fiber bundle connected to a grating spectrograph. With the described setup a continuous monitoring of living tissue optical properties is possible. As water is the main absorber for light in biological tissue in the wavelength range from 950 nm to 1000 nm, human skin shows a peak in the absorption spectrum between 959 nm and 965 nm [7]. The continuous measurement of μ_a provides information on the evolution of water content in subcutaneous skin. This information could supply useful support for diagnostic decisions. Body water is an important indicator for renal disease and dialysis patients as well as for obese and elderly people. Established methods for body water estimation like bioelectrical impedance analysis (BIA) [8] and deuterium dilution method are considering the overall body water level. Continuous estimation is not possible [9]. The method introduced in this paper provides local and continuous information on the skin water content by estimating the absorption coefficient μ_a in the range between 959 nm and 965 nm.

2   Material and Methods

In this section the model assumed for human tissue and the calibration phantoms, is described. Furthermore, sensor device, setup and the methods for calibration and estimation of the optical properties are explained.

2.1   Material

Human skin consists of three main layers. Epidermis is the exterior layer that forms a protective barrier, to the environment. Its thickness is about 100 μm and contains no blood vessels. In this layer the main part of skin melanin is located. Beneath the epidermis is a 1-4 mm thick layer called dermis, containing connective tissue that dominates the scattering properties. Main absorber is haemoglobin in the visible range with absorption peaks at 546 nm and 578 nm for oxygenated haemoglobin and water in the near infrared range with the absorption peak at 960 nm. Deepest skin layer is the subcutis with a thickness of 1-6 mm and haemoglobin, lipids and water as main absorbers. A fat layer is following beneath the skin. Regarding our measuring scenario described in this paper, the tissue layers penetrated by the incident photons are skin, fat and, depending on the embodiment, muscle. In our investigations we have chosen a measurement point, where the fat layer could be assumed as semi-infinite in both dimensions, such as the upper hip. This approach suppresses the influence of high absorbing muscle tissue. In a first theoretical model, we approximate the three skin layers and the fat layer as a homogeneous medium, however the measurement point consists of more layers in practice.

For calibration, we used epoxy phantoms as shown in Fig. 1. The optical properties were evaluated with absolute reflectance measurements [10]. Phantom matrix is made of epoxy with titanium dioxide as scattering particles and pigments as absorber. The epoxy phantoms are homogeneous and satisfy the assumption of a semi-infinite sample.

Figure 1: Sensor device placed on epoxy phantom made with titanium dioxide as scattering particles and pigment absorbers [10].

2.2   Method

Measurement setup
A sensor device is developed which uses five near-infrared LEDs (OSA Opto Light, OCI-440 970-X-T) emitting in the wavelength range between 900 nm and 1000 nm with 973 nm peak wavelength. The LEDs are mounted on a 3×5 cm base plate. Between LED and sample a filter glass (Schott, BK7) with diameter 3.5 mm is placed. The distances between LEDs and detection fiber bundle are 4, 4.75, 8, 11 and 14 mm arranged as shown in Fig. 3. The LEDs can be switched on sequentially by a control board. An optical fiber bundle with diameter 2 mm and numerical aperture (NA) of 0.22 is used to capture the diffuse scattered light leaving the sample at detection position. The fiber bundle is connected to a grating 2d-spectrometer (inno-spec, IST 5010) which uses a 30 μm slit, grating constant 3.5 μm. It enables measurements in the spectral range from 500 nm to 1000 nm. For image detection a CMOS digital image sensor (MT9V024) from Aptina with 8-bit 752×480 pixels full resolution and a frame rate of 60 fps is implemented. For each wavelength in the range from 900 nm to 1000 nm the intensity of remitted light at each LED-detector distance is measured. Sample optical properties μ'_s and μ_a are estimated by applying a neural network (NN) which was trained with reflectance data calculated by use of a MC simulation as described in [11]. The MC simulation represents a solution of radiative transfer theory and takes into account the probe geometry of our setup. In Fig. 2 the measuring setup is shown.

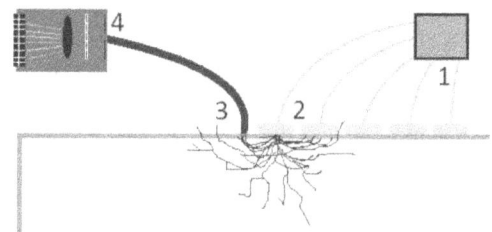

Figure 2: (1) LED control, (2) near-infrared LEDs, (3) optical fiber bundle, (4) grating spectrometer with camera chip.

Data acquisition
The image is processed by an image acquisition software developed in LabView for this implementation. Every light image follows a dark image which is subtracted from first one to correct ambient light impact. The average over all columns of the image gives the intensity for each wavelength. Repeating this with increasing integration time for each distance results in a complete data set. Higher integration time at larger LED-detector distances produces better signal to noise ratio (SNR) and is corrected after image acquisition to get the according intensity. Then spectral mean of 2 nm and time mean over 60 data sets is calculated. This provides us decrease of remitted light intensity curves over LED-detector distance in 2 nm intervals from 900 nm to 1000 nm.

Calibration

For calibration of the setup, a calibration phantom with known optical properties, obtained by measurements with well evaluated optical setups are used [10]. The expected reflectance data for the calibration phantom is calculated by MC simulation, that takes into account the geometry of our setup. These reflectance data are used as forward data. The sensor is placed on the phantom and measured data is divided by MC simulated forward data. The resulting data set gives the calibration value for each LED-detector distance depending on the wavelength. Different epoxy phantoms with known μ'_s and μ_a, similar to biological tissue, were tested for evaluation of the setup.

Figure 3: Sensor device, left: plate top with circular fiber bundle mount, right: plate bottom with LED sources and entry hole for remitted light (marked with arrow).

Inverse problem

To derive the optical properties μ'_s and μ_a from the measurement a NN is used. Two NN were trained with MC simulated forward data representing the sensor setup and the measuring scenario which is on the one hand the measurement on epoxy phantom and on the other hand the measurement on human tissue. The range of the optical properties for the forward data sets varied from 0.5-2.4 mm^{-1} for μ'_s and from 0.005-0.1 mm^{-1} for μ_a. The forward data are calculated by MC simulation assuming a single layer and semi-infinite model. We used a NN with three layers as described in [11]. The measured intensities are normalised to average intensity of first and second LED-detector distance. For each wavelength the normalised values are used as input to the five neurons of the input layer. The hidden layer contains five neurons and the two neurons of the output layer represent μ'_s and μ_a.

3 Results and Discussion

For evaluation the measuring setup was calibrated with epoxy phantom MC simulated forward data. Spatial and spectral resolved diffuse reflectance were measured for three different epoxy phantoms with known optical properties and μ'_s and μ_a were estimated by applying a NN. Spectral mean of 2 nm and time mean over 60 data sets were calculated. In total five measurements at each epoxy phantom were performed. The sensor was reapplied for each mea-

surement. In Fig. 4 the results of the evaluation for estimating μ'_s are shown. Orange and marron phantom estimations are shown from 900-980 nm, ferrite phantom is displayed from 920-980 nm because of low LED emission outside of this range. Reference graphs are plotted for ferrite and marron phantom. Deviation between estimated ferrite phantom μ'_s and its reference is less than 7 % from 920-975 nm. For orange and marron phantom the deviation from reference is less than 4 % from 900-990 nm. Results for the estimation of μ_a are displayed in Fig. 5. Ferrite phantom deviation of μ_a in the spectral range from 920-975 nm and for marron and orange phantom from 920-990 nm is always below 15 %.

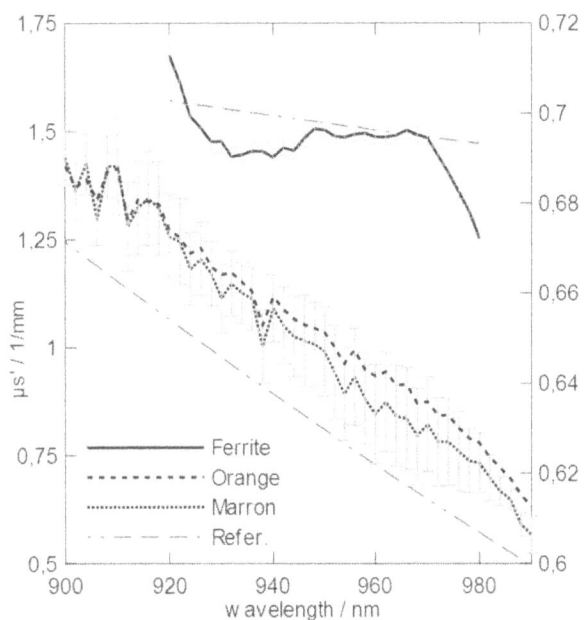

Figure 4: Estimation of reduced scattering coefficient μ'_s for 3 different epoxy phantoms. Reference graph is μ'_s for particular epoxy phantom measured as shown in [10]. Ferrite phantom is represented on left y-axis, orange- and marron on the right y-axis. Because orange phantom reference graph would be covered by the orange phantom measurement, it is not plotted.

Although epoxy phantom optical properties were obtained by measurement with well evaluated optical setups, there are questionable μ_a measurement values in the range of 956-964 nm because of detector change. This results in questionable calibration values and therefore higher deviation from reference μ_a in this wavelength range. Estimation of marron phantom μ_a is more accurate outside this range. For measuring skin optical properties a two or more layer model could be used. This would lead to a better fit of the three layer structure of human skin and subjacent fat tissue. Including prior knowledge like thickness of skin and fat layer could improve the method.

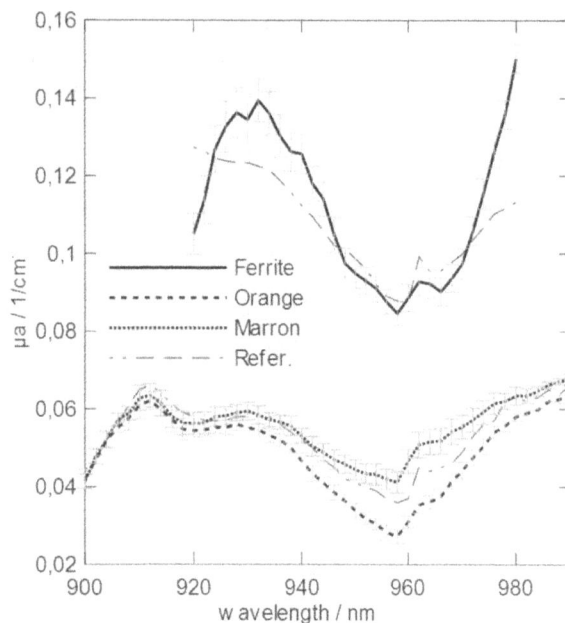

Figure 5: Estimation of absorption coefficient μ_a for 3 different epoxy phantoms. Reference is μ_a for particular epoxy phantom measured as shown in [10]. Orange phantom reference graph is not plotted.

4   Conclusion

In this paper a method for estimating the absolute subcutaneous water content of human skin by noninvasive spatial resolved diffuse reflectance measurement is presented. Sensor calibration can be done with MC simulated forward data for epoxy phantoms with known optical properties. NN trained with MC simulated forward data are applied on measured intensities of remitted light at different LED-detector distances. We presented the estimation of epoxy phantom optical properties with a small sensor device and comparable low equipment costs in the wavelength range from 920-980 nm with an inaccuracy below 15 %.

Further investigations on estimating μ_s' and μ_a with multi layer model trained NN are needed. The influence of skin and fat layer thickness likewise the point of measurement on the human body are part of more research on this method. Next steps also including test series with different probands at normal, low and high hydration levels. This would enable to study relation between total body water and subcutaneous water content represented by skin μ_a.

Acknowledgement

The work has been carried out at Senspec GmbH, Rostock. This research was supported by employees of Institut für Lasertechnologien in der Medizin und Meßtechnik Ulm, who provided the MC simulation, training of the NN and epoxy phantoms.

5   References

[1] M. Hammer, A. Roggan, D. Schweitzer and G. Muller, *Optical properties of ocular fundus tissues-an in vitro study using the double-integrating-sphere technique and inverse Monte Carlo simulation.* Physics in Medicine and Biology, vol. 40, no. 6, pp. 963, 1995

[2] M. T. Heine, F. Foschum, and A. Kienle, *Determination of the optical properties of anisotropic turbid media using an integrating sphere.* Proc. SPIE, vol. 7368, pp. 1-6, 2009

[3] S. H. Tseng, A. Grant and A. J. Durkin, *In vivo determination of skin near-infrared optical properties using diffuse optical spectroscopy.* Journal of Biomedical Optics, vol. 13, no. 1, pp. 1-7, 2008

[4] L. Spinelli, A. Torricelli, A. Pifferi, P. Taroni, G. M. Danesini, et al., *Bulk optical properties and tissue components in the female breast from multiwavelength time-resolved optical mammography.* Journal of Biomedical Optics, vol. 9, no. 6, pp. 1137-1142, 2004

[5] A. Kienle, L. Lilge, M. Patterson, R. Hibst, R. Steiner, et al., *Spatially resolved absolute diffuse reflectance measurements for noninvasive determination of the optical scattering and absorption coefficients of biological tissue.* Appl. Opt., vol. 35, no. 13, pp. 2304-2314, 1996

[6] M. G. Nichols, E. L. Hull and T. H. Foster, *Design and testing of a white-light, steady-state diffuse reflectance spectrometer for determination of optical properties of highly scattering systems.* Appl. Opt., vol. 36, no. 1, pp. 93-104, 1997

[7] S. L. Jacques, *Optical properties of biological tissues: a review.* Physics in Medicine and Biology, vol. 58, no. 11, pp. 37, 2013

[8] U. G. Kyle, I. Bosaeus, A. D. De Lorenzo, *Bioelectrical impedance analysis part I: review of principles and methods.* Clinical Nutrition, vol. 23, no. 5, pp. 1226-1243, 2004

[9] D. Halliday, A. G. Miller, *Precise measurement of total body water using trace quantities of deuterium oxide.* Biomedical Mass Spectrometry, vol. 4, no. 2, pp. 82-87, 1977

[10] P. Krauter, S. Nothelfer, N. Bodenschatz, E. Simon, S. Stocker, et al., *Optical phantoms with adjustable subdiffusive scattering parameters.* Journal of Biomedical Optics, vol. 20, no. 10, pp. 105008, 2015

[11] M. Jäger, F. Foschum, A. Kienle, *Application of multiple artificial neural networks for the determination of the optical properties of turbid media.* Journal of Biomedical Optics, vol. 18, no. 5, pp. 057005, 2013

Design and development of a steering platform for an innovative infusion system

M. Al Msalma [1], S. Abdul-Karim [2], J. Schroeter [2], and B. Nestler [2]

[1] Medizinische Ingenieurwissenschaft, Universität zu Lübeck, momen.almsalma@student.uni-luebeck.de

[2] Medical Sensors and Devices Laboratory, Luebeck University of Applied Sciences,{saif.abdul-karim, joerg.schroeter, bodo.nestler}@fh-luebeck.de

Abstract

Patients in intensive care units may receive several medications intravenously and simultaneously. When 6 different medications are needed for example, they are delivered by 6 syringe pumps via infusion lines. However, a net of tubes must be established which in turn leads to a complex and vague scene of the devices and causes hygiene problems to patients. To solve this problem, a new model for multiple infusions was developed to reduce the number of tubes that reach patient to only one. The module integrates the syringe pumps into one device. It mainly consists of a rotary selection valve and 6 compartments for medications supply. The selection valve and the compartments are driven by stepper motors. This paper presents a preliminary design of a platform for steering the stepper motors. It also presents a tested algorithm for synchronized work of the motors in the module.

1 Introduction

Multiple intravenous infusions are used in Intensive Care Units (ICU) for simultaneous application of medications to patients [1]. In these setups several syringe pumps, extension tubes, three-way valves, and valve banks are joined together. The structure of these systems imply errors, particularly because the tubes appear very complex and confusing to medical staff. It may cause hygiene problems when transporting the patient from operation room to ICU [2]. Another potential problem in the delivery of several drugs is that they can interact with each other in the tube. This can lead to incompatibility reactions and thus loss of efficacy of the drugs [3]. References [4]-[7] show that the potential infection risk to patients increases, if more than one catheter is placed. Fig. 1 shows a drug delivery station at an ICU of the University clinic Schleswig-Holstein in Luebeck. The project has the basic aims to make the drug delivery station simpler. A new multiple infusion model (MIM) is being developed. It shall reduce the potential risks to patients. A 3D print of the first MIM is shown in Fig. 2.

Figure 2: First manufactured MIM with 50 ml syringes loaded.

2 Materials and Methods

The MIM mainly consists of 6 compartments, i.e. syringe pumps, and a rotary selection valve that orders the delivery

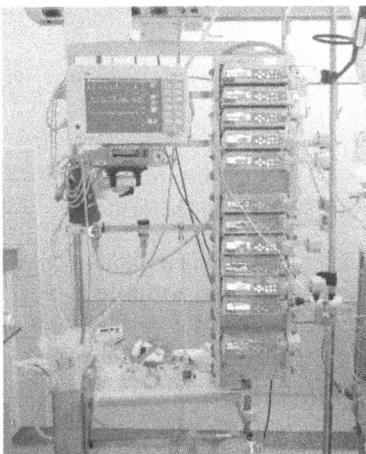

Figure 1: Multiple infusion station in ICUs is a source of errors due to chaos and complexity of infusion lines.

of the drugs' segments. For flow generation, the syringe pumps are driven by 6 stepper motors (Nanotec Electronic GmbH & Co. KG, Germany), and the selection valve is driven by another model of stepper motor (VICI AG, USA). This is because the stepper motor comes as a built-in unit with the selection valve. In the context of this work, the stepper motors of the syringe pumps are named syringe motors, and the selection valve motor is named valve motor. This work is divided into 2 parts: the first part is configuring a control unit for the valve as a main controller for the supply of liquids. The second part is configuring the syringes control unit for the 6 syringe motors. In the future work, there must be 1 control unit for both parts.

2.1   Selection valve unit

The selection valve is responsible for the order of medications' flow. The provided control unit for the valve is bulky in comparison to the size of the available room inside the MIM. Therefore, hardware and software modifications must be done.

2.1.1   Hardware of valve motor

The selection valve unit, which can be seen in Fig. 3, consists of the valve head and a stepper motor. It can be operated manually via serial interface or USB interface by a multi-position actuator control module which can receive commands from a peripheral device, send them to the valve motor and give a feedback of the current valve motor position. Sending the commands must be done by *PSoC4* microcontroller (Cypress Semiconductor, USA). The communication between *PSoC4* and the control unit is performed by the serial interface Universal Asynchronous Receiver Transmitter (UART). Fig. 3 schematically shows the structure of the communication between the individual parts.

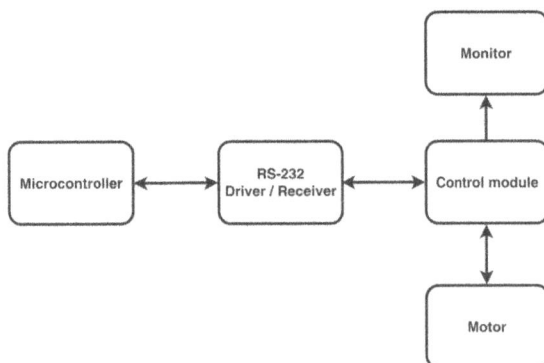

Figure 3: Block diagram of the developed communication between the individual parts of the valve control unit. The RS-232 Driver/Receiver represents the major designed part in this diagram.

2.1.2   Software for selection valve

The microcontroller has the function of sending the commands to the control module, receiving messages from it,

handling these messages and responding to them. There is a prescribed set of commands available for operating or configuration of the valve motor. These commands will be implemented in a control program. Using *PSoC creator* software (Cypress Semiconductor, USA) as a programming environment, a control program is written and transmitted directly to the microcontroller via USB interface.

2.2   Syringe pumps control unit

The syringe motors are stepper motors that are used for delivery of drugs in the appropriate channels. In the developed instrument 6 stepper motors should be used. But for test purposes in this work and to reduce the material costs, only 3 motors are used.

2.2.1   Hardware of syringe motors

Each syringe motor is provided with 1 driver model *SMCI33-2* (Nanotec, Germany) with 3 outputs. Each driver contains a programmable integrated controller. The 3 outputs enable 8 logic states, and each state can be used to operate a particular motor. Because the device has a small room for electrical boards inside, using the 8 logic states will reduce the number of drivers from 6 to 1. The driver has 6 inputs that can be exploited. The syringe motor is connected via a four-core cable with the *SMCI33-2*, and in the encoders 3 channels must be connected to the *SMCI33-2* to perform a position monitoring. The *SMCI33-2* takes over the transmission of the commands to the motor and monitors them. *PSoC4*, with the help of a constructed relay board, decides which motor or encoder should be used. Fig. 4 shows schematically the communication between the major components of the syringes control.

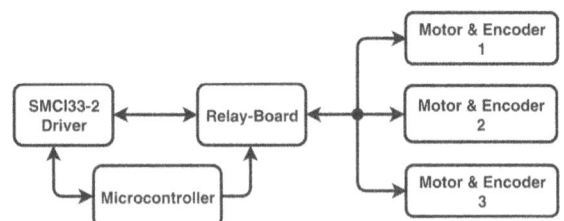

Figure 4: Block diagram of the designed syringe pumps control unit.

2.2.2   Software of syringe motors

The programming of the stepper motor controller was carried out using the Windows software *NanoJEasy*. It is a programming environment for developing Java programs, which are executable on Nanotec stepper motor controller and enable advanced programming of controllers. The control system includes a Java Virtual Machine with integrated functions, which can be used in the program. These functions can be used for communication with the motor. But appropriate classes at the beginning of the program must be implemented.

3   Results and Discussion

3.1   Hard- and software of valve motor

For the communication between the microcontroller and the control module a converter board was developed because the microcontroller has only TTL level (0 to 5 V) available, but should still communicate with the control module in accordance with RS232, which works with 12 V signals. This was accomplished by using level converter *MAX232* as shown in Fig. 5.

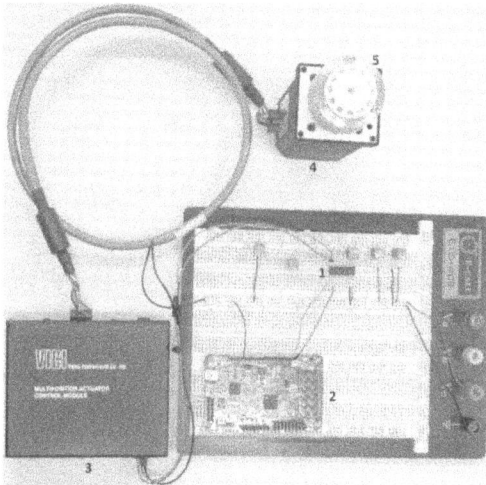

Figure 5: Control of the valve unit: (1) MAX232, (2) microcontroller, (3) control module,(4) stepper motor, (5) valve.

A 5 V supply voltage is sufficient for the *MAX232* to convert from TTL level to RS232 level. The working method of the *MAX232*, the drivers and the connections is illustrated in Fig. 6. The next step is to test the UART communications. For this, an interrupt-based program was implemented, which is presented in the flow chart in Fig. 7. At the beginning of the program the UART communication will be initialized and started. A Light Emitting Device (LED) helps to confirm the state of the switcher. Each time the button is pulled down, the LED turns off. At the same time the motor changes in the operating state. The valve rotates the channel to the desired position with each button press. Finally, the microcontroller sends a command for position detection to the control module, which returns on the position as a string like "position is xx r" in 8 bit packages. Since the microcontroller can read faster than the control module can send the data, an interrupt was used which is triggered every time a packet is received. The microcontroller reads the packet and waits while the control module sends the next packet, otherwise the microcontroller reads the same package repeatedly. The received string is stored

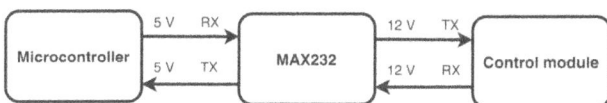

Figure 6: Working method of the converter board.

in a variable, after the control module finishes sending the data. The position number is thereafter cut out of the string and stored in a variable that is ultimately shown on a screen. The test of the implemented program has demonstrated that the developed control unit worked successfully. The hardware in Fig. 5 can now be used for any algorithm.

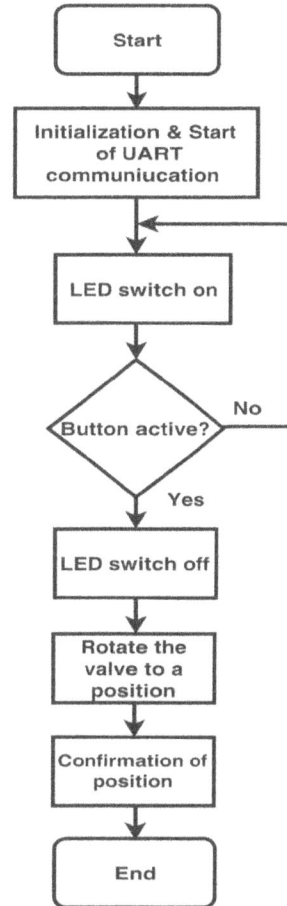

Figure 7: Sample algorithm as flow diagram of the test program for the valve motor.

3.2   Hard- and software of syringe motors

The challenge in the syringes control unit was to reduce the number of drivers, although a driver is intended only for 1 motor. In order to realize this, a relay board was designed as mentioned before which with the aid of the *PSoC4* and the *SMCI33-2* makes it possible to switch between the motors and the encoders. This board includes a four-pole relay for each motor and a three-pole for each encoder. When the current flows through a relay, the corresponding cable of the motor or encoder can be connected to the *SMCI33-2*. The current flow is however controlled by optocouplers. Fig. 8 shows the structure of the relay boards and Fig. 9 shows the developed syringe pumps control unit. So the optocouplers operate as switches that close the electrical circuit when they receive a signal from the microcontroller. For the generation of the signals a program has been implemented and stored in *PSoC4*. The program contains a counter, which was first initialized to the zero value. Depending on the counter value, a motor and its encoder can be connected with the *SMCI33-2* and the other 2 motors will be separated. With each keystroke the counter value is increased by 1 until it exceeds the value 3, then it is again reset to value 1. The connected motor will then carry out the command list which is stored on the *SMCI33-2*. The hardware was tested with the original software *NanoPro* from Nanotec. It successfully functioned and responded to the sent commands without errors.

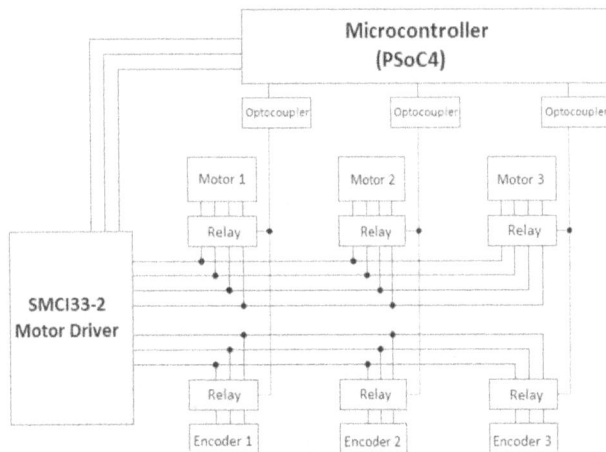

Figure 8: Relay board to switch between motors and its encoders.

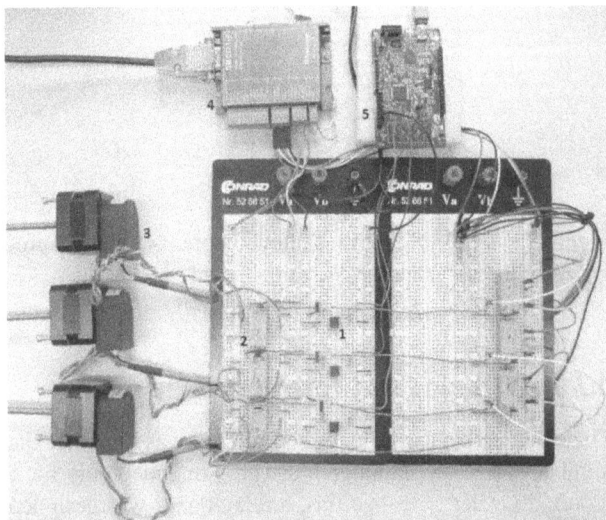

Figure 9: Control of 3 motors: (1) optocoupler, (2) relay, (3) stepper motor, (4) motor driver, (5) microcontroller.

4   Conclusion

The aim of this work was to develop a control unit for a number of motors, which could be used in an innovative infusion system later. The platform will enable to set an ordered medication supply, by synchronization the work of the motors. For this reason, 2 electrical circuits, a converter board, and a relay board were designed and tested, which enabled the conversion of the voltage level between the microcontroller and the control module and switching between the motors. It was checked how commands are sent via a microcontroller to the motor controls, which then rotate the syringe motors with a determined driving profile or turn the valve motor to a specific position.

Acknowledgement

This work was carried out at the Medical Sensors and Devices Laboratory, Luebeck University of Applied Science, Germany.

5   References

[1] Hintzenstern, *Infusion-Transfusion-Parenterale Ernährung*. Urban & Fischer, 2004.

[2] S. Abdul-Karim, *Design and development of a test bench to separate drugs in a multi-infusion system using gas bubbles*. Master Thesis, Lübeck University of Applied Sciences, 2013.

[3] A.A. Ameri, DIVI 2008 Hamburg, *Modernes Infusionsmanagement auf der Intensivstation Was ist sinnvoll, was ist notwendig?*. Available: http://vs1.exploredoc.com/store/data/001881221.pdf?key=02a0c246df297770fcc41b61963f68cd&r=1&fn=1881221.pdf&t=1455106519681&p=86400 [last accessed on 16.01.2016].

[4] S. Weiland, *Koalition prüft schärfere Infektionsschutzregeln*. Available: www.spiegel.de/politik/deutschland/0,1518,713482,00.html [last accessed on 16.01.2016].

[5] H. Haselmayer, *Hygiene invasiver Katheter - Umgang mit Infusionssystemen*. Available: http://www.medicom.cc/medicom-de/inhalte/intensiv-news/entries/1907/entries_sec/1917.php?WSESSIONID=e80f538113f19cc49fec86e5aaa5ea75 [last accessed on 16.01.2016].

[6] Universitätsklinikum Tübingen, *Hygienemaßnahmen bei Injektionen, Punktionen, Infusionen und bei praeoperativer Hautantiseptik*. Available: http://www.medizin.uni-tuebingen.de/uktmedia/EINRICHTUNGEN/Institute/Institut+fuer+Medizinische+Mikrobiologie+und+Hygiene/PDF_Archiv/Hygiene/HygienemassnahmeBeiInjUndInf-port-10011.pdf [last accessed on 16.01.2016].

[7] Pflegewiki, *Zentralvenenkatheter*. Available: http://www.pflegewiki.de/wiki/Zentralvenenkatheter [last accessed on 16.01.2016].

Characterisation, installation and functional testing of a collimator at a research linear accelerator

J. Beer [1]

[1] Medizinische Ingenieurwissenschaft, Universität zu Lübeck, julian.beer@student.uni-luebeck.de

Abstract

The combination of a Magnetic Resonance Imaging (MRI) system and a linear accelerator in a hybrid device (MRI-Linac) is a new opportunity in radiotherapy. One aspect in research is the dosimetry at MRI-Linacs. In this regard an experimental set-up is in preparation at a research electron accelerator for radiotherapy dosimetry. Since geometrically defined radiation fields are required in the experiments, a collimator had to be installed. This paper presents the implementation of a collimator system added to PTB's research linear accelerator and the results of radiation field measurements used to determine the resultant field dimensions. With the newly installed collimator system it was possible to create reproducibly radiation fields of various sizes. The radiation fields vary from $9{\times}9$ cm^2 up to $20{\times}20$ cm^2, with a maximal deviation of 3.64 % in horizontal and 2.25 % in vertical size.

1 Introduction

In radiation therapy diagnostic imaging is used to perform irradiations more precisely. This technique is called image guided radiation therapy (IGRT) and is typically done sequentially, i.e. the irradiation is applied after image acquisition, and at stand-alone devices. A new development is the combination of image acquisition and radiotherapy within one hybrid device which is realized, e.g., by combining an MRI system and a linear accelerator [1]. This MRI-Linac makes it possible to adapt the irradiation almost in real-time. At the same time it utilizes the benefits of MRI, like good soft-tissue visualization, a high temporal resolution. Furthermore MRI does not lead to additional radiation exposure [2]. At MRI-Linacs the strong magnetic field of the MR is always present. Investigations have shown that the response of dosimeters is affected by static magnetic fields [3]. This gives rise for further research and PTB (Physikalisch-Technische Bundesanstalt) is preparing experimental investigations regarding dosimetric measurements inside the magnetic field of an electromagnet. During the irradiation process, the measurements between the pole shoes should not be affected due to additional scattering, magnet materials should not be activated and the electronics should be preserved from damages due to radiation. Therefore, radiation fields have to be confined to the space between the pole shoes. This was realised by a newly installed collimator system which was geometrically characterised and installed at a research linac. The geometrical information was of interest to get a (theoretical) estimation of field dimensions and to evaluate the difference to the measured fields.

2 Material and Methods

Fig. 1 shows the three main parts of the set-up: The research accelerator, the newly installed collimator and the measuring system which was placed inside a phantom. The parts are described in the sections 2.1 till 2.3. The distance between the source of the radiation field and the phantom surface is called the source to surface distance (SSD). In this set-up it is 1 m. Using the room lasers, collimator and measuring system were positioned in line with the central axis of the beamline.

2.1 Research accelerator

The investigations were executed at PTB's research electron linear accelerator for radiotherapy dosimetry, custom-designed for PTB, manufactured by ACCEL Instruments (ACCEL Instruments GmbH, Bergisch-Gladbach, Germany) [4]. At the high-energy section electron beams from 6 MeV up to 50 MeV are possible. In this work a nominal electron energy of about 12 MeV was used. The electron beam leaves the beamline at the end by passing a thin exit window made of copper (Fig. 1). With respect to the exit window, the beam was set up centrally. The beam diameter near the exit was about 10 mm Full Width at Half Maximum (FWHM) as measured by a beam profile monitor at the beamline. After the exit window the electron radiation expands through the air and enters the phantom. The spatial distribution of this radiation is the radiation field [5].

2.2 Collimator system

The used collimator is a home-built system, the schematic layout is shown in Fig. 1. The collimator causes a restric-

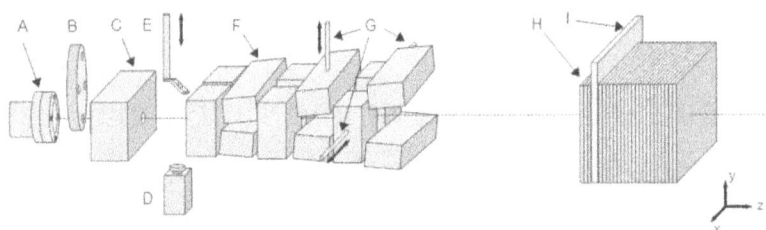

Figure 1: Schematic drawing of the set-up: The end of the beamline (A), the collimator system which consist of a filter wheel (B), fist collimator (C), video camera (D), mirror (E), second collimator (F) and limit stops (G) and the phantom (H) with detector array inside (I) to measure radiation field dimensions. The aperture axis is also shown (dashed line).

tion of the field size perpendicular to the direction of the central beam. The first collimator is a fixed block made of lead with a cone-shaped aperture and it provides the first beam forming. This aperture has a widening diameter from 1.3 cm to 4.4 cm and the collimator installation and positioning should have caused the aperture axis to be in line with the beamline central axis. The second collimator is made of twelve lead blocks mounted behind the primary collimator; in x- and y- direction three in a row are connected. Since the block triplets are movable, rectangular radiation fields of different sizes are feasible. To allow the reproducibility of certain radiation field dimensions, three aluminium bars with a ruler (accurate to 1 mm) were mounted at the outside of the collimator system. These bars operate as limit stops. Since they are movable, the distance to the blocks and consequently the maximum collimator opening is adjustable (Fig. 1). Contrary to the horizontal blocks the vertical blocks are coupled and just one stop, instead of two, is necessary there.

In addition to the collimating parts, a motorised filter wheel is located between beam exit window and primary collimator. Up to ten round inserts with a diameter of 18 mm each are placeable inside the holes of the wheel. As shown in Fig. 1 the insert is placed inside the beam path. In this work, two different inserts were used:
1. A luminescent screen to verify the beam position inside the primary collimator. This screen is coated with a phosphorescent powder which emits light during irradiation, observable via a mirror-camera system (Fig. 1).
2. A lead foil, 1 mm thick, for an increase of electron scattering and consequently widening of the radiation field (section 3.1).

2.3   Field measurements

The determination of the field distribution is based on dose measurements across the radiation field. Common practice for dosimetry in radiotherapy is the use of ionisation chambers (IC) placed inside a water phantom [6]. Since the used measurement systems are not waterproof, a solid slab phantom was used. The phantom material was PMMA (polymethyl methacrylate) which is a water-equivalent plastic according to E DIN 6800-1 [5]. The slab phantom is about $30 \times 30 \times 30$ cm^3 (W$\times$H$\times$D). The surfaces of the slabs were orientated parallel to the x-y-plane defined by the coordi-

nate system in Fig. 1. The central beam axis which coincides with the aperture axis in Fig. 1 passes centrally through the slabs. With this set-up it is possible to place dosimetric detectors in various depth inside the phantom. In this work two different detectors were used:
1. A cylindrical IC (NE2571, Nuclear Enterprises, Wichita, KS, USA) fixed inside a special slab was placed in increasing depths inside the phantom and its signal at each depth was measured. The position of the chamber was manually changed from a depth of 9 mm to 59 mm in steps of 5 mm; the IC reference point always lying on the central beam axis. The resulting signal curve was analysed regarding an appropriate depth for the measurements with the OCTAVIUS detector. For these measurements the nominal field size was 20×20 cm^2 according to the recommendations in DIN 6800-2 [6]. Here, the field size was adjusted using the geometrical characterisation of the collimator. Based on this characterisation, the radiation field expansion based on the intercept theorems was assumed, i.e. after the first and the second collimator the field size of the radiation field increases linearly with distance from the source.
2. A two-dimensional array of ionisation chambers (OCTAVIUS Detector 1500, PTW, Freiburg, Germany) is placed at a specific depth inside the phantom to determine the dose distribution. The detector has 1405 evenly distributed ionisation chambers which cover the maximum detectable field size of 27×27 cm^2 [7]. It is appropriate to measure relevant fields up to 20×20 cm^2. In conjunction with the software Verisoft provided by PTW dose distributions and therefore radiation field dimensions are measurable and representable with a resolution of 0.5 cm. Like the collimator, the phantom is aligned in axis with the beamline and so is the detector centre.

3   Results and Discussion

As preparatory work the dose distribution across the whole detector was measured with and without a scattering foil to determine the influence of the foil on the radiation field distribution (section 3.1). Furthermore the depth dose curve at central axis was measured to define an appropriate depth inside the phantom for field measurements (also section 3.1). Finally, field distribution measurements were performed and the field dimensions determined (section 3.2).

3.1 Preparatory work

Fig. 2 shows two profiles along the x- axis of dose distributions with and without a scattering foil (see section 2.3). The profile with scattering foil belongs to the dashed line in Fig. 3, positioned at the dose maximum of the field. Both fields are restricted by the collimator, adjusted at a nominal field size of 20×20 cm^2. Without the scattering foil the dose distribution is unsymmetrical near the centre. In comparison, the profile created from a measurement with a scattering foil is flatter at dose maximum. This relation is found for the vertical axis, too. The ideal profile is formed like a rectangular function - evenly distributed and a steep increase and decrease at specific values. Since the profile with scattering foil is more similar to the ideal distribution (rectangular function) the measurements were executed with scattering foil.

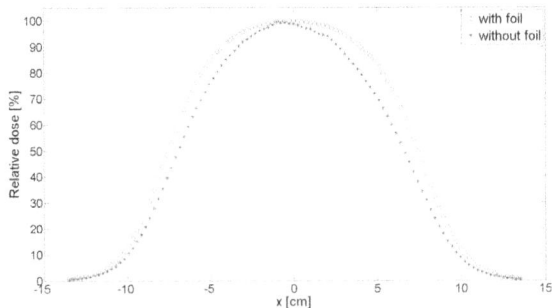

Figure 2: Relative dose distribution of a $20 \, \text{cm} \times 20 \, \text{cm}$ radiation field with and without lead foil at the dose maximum of both fields.

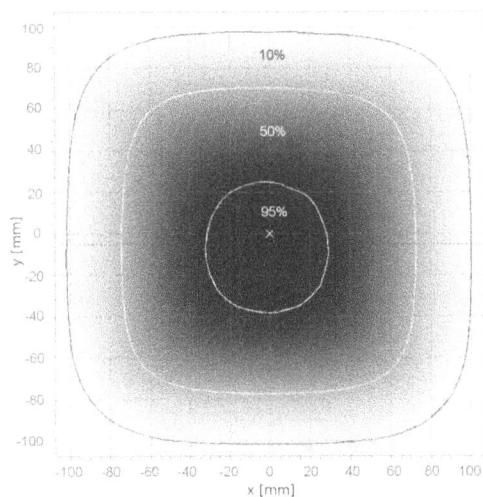

Figure 3: Relative dose distribution of a $20 \, \text{cm} \times 20 \, \text{cm}$ field with lead foil. The white cross indicates the middle of the detector array.

The results of the measurements of the IC signal in increasing depth are shown in Fig. 4. With rising depth the measured signal increases to the maximum at about 25 mm depth. At a depth of 45 mm it is nearly 50 % of the maximum dose. Fig. 4 reflects characteristics of a depth dose curve for electron beams: There is a slight increase in the

dose up to the maximum which is located near the surface and a rapid decrease after the maximum [8]. Based on Fig. 4, the measurements with the OCTAVIUS detector were executed at a depth of 46 mm, at which the signal is decreased to 50 % of the maximum. The 50 % signal level was chosen since the corresponding depth can be evaluated well for different electron energies because of the almost linear decrease of the signal curve at the 50 % level. This makes it possible to investigate field sizes for other energies beyond or above 12 MeV with the same method described in this work.

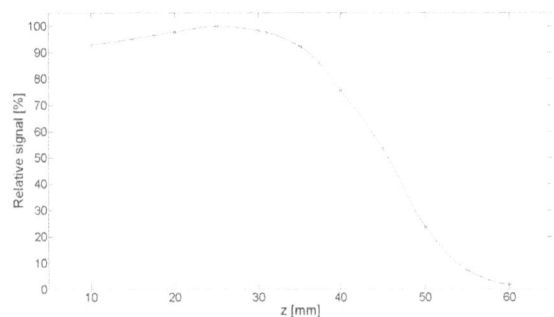

Figure 4: Signal of an ionization chamber (NE2571) at different depths of a PMMA phantom for a 12 MeV electron beam collimated to a $20 \, \text{cm} \times 20 \, \text{cm}$ field.

3.2 Radiation fields

The determination of the field sizes was executed based on dose distributions across the detector array. An example of the dose distribution for a field with a nominal size of 20×20 cm^2 is shown in Fig. 3. The highest dose is near the detector centre. This is indicated by the light grey isoline which represents 95 % of maximum dose. Contrary to common practise [5], the field size is not estimated at FWHM of maximum dose. With respect to the experiments with the electromagnet, portions of the field hitting the magnet and its pole shoes should be small for reasons already mentioned in section 1. An irradiation of the magnet with doses ≤ 10 % was chosen as limit and therefore the field sizes are determined with respect to the criterion of 10 % of maximum dose. As mentioned in section 3.1 the radiation field measurements were executed at a depth of 46 mm inside the slab phantom. First investigations were based on the geometrical characterisation of the collimator. The adjusted collimator opening should result in an expected field size which was predicted by a linear size-distance dependence. The investigations showed that the radiation fields were not collimated as predicted; the measured sizes were larger than expected (Table 1).

Because of this and based on these measurement results, a method was developed to obtain more realistic field size predictions. The results are presented in Table 2. All measured field sizes are analysed with respect to their nominal field size and the shift of the dose maximum to detector centre. Since all fields are shifted to negative values, the field sizes and deviations were determined with respect to the

Table 1: Nominal field sizes with regard to the measured sizes with the deviation in horizontal and vertical dimension (width / height).

Nominal [cm^2]	Measured [cm^2]	Deviation [%]
5×5	10.22×10.69	+104.40 / +113.80
8×8	13.20×12.90	+65.00 / +61.25
12×12	17.93×20.38	+49.42 / +69.83
14×14	20.94×22.22	+49.57 / +58.71

field centre (dose maximum) instead of the detector centre. The occurring shift to negative values is probably caused by the positioning of the collimator or the slab phantom.

Table 2: Nominal field sizes and the measured sizes with the deviation in horizontal and vertical dimension (width / height). Regarding to the detector centre, the spatial shift from field centre is determined.

Nominal [cm^2]	Measured [cm^2]	Deviation [%]	Shift [mm]
9×9	9.28×8.98	+3.11 / -0.22	-0.5 / -3.5
10×10	10.19×10.12	+1.90 / +1.20	+1.0 / -4.0
11×11	11.40×11.15	+3.64 / +1.36	+0.5 / -3.0
12×12	11.90×12.27	-0.83 / +2.25	+0.5 / -5.0
13×13	12.80×13.04	-1.54 / +0.31	0.0 / -5.5
14×14	13.83×14.12	-1.21 / +0.86	0.0 / -5.5
15×15	14.82×14.85	-1.20 / -1.00	+0.5 / -6.0
16×16	15.54×15.81	-2.87 / -1.19	-1.0 / -6.0
17×17	17.00×17.10	0.00 / +0.59	-1.0 / -6.0
18×18	17.93×18.39	-0.39 / +2.17	+0.0 / -7.0
19×19	19.09×19.15	+0.47 / +0.79	-0.5 / -7.0
20×20	20.05×19.66	+0.25 / -1.70	-1.5 / -7.0

The measured field sizes deviate maximally by an absolute value of +3.64 % (11×11 cm^2 field) in horizontal direction and +2.25 % (12×12 cm^2 field) in vertical dimension. Since the largest deviations occur at the smaller fields, a possible explanation is the limited precision of the limit stops (1 mm). The measurements indicate that a deviation of ±0.15 cm of the collimator blocks in vertical dimension leads to a changed field size of approximately 1 cm. Additionally, the shape of the fields depends on the nominal field size. With decreasing size the fields appear rounder, but large fields like the 20×20 cm^2 field in Fig. 3 have an approximately rectangular shape. The smallest field that is possible is 9×9 cm^2. At this adjustment the collimator is almost closed completely.

4 Conclusion

With the installed collimator system, it is possible to create fields at a distance of 1 m SSD with different sizes and shapes from 9×9 cm^2 up to 20×20 cm^2. Smaller fields are not possible and larger fields not of interest. The fields deviate maximally by an absolute value of 3.64 % (horizontal) or 2.25 % (vertical), respectively. With the newly installed limit stops the field sizes can be adjusted reproducibly. The

performed geometrical characterisation was helpful for first approximation for the dimensions of the fields.

In the future, experiments with x- ray targets inside the filter wheel to create photon radiation fields will be performed. The different scattering characteristics will probably cause other field distributions. With respect to experiments with the electromagnet, a shorter SSD could be necessary, e.g. to create fields smaller than 9×9 cm^2. Additionally, measurements at another depth inside the slab phantom might be of interest. For all the changes in the experimental set-up mentioned above evaluations of the field sizes will be necessary.

Acknowledgement

This work has been carried out at the Physikalisch-Technische Bundesanstalt Braunschweig, Department 6.2: Dosimetry for radiation therapy and diagnostic radiology. Thanks for support by the Universität of Lübeck to Prof. Martin Koch from the Institute of Medical Engineering.

5 References

[1] B. W. Raaymakers et al., *Integrating a 1.5 T MRI scanner with a 6 MV accelerator: proof of concept*. Phys. Med. Biol. 54 N229, 2009.

[2] Jan J. W. Lagendijk et al., *MR guidance in radiotherapy*. Phys. Med. Biol. 59 R349, 2014.

[3] S. Stefanowicz et al., *Dosimetry in clinical magnetic fields using plastic scintillation detectors*. Radiation Measurement 56, pp. 357-360, 2013.

[4] K. Derikum, *A Dedicated Irradiation Facility for Radiotherapy Dosimetry*. Volume 25/1 of IFMBE Proceedings, pp. 53-55, World Congress on Medical Physics and Biomedical Engineering, Munich, 2009.

[5] E DIN 6800-1:2013-04, *Dosismessverfahren nach der Sondenmethode für Photonen- und Elektronenstrahlung - Teil 1: Allgemeines*. Beuth Verlag, Berlin, 2013.

[6] DIN 6800-2:2008-03, *Dosismessverfahren nach der Sondenmethode für Photonen- und Elektronenstrahlung - Teil 2: Dosimetrie hochenergetischer Photonen- und Elektronenstrahlung mit Ionisationskammern*. Beuth Verlag, Berlin, 2008.

[7] PTW, *OCTAVIUS 1500 Specifications*. Available: http://www.ptw.de/3088.html?&cId=9212 [08.02.2016].

[8] Hanno Krieger, *Strahlenphysik, Dosimetrie und Strahlenschutz - Band 2: Strahlungsquellen, Detekttoren und klinische Dosimetrie*. B.G. Teubner, Stuttgart/ Leipzig/ Wiesbaden, 2001.

Development of a Test Facility for non-invasive Ventilation Equipment

M. Apostel [1], S. Schimpf [2]

[1] Medizinische Ingenieurwissenschaft, Universität zu Lübeck, Martinapostel@gmx.de
[2] Drägerwerk AG & Co. KGaA, Lübeck, stefan.schimpf@draeger.com

Abstract

The Infinity® MCable™-Mainstream CO_2 sensor is an accredited medical product, developed by the companies Drägerwerk AG & Co. KGaA and Siemens to measure the carbon dioxide (CO_2) concentration in the patient respiratory flow. This sensor is installed in an automated test bench and measures the CO_2 concentration during ventilation in the dead space between the patient's face and a breathing mask. The integrated components are regulated by the DIN ES ISO 9001 norm, although they are commercially available medical devices and have to perform an accuracy test in a specific time-frame. This paper describes the development of a test- and calibration software written in the application development software LabVIEW. A gas mixture of 95% nitrogen (N_2) and 5% carbon dioxide is used to validate the accuracy test by comparing the measured CO_2 concentration and the 5% CO_2 gas mixture concentration. The developed software realizes the aimed accuracy test.

1 Introduction

Industrial manufacturers are forced to provide evidence of realized procedures to guarantee high quality products. Not only because of the international jurisdiction of product liability, but also because of consumer demands. These evidences were provided by quality management systems which are accredited by the Deutsche Industrie Norm (DIN) and the International Standard Organization (ISO). These comprise a description of guidelines to keep an internationally defined quality standard for companies. The DIN and ISO norms are valid for a classic manufacture company as well as banks and insurance companies to certify their quality assurance system by these guidelines [1]. In a production chain the quality control is given by measurement equipment or whole test facilities which certify the specification durability of a product. Even the best-engineered measuring devices may have inaccuracies due to environmental influences or material transformations. DIN/ISO quality standards counteract these inaccuracies of measuring equipment by dictating calibration tests in regular intervals. As soon as they are installed in test benches, their classification changes to measuring devices and underlies the guidance of DIN EN ISO 9001:2008 norm. The devices must be calibrated or undergo frequently accuracy tests [2].

Among a variety of other products the company Drägerwerk AG & Co. KGaA manufactures breathing masks for domestic use, that have to pass a CO_2 washout test. This test is performed by a new automated test bench and measures the concentration of CO_2 that accumulates in the dead space

between mask and patient's face. A concentration above 5% is unhealthy and can lead to death, by increasing over 10% and longterm inhalation [3]. Therefore the danger of re-inhalation of the CO_2 concentration in the dead space must be minimized. In this test facility the CO_2 concentration is measured by using the Infinity® MCable™-Mainstream CO_2 sensor (Fig. 1).

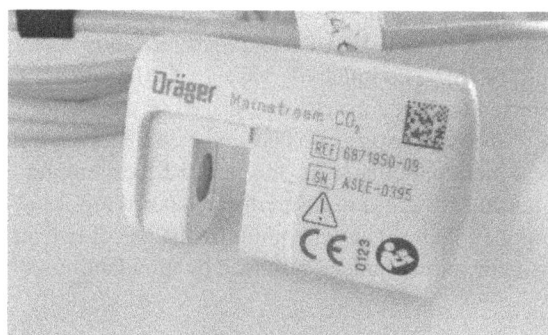

Figure 1: The Infinity® MCable™-Mainstream CO_2 sensor

Due to the fact that the previous test bench worked with a entirely different CO_2 sensor, there is no actual procedure to realize a calibration or an accuracy test for the Infinity® MCable™-Mainstream. To obey the DIN it is necessary to develop a device specific accuracy test. This paper describes the development of a calibration and control software that is able to perform the accuracy test. The software is written in LabVIEW to simplify the handling and to keep the mobility high.

2   Material and Methods

The Infinity® MCable™-Mainstream CO_2 sensor (Drägerwerk AG & Co. KGaA, Lübeck, D) is connected to a laptop's RS232 port by using a null-modem cable. The transmission uses a baudrate of 19200 $\frac{symbols}{second}$ and a 8N1 (8 Data Bit, none Parity Bit and 1 Stop Bit) single ended serial connection. The operation voltage of 5 V DC is supplied by the laptop's USB port. The sensor measures the concentration by using the mainstream method and non-dispersive infrared (NDIR) method (Fig. 2). The measurement range

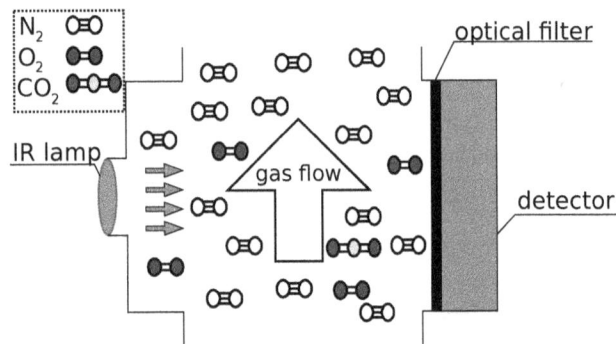

Figure 2: Schematic diagram of a non-dispersive infrared spectroscopy

is from 0 to 13,2 Vol.% with a precision of 5 % of the measured value [4]. To control the gas flow, a flow meter (TSI Incorporated, MN, USA) was installed between the gas cylinder and the CO_2 sensor. The flow meter's accuracy is 2 % of the measurement value and is necessary to keep a constant flow and not overstep the maximum flow of 0,1 $\frac{l}{min}$. The flow meter's baudrate is 38400 $\frac{symbols}{min}$ with a 8N1 data transmission [5]. To check the sensor's accuracy we used a test gas mixture by Linde AG with a concentration of 95% N_2 and 5% CO_2 and a manufacturer tolerance of $\pm$ 5 % [6]. The Infinity® MCable™-Mainstream CO_2 sensor is fitted on a cuvette and placed in the mainstream flow of the gas mixture. This cuvette guarantees a constant optical path length between the infrared light emitter and the detector. It is also to ensure that a tapered top fits at the end of the cuvette. This will minimize the risk of noticeable measurement inaccuracies provided by turbulent flow (Fig. 3).

2.1   Developing a LabVIEW Driver

A transmission protocol is integrated into the CO_2 sensor that allows to transmit, receive and process commands given by the host. This protocol is called STORM and stands for *Serial Transmission Optimized for Real-time Measurement* and has been developed by Drägerwerk. The network structure is a typical host-to-slave structure. The device only reacts if the host transmits a correct command. The transmitted data is separated in three parts. The first part is the header, followed by the parameter or data part, and the last part, is the checksum. The header part includes the identification numbers and the start byte. The data or

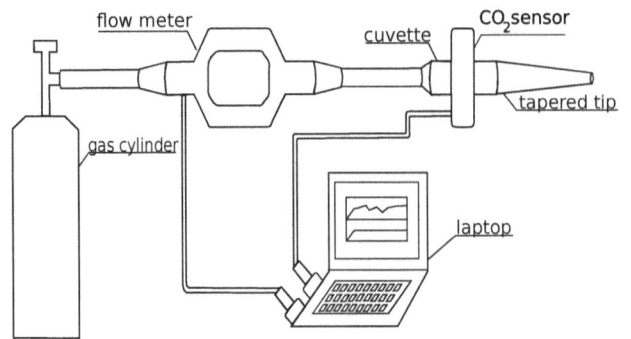

Figure 3: Schematic test structure with connected CO_2 sensor, gas cylinder and flow meter

parameter part consists of the requested data or parameter value, depending on the direction of the transmission. The checksum acts as a failure identification. In general a transmission begins with a start byte that is given by the hexadecimal 0x10 followed by the length byte, which describes the whole command length. The identification byte is the last part of the start sequence and represents the ID of every single command of this protocol. The calculation of the length byte is given by the equation

$$LB = x + 1, \ x \in [0, 255], \tag{1}$$

where x describe the return value of the data parameter. The checksum is calculated by adding 1 to the one's compliment of the parameter value sum and is given by:

$$CS = ((-1) * (\sum_{n=0}^{LB} PV)) + 1, \ CS \in [0, 255]. \tag{2}$$

The LB is the length byte value and PV the parameter value. If the calculation fails, the sensor will not accept the commands given by the host.

The driver to control the device over LabVIEW is separated into two parts (Figure 4). The commander links the selected

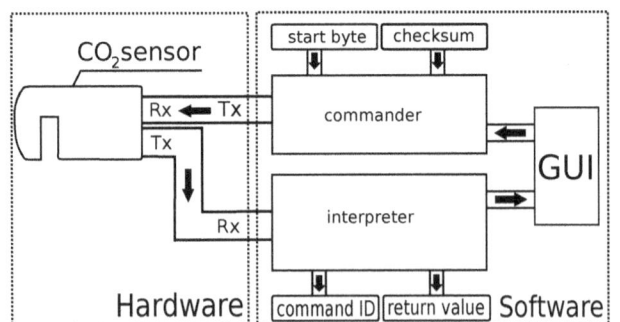

Figure 4: The commander transmits the linked command to the sensor's Rx wire. The interpreter gets the return value from the Tx wire.

command with the start byte and calculates the checksum of the host command. The interpreter waits until the sensor return value bytes arrive at the laptop's RS232 port. After arriving the command ID and the start byte will be assigned to the corresponding command.

2.2 Software

The queued state machine architecture was chosen as program structure for the control software. Generally, this architecture allows to send commands and data from multiple source points and is handled by only one event handler [7]. Another advantage is that the program possesses an idling cycle and the user is able to start, stop or repeat tasks dynamically. The sequential task work-off may cause a blockage in the data flow and the dynamic control. This happens when tasks in the queue comprise a loop. The blockage gets bypassed by implementing a second queue that executes the other tasks (Fig. 5). This kind of program

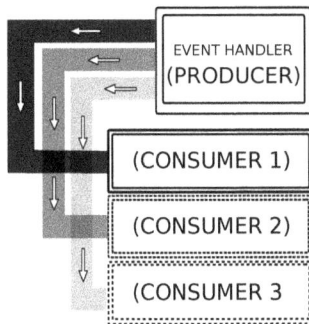

Figure 5: Schematic structure of a queued state machine, with one event handler controls more than one consumer

structure still guarantees a dynamic user control. The software accesses the LabVIEW implemented I/O application programming interface Virtual Instrument Software Architecture (VISA). This realize the data acquisition and the component control. To minimize the emergence of condensed water, the Infinity® MCable™-Mainstream CO_2 sensor has a 2 minute warm-up phase. During the warm-up phase the user is able to check the information package that includes the serial key, manufacture date and the actual sensor status. After the warm up the cuvette selection follows. The selection of a proper cuvette is important for the calibration check because variations of resistance, optical density and optical path length could cause a wrong CO_2 concentration. After selecting a cuvette the actual ambient pressure must be transmitted. The interdependency between CO_2 concentration measurement and the ambient pressure is the reason why the user must transmit the correct pressure before each calibration. The infrared-sensor will detect more molecules if the pressure is higher than its actual value or vice versa (Fig. 6). After transmitting the

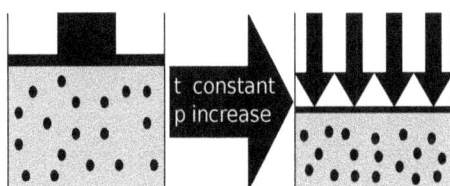

Figure 6: Infrared sensor measures higher CO_2 concentration with higher ambient pressure (cp. [8], S. 2)

choice of cuvette and the ambient pressure the user is able to start the sensor zeroing. This zeroing task includes a sensor measurement of the actual CO_2 concentration and saves this value as an offset. Assuming the zeroing passes without any errors the new sensor value is $0,0 \pm 0,05$. The user is now able to start an accuracy test. The structure chart shows the program flow (Fig. 7). During the measurement the ac-

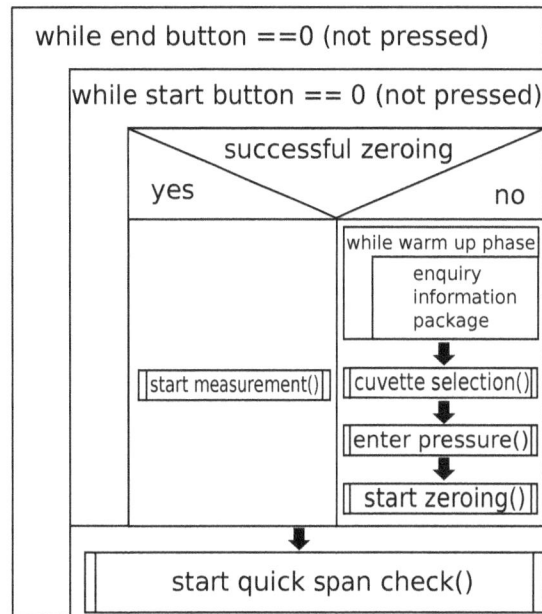

Figure 7: A Nassi-Shneiderman-Diagram of the calibration program

tual CO_2 concentration and the gas flow are shown on the control software's front panel. The synchronization is realized by similar time base. At the end of the accuracy test a report gives a summary of the sensor's parameter. The information includes the ambient pressure, the measuring time and the last stable CO_2 concentration. If the measured CO_2 concentration is smaller than 5 % of the CO_2 concentration of the used carrier gas, the accuracy test passed. It may also be possible that the CO_2 sensor's hardware is broken.This is the reason why another test was installed. This test is able to verify the hardware of the CO_2 sensor. Instead of placing a cuvette between the light source and the detector of the sensor, it is possible to use a quartz glass. The optical density value of the quartz glass is saved in the sensor's firmware. If the actual quartz glass value differs from the value that is deposited in the firmware, a hardware failure will be detected.

3 Results

The validation of the test procedure comprises 20 test measurements. These were realized with two different cuvettes (disposable and reusable) and ten measurements were performed for each kind (Table 1). The successful communication between the software and the sensor allows to enter the ambient pressure and the selection of the used cuvette.

Table 1: Accuracy test results

Testnr.	gas mixture	results	cuvette
1	5 % CO_2 and 95 % N_2	5,13 %	reusable
2	5 % CO_2 and 95 % N_2	5,01 %	reusable
3	5 % CO_2 and 95 % N_2	4,88 %	reusable
4	5 % CO_2 and 95 % N_2	5,11 %	reusable
5	5 % CO_2 and 95 % N_2	4,87 %	reusable
6	5 % CO_2 and 95 % N_2	5,05 %	reusable
7	5 % CO_2 and 95 % N_2	5,11 %	reusable
8	5 % CO_2 and 95 % N_2	4,97 %	reusable
9	5 % CO_2 and 95 % N_2	4,86 %	reusable
10	5 % CO_2 and 95 % N_2	4,90 %	reusable
11	5 % CO_2 and 95 % N_2	5,01 %	disposable
12	5 % CO_2 and 95 % N_2	5,10 %	disposable
13	5 % CO_2 and 95 % N_2	4,88 %	disposable
14	5 % CO_2 and 95 % N_2	4,76 %	disposable
15	5 % CO_2 and 95 % N_2	5,20 %	disposable
16	5 % CO_2 and 95 % N_2	5,11 %	disposable
17	5 % CO_2 and 95 % N_2	5,03 %	disposable
18	5 % CO_2 and 95 % N_2	5,05 %	disposable
19	5 % CO_2 and 95 % N_2	5,10 %	disposable
20	5 % CO_2 and 95 % N_2	4,84 %	disposable

The big value spreading by $\pm 0,24$ arises from the Infinity® MCable™-Mainstream CO_2 sensor's precision. Based on the measured values a reproducibility is given. At the end of an accuracy test a PDF report is prepared that shows the manufactured components and certifies the results.

4   Conclusion

This paper shows the development of a test facility for non-invasive ventilation equipment with a control software written with the application development system LabVIEW. The purpose of developing this test procedure was to realize a calibration or accuracy test of the Infinity® MCable™-Mainstream CO_2 sensor. This sensor is a commercially available medical device and is installed in a test bench. Therefore the sensor is categorized as a measuring equipment and underlies the DIN EN ISO 9001:2008 guideline that demand a calibration or accuracy test in specific time frames. At the end of the calibration a report attests the accuracy of the Infinity® MCable™-Mainstream CO_2 sensor by showing components parameter details and a all-over documentation of the calibration data values. Based on the results a 100 % test accuracy can be certified to the test procedure. As an outlook and to optimize the handling and minimize the human failure by entering a wrong pressure, an ambient pressure sensor could be implemented. Further, an implementation of a mass flow controller instead of a flow meter would increase the automation of the test facility and decrease the human failure to a minimum.

Acknowledgement

The work has been carried out at Drägerwerk AG & Co. KGaA, 23558 Lübeck.

5   References

[1] Testo Industrial Services GmbH, *Prüfmittelmanagement und Kalibrierung.* Available: http://download.testotis.de/files/DE/Fibeln/ Pruefmittelmanagement_und_Kalibrierung_2013.pdf [last accessed on 28.01.2016]

[2] S. Sieben, *Qualitätsmanagement und die DIN EN ISO 9001:2008: Eine Einführung: mit Trainingsleitfaden.* BoD – Books on Demand, 2013

[3] EUROPEAN INDUSTRIAL GASES ASSOCIATION, *Physiologische Gefahren durch Kohlendioxid (CO2).* Available: http://www.oeigv.at/PDFs/EIGA-IGV-SI24-11-D.pdf [last accessed on 27.01.2016]

[4] Drägerwerk AG & Co. KGaA, *Infinity® MCable™-Mainstream CO.* Available: http://www.draeger.com/ sites/assets/PublishingImages/Products/mon_infinity_ acute_care_system/US/Accessories/infinity-mcable-mainstream-co2-ds-9066255-us.pdf [last accessed on 26.1.2016]

[5] TSI Incorporated, *TSI SERIES 4000/4100 Datenblatt.* Availble: http://www.tsi.com/uploadedFiles/ _Site_Root/Products/Literature/Manuals/1980430-4000-4100-Design-Guide-web.pdf [last accessed on 23.1.2016]

[6] Linde AG, *Produktdatenblatt Prüfgas in HiQ® MINICAN.* Available: http://produkte.linde-gase.de/db_neu/ 0,05_co2_n2_minican.pdf [last accessed on 27.1.2016]

[7] A. Lukindo, *LabVIEW Queued State Machine Consumer Producer Architecture.* Mezintel Incorporated, Canada, 2007. https://decibel.ni.com/content/docs/ DOC-32964 [last accessed on 26.01.2016]

[8] VAISALA, *Wie misst man Kohlendioxid?.* Available http://www.vaisala.de/Vaisala%20Documents/ Application%20notes/CEN-TIA-Parameter-How-to-measure-CO2-Application-note-B211228DE-A.pdf [last accessed on 10.02.2016]

Validation of non-invasive, neonatal blood pressure cuffs

B. Al-Tashi [1], Thomas Graßl [2]

[1] Medizinische Ingenieurwissenschaft, Universität zu Lübeck, basheer.al-tashi@student.uni-luebeck.de

[2] Drägerwerk AG & Co. KGaA, thomas.grassl@draeger.com

Abstract

Apart from measuring blood pressure directly and hence invasive the non-invasive indirect blood pressure measurement represents the conventional approach to evaluate blood pressure. It has been shown that a detailed characterization of a blood pressure cuff's transfer function requires alternative concepts to rate its performance in addition to common clinical studies. In this paper a new approach basing on a limb-simulator is presented. Referring to the non-invasive oscillatory blood pressure measurement, blood pressure cuffs are widely used as indicators or rather sensors that translate oscillations of an arterial wall located in a patient's limb to the NiBP -measuring instrument. Usually, the brachial artery is chosen to derive the blood pressure. The validation of compliant NiBP-meters is defined in related standards published by the International Organization for Standardization (ISO) whereas effects of the applied cuffs are not covered by any standard [1]-[3]. Side effects caused by cuffs are even more crucial in case of neonatal patients. Therefore, in the scope of this paper the implementation of a standardized validation process for neonatal blood pressure cuffs utilizing a limb-simulator is discussed. More precisely, completion of an existing test rig as well as the implementation of the necessary control software are described before discussing results obtained through validation and verification of the test bench.

1 Introduction

A blood pressure cuff consists of either a self-contained bladder that is encapsulated by an inelastic part or an integrated bladder forming one coherent unit. In accordance with (IEC 80601-2-30 2009-01)[1] the bladder itself is defined as the part of a cuff that is inflated during application. Many different characteristics of blood pressure cuffs are key for a proper transmission of blood-oscillations further depending on the particular algorithm used during measurement. Neonatal blood pressure cuffs differ from their adult versions regarding not only shape but also bladder, connected hose or cuff-tube interface. Even though, detecting an oscillatory blood pressure is theoretically possible by utilizing a bicycle tube, reproducible results can only be guaranteed by using special designed materials suited for blood pressure measurements. Thus the selected cuff-materials are of great importance for obtaining proper blood pressure values.

In the present paper a limb-simulator is proposed for both the analyzation and evaluation of the transfer characteristics regarding a neonatal cuff due to the fact that clinical studies are impossible or at least very difficult to perform in the face of neonates. The presented results mainly focus on the utilized limb-simulator that was finalized in a first step. Subsequently, the necessary control unit was implemented in software using National Instrument's LabView [2]. Finally test manoeuvres were carried out for both validating and verifying the test bench.

2 Material and Methods

The test bench considered in this paper is presented in Fi. 1. The test bench consists of a microcontroller, a linear

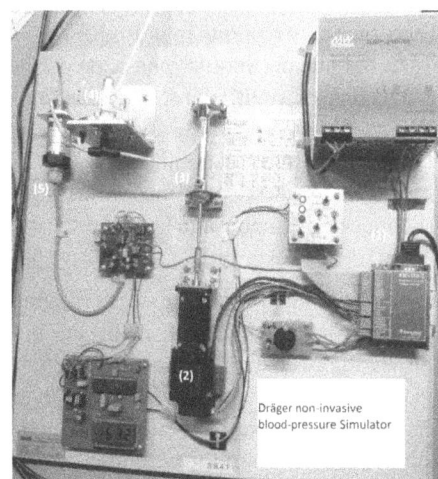

Figure 1: Dräger non-invasive Blood Pressure Simulator (1) Microcontroller, (2) Linear-Motor, (3) Pneumatic Cylinder, (4) Artifical Limb, (5) Pressure Sensor

---

[1] Particular requirements for the basic safety and essential performance of automate non-invasive sphygmomanometers

[2] (Laboratory Virtual Instrument Engineering Workbench

motor, a pneumatic cylinder and an artificial limb. The used stepper motor is controlled through the microcontroller that interfaces with the control unit represented in hardware by a constant current power output stage. The final output stage is implemented through an inexpensive and compact module, namely Nanotec's SMCI33 unit, comprising an integrated closed loop current feedback control. The microcontroller offers an easy way to fulfill a plurality of requirements regarding the drive. With the possibility to set motor-related parameters such as applied phase current, step- resolution or used adaptive micro step the operational behavior of the stepper motor can be altered and optimized according to requirements imposed by a test-scenario aiming at a simulated artificial blood pressure. Some parameters can be preset with the software NANOPRO provided by Nanotec.

Forward and backward movements of the stepper motor are translated into pressure waves by the connected pneumatic cylinder. Depending on the applied pulse duration and frequency different blood pressure oscillations can be simulated that are directly fed to the artificial limb. The recreated arm has to fulfill certain rules regarding not only shape but also inner pressure-volume relation to allow for proper measurements. Furthermore a suited blood pressure monitor is connected via a separated hose and a Y-piece to the pressure sensor to enable both realistic test conditions including cuff-inflation and a comparison between conventional derived and test bench related pressure values. Regarding setup it is important to activate the neonatal inflation modus in the monitor's operational settings as in case of disregard the measurements can be distorted by incorrectly applied peak pressures during cuff inflation. When applying too high pressures via the patient-connected cuff the test bench should even abort measuring due to patient safety taking into account the neonatal application case. After initializing, the test bench is able to automatically detect pressure exerted on the artificial limb by the wrapping round cuff. At a certain peak pressure the stepper motor runs a predefined script generating a series of oscillations. As the pulses have different characteristics concerning amplitude and duration the human blood pressure is mimicked. Vice versa the neonatal cuff detects these pressure waves transmitting them to the connected monitor and pressure sensor.

Within the artificial limb pressure wave propagation is superimposed due to pulses impressed by the inflation of the cuff and via the pneumatic cylinder by the movements of the stepper motor. These interactions establish an airflow to pressure ratio that is key for a proper measurement regarding the connected monitoring units. Therefore, an artificial limb has to meet certain requirements regarding not only its shape but also its compliance. The latter determines to which extend the arm can fallow smallest pressure variations due to its inertia. To achieve an artificial limb with a high sensitivity an ultra-thin elastic, yet hermetic material has to be used for construction. Without proper

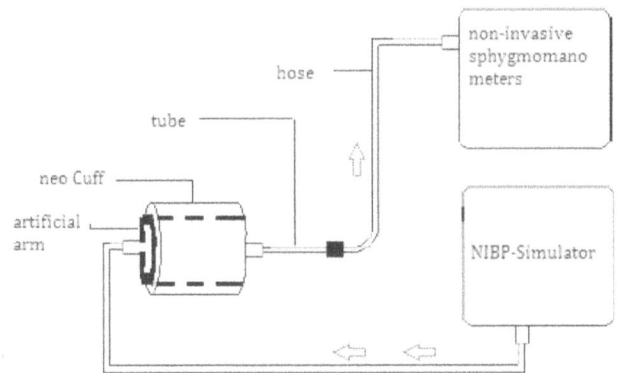

Figure 2: Operating schematic of the NiBP Simulator

impermeability the pressure wave propagation would be affected as no stationary pressure level could be arrived in the first place. Different materials were examined and assessed regarding their applicability. At last a section of a latex glove was chosen that was fixated with an adhesive polyurethane-foil.

To validate the test bench an additional pressure sensor that directly tracks the pressure level at the neonatal blood pressure cuff was necessary. The integration into the present test bench is pictured in Fig. 3. The parallel derived pressure

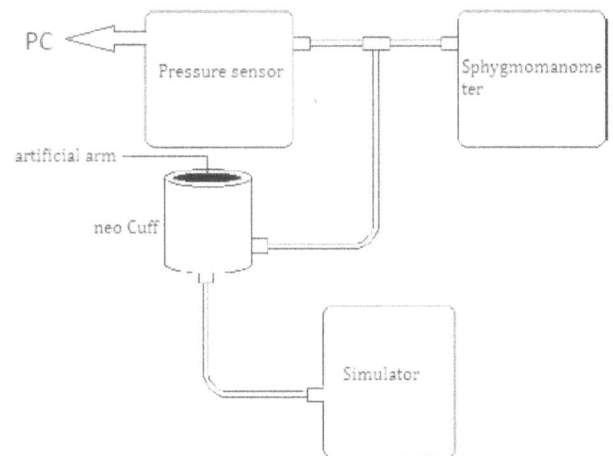

Figure 3: NIBP with two pressure sensors

values are recorded and plotted using a self-implemented software application containing a graphical user interface. The application was created using NI's LabView. The connected patient monitor incorporates a pressure sensor so that pressure values can be sampled and moreover be transmitted to LabView. However, interfacing the internal sensor may impair the measurement of the monitor which would require complex calibration procedures for every further measurement. The parallel derived pressure values are recorded and plotted using a self-implemented software application containing a graphical user interface. The application was created using NI[3]'s LabView. The connected pa-

---

[3]National Instrument

tient monitor incorporates a pressure sensor so that pressure values can be sampled and moreover be transmitted to Lab-View. However, interfacing the internal sensor may impair the measurement of the monitor which would require complex calibration procedures for every further measurement. The subsequent Fig. 4 shows different pressure curves obtained by the measuring instruments described above.

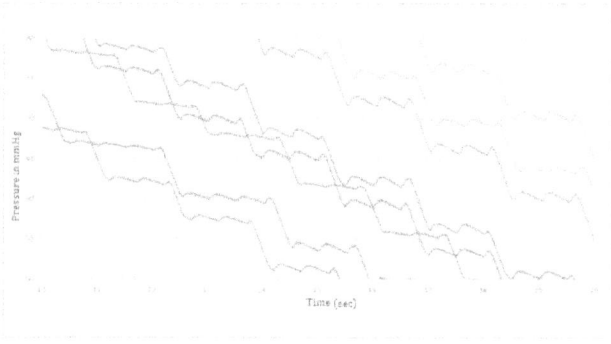

Figure 4: Different obtained pressure curves for simulated blood oscillations (various algorithms)

2.1 Software

A java program was written on the simulators micro-controller. The simulator has a certain driving profile that simulates the hearts oscillation and is used for test purposes. The program reacts to the measured pressure signal in the simulated upper arm and adjusts the linear motors movement.

A LabVIEW program was developed to display the pressure sensors signal on the display of a computer. LabVIEW programs are called VIs (Virtual Instruments). It is possible to include SubVIs in a VI. A SubVI that automatically stores the values at the end of each measurement in an Excel (by Microsoft Windows) sheed is included in the program. The user can select the duration and the sampling rate for pressure values derived at the simulated upper arm or at the blood pressure cuff for each test series independently. With a PC that runs the LabVIEW program the pressure course in the simulated arm as well as the blood pressure cuff is stored.

3 Results and Discussion

A direct detection of the blast wave that is initiated by the blood pressure and detected by the neonatal blood pressure cuff is necessary to determine the blood pressure cuff's transfer function and its validation. Our test bench simulates an upper arm that allows the validation of neonatal blood pressure cuffs.

While using a NIBP-instrument an error can be observed during evaluation. The instrument detects the pressure variations in the blood pressure cuff and uses those values

to calculate the blood pressure. Blood pressure cuffs are medical products. However there are no standards for their validation. The (ISO 80601-2-30:2009-01) defines a method to check the complete system. The pressure wave is directly injected to the measuring line by using a T-piece. This method enables determination of the repeatability regarding the results derived by the pressure sensor on the Monitor. However, it is not possible to determine the transfer function from the patient's blood pressure wave to the measured pressure wave via the blood pressure cuff (see Fig. 5).

The cuff is replaced by a metal container that is shown in Fig. 5. The concept of the ISO-80601-2-30 is adapted to the present test bench to determine the pressure curve defined by the blood pressure cuff by including a Y-piece between the cuff and the blood pressure monitor.

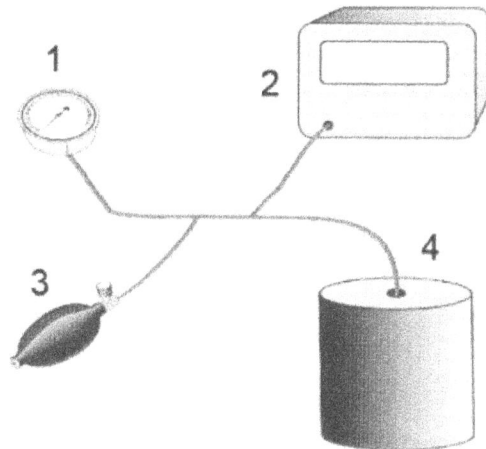

Figure 5: Test construction to determine the cuff pressure indicator's tolerance. (1) Calibrated reference manometer, (2) Not invasive blood pressure measurement under test, (3) Pressure generator, (4) Rigid metal container.

The Y-piece was chosen regarding to the diameters of the connecting hoses to minimize leakage and improve the results. Different procedures to supress the noise were tested. The results of the RMS-filter (Root Mean Square) were satisfying Applying a RMS filter provides satisfying results. An example can be seen in Fig. 6.

4 Conclusion

A NIBP blood pressure simulator to validate neonatal blood pressure cuffs was presented in this paper. It enables a validation without testing on neonates that are fraught with risks and complications. Blood pressure simulators that are available on the marked, simulate blood pressure oscillation with the help of a blood pressure monitor. Those simulated pressure waves are not suitable to validate blood pressure cuffs even if it is often said otherwise. They are

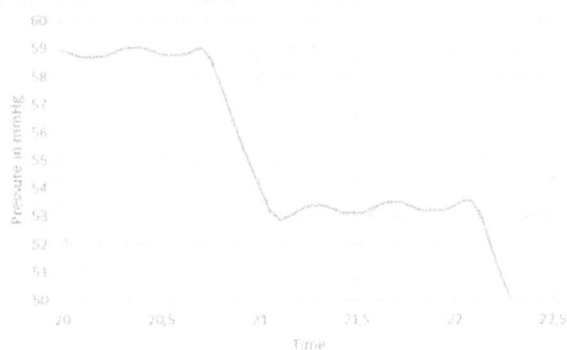

Figure 6: Pressure curve after RMS

suitable to test NIBP monitoring systems for adults and children as well as neonates. To validate blood pressure cuffs a suitable artificial limb as well as two sensors for the control of the simulated arm's pressure are needed to detect pressure oscillations within the cuff. It is important to use the cuff as a sensor instead of measuring past the cuff in twhich case the cuff would function as a buffer capacity.

During the blood pressure cuff's verification it is probably an advantage to choose a test setup without NiBP because the build-up of pressure due to the measuring device itself could falsify the measurement results. A comparison of measurements with and without NIBP monitor is shown in Fig. 7.

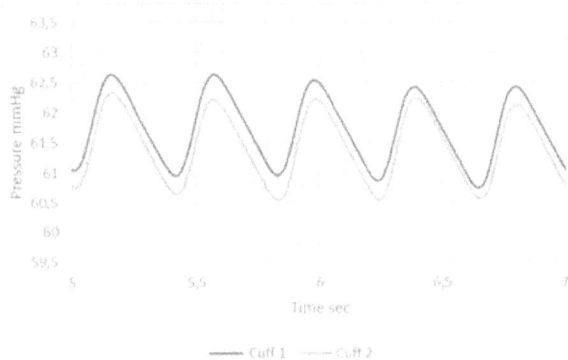

Figure 7: Pressure curves of the cuffs with and without the usage of the NIPB monitor

Acknowledgement

The work has been carried out at Drägerwerk AG & Co.KGaA, Moislinger Allee 53-55, 23558 Lübeck. I would like to express our thanks to the colleagues of the department Accessories & Consumables for the great support in so many different ways.

5   References

[1] Association for the Advancement of Medical Instrumentation, *Manual, electronic, or automated sphygmomanometers*, 2002.

[2] Deutsches Institut für Normung e.V., *Nichtinvasive Blutdruckmessgeräte – Teil 1: Allgemeine Anforderungen, Deutsche Fassung EN 1060-1:1995+A2:2009*, 2010.

[3] Deutsches Institut für Normung e.V., *Non-invasive sphygmomanometers – Part 3: Supplementary requirements for electro-mechanical blood pressure measuring systems* ;German version EN 1060-3:1997+A2:2009

[4] H. Groß, S. Mieke, and M. Ulbrich, *An arm phantom: a digital simulation system for testing sphygmomanometers* in Journal of Medical Engineering & Technology, vol.20 no.2, pp. 75–83, 1996.

[5] E. C. Walker, *Blood pressure cuff calibration system*, 1984.

[6] P. E. G. Kitzenmaier, *Oberarmsimulator für nichtinvasive Blutdruckmessgeräte*, 1998.

[7] R. Padwal, J. Millay, S. Babwick, M. Neil, L. Langkaas, J.Ringrose, *Effect of the accuracy of oscillometric blood pressure measurements*, University of Alberta, Edmonton, 2015.

Optimization and Evaluation of a Positive End-Expiratory Pressure Valve for Dräger Anesthesia Machines

Lucas Davenport [1] and Roman Weikert [2]

[1] Medizinische Ingenieurwissenschaft, Universität zu Lübeck, lucas.davenport@draeger.com
[2] Drägerwerk AG & Co. KGaA, Lübeck, roman.weikert@draeger.com

Abstract

The positive end-expiratory pressure valve of current Dräger anesthetical devices can cause oscillations under certain circumstances. This influence on artificial respiration has no hazard for the patients. However, it may cause an audible noise, so design changes have been developed and investigated [4]-[5]. The new design has been compared to the series valve referring to oscillation and expiratory resistance. A setup was built to make different measurements possible. The expiratory resistance was measured between ten series valves and ten of the new design. For oscillation analysis a combination of 15 valves with 17 tubes of different material and length was analyzed. Following this, an investigation of airway pressure curves was performed in MatLab software. The results show that the new design of expiratory valve has potential to reduce oscillations considerably.

1 Introduction

Ventilation with positive end-expiratory pressure (PEEP) improves the human oxygenation, especially for patients which suffer under lung diseases. It has become more and more popular during the last years for anesthesiologists and in intensive care [1]-[2]. A special valve in the breathing system makes this mode of ventilation possible. This so called PEEP valve can oscillate under certain circumstances. This can cause typical noise, which reduces the reliability of anesthesia machines from customer view. Due to this the improvement of the current PEEP-valve is an important topic for Dräger as a company of high medical standards. Therefore, a new design with changes on valve part level and material was developed [4]-[5]. To prove the capability of the new design, different measurements have been defined and performed in order to characterize and compare the current series and the new optimized design. Two of these measurements and their results will be discussed in this paper.

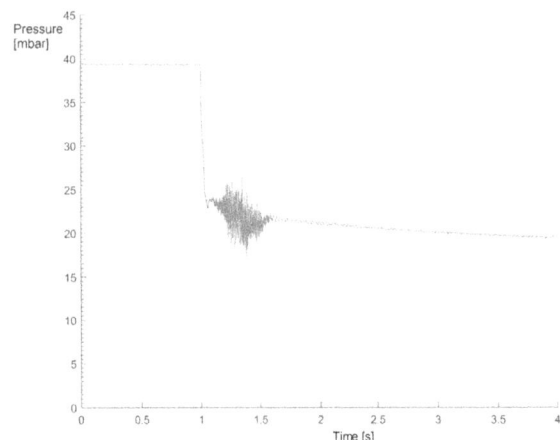

Figure 1: Breathing cycle with an oscillation detected at the PEEP-valve for an expiratory pressure drop from 40 mbar to 20 mbar.

2 Material and methods

Oscillation noise influences the customer perception of the anesthesia machines. In Fig. 1 an oscillation of an airway pressure signal for a breathing cycle with PEEP of 20 mbar is shown. For comparison Fig. 2 represents a typical breathing cycle with a PEEP of 0 mbar. One breathing cycle contains the time for inspiration and expiration. The upper part illustrates the pressure curve whereas the lower signal shows the corresponding flow curve.

The positive peak in the flow curve represents the inspiration flow and the negative peak shows the expiration flow. In Fig. 1 only the expiration part of the breathing cycle is visible, because oscillations only occur in this stage. At the beginning of expiration the pressure drops from inspiratory pressure of 40 mbar to PEEP 20 mbar. In the early phase of expiration it shows the typical oscillation behaviour. The frequency of this oscillation is about 200-250 Hz. The amplitude of 6-10 mbar leads to an audible noise. Although there is an oscillation, the PEEP can be hold properly by the valve. Generally oscillations increase at higher PEEPs whereas lower PEEPs like 0-5 mbar usually do not generate oscillations. However the highest adjustable PEEP is 20 mbar, which is not used frequently.

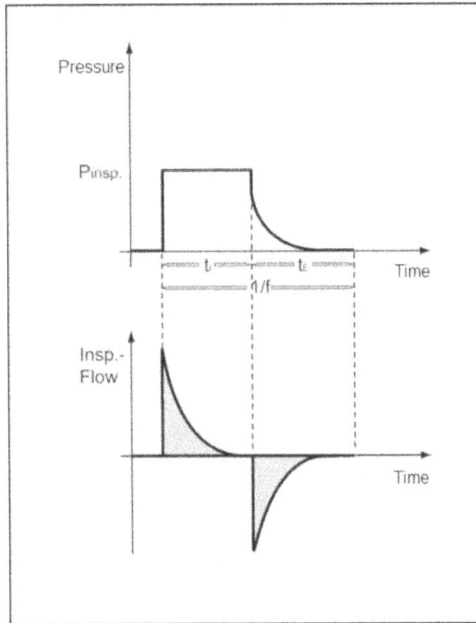

Figure 2: Pressure and flow curve from a breathing cycle for a PCV [3].

In order to get a short view on the reasons for oscillation, the function of an anesthesia machine's breathing system shall be introduced. Fig. 3 shows the circuit of an anesthesia device's breathing system. At the top you can see the patients lung. When the patient expirates, the inspiratory valve is closed and the gas flows through the right path (expiratory path), passes the flow sensor and reaches the PEEP valve. This generates a blockade for patients breath until the pressure of his breath is higher than the control pressure of valve. That is why a residual pressure stays in the lung, whereby a good oxygenation is assured and a collapse of the lung is prevented [2]. Gas which passes the PEEP valve flows through another expiration valve and the absorber or leaves the circuit by the exhaust valve [6]. Passing gas can cause parts of PEEP valve to oscillate, therefore a closer investigation is necessary.

2.1   PEEP valve

In Fig. 4 a model of the new PEEP valve design is shown. The membrane is connected with a disc. It is fixed between the breathing gas block and a lid which contains a connection to a control pressure. The membrane is made of a specific elastomer compound, so axial movement of the valve body is possible. If there is a greater control pressure than airway pressure the valve seals the crater opening. Hereby no gas can enter the expiratory path. In the opposite case the valve opens and air can escape. Reaching a certain combination of flow and pressure, oscillations can occur. The valve opens and closes many times with small gaps. The oscillations can be enhanced or weakened by different system parameters. The new design's membrane provides optimized properties to damp oscillation amplitude. Moreover

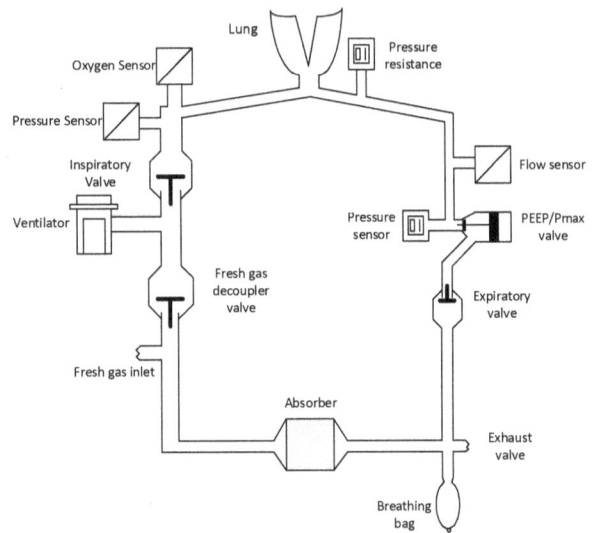

Figure 3: Breathing system circuit of common anesthesia machines [6]-[7].

the lid was new designed to optimize the volume control pressure circuit.

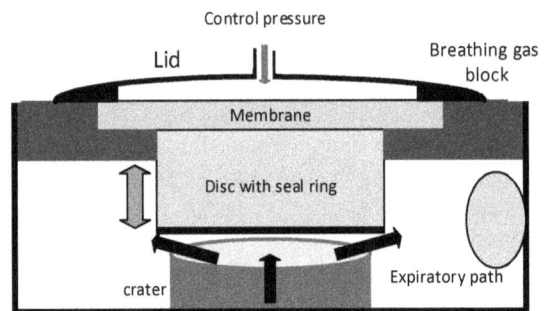

Figure 4: Cross section view of the new designed positive end-expiratory pressure valve.

2.2   Measurement setup for the oscillation analysis

To compare series valve with the new design, a set of measurements regarding the current oscillation behaviour was performed for a system similar to the one shown in Fig. 3. Instead of human subjects lung models consisting of a copper wool filled glas reservoir have been used, which show comparable values for compliance and resistance to human lungs. The formulas (1) and (2) tell us how to calculate these values [8]. Equation (1) shows how to calculate the compliance of a lung. V stands for the volumina of a lung, p_{pul} gives the intrapulmonal pressure and p_{pleu} the intrapleural pressure.

$$C_{lung} = \frac{\Delta V}{(p_{pul} - p_{pleu})} \quad (1)$$

Formula (2) is modified by the Hagen-Poiseuille-principle. Herein r stands for radius and L for the length of the tube[8].

$$C = Viskosity(\varphi)\frac{8 \times L}{r^4} \qquad (2)$$

We took two different lungs for oscillation analysis, a greater lung comparable to an adult lung and a smaller one representing a child lung. The compliance values are 50 mbar/ml and 11 mbar/ml. Their resistances are about 5 mbar and 50 mbar at 60 l/min. For detection of signals a dSPACE station connected with different pressure and flow sensors was used (Fig. 3). This hardware makes a regulation and detection of breathing cycles possible. In order to test compatibility of all possible tube constellations, 17 different tubes have been tested. For every combination of valve and tube a measurement of pressure values was done. This contains eight breathing cycles with two different pressure jumps. The first four cycles reach a PEEP of 20 mbar, the last four a PEEP of 15 mbar. The inspiratory pressure was 40 mbar in both cases. We have used 15 valves for oscillation analysis, five of them were series valves and seven were valves of new design type. Moreover we applied a microbic filter to check whether there is an influence on oscillation. In conclusion this leads to $(17 \cdot 8 \cdot 12 \cdot 2)$ breathing cycles. The analysis of the measured signals was performed in MatLab. Different algorithms were used to edit, sort and filter the data. The pressure signals have been split for each jump, so we got eight curves for every valve and tube combination. The airway pressure signals were arranged with a highpass filter and a butterworth filter. After that step a fast fourier transformation was performed to allow a spectral analysis. The highest amplitude of this frequency band was determined.

2.3 Expiration resistance

The higher the expiration resistance, the harder is the expiration for patients. Therefore it is necessary to be in specified international medical standards [9]. In order to meet the system requirements the new design must not generate a higher expiratory resistance compared to the series valve. The structure of this measurement is based on the first measurement. However there are some variations: The airway pressure is recorded close to the lung. The PEEP valve control pressure is set constant to 0 mbar. A flow of 60 $l \times min^{-1}$ air runs through the system. The expiration resistance is equivalent to the airway pressure. System specific parameters in combination with the high flow influence the pressure. We have used ten series valves and ten of the new design. In the following data was plotted to compare both valves.

3 Results and Discussion

In this chapter the results of both measurements will be presented. In the first part the oscillation analysis, after that the expiration resistance will be shown. At the end a summary

of the results and a conclusion for the future of this project will be presented.

3.1 Analysis of oscillation

At the beginning every single breathing cycle was analysed as illustrated in Fig. 5. The upper diagram shows three signals: The first signal stands for the measured, unfiltered airway pressure signal and starts as the highest signal in the beginning, the second signal represents airway pressure signal filtered with highpass and butterworth filter and starts at 0 mbar pressure and the last signal shows the control pressure. The lower diagram show the result of a fourier analysis with a time interval of 32 ms. For Fig. 5 and Fig. 6 a combination of reusable breathing tubes with microbic filter was used.

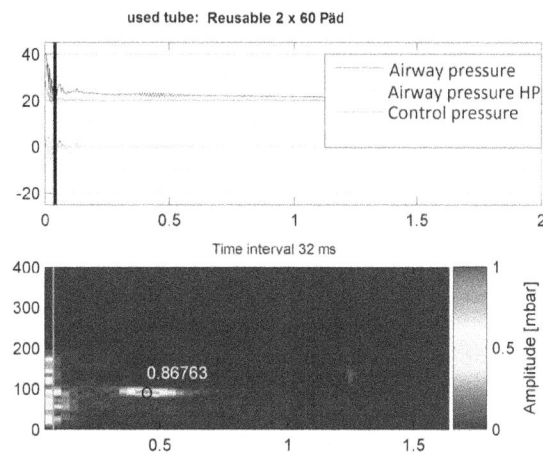

Figure 5: Analysis of one expiration cycle with a pressure jump from inspiration pressure 40 mbar, to a PEEP of 20 mbar for the new designed valve.

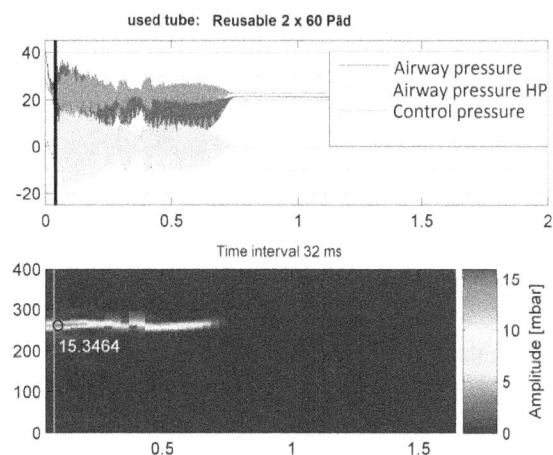

Figure 6: Analysis of one expiration cycle with a pressure jump from inspiration pressure 40 mbar, to a PEEP of 20 mbar for the series valve.

In comparison to Fig. 6, which shows the series valve under same conditions, the signal of optimized design contains no oscillations. Because one signal comparison has limited predictive value, Fig. 7 displays the arrangement of all possible combinations. The frequency on the x-axis is plotted against the amplitude on the y-axis. Series valve pressure oscillation maxima are plotted in black and new valves maxima are plotted in grey. The amplitudes of oscillations, which correlates with a louder noise, are generally higher. The frequencies compared to new PEEP valve design are higher, too. Because of the higher frequencies the noise is more annoying for the customer. In conclusion the general oscillation behaviour of the new designed valve is much better than the behaviour of the series valve.

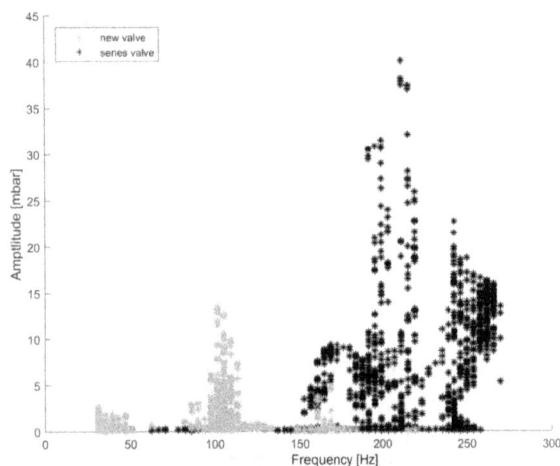

Figure 8: Analyses of one expiration cycle with a 20 mbar PEEP. Black bars are series valves, grey bars represent new valves.

Acknowledgement

The work has been carried out at the life cycle engineering of Drägerwerk AG & Co. KGaA, Lübeck.

5   References

[1] J. Rathgeber, *Grundlagen der maschinellen Beatmung: Einführung in die Beatmung für Ärzte und Pflegekräfte.* Thieme, 2010, pp. 103–166.

[2] M. Prella, F. Feihl and F. Domenighetti, *Effects of short-term pressure-controlled ventilation on gas exchange, airway pressures, and gas distribution in patients with acute lung injury/ARDS: comparison with volume-controlled ventilation.* Chest, 2002.

[3] R. Kramme and E. Siegel, *Medizintechnik.* Springer, Berlin/Heidelberg, 2002, pp. 427–442.

[4] S. Schröter, *Numerische Simulation der Strömung eines pneumatisch gesteuerten Beatmungsventils*, 2013, Universität Siegen.

[5] S. Schröter, *Experimentelle Schwingungsanalyse des Expirationsventils (PEEP-Ventil) des Anästhesiegeräts Fabius GS*, 2012, Universität Siegen.

[6] J.M. Yoder, *Understanding modern anesthesia systems.* Dräger Medical Inc., 2009.

[7] Dr. med. M. Heck and Dr. med. M. Fresenius, *Repetitorium Anästhesiologie.* Springer, Berlin/Heidelberg, 2007, pp. 109–110.

[8] Dr. med. M. Heck and Dr. med. M. Fresenius, *Klinikmanual Anästhesie.* Springer, Berlin/Heidelberg, 2008.

[9] DIN EN ISO 8835-2, *Systeme für die Inhalationsanästhesie - Teil 2.* 2009-08.

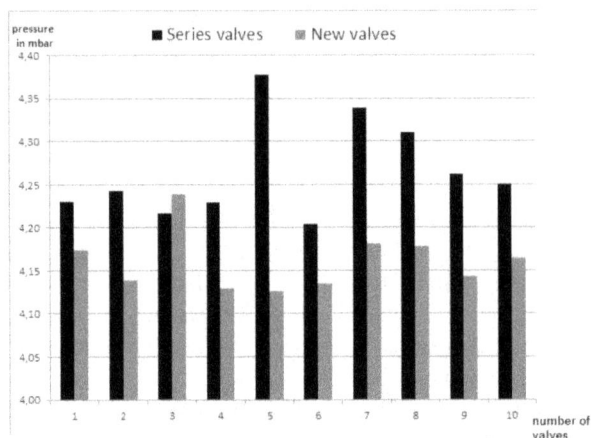

Figure 7: Analyses of frequencies for two different valve designs and two different PEEPs.

3.2   Expiration resistance

Fig. 7 shows the resistances of the breathing system with ten values of series valves and new design values. The average value of the new design is lower than the average of series valve values. This fact proves, that the new PEEP valve has better properties regarding expiration resistance and it fulfills the medical standard [9].

4   Conclusion

To approve the new model first steps were taken with the two measurements. By the performed measurements the first steps were taken to approve a new designed PEEP valve. It has better resistance properties. The oscillation analysis showed, that the frequency has been moved to lower values, which are more pleasant for the customers. Moreover the amplitudes have been reduced. All in all the new valve shows the potential to improve current anesthesia machines. For an authorization it is necessary to do more specific norm tests in the future, such as integration tests, reaction times and tests of lifetime.

5

Safety and Quality

Determination of action forces applied by users
– To make a medical device more reliable –

K. Steppke [1], A. Kunath [2], F. Grüner [2], J. Ingenerf [3] and H. Handels [3]

[1] Medizinische Ingenieurwissenschaft, Universität zu Lübeck, katja.steppke@student.uni-luebeck.de
[2] Research & Development, Drägerwerk AG & Co. KGaA, Lübeck {andre.kunath, fabian.gruener}@draeger.com
[3] Institute of Medical Informatics, Universität zu Lübeck, {ingenerf, handels}@imi.uni-luebeck.de

Abstract

The scope of this paper is to determine maximum forces applied by users of medical devices. Medical devices are often handled with different forces by the operator. For development and construction process there is a need of knowledge what dimension of human actions forces a medical device must withstand. The determined force values are used in static load tests in order to fulfil the given safety requirements. Several studies with similar measurement of isometric static force in different application fields were collected and evaluated. To verify the data, a study collecting pushing and pulling forces at 80 and 130 cm for medical devices was performed. Men achieved 310 ± 76 N, women 237 ± 48 N while pulling in standing position. The aim in the future is to provide a guideline regarding human action forces applying to medical devices for reliable product design.

1 Introduction

Human action forces are ubiquitous in the clinical daily work environment. Bed sides are pulled and pushed. Interactive control panels for patient data monitoring are installed on movable extension bars. A force can be applied to the pivotal point which must withstand the force. In order to cope with applied forces it is important to supply knowledge of human action forces to engineers, especially during the development of a product.

There are two point of views. Firstly the ergonomic perspective: a certain force value is not allowed to be exceeded in order to protect the physical well-being of the operator when the mechanical task has to be repeated several times. Usually there are specification how the maximum ergonomic force value has to be chosen to fulfill the ergonomic requirements. On the other hand the operator of the medical device is able to apply forces larger then the ergonomic force value (Fig. 1). This project's investigation aims for a determination of maximum force value a human can perform.

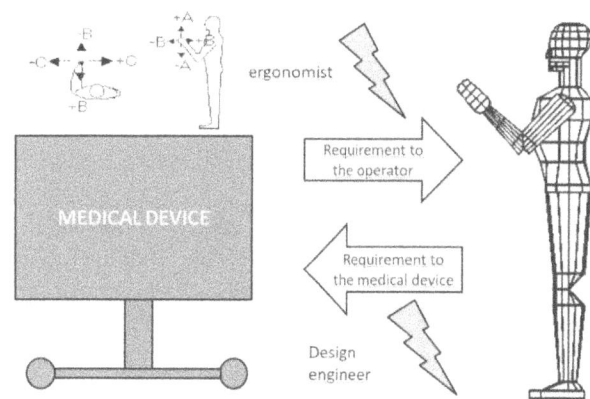

Figure 1: Two perspectives of human forces applied to medical devices: ergonomic view vs engineering view. The sketch of the human is taken from [3]. Standardised names of action force directions used in other studies and this project, referring to spatial dimensions [3], are indicated in upper left corner.

1.1 Context of Reliability Engineering

In order to increase quality and reliability it should be possible to specify more test cases by the means of knowledge of human forces provided in this project, which are not covered by the norm yet. Reliability is *"the probability that an item will perform a required function without failure under stated conditions for a stated period of time"* as defined in [1]. It shall be a measure to decrease the number of it-

eration in the product development process. Referring to the V-Modell [2] the relevance of this project is shown in Fig. 2. At the end of a product development process the Product Requirements (PR) are validated. Based on PR the Technical System Requirements (TSR) are deduced, which are later on verified before the validation. TSR cover e.g. the desired maximum Field Failure Rate (FFR) of medical devices. In the TSR additionally safety requirements are included, which are explained in details in [4]. The testcenter documents Certified-Body-Reports based on passed tests, which are important for the approval of medical devices in the international markets [5]. Safety factors as a part of safety reqirements in the TSR can be reduced by the knowledge of all external forces as mentioned e.g. in Table 21 of [4]. This project shall provide reliable data for mechanical safety tests.

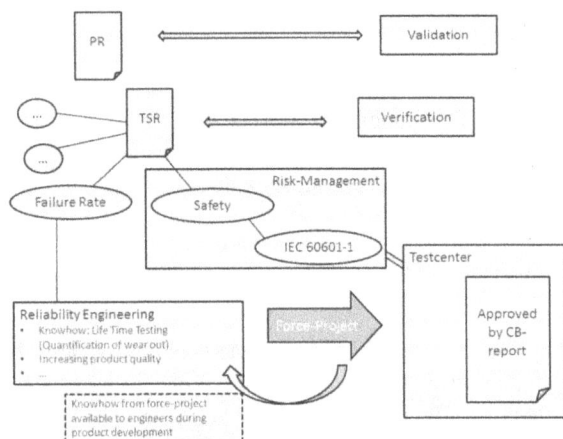

Figure 2: Referring to the V-Model [2] this project shows relevance in terms of fullfilling parts (approval for international market) of the TSR. By the knowledge of human action forces at an early stage of product development the number of iterations in the development can be reduced as well as the FFR.

1.2   Short overview of human action forces

There have been studies in the past, which investigated the magnitude of human action forces [6]-[9]. Most of the studies were brought up by ergonomic researchers demanding for more ergonomic functionality in various work environments. In order to derive the ergonomic force value they often measured in a first step the *maximum action force*. Secondly they reduced the force value by different factors such as repetition, age, gender to gain the *ergonomic force value*. For this project the first step is of relevance: the maximum action force. Human forces are divided threefold: to muscle force, mass force and the action force. The action force has an effect not inside but outside of the human body and is thus measurable. The action force can be exercised in various ways and [3] provides a systematisation. In order to provide a consistent wording the directions of forces are standardised as displayed in Fig. 1 and used in different studies as well as in this project.

The *isometric static action force* has been measured in different application fields and thus provides a reasonable measurement for this project. An isometric force measurement requires a constant muscle length during exercise. Due to the tension of the muscle the supply of blood is reduced leading to fatigue of the muscle. It has been proven that at a level of 15% of the maximum action force the exercise can be done without exhaustion [6], [10]. The measurement of isometric static action forces follows the Caldwell-Regimen, with building up the maximum force within 1 s

and maintaining either 3 or 4 s following the shape of a plateau [11].

2   Material and Methods

This project consists of two parts: the first is based on literature research (assembly [6], stewardess [7] and farm workers [8]) in order to set up a data base. They surveyed detailed isometric action forces in different positions and points of application forces (heights). They all used the Caldwell-Regimen. The second part is a survey what dimension of human action forces are applied to two typical heights of Dräger Medical Products. In the following section the evaluation of the data from literature research is explained. A short description of the experimental set up then follows.

2.1   Maximum Action Force data base

The results of several studies were collected in a database implemented in MS Excel®. This allows to filter force cases by following characterizations

- force application height [mm]

- force direction [A+, ... C-]

- force value [N]

Further categories are subject's posture, single-handed vs. both hands and gender. Depending on the source the isometric force values were given as a percentil (50., 95. or 99.) or as a mean value with standard deviation. Usually the size of the random sample is given. Additionally in some sources the minimum, maximum and the interquartils-range are given.

For initial investigation two force cases were chosen: The filter was set to a height of 1200-1500 mm (standing posture, case 1) and to 600-800 mm (sitting posture, case 2). The next step was sorting according to the force direction as in Fig. 1. In [4] the norm patient's weight ist defined as 135 kg according to the 99. percentile. Thus the recommendation is to take the force value from the 99. percentile as well. Reference [3] recommends on the other hand to orientate to the 95. percentile. Accoording to safety standards the highest available given force value should be taken and if not available or for any reason is not applicable, there must be a logical, argumentically safe downgrading to 95. or 50. percentile. The safety can be maintained by choosing the factor suitable to the case and normative standards.

For each force category (99., 95. and 50. percentile) the maximum and minimum was taken and the mean over the given values in a percentile group was calculated. For the given mean values the maximum and minimum were noticed and the mean over means was calculated by (1) given by [12] with $\bar{\bar{x}}$ the mean of means, n_i the sample size of sample i, $\overline{x_i}$ the mean value of the sample i and s_i the standard deviation of the mean of the sample i.

$$\overline{\overline{x}} = \frac{\sum_{i=1}^{k} n_i \overline{x_i}/s_i^2}{\sum_{i=1}^{k} n_i/s_i^2} \quad (1)$$

2.2 Experimental Setup

The two cases described above pulling and pushing in two different heights shall be validated in an experimental set up. A bidirectional load cell (FMI-B30K5) by Alluris GmbH was used for force measurement. It provides a measurement up to 5000 N with a precision of ± 1 N. Two handle bars (length 80 mm) were attached to the load cell which was fixated horizontally on a carrier. The carrier could be moved between two heights (80 and 130 cm). During the measurement the force data were transferred realtime via USB-cable to a laptop using the Excel® AddIn *FMI_Connect 3.02* provided by Alluris GmbH.

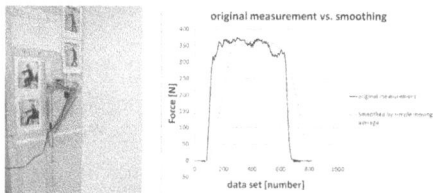

(a) Set up   (b) Original/smoothed force data

Figure 3: Experimental set up and a typical force-time-curve are shown here.

The participants (Dräger - employees) took part voluntary during an "Open Day" of the testcenter for mechanical safety. The 131 participants (30 W, 101 M) were advised to build up the maximum force within 1 s and maintain it for 4 s. The order of the four tasks was randomised to eliminate the influence of order. Furthermore information of body height, body weight, age, gender, dextrality and level of fitness were collected via questionaire. The maximum force value for each force case is obtained by taking the maximum of the simple moving average over 1,5 s - interval of the 5 s measurement duration, see (2), as in [6], [7].

$$m_{SMA}^{n} = \frac{1}{n} \sum_{i=0}^{n-1} x(t-i) \quad (2)$$

3 Results and Discussion

By the help of experimental set up the data base could be expanded by four force cases for male and female. In the data base 568 force cases were collected from four sources and are available for supplying multiplexed information to engineers. In the data base 52 force cases were found for case 1, with the additional restriction that the horizontal distance is no larger than 80 cm. For case 2 there are 21 force cases available. As an example Table 1 shows the average force [N] for pushing (B-) and pulling (B+) at a height of

120-150 cm according to the four mentioned force value types as mentioned in section 2.1. The two right columns show the values for pushing and pulling while standing of the male participants from the own study. They are slightly lower than those from the studies. The average height is 1,69±0,06 m (W), 1,82±0,06 m (M), the age 35,5±11,48 yrs (W), 42,25± 11,18 yrs (M) and weight 71,21±10,83 kg (W) and 85,92±14,12 kg (M).

Table 1: For the 4 categories of force values (s.a.) the maximum, minimum and average of resulting force values from literature research were calculated. This table gives an example of pushing (B-) and pulling (B+) force values at measured heights 1200-1500 mm (standing position). The two right columns show results from field study for male participants at height of 1300 mm.

Force Value type	B+	B-	B+	B-
99. Average [N]	619	723	580	671
95. Average [N]	520	730	461	519
50. Average [N]	352	489	302	341
Mean of Mean [N]	299	348	310	343

With a t-Test and a significance of 5% it could be proven, that the strength between male and female are different as mentioned in studies. All combination of the four cases were tested within one gender group and have to be assumed to be different to each other with a significance of 5% with the dependent t-Test, except pushing and pulling in sitting posture for the female. The frequency of force values for the four cases are shown in Fig. 4, nearly normal distribution can be assumed. The small images in Fig. 4(a) and 4(c) show posture of a participant. The averages and the standard deviation are given in Table 2.

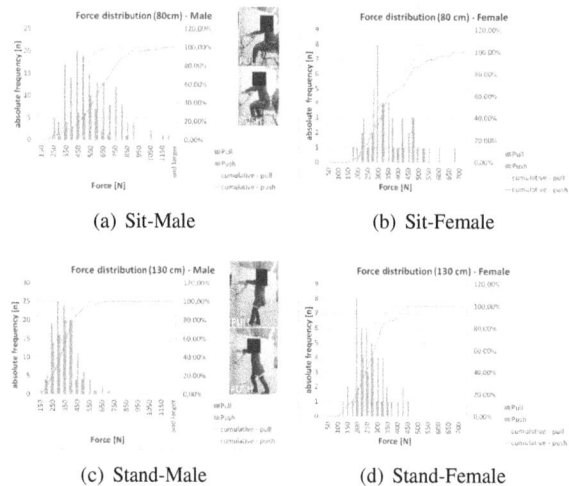

(a) Sit-Male   (b) Sit-Female

(c) Stand-Male   (d) Stand-Female

Figure 4: Distribution of forces for fe/male participants of the four cases.

The test condition were not as good as mentioned in literature due to operational reason: it was not possible to invite the participants exclusively for 20 minutes in order to supply the best reproducibility. For that a quiet environment, enough time to fully introduce the participants indiviudally,

Table 2: Averages and standard deviation for four test cases.

Action	Women	Men
Stand - Pull [N]	237±48	310±76
Stand - Push [N]	264±82	343±103
Sit - Pull [N]	334±85	437±98
Sit - Push [N]	355±122	610±188

no competition were required. Instead of that the surrounding was fluctuant noisy due to the set up of testcenter's "Open Day" and people coming and leaving. One aim of the study was to increase the developer's awareness towards human action forces. Thus our priority was set to reach more people instead of measuring a few very precisely and extensive. The available time window was seven hours thus there were just 3-4 minutes per participant. They were introduced to their task in groups by a standardised text. At their turn they received a short summary of the task to perform. Due to the queue of waiting people it was not possible to repeat the task in order to eliminate uncertainty or underestimation of their personal maximum force. Neither it was possible to allow much time of training or getting confident with the experimental set up. It was possible to observe previous participants. In comparison to the values in the literature our gained values are lower except in one case: pushing in sitting posture. In this case they are higher due to the extensive usage of the fixed back seat as resistance, which did not exist in the existing studies. An explanation for the lower values in the other cases is that the Dräger-sample consists of office workers, whereas the workers from the studies are employed in physical work environments. Irregardless the values of the 131 participants give important information.

4   Conclusion

There was a need for a data base supplying human action forces collecting information provided by different studies. The data base was set up successfully in Excel®, a future option may be to transfer it to an interactive web interface in order to provide the accessability to other engineers. In a laboratory study pulling and pushing values were maintained at 80 and 130 cm. Thus 2x4 cases could be added to the data base with typical height of handle bars of Draeger Medical products. It is important that there is not a linear relation between height and force thus each height has to be examined indivually. Here can be an option for further investigation and a transfer of the measurements to the population of clinical health workers.

Acknowledgement

The work has been carried out at the company Drägerwerk AG & Co. KGaA, Lübeck. I would like to thank my supervisors André Kunath and Fabian Grüner for their support, assistance and for giving the possibility to carry out this project.

5   References

[1] P. D. T. O'Connor and A. Kleyner. *Practical Reliability Engineering*. 5th Edition. John Wiley & Sons, Ltd., 2012.

[2] B. Bertsche, P. Göhner, U. Jensen, W. Schinköthe and J. Wunderlich, *Zuverlässigkeit mechatronischer Systeme – Grundlagen und Bewertung in frühen Entwicklungsphasen*. Springer Verlag, Berlin/Heidelberg 2009.

[3] DIN 33411-5:1999-11, *Physical strength of man - Part 5: Maximal isometric action forces, values*. Beuth-Verlag GmbH, 1999.

[4] DIN EN 60601-1 (VDE 0750-1):2013-12, Medical electrical equipment – Part 1: General requirements for basic safety and essential performance. Beuth-Verlag GmbH, 2013.

[5] IEC, *About the CB Scheme*, 2014. URL: http//www.iecee.org/cbscheme/pdf/cbfunct.pdf (visited 24.02.2015)

[6] J. Wakula, K. Berg, K. Schaub, R. Bruder, U. Glitsch and R. Ellegast, *Der montagespezifische Kraftatlas*. BGIA-Report 3/2009. Deutsche Gesetzliche Unfallversicherung (Hrsg.), 2009.

[7] U. Glitsch et al., *Untersuchung der Belastung von Flugbegleiterinnen und Flugbegleitern beim Schieben und Ziehen von Trolleys in Flugzeugen*. BIA-Report 5/2004. Berufsgenossenschaftliches Institut für Arbeitsschutz, 2004.

[8] A. Klußmann, P. Serafin, A. Schäfer, M. Keuchel, B. Neumann and K. Lang, *Betätigungskräfte an Landmaschinen. Analyse und Messung von Handbetätigungskräften und abgeleitete Empfehlungen*. Verein zur Förderung der Arbeitssicherheit in Europa e.V., 2013.

[9] R. Cutlip, H. Hsiao, R. Garcia, E. Becker and B. Mayeux, *Optimal hand locations for safe scaffold-ende-frame disassembly*. in: Applied Ergonomics, vol. 33, pp. 349-355, 2002.

[10] F. Marschall and M. Fröhlich, *Testing the correlation between maximal strength and repition maximum for deduced lower levels of intensity*. in: Deutsche Zeitschrift für Sportmedizin, vol. 50, nr. 10 pp. 311–315, 1999.

[11] L. S. Caldwell, D. B. Chaffin, F. N. Dukes-Dobos, L. L. Laubach, K. H. E. Kroemer, S. H. Snook and D. E. Wassermann, *A Proposed standard procedure for static muscle strength testing*. in: American Industrial Hygiene Association Journal, vol. 35, no. 4, pp. 201–206, 1974.

[12] L. Sachs, *Angewandte Statistik*. 10th Edition. Springer Verlag, Berlin/Heidelberg 2002.

Completion of a Test Bench to Verify the Peel Adhesion of Medical Devices: Software Development and Validation

A. Riebesel [1], M. Vienenkoetter [2], D. Dreffkorn [3], and P. Rostalski [4]

[1] Medizinische Ingenieurwissenschaft, Universität zu Lübeck, anke.riebesel@student.uni-luebeck.de
[2] Biomedizinische Technologie, Hochschule Hamm-Lippstadt, maxie.vienenkoetter@stud.hshl.de
[3] Drägerwerk AG & Co. KGaA, Lübeck, dirk.dreffkorn@draeger.com
[4] Institute for Electrical Engineering in Medicine, Universität zu Lübeck, philipp.rostalski@uni-luebeck.de

Abstract

Adhesive layers are important to guarantee secure placement of various medical devices like biosensors on the patient's skin. A proper device-patient connection has to be assured during treatment. Therefore information on peel adhesion regarding skin is required. To measure the peel adhesion on skin, Drägerwerk AG & Co. KGaA developed a test bench with different setups. In this task the requirements for the software were determined and the software was implemented as well as verified. Validation measurements using tape with and without carrier material for the different setups under various conditions were carried out. The measurement delivered promising results: Peel adhesion on steel at immediate pull off were five times higher than on the skin equivalent. The peel adhesion on skin increased with longer dwell times. Furthermore the peel adhesion observed on actual human skin differed less than $0{,}2\,\frac{\text{N}}{\text{cm}}$ from the peel adhesion observed on the skin equivalent.

1 Introduction

To guarantee a secure placement of various medical devices like biosensors on the patient's skin, adhesive layers are essential. An example is the Tcore™ (Drägerwerk AG & Co. KGaA). It non-invasively detrmines the patient's core temperature while stuck to his forehead [1]. The information on peel adhesion regarding skin is necessary to ensure the fixated device-patient connection.

Manufacturers are required to specify the peel adhesion of their products. The DIN (German Institute for Standardization) EN (European Committee for Standardization) 1939:2003-12 prescribes how to evaluate peel adhesions: An adhesive strip with a width of 24 mm and a minimum length of 125 mm has to be pulled off with $5\,\frac{\text{mm}}{\text{s}}$ from a standard steel panel either when peeled at $180°$ or $90°$. Neither a broadly accepted method to test peel adhesion by using a skin equivalent nor a method to transfer the values from steel to skin exist. The goal of this assignment is to develop a test apparatus to determine the peel adhesion of adhesives from human skin without requiring a clinical study involving test persons.

2 Material and Methods

To evaluate the peel adhesion on human skin the test bench has to be put into operation which requires software to expand the capabilities of the existing test bench. Therefore a software is required. The completed test bench has to be verified and validated.

2.1 Hardware

The test bench considered in this paper is presented in Fig. 1. Due to the modular scheme, the test bench enables

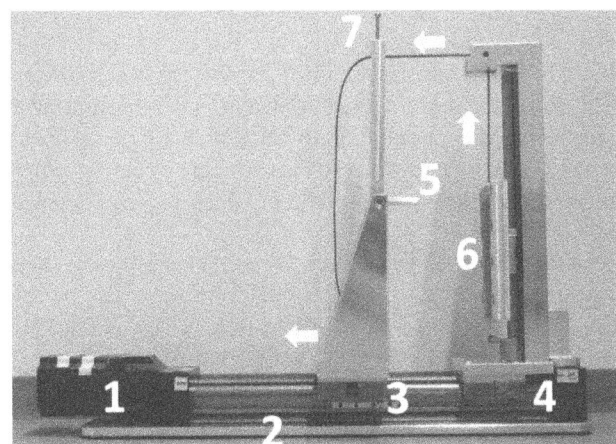

Figure 1: The test bench for the $90°$ peel test. 1) stepper motor 2) spindle lift table 3) with the stepping motor controlled slide 4) fixed slide 5) fastening for the sensor 6) skin equivalent on carrier plate 7) fastening mechanism to tension the cable, arrows) movement during pull off

two different test setups for both $180°$ and $90°$ pull off manoeuvres. Moreover, it supports two different test-surfaces:

steel and a skin equivalent. Both test surfaces can be fixed on a carrier plate. The skin equivalent is a polyurethane rubber simulating a forearm's surface. In Fig. 1 the 90°-setup with the skin equivalent (5) is presented. The test bench is placed on a spindle lift table (2). To pull the tape off, one part of the setup is placed on a slide that can be moved (3) by a stepper motor from Nanotec Electronic GmbH & Co. KG (1). Nanotec provides a control system for the stepper motor. By using the rpm (revolutions per minute) mode, the motor accelerates to the given maximal rpm [2]. The carrier plate is mounted to a fixture opposite the stepper motor (4). For the 90°-setup the carrier plate (6) for either the skin equivalent or steel is placed on a slide that has to move upwards during the pull off to maintain the 90° peel angle. The upward movement is ensured by a cable between the carrier plate and the slide controlled by the stepper motor. The cable can be fixated with a fastening mechanism (7). The upper end of the adhesive tape has to be clamped to a clamp which is attached to a force sensor from Burster Präzissionstechnik GmbH & Co. KG that measures the pull off forces. The force sensor itself is affixed to a holder (5). The sensor can be connected to a PC via its USB (Universal Serial Bus) -sensor-interface. The sensors accuracy is $\pm 0,3$ N [3].

2.2 Requirements for the Software

The DIN EN 1939:2003-12 regulates the determination of peel adhesion properties of self adhesive tapes on steel. Due to variations of the adhesive force for different pull off speeds, it dictates the pull off speed $5\,\frac{mm}{s} \pm 0,2\,\frac{mm}{s}$. The force values have to be sampled at a frequency of 5 Hz or more. From the pull off path of 125 mm, the force values from 50 mm to 75 mm should be averaged and divided by the width of the tape to get the peel adhesion in $\frac{N}{cm}$. However, the test bench's length is limited by the spindle lift table and the size of the adhesive medical devices. Therefore the peel adhesion is derived by averaging over the force values from 20 % to 60 % of the peel off path. Additionally, the norm prescribes a report including determined information, an example of which that can be seen in Table 1 (A6).

There are also requirements that are dictated by the user. Instructions to guide the user within the software, a clear user interface and a data output in Excel are important for the simple handling of the test bench. The test bench's software requirements resulting from the available hardware, the standard, and the user are composed in the product requirement specification in Table 1.

2.3 Software Realisation

To fulfil the requirements a user interface to control the program and enable communication with the sensor as well as the stepper motor of the test bench are needed. Such an interface can be implemented with LabVIEW

Table 1: Product requirement specification: Requirements and ID (Identification)

ID	Requirement
A1	Accuracy sensor: $\pm 0,3$ N
A2	Pull off speed: $5\,\frac{mm}{s} \pm 0,2\,\frac{mm}{s}$
A3	Sampling the peel adhesion at least every 0,2 s
A4	Averaging between 20 % and 60 % of the path
A5	Peel adhesion: in $\frac{N}{cm}$
A6	Report with tester, date, test method, tape, dwell time, anomalous behaviour, tape width, peel adhesion, temperature, pressure, humidity, angle
A7	Usable for the different angles and surfaces
A8	Instructions to guide the user
A9	Clear user Interface
A10	Possibility of data output in Excel

(Laboratory Virtual Instrument Engineering Workbench), a development environment from National Instruments that provides a visual programming language. A LabVIEW program, called VI (Virtual Instrument), consists of both a frontpanel and a backpanel. The front panel represents the user interface which consists of both control and display elements. These elements can be freely arranged and are mapped to the backpanel which reflects the program's block diagram. It is possible to integrate multiple Sub-VIs in an overlaying VI. Burster and Nanotec provide matching VIs to communicate with the sensor and the stepper motor. Those VIs use Sub-VIs provided by LabVIEW to interface via USB interfaces [4]. A Software that meets the requirements and uses the SubVIs provided needed to be developed.

A queued state machine was used to realize the peel adhesion measurement. The queue sequentially stores values until they are requested allowing data sharing between different loops within the program. There are two main loops: one comprising the state machine that organizes partial processes of the program in the different states and the other providing an event handler that adds elements to the queue due to user interaction like clicking a button at the frontpanel.

The front panel consists of different buttons in each state to simplify the handling. During first state "INIT", the COM (Communication) -Ports to which the motor and sensor are connected are searched. The user has to fill in information about the test scenario like the product under consideration, the angle and whether steel or the skin equivalent is used. The motor and sensor are then opened. After the user confirms the information with a click on a button at the frontpanel the program calls the "Positioning" that enables the user to drive the slide to the desired position to first adhere and then clamp the adhesive tape. The selected mode and its speed are sent to the motor automatically. After confirmation that the tape is clamped the state changes to "Measurement" and the slide (6) drives to the left side. If the measured force exceeds the threshold 0,1 N the signal is sampled with 10 Hz and displayed in an integrated graph according to its time stamp. Signal acquisition and the motor stop as soon as the average of the last three values is lower than the threshold. Then the adhesive force is evaluated between 20 % and 60 % of the actual path and normalized to $\frac{N}{cm}$. Finally the user can decide whether the program changes to the "Report" or "Excel" as well as the "End" state by clicking the corresponding button. The "Report" state creates and opens the corresponding report. In "Excel" state the user can open the data in Excel. By switching to the end state, the motor and sensor are closed and LabVIEW is terminated. For safety reasons, there is a button allowing the user to change to the "Stop Motor" state at any time.

2.4 Verification

The verification examines whether the product complies the product requirement specification (see Table 1). The sensor is calibrated every year and has an accuracy of $0,3\,N$ (A1) [3]. A light barrier is used to confirm the adhesive tape is pulled off at a rate of $5\,\frac{mm}{s} \pm 0,2\,\frac{mm}{s}$ (A2) using the LabVIEW program. By evaluating the data output of 20 measurements using Excel a sampling rate above $9\,Hz$ (A3) was confirmed. The Excel data was used to confirm the correct averaging of the force measurements over 20 % to 60 % of the pell-off path (A4) as well as their conversion to $\frac{N}{cm}$ (A5). A6 - A10 were confirmed by using the test bench. Thus, the test bench's software fulfils its product requirement specification and the verification was completed successfully.

2.5 Validation

Following the verification of the software the validation of the test bench was performed by various measurements. Any test bench must prove it can provide reproducible test results under identical test conditions. The reproducibility of the pull off force was tested with 20 industrial tapes that had nearly the same width and adhered length on the carrier plate. In order to check the test bench's behaviour, tapes having higher and lower peel adhesions were tested. Additionally a test series was performed using an industrial double adhesive MH-tape from ATP adhesive systems GmbH. One side of the tape is certified for the adhesion on human skin. The other side can be used to adhere an optional carrier material. In medical applications PE (Polyethylene) foam is often used as carrier material to shield the sensor from the environment. The tape was tested both with and without PE foam from PANA Foamtec GmbH as a carrier material considering both test angles and surfaces. The single measurements were repeated at least five times. The average peel adhesions are presented in the paper and are presented in charts created with Excel. The immediate pull off required by the DIN EN 1939:2003-12 was tested in the first test scenario. In the second test scenario a longer adhere period of one hour was tested to examine the consequences of a rising adhesion time. In clinical practice adhesive products are adhered much longer than a minute. In a final test

series the skin equivalent's surface was preheated to $32°\,C \pm 1°\,C$ as $32°\,C$ is the temperature of the skin's surface at room temperature where a naked person still feels comfortable [5]. To compare the results from the skin equivalent with the peel adhesion on actual human skin, the adhesive force at the arm of a staff member was measured under equivalent test conditions.

3 Results and Discussion

The results of the repeatability test were the peel adhesions of $3,66\,\frac{N}{cm} \pm 0,42\,\frac{N}{cm}$ for 90° and $3,32\,\frac{N}{cm} \pm 0,35\,\frac{N}{cm}$ for 180°. However, it must be noted that the tests performed were afflicted by some uncertainties. There are uncertainties due to the manual cutting, sticking on and clamping of the tapes as well as slightly differences in the environmental conditions.

The selected measurement threshold imposes the restriction that the peel adhesion has to exceed $0,1\,N$. This limitation is acceptable because a stripe with a peel adhesion as low as that may peel away easily and is not suitable as a medical device. The test bench was verified for a peel adhesion up to $70\,\frac{N}{cm}$ which exceeds the adhesive force of adhesive medical devices.

Measurements of the same tape under different application

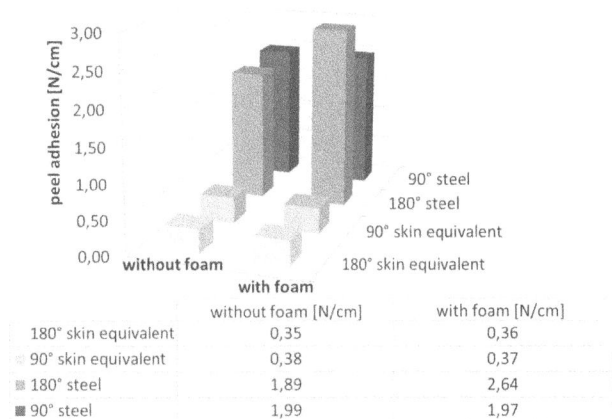

	without foam [N/cm]	with foam [N/cm]
180° skin equivalent	0,35	0,36
90° skin equivalent	0,38	0,37
180° steel	1,89	2,64
90° steel	1,99	1,97

Figure 2: Peel adhesion of MH-tape for immediate pull off

scenarios were performed and compared. The results of first test scenario with the immediate pull off imposed by the DIN EN 1939:2003-12 are presented in Fig. 2). The peel adhesion on steel is more than five times as high as on the skin equivalent. In contrast to the skin equivalent the surface of the steel is smoother and not deformable. Peel adhesions with and without foam differ by less than $0,03\,\frac{N}{cm}$ excluding the 180° test series using foam on steel: there the value with foam is $0,75\,\frac{N}{cm}$ higher than the measurement without foam. Furthermore, the measurement on steel with foam is the only measurement were the difference between 90° and 180° is higher than $0,1\,\frac{N}{cm}$.

For the next test scenarios only the results for the 90° test angle are regarded. The influence of a rising adhesion time can be seen in Fig. 3. The peel adhesion on the skin equiva-

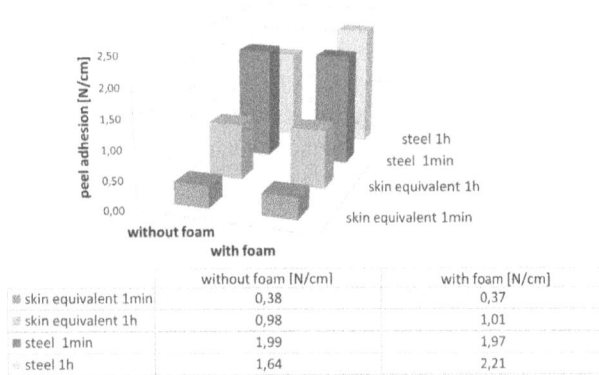

	without foam [N/cm]	with foam [N/cm]
skin equivalent 1min	0,38	0,37
skin equivalent 1h	0,98	1,01
steel 1min	1,99	1,97
steel 1h	1,64	2,21

Figure 3: Peel adhesion of MH-tape using the 90°-setup and different dwell times

lent has more than doubled due to the extended dwell time. One reason for the measured increase might lie within the evaporation of water from the tape [6]. An equilibrium between tape and skin equivalent based on a uniform distribution of water molecules within the contact area causes higher adhesive forces [6]. On steel the foam had a large influence on the results: the peel adhesion without carrier material decreased by $0{,}25\,\frac{N}{cm}$ while the peel adhesion with foam increased by $0{,}24\,\frac{N}{cm}$.

The peel adhesion on the skin equivalent for different temperatures and the peel adhesion on the test person's skin are depicted in Fig 4. It should be considered that the peel adhe-

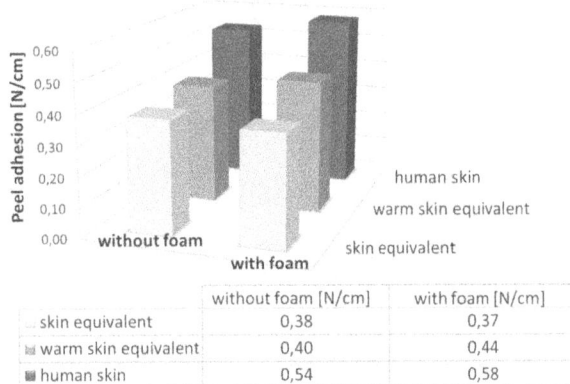

	without foam [N/cm]	with foam [N/cm]
skin equivalent	0,38	0,37
warm skin equivalent	0,40	0,44
human skin	0,54	0,58

Figure 4: Peel adhesion of MH-tape with 90°-setup for immediate pull off

sion for human skin was only tested on one person's forearm. Differences in the mechanical properties may occur according to gender, age or tested body region [7]. The peel adhesion for the heated skin equivalent is slightly higher compared to the skin equivalent at room temperature. Depending on the use of carrier material it increased by about $0{,}07\,\frac{N}{cm}$ in cases with foam and by $0{,}03\,\frac{N}{cm}$ otherwise. The values from the measurement on human skin are $0{,}14\,\frac{N}{cm}$ (no carrier material) and $0{,}13\,\frac{N}{cm}$ (foam) higher than the results obtained for the preheated skin equivalent. As the skin equivalent provides a better approximation of real skin than steel regarding the peel adhesion, especially a heated skin model, the proposed test bench is suited to classify adhe-

sive medical devices.

4   Conclusion

A new test bench to measure the peel adhesion on a skin equivalent to guarantee the safety of the fixed patient-device connection for adhesive medial devices like biosensors was put into operation. The test bench determines the peel adhesion on human skin by using a skin equivalent which is a promising result. If further investigations confirm the presented results, the peel adhesion on skin can be easily determined in a laboratory. Future studies should investigate whether similar results can be obtained for further adhesive tapes. Furthermore, the peel adhesions on different test persons should be measured.

Acknowledgement

The work has been carried out at Drägerwerk AG & Co. KGaA, Moislinger Allee 53-55, 23558 Lübeck. We would like to thank the department Accessories and Consumables for financing and supporting the project.

5   References

[1] Drägerwerk AG & Co. KGaA, $Tcore^{TM}$ *Temperaturmonitoring-system sicher, präzise, nicht invasiv.* Drägerwerk AG & Co. KGaA, Lübeck, 2015.

[2] Nanotec Electronic GmbH & Co. KG, *NanoPro Steuerungssoftware für Schrittmotorsteuerungen und Plug & Drive Motoren.* Available: http://de.nanotec.com/download/handbuecher/ [last accessed on 21.01.2016].

[3] burster präzisionsmesstechnik gmbh & co kg, *Precision load cell.* Available: http://www.burster/bmt-20.de/de/produkte/p/detail/8431-8432/ [last accessed on 18.12.2015].

[4] W. Georgi, E. Meti, *Einführung in LabVIEW.* Carl Hanser, München, 2011.

[5] G. Neuroth, *Hauttemperatur im Dienste der Wärmeregulation* in Pflügers Archiv European Journal of Physiology, vol. 250, no. 3, Springer, Berlin/Heidelberg/New York, p. 396, 1948.

[6] A. Nussinovitch, A. Gal, C. Padula, P. Santi, *Physical characterization of a new skin bioadhesive film* in AAPS PharmSciTech, vol. 9, no. 2, Springer, Berlin/Heidelberg/New York, pp. 458-463, 2008.

[7] J. Renvoise, D. Burlot, G. Marin, C. Derail, *Adherence performances of pressure sensitive adhesives on a model viscoelastic synthetic film: A tool for the understanding of adhesion on the human skin* in International journal of pharmaceutics, vol. 368, Elsevier, pp. 83-88, 2008.

Qualification of a new Cutplotter for ULTRAPRO COMFORT PLUG™

A. Boy [1], T. Licht [2], J. Haidar [2], and J. Völkel [2]

[1] Medizinische Ingenieurwissenschaft, Universität zu Lübeck, alexander.boy@student.uni-luebeck.de
[2] Johnson & Johnson Medical GmbH, Norderstedt, {TLicht, JHaidar, JVoelkel}@ITS.JNJ.com

Abstract

ULTRAPRO COMFORT PLUG™(UCP) is a partially absorbable device for open inguinal hernia repair [1] marketed by Ethicon, which among others will be produced with a cutplotter. From a lean manufacturing point of view there is a discussion about a transfer of a manufacturing process of the UCP, the so called cutplotting, to a secondary cutter. To explore the suitability of a secondary cutter and to check if a new cutplotter is necessary different kinds of analysis have been performed, for example capacity analysis or flatness measurements. The first step was to build an overview about the machine workloads based on sales figures provided by the marketing department. After that some parameters of the cutplotting process were adjusted (e.g. cutting depth, vacuum). Based on this data the development team may take a decision to the following validation strategy.

1 Introduction

A hernia can appear at every part of the abdominal wall. A bulge occurs, because of tissue that leaks through an open section of the muscles. Often a hernia grows in navel (umbilical), in groin (inguinal) or in places after surgical cutting [2]. To repair such lesions a lot of devices were developed and one of these is the ULTRAPRO COMFORT PLUG™ (Fig. 1) by Ethicon for open inguinal hernia healing, which is available in two different sizes (40 mm and 55 mm). It consists of an onlay, a partially absorbable macroporous mesh and a rib consisting of Polydioxanon (PDS) [1]. An initial step of the production is cutting out the components with correct dimensions using the cutplotter equipment. Afterwards mesh and rib will be connected by welding. Other processes follow till the product is ready for packaging and sterilization.

Befor UCP was launched all manufacturing processes have been validated. The Food and Drug Administration (FDA) defines the validation „as the collection and evaluation of data, from the process design stage through commercial production, which establishes scientific evidence that a process is capable of consistently delivering quality product" [3]. This validation process is divided into three main parts: installation qualification (IQ), operational qualification (OQ) and performance qualification (PQ). The IQ proves evidence, that all components of the product (equipment and ancillary system from the development process) agree on the manufacturer's specification. Up next is the OQ, which shows, that the product produced under worst case conditions is as good as a product under normal conditions. Therefore different parameters can be set up, e.g. temperature and pressure. The end of the validation process is the PQ after successfully finishing IQ and OQ. This step demonstrates, that the product agrees to all predetermined requirements [4].

This project deals with the problem that the cutting process should be progressed closer to lean manufacturing. Because of that it has to be clarified whether a transfer from cutter A to B (next to the ensuing steps) is possible and which qualifications - and validation steps are necessary. In the following the questions "Is there enough workload on cutter B?", "Is the flatness of the cutting ground sufficient?" and "Which impact have different cutting parameters to product properties?" will be answered.

Figure 1: The ULTRAPRO COMFORT PLUG™consisting of onlay, mesh and rib.

2 Material and Methods

In the following section the methods to analyse the cutting parameters are described. At first a capacity analysis was made to achieve an overview of the cutters workload. Due to a new cutting ground of cutter B a flatness measurement was performed afterwards. Then different parameters were used to cut the PDS rib. The results can be compared regarding form, size and detachment.

2.1 Capacity Analysis

The capacity analysis based on planned sales from marketing for the next years serves to acquire the workload of production machines. Those sales were given in ea per year [ea/year]. Based on one sheet a certain amount of ea (implants) will be produced depending on their size. In addition to the sales the time [h] to produce 100 ea was given. This value was determined by previous experiments and faciliates the calculation. Finally the rolled throughput yield (RTY) [%] is necessary. It is based on the yield value, which describes the outcome from a single step of the process. However the RTY is the product of all yields throughout all processes. With the formula

$$workload = \frac{\frac{ea}{year} \cdot \frac{h}{100ea}}{RTY} \qquad (1)$$

the workload [hours/year] of the production machines can be charged for the next years. The working days per year are also given. It is important to calculate each step separately, because of three plug components (mesh, rib, onlay). Therefore three different times per 100 ea and RTYs are given.

2.2 Measurement of Flatness

The flatness was measured because of a new metal plate under the cutting ground. Fig. 2 shows the old and new design.

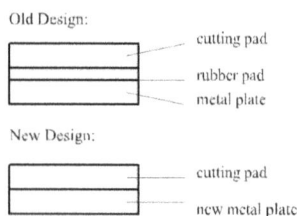

Figure 2: Old and new design of the cutting ground of the cutplotter.

For flatness measuring a dial gauge (Mitutoyo Absolute Meter) was mounted on a tripod by using an adapter on the cutplotter. After that a 10 μm monocryl foil was placed on the cutting pad to get a high vacuum. A ruler showed measurement points that have been defined in advance. To prevent that the dial gauge glides into a hole inside the cutting pad, a regular piece of metal was placed under the gauge. The setup of the measurement can be seen in Fig. 3.

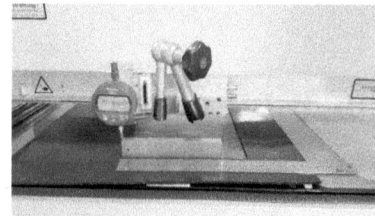

Figure 3: Experimental design for flatness measurements.

Now the measurement was taken at the points defined before. For that the dial gauge was moved by the control panel of the cutter. According to earlier validations the total variance of flatness should not exceed $\pm$ 0.2 mm, laid down in earlier validations.

2.3 Analysis of the PDS-Rib manufacturing process

To compare the rib quality of both cutplotters with regard to different characteristics (e.g. cutting edge, foreign matter, size, holes) some experiments were realized. With the help of different input variables, e.g. cutting depth, oscillation and vacuum, the cutting results can be compared in some respects. In addition, the removeability of all 24 ea per sheet was valued, if necessary, to determine more suitable setup parameters.

At first a Polydioxanon foil was laid on the cutting ground. PDS is a polymer and was described at first in 1980. In ground state it is colourless, but PDS can be enriched with a violet dye for better visibility and is applied in medical suture material and hernia supply because of its special tissue compatibility and flexibility. Additionally PDS can be easily reduced by the human body [5]. After foil placing the parameters of cutting were adjusted with the help of the control panel, that can be seen in Fig. 4. Therefore a zero position was defined visually (LC A: 2994 mm; LC B: 2860 mm). Then experiments were carried out with different cutting depths.

Figure 4: Control panel of the cutplotter.

Afterwards each of the 24 ribs was numbered to evaluate them in terms of extraction (1: very good, 2: easy, 3: complex, no extraction possible), cutting edge (1: very good, 2: good, 3: insufficient), foreign matter (yes/no), size (Ok/NOk) and holes (yes/no). A drawing of the previous described rib can be seen in Fig. 5.

Then a visual inspection was made using pre-defined positions under a microscope. Differences in quality may oc-

Figure 5: Drawing of a PDS-Rib.

cur, because of different oscillations of the cutplotters (A: 12000 oscillations/minute, B: 18000 oscillations/minute).

3   Results and Discussion

First a capacity analysis was carried out to calculate the workload of the cutplotter. Subsequently measurements of cutter B's cutting ground were started. Based on this data the depth of blade and vacuum were varied to find the best settings for PDS-rib cutting.

3.1   Capacity Analysis

For cutplotter B a maximal available annual capacity of approximately 5000 hours/year was determined. This includes 15 shifts per week. In addition there is a buffer (20 shifts/week), which can be included. In case of need it enables additionally production time of 1500 hours/year. In the diagram below the actually workload with all products (dotted line), the possible capacity and the additional buffer are shown. A transfer of this PDS-rib manufacturing process seems possible because of enough free capacity of cutter B.

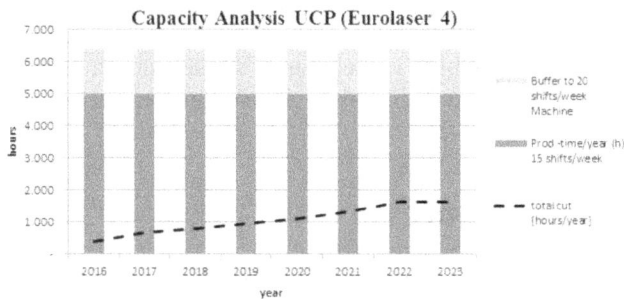

Figure 6: Workload of cutter B.

3.2   Measurement of Flatness

The total variance of flatness should not exceed $\pm$ 0.2 mm. The measurement showed that this condition can be met. The cutting ground has its maximum at 0.077 mm in the upper left corner and the minimum at -0.161 mm on the right side, which is the maximum deviation to zero position.

At other points on the cutting ground there are only little fluctuations. To obtain a better overview the values were presented in a 3D-plot in Fig. 7. In the lower left corner there is the zero position at point (0.0).

In addition it becomes apparent that the new cutting ground (maximum difference: 0.238 mm) is more uniform than the old one (maximum difference: 0.25 mm). Consequently

the new plate is more suitable for the cutting process. Summarising from flatness perspective there seems to be no possibility why the transfer of the cutplotting process could fail.

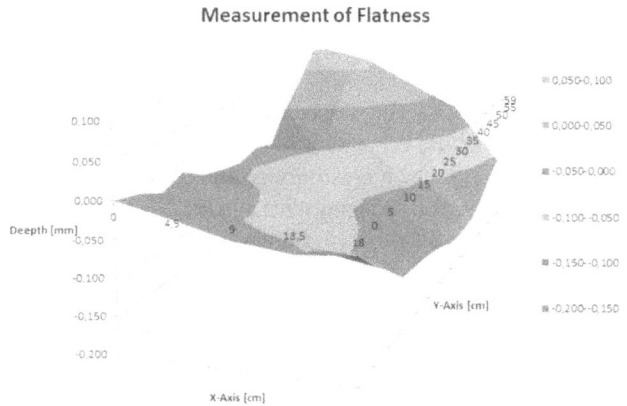

Figure 7: 3D-plot of cutting ground's flatness.

3.3   Analysis of PDS-Rib (Variation of cutting depth and vacuum)

Eleven cutting runs were carried out, in which either the cutting depth or vacuum has been varied. The speed of cutting was kept constant (150 mm/s).

The runs 1-5 were performed at the new cutter B with a vacuum of approximately 55 mbar. The first cut with a depth of 2860+90 (zero position +0.9 mm) resulted in very good ribs with very good removal (each of the 24 ribs were rated with "1"). That means the PDS foil can be lifted without falling down the ribs. Their subsequent removal requires no energy at all.

After that the trial with a depth of 2860+15 (run 2) according procedure followed. With this parameters the foil wasn't cut, because the blade didn't touch the cutting ground. Consequently an evaluation was not possible.

The area around +90 was scanned in 0.15 mm stages. The desired settings could not always be complied because of touch panel sensitivity, but the actual values slightly differed from required values ($\pm$0.01 mm). Run 3 with addition of 0.61 mm to cutting depth did not provide good results, which were unsatisfactory (rated with "3").

The next cutting step followed with +74 to zero position (run 4). Using this settings good results have been obtained at all 24 ribs. They could be removed with relative ease and their material was clean and intact. Because of that these parameters represent a good alternative for future production.

The fifth run was cut with +1.34 mm. The ribs peeled off from the foil when lifted. This is not designed for future use and therefore it presents no good results.

In order to compare both, run six and seven were carried out at cutter A with the settings (2994+15 and 2994+90). It was necessary to adjust the zero position, because the metal plate under the cutting ground was different to cutter B, but the vacuum was nearly the same. The results of both cuttings were similar to runs 1 and 2. The addition of 0.15 mm (run 6) yielded nothing and the results of run 7 (+0.9 mm) were also very good. As a consequence an optimal adjustment for cutting depth was found.

To find possible cutting limits +1.05 mm was tested in trial run 8 at cutter B. All ribs showed a very good removing. As a consequence this value was set as the high value of cutting interval.

Summarising by varying the cutting depth, an ideal setting was found. The old value with addition of 0.15 mm did not yield to foil cutting, it was only scratched. On the other hand the setup 2860+90 represents a better alternative. The ribs, adhering to PDS sheet, did not detach on the cutting ground while lifting the foil. Nevertheless they could be easily peeled up. Further an interval with its maximum = 2860+105 and minimum = 2860+75 for a good production outcome around the optimum exists. Otherwise either the edge of the PDS rib becomes wavy, because of pushing the foil ahead of itself by too deep blade, or the ribs won't be cut or the removal is too work-intensive. By extracting the ribs can be stretched or teared up.

Trials 9 and 10 were carried out with constant cutting depth (2860+90) and variable vacuum. At first the upper limit was tested at approximately 80 mbar, which produced very good results. All checking parts were marked with "1".

The last tenth cut was started with 12 mbar (run 10 a), but it was aborted after the fifth rib because of destruction of the foil. This test illustrated, that 12 mbar are not enough to fix the PDS on cutting ground. It wasn't sucked in by vacuum. As a consequence the ribs 1 till 5 were cut apart.

In the following the PDS was cut with 20 mbar (run 10 b), resulting in very good outcomes (all parameters have been rated with "1"). Because of these results a vacuum interval with maximum = 80 mbar and minimum = 20 mbar could be set for the future.

These findings are shown in Table 1. All ribs neither did show any holes nor foreign matter. Additionally their sizes were within the bounds and the cutting edges were also very good. Consequently this paramters are not shown in the table.

Table 1: Overview of the cutting results

Run	Lasercutter	Extraction	Depth	Vacuum
1	4	1	+90	55 mbar
2	4	no extraction	+15	55 mbar
3	4	3	+61	55 mbar
4	4	2	+74	54 mbar
5	4	1	+134	54 mbar
6	2	no extraction	+15	50 mbar
7	2	1	+90	50 mbar
8	4	1	+105	56 mbar
9	4	1	+90	80 mbar
10 a	4	foil destruction	+90	12 mbar
10 b	4	1	+90	20 mbar

3.4   Comparison between oscillations

To reveal possible non-detectable faults one rib per cutter was considered under the microscope. For this the two PDS-ribs were compared at prominent positions. It seems obvious, that higher oscillations lead to finer cutting edges, but no major differences in quality could be found. Both showed neither any faults nor bad rib edges. The consequence is that the oscillation frequency of cutter B is also suitable for rib cutting. An example of one comparison is demonstrated in Fig. 8.

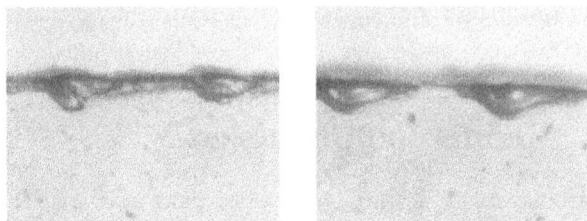

Figure 8: Example of comparison of two ribs (The dark area is the cutting edge). Left: Rib from cutter A, Right: Rib from cutter B.

4   Conclusion

Summing up, the results do not argue against a transfer of the rib manufacturing process to cutter B. The flatness as well as the cutting results show, that the new cutplotter seems well suited. Based on the determined data the project team has to plan the further action and develop a validation strategy. Maybe additional experiments are necessary before making a final decision for the next steps.

Acknowledgement

The work has been carried out in the department of Product Management at Johnson & Johnson Medical GmbH, Norderstedt.

5   References

[1] D. Krause, *510(k) SUMMARY*. ETHICON. Harlow, Sommerville, 2014.

[2] K. Heneghan, M. Malangioni and R. Fitzgibbons, *Groin Hernia: Inguinal and Femoral Repair*. American College of Surgeons, Chicago, 2009.

[3] U.S. Department of Health and Human Services, *Guidance for Industry*. Food and Drug Administration, Rockville, 2011.

[4] Der Arbeitskreis Dentalinstrumente (AKDI), *Validierung von Dampfsterilisationsverfahren*. Statement zur Validierung von Dampfsterilisationsverfahren, Berlin, 2010.

[5] K. E. Rehm, K. H. Schultheis, *Bandersatz mit Polydioxanon (PDS ®)*. Unfallchirurgie 11, 264-273 (Nr. 5), Unfallchirurgische Klinik und Poliklinik am Zentrum für Chirurgie der Justus-Liebig-Universität Gießen, Gießen, 1985.

Creation of a new test method and a validation strategy for the visual inspection of LDPE-bags

Q. Kelmendi [1], C. Maser [2], and S. Schoenfisch [2]

[1] Medizinische Ingenieurwissenschaft, Universität zu Lübeck, qendresakelmendi@student.uni-luebeck.de
[2] Johnson & Johnson Medical GmbH, Norderstedt, {CMaser,SSchoenf}@its.jnj.com

Abstract

A new test method for the identification of defect LDPE-bags (Low Density Polyethylene-bags) will be introduced for the raw mesh production as well as in the incoming goods inspection of the bags. With this method it is possible to see polluted or damaged bags with a visual inspection, so that the employee can sort out the faulty bags during the production process rather than checking each bag before production. The new method is time-saving for the employee and therefore more cost-efficient for the company. A validation of the new test method will confirm its effectiveness. The test method validation establishes a high degree of assurance that the test method is appropriate to the sample being tested and will consistently yield results that accurately reflect the quality characteristics of the product tested.

1 Introduction

LDPE-bags play an important role in the production of surgical meshes. At first single fibers are reeled of the coil and braided into one another with the same thread tension [1][2]. During this process it is possible for the coils to get tangled up with each other. To prevent this, LDPE-bags are used. A major requirement for the use of LDPE-bags during production is a flawless bag, therefore the bags have to be tested with a certain test method. At first a new test method has to be developed in such a way that it fits into the production steps. In order to standardize the test method for the production process it has to be validated accordingly. Once the test method has proven its reliability, the next step is to explore an automated process.

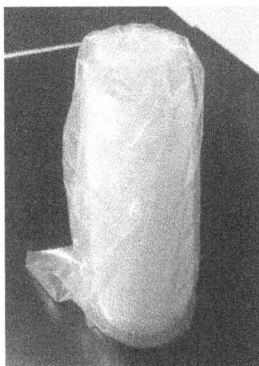

Figure 1: Coil with LDPE-bag

2 Material and Methods

In this section the materials used in the production and the developed test method are presented. At first the production of meshes and the application of LDPE-bags is illustrated in detail. Afterwards the newly developed test method is described.

2.1 Production Process of Meshes

Currently surgical meshes are commonly used as a replacement of a functional abdominal wall. In order to meet the requirements for this application, the meshes have to fullfill certain properties like stability and flexibility, which are gained through the production process [2]. To achieve these properties, single threads from coil with monofilament and multifilament are combined with the same thread tension on sectional beams [4]. During this process this sectional warping, the single fibers are combined to an extensive mesh [5]. Before the sectional warping the mesh coils are placed in the sectional warping creels, as seen in Fig. 2 [4]. Every single coil is wrapped into a LDPE-bag to prevent the filaments from getting tangled up with each other. Afterwards the fibers pass through the warping head[5]. This process is repeated until the suture coils become meshes. Afterwards the meshes are processed further and need to run through more production steps until they are ready to be used in surgical procedures.

2.2 Test method

For the development of the test method it is important to get an overview over the test criteria as well as the corresponding test method. This is called a raw material

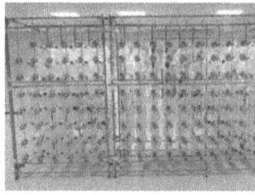

Figure 2: Sectional warping creel with the coils

specification. This specification shows which test methods are already existing and applicable. In addition to this it contains all the information that is needed for a detailed description of the raw material.

In the following the development of a raw material specification is presented. The polyethylene bags are checked with the following test instructions:

1. Fourier-Transform-Infrared-spectrometry

2. Dimensions

3. Item number

4. Appearance

A distinct identification of the item number is done by a visual inspection. The number on the box and on the delivery note are compared to each other. However, no explicit description of how the bags have to be checked for dirt and damages during the visual inspection is provided. In order to provide clear instructions for the checking, an explicit test method needs to be developed.

2.3 Validation strategy

The validation of the test method ensures that all the necessary requirements, including possible corrective actions, are met in order to introduce the method as a standard procedure. The validation strategy has been based on a non-destructive attributive examination – a visual quality test that divides the untis in good and bad bags[4]. This is called a qualitative test method and the sample is not destroyed during the inspection. At first the testers are divided into four groups – a master tester and three additional testers. The testers are employees of the raw mesh manufacturing who will be using this method during production. The master tester should have extensive knowledge of the classification of the concerned failures. In this case the head of the raw mesh manufacturing is chosen. For the test 20 LDPE-bags have been used – 6 bad and 14 good samples. The minor number of 20 bags is sufficient because the bags are not destroyed by the testers during the validation process and can be reused. Therefore it is also ensured that each tester has the same precondition. Afterwards each sample is rated by the criteria „ok" or „not ok". This evaluation is done separately from all testers in two cycles.

2.4 Classification of the samples

A requirement for the validation strategy is the classification of the samples into defect classes, shown in Table 1. These are defined by Johnson & Johnson for validation process [6]. A non-compliant sample is set as Ic - it is a critical defect that can have significant impact on the health and can worsen the condition of the patient [4]. To reach a universal validity of the test method the highest defect class is taken as a basis – meaning every failure is assumed as a failure of class Ic. The tester only decides if it is a compliant or a non-compliant sample. Since all occurring failures are assumed to be of class Ic, all of the non-compliant samples have to be identified. Therefore all acceptance criteria have to reach 100%. The acceptance criteria include agreement for all the evaluations among the tester as well as compliance with the specification of the good and bad bags, from now on referred to as Standard. In Addition the κ-values have to be met by all testers. The κ-value is a measurement of the compliance by different or the same testers on certain samples. Since the evaluations of the testers have to agree among each other the κ-value has to be 1, meaning 100% conformity, for the test method to be reliable.

Table 1: Defect classes

Class	Description
Critical (Ic)	A non-conformance to specifications, label claims, regulatory, customer or other requirements that could cause serious injury or illness to the customer.
CLASS I	A non-conformance to regulatory requirements or a non-conformance to specifications, customer or other requirements that will not cause serious injury or illness to the customer but could render the product unusable or difficult to use.
CLASS II	A non-conformance to specifications that will not cause injury or illness to the customer and is unlikely to render the product unusable. It will likely be noticed by the customer and may cause slight customer annoyance.
CLASS III	A non-conformance to specifications that will not cause injury or illness to the customer and is unlikely to render the product unusable. It will generally be noticed by the customer and affect perception of quality.

3 Results and Discussion

The results of the test method validation are shown in Table 2. In the table both validation cycles of the 20 bags by each tester are shown. The last column is indicating the predetermined condition of the bag. It can be seen that all testers identified the bags in agreement among each other and also in compliance to the Standard. Additionally the κ-values have been calculated for conformity among the testers and with the Standard. For the calculation some additional values have to be determined. In Table 3 the number of identi-

Table 2: Results of the test method validation

Sample Nr.	Tester1 Exp. 1	Tester1 Exp. 2	Tester2 Exp. 1	Tester2 Exp. 2	Tester3 Exp. 1	Tester3 Exp. 2	Standard
1	OK	OK	OK	OK	OK	OK	OK
2	OK	OK	OK	OK	OK	OK	OK
3	Not OK	Not OK	Not OK	Not OK	Not OK	Not OK	Not OK
4	OK	OK	OK	OK	OK	OK	OK
5	OK	OK	OK	OK	OK	OK	OK
6	Not OK	Not OK	Not OK	Not OK	Not OK	Not OK	Not OK
7	Not OK	Not OK	Not OK	Not OK	Not OK	Not OK	Not OK
8	OK	OK	OK	OK	OK	OK	OK
9	Not OK	Not OK	Not OK	Not OK	Not OK	Not OK	Not OK
10	OK	OK	OK	OK	OK	OK	OK
11	OK	OK	OK	OK	OK	OK	OK
12	OK	OK	OK	OK	OK	OK	OK
13	OK	OK	OK	OK	OK	OK	OK
14	Not OK	Not OK	Not OK	Not OK	Not OK	Not OK	Not OK
15	OK	OK	OK	OK	OK	OK	OK
16	OK	OK	OK	OK	OK	OK	OK
17	OK	OK	OK	OK	OK	OK	OK
18	OK	OK	OK	OK	OK	OK	OK
19	OK	OK	OK	OK	OK	OK	OK
20	Not OK	Not OK	Not OK	Not OK	Not OK	Not OK	Not OK

Table 3: Number of identification per bag

Sample Nr.	Good	Failure Class III	Failure Class II	Failure Class I	Failure Class I c	Square sum of the row
1	6				0	36
2	6				0	36
3	0				6	36
4	6				0	36
5	6				0	36
6	0				6	36
7	0				6	36
8	6				0	36
9	0				6	36
10	6				0	36
11	6				0	36
12	6				0	36
13	6				0	36
14	0				6	36
15	6				0	36
16	6				0	36
17	6				0	36
18	6				0	36
19	6				0	36
20	0				6	36
Sum of Column	Sum 1: 84	Sum 2: NA	Sum 3: NA	Sum 4: NA	Sum 5: 36	Sum 6: 720

fication (good or failure class Ic) per bag is shown. Taking the first row as an example, the 6 good identified bags result from the three testers on both validation cycles for the bag. Furthermore the number per row have to be added and squared.

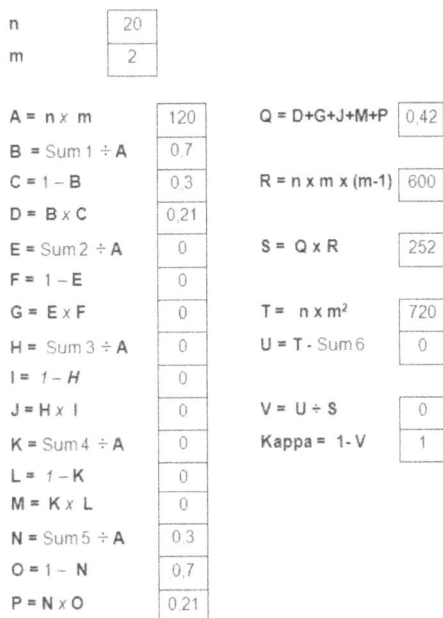

n	20
m	2

A = n x m	120	Q = D+G+J+M+P	0,42
B = Sum 1 ÷ A	0,7		
C = 1 − B	0,3		
D = B x C	0,21	R = n x m x (m-1)	600
E = Sum 2 ÷ A	0	S = Q x R	252
F = 1 − E	0		
G = E x F	0	T = n x m²	720
H = Sum 3 ÷ A	0	U = T - Sum 6	0
I = 1 − H	0		
J = H x I	0	V = U ÷ S	0
K = Sum 4 ÷ A	0	Kappa = 1- V	1
L = 1 − K	0		
M = K x L	0		
N = Sum 5 ÷ A	0,3		
O = 1 − N	0,7		
P = N x O	0,21		

Figure 3: Calculation of κ

The κ-value is then calculated as shown by the schematic

outline in Fig. 3. As input values the number of samples (n) and the number of bag classification categories (m) in Table 3. Since the categories are only "Good" and "Ic" m is two in this case. Executing the calculation steps yields to a κ-value of 1 for the conformity of the tester among each other. Repeating this calculation for the conformity of the tester with the Standard results in a value of 1 equally. Furthermore the detection time of a damaged or polluted bag has been measured in order to investigate the time effectives of the method. The results are shown in Table 4. In addition the data illustrated in Fig. 4.

Table 4: Detection time of damaged or polluted bags

	Detection Time [sec]						
	Bag 3 Damage	Bag 6 Rift	Bag 7 Rift	Bag 9 Rift	Bag 14 Damage	Bag 20 Damage	All Bags Average
Mastertester	2.5	5.0	4.9	3.1	2.2	2.9	3.4
Tester 1	3.5	6.3	12.0	10.0	3.3	3.0	6.4
Tester 2	2.0	7.6	6.8	8.9	5.6	3.3	5.7
Tester 3	3.7	9.9	5.6	6.8	6.2	4.0	6.0
Average	2.9	7.2	7.3	7.2	4.3	3.3	5.4
STDEV	0.8	2.1	3.2	3.0	1.9	0.5	1.3

It can be seen that the master tester has the fastest detection time with an average of 3.44 sec. Also the other three testers show solid detection times averaging around 6 sec per failure. The data also indicates that polluted bags are detected faster than damaged bags. In order to enhance the interpretation of the results, the pollutions and damages have been

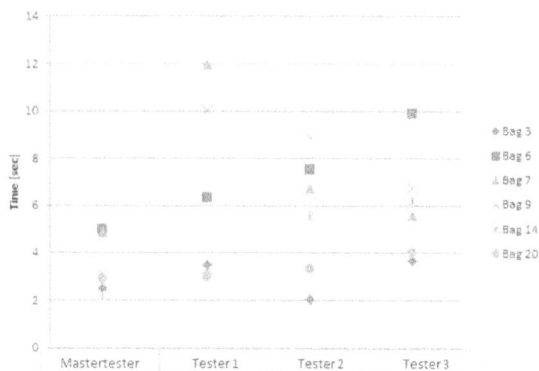

Figure 4: Graph of the detection time of damaged or polluted bags

measured with a three dimensional measurement machine (DeMeet 400) with an accuracy class of 4 μm. The results are shown in Table 5.

Table 5: Measured polluted and damaged bags

	Type of error	Length
Bag 3	Damage	5.233 mm
Bag 6	Rift	50.55 mm
Bag 7	Rift	45.55 mm
Bag 9	Rift	24.33 mm
Bag 14	Damage	3.23 mm
Bag 20	Damage	4.22 mm

From the data it can be seen that there is difference in the size between a pollution and a damage. The size of the pollution on bag 14, for example, is 3.23 mm (Fig. 6) whereas the fissure in bag 9 has a size of 24.33 mm and a diameter of 11.21 mm as seen in Fig. 5. The results show that a quantitative method is not required in the introduction phase of the test method since all failures can be detected by the human eye. In addition the test is not very time consuming for even the smallest failures were detected in a reasonable time.

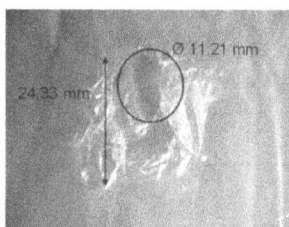

Figure 5: Rift

4   Conclusion

The test method for LDPE-bags will ensure the production of effective materials and proven manufacturing techniques

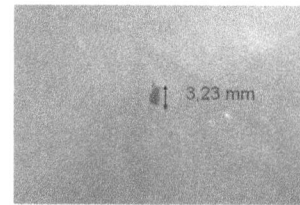

Figure 6: Dirt

at Johnson & Johnson Medical GmbH in Norderstedt. According to the validation results the test method is a reliable procedure to identify faulty LDPE-bags. Initially the test stage of the test method is carried out by visual inspection by the employee. In accordance with the presented data the method is effective and not time-consuming. At this stage it is more cost-efficient than the acquisition of a new machine to automate this procedure. Once the test method is successfully introduced into the production process an enhancement to an automated process based on camera technology will be investigated. Such a machine works with scanners and surface sampling. Possible systems are currently available like Cognex Insight System, Keyence Lumitrax System or the Keyence LJ-v7000-2D/3D-Laser-Profile Measurement System

Acknowledgement

The work has been carried out in the department of Manufacturing Engineering at Johnson & Johnson Medical GmbH in Norderstedt.

5   References

[1] A. Hirner and K. Weise , *Chirurgie*. Georg Thieme, Stuttgart, 2008.

[2] Johnson & Johnson Medical GmbH, *Prolene Netz – Das Netz für die dauerhafte Hernienreparation*. Production brochure, Norderstedt, 2007.

[3] Johnson & Johnson Medical GmbH, *Ethicon complete catalogue 2015*. Available: http://de.ethicon.com/healthcare-professionals/Unsere-Produkte/Die-Produkte/Die-Produkte [last accessed on 02.01.2016].

[4] Johnson & Johnson Medical GmbH, *QSV-10-03 and PR-0011015 Methodenvalidierung*.Internal documents by Johnson & Johnson Medical Norderstedt, 2016.

[5] Johnson & Johnson Medical GmbH, *100093860 Schären und Wirken von Ultrapro Netzen*. Internal documents by Johnson & Johnson Medical Norderstedt, 2016.

[6] Johnson & Johnson Medical GmbH, *SOP 10005 Defect Classifications* . Internal documents by Johnson & Johnson Medical Norderstedt, 2016.

Determining the proportionality of the right and left pupil distance using the video centration system Zeiss i.Terminal® 2 and establishing a measuring method with high accuracy and high reproducibility for determining the pupil distance (PD)

P. Flesch [1], O. Schmidt-Kiy [2], and A. Ritsche [2]

[1] Medizinische Ingenieurwissenschaft, Universität zu Lübeck, pia.flesch@student.uni-luebeck.de
[2] Fielmann Akademie Schloss Plön, {o.schmidt-kiy,a.ritsche}@fielmann-akademie.com

Abstract

In the presented study the right and left pupil distance of 39 subjects is measured twelve times. These values describe the distance between the right or left centre of pupil and the nasal root. Adding both values together the pupil distance is retained. In total 468 values for each side are processed and statistically analysed. The statistical evaluation is implemented by a program called *STATISTICA*. The measurement of the pupil distance is performed by video centration. The video centration system used in this study is named Zeiss i.Terminal® 2. The purpose is to analyse the proportionality of the right and left pupil distance and to investigate pupil distance measuring methods. Within this paper the examined cohort continuously exhibits a larger pupil distances on the left side. Based on the 95 percent confidence interval, the correlation coefficient as well as on the interquartile range conclusions drawn from these measurements are valid.

1 Introduction

Measuring the pupil distance (PD) is important for grinding individual and correctly centred ophthalmic lenses. Complex and expensive optical glasses such as progressive lenses and single vision lenses require an accurate determination of the pupil distance. Reclamation of recently adjusted glasses is mostly due to imprecise measuring values. This study aims to establish pupil distance measuring methods with minimal spreading width and high reproducibility. Furthermore the aim is to analyse the proportionality of the distance from the pupil middle to the nasal root on both sides.

Pupil distance means the distance between the right and the left pupil centre. Because the single distances from one eye to the nasal root are not identical, separate measurements for right and left pupil distance (pR, pL) must be determined. Adding both single pupil distances pR and pL together, the PD value is obtained. The PD is a highly individual value. It is known to vary "[...] with respect to age, gender [...]" [1] and race. It ranges from 58 mm to 70 mm on average while the vast majority of adults have PD values "[...] in the range of 50 mm to 75 mm" [1]. The mean adult PD value measures 63 mm. There are three options for determining the right and left pupil distance [2]:

1. *Manual methods* are the oldest alternative. The single PD is determined manually by means of a PD measuring bar.
2. *Pupilometers* are the first electronic instruments which measure the pupil distance.

3. The introduction of *video centration systems* promises time-saving and more accurate results than manual processes. Another benefit is the output of additional parameters.

Furthermore two possible centration points exist: centration on pupil middle or centration on cornea reflex. While the user defines the centre of the pupil middle himself the cornea reflex is caused by the illumination of pupilometers or centration systems. Studies [3] – [4] show that the PD measured on cornea reflex is about 0.5 mm smaller than the PD measured on pupil middle. Consequently these centration points are located closer to the nose. The reproducibility by centration on cornea reflex is higher than by centration on pupil middle. According to Wesemann possible reasons for this might be the easier centration on cornea reflex while the pupil middle is estimated by the user [3] – [4].

Video centration systems are measuring the PD the most accurate and with the lowest spread width followed by pupilometers and finally by PD measurement bars [2] – [6]. According to Brunner and Schlemmer [5] the standard deviation (sd) averaged over all PD values measured by video centration system is sd = 0.32 mm. Bodenstab is comparing video centration systems [6]. The averaged standard deviation over all systems displays sd = 0.24 mm. Wesemann [3] compares all three different methods. By centration of pupil middle video centration systems display the smallest standard deviation sd = 0.31 mm, followed by the standard deviation averaged over all pupilometers sd = 0.55 mm and final by the standard deviation averaged over all pupil measuring bars sd = 0.71 mm. By centration on cornea reflex the standard deviation averaged over all pupilometers displays sd = 0.51 mm. Within the study [4] video centration systems „[...] measure the PD more accurate [...]" [4] than digital pupilometers. The averaged standard deviation of pupilometers sd = 0.34 mm is almost twice the size of the standard deviation averaged over all video centration systems sd = 0.18 mm.

2   Material and Methods

Adjusting glasses and lenses plays a key role in maximizing visual comfort. During the process of adjustment fitting errors can occur. Because these fitting errors can cause up to 40 percent loss in glasses or lens performance adjusting processes aim to ensure optimal adaption. With the Zeiss i.Terminal® 2 in Fig. 1 "fitting parameters are captured digitally for advanced lens customisation" [7] with an accuracy of 0.1 mm. Using the Zeiss i.Terminal® 2 is according to the manufacturer 60 percent faster than using manual measuring methods and "[...] 84 percent more accurate than a manual process" [7].

Before starting the measurement the eyeglass frame is anatomically adjusted on the customer and has to be clamped into a certain calibration tool. This calibration tool is equipped with front and side markers. The markers will be detected of the video centration system for the determination of the customer's individual face parameters. During the measuring process both supporting glasses have to be removed to avoid disturbing reflexes. Afterwards the customer places the tool including frame on his own face to simulate the prospective frame fit. Next he places himself in a natural body posture directly in front of the Zeiss i.Terminal® 2. According to the manufacturer the optimum distance is 75 cm [7]. Subsequently the optician is adapting the camera to the customer's size. The customer is now able to recognize the red fixation cross without shading and is referred to look straight ahead in the distance. To capture the customer's individual fitting parameters two pictures are required. For the frontal shot the customer has to fix the red cross, for the following side shot he has to rotate by 90 degree. The side picture is necessary to establish further individual face parameters which are also required for optimal glasses or lens fitting. Subsequently the user "[...] can adjust the markers on the frame directly on the pictures and the system will calculate the data automatically [...]" [7]. Based on the results of the customer's individual parameters lenses will be determined which suit his face the best.

The functionality of the calibration bracket as well as the functionality of the front markers has to be ensured. Additionally the customer has to adopt a natural posture to simulate his ordinary tone of the body.

39 subjects are measured twelve times by the Zeiss i.Terminal® 2. In total 468 measured values are recorded. 56 percent of the subjects are male aged between 35 and 76. 44 percent of the subjects are female aged between 48 and 80. The average age of male and female is 63.

For statistical evaluation the program *STATISTICA* will be

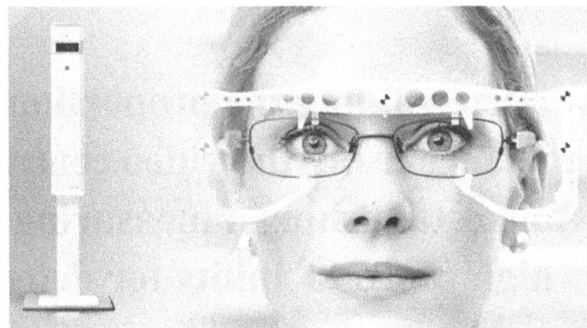

Figure 1: Zeiss i.Terminal® 2 with the calibration tool.

used. This program is designed by StatSoft (Europe) GmbH especially for data analysis.

3   Results and Discussion

In the following the results of the study meaning the proportionality of right and left pupil distances are presented. They are justified by statistical evaluation while graphical representations are intended to clarify this. Subsequently these results are critically discussed. The discussion also covers the results achieved from the underlying studies [2] – [6] about the accuracy of the three different pupil distance measuring methods.

3.1   Distribution of pL-pR over PD

In Fig. 2 the axis of abscissa contains all PD values of each subject while the ordinate contains the difference between pL and pR values. Positive data points mean a larger pL than pR, negative data points mean a smaller pL than pR. As can be obtained from Fig. 2 most PD values are located in the average range of 58 mm to 70 mm.

A mean value of 0 mm would result from identical pR and pL values and therefore from a perfectly symmetric face. Fig. 2 contains a mean value of 1.335 mm shifted to the positive range. This indicates that pL values are larger than pR values. The confidence interval containing 95 percent of all measured values is also shifted to the positive range and makes this even clearer. The positive shifting is apparent from the confidence range of 5.065 mm (mean value + 1.96*SD) and -2.395 mm (mean value – 1.96*SD).

With rising PD values the difference between pL and pR is not proportionally increasing. On the contrary the slope of the regression line is slightly negative. There is no dependency between PD and pL values because of the small correlation coefficient of r = -0.091. The small slope m = -0.042 mm of the regression line indicates the same.

All correlation coefficients in this study are calculated by *STATISTICA*.

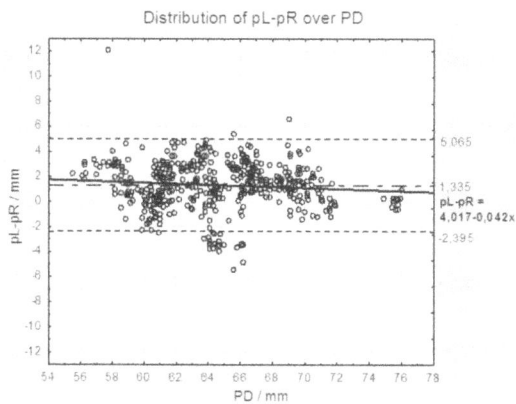

Figure 2: Display and distribution of the difference between pR and pL plotted over PD. Mean value, regression line and 95 percent confidence interval (mean value ± 1.96*SD) are marked.

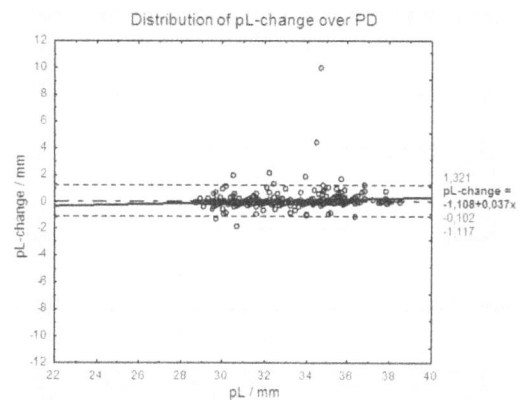

Figure 3: Display of pR-change in dependence on average pR. Mean value, regression line and 95 percent confidence interval (mean value ± 1.96*SD) are marked.

3.2 Distribution of pR-change over PD / pL-change over PD

The ordinate of Fig. 3 (or Fig. 4) records the pR-change (or pL-change). The axis of abscissa records the measured values of pR (or pL). pR-change means the difference between pR and its corrected value, same with pL-change. Corrected values result out of a manual post control which is implemented after the centration process.

Gathering from Fig. 3 as well as from Fig. 4 it is obvious that pL values are larger than pR values. As Fig. 3 shows most pR values lie between 27 mm and 36 mm. One can gather from Fig. 4 that pL values are sharply shifted at about 2 mm to larger values from 29 mm to 38 mm. The confidence interval of pR is also slightly more negative while the confidence interval of pL is slightly more positive. On the right side the confidence range lies in between 1.751 mm (mean value + 1.96*SD) and -2.021 mm (mean value – 1.96*SD), on the left side the range is restricted by 1.321 mm (mean value + 1.96*SD) and by -1.117 mm (mean value – 1.96*SD). This underlines even more that pL values are larger. Furthermore the mean value of pL which measures 0.102 mm is more positive than the mean value of pR with -0.135 mm.

On the right side the correlation coefficient r = 0.237 and the slope m = 0.097 mm of the regression line shown in Fig. 3 are higher than on the left side. Nevertheless both values are that small, that there is no relation between pR-change and pR. The similarly small correlation coefficient of r = 0.130 on the left side as well as the small slope of m = 0.037 mm shown in Fig. 4 give no indication for the existence of a relation between pL-change and pL. This entails that increasing pR values or increasing pL values may not be indicative of a rising pR-change or a rising pL-change.

3.3 Interquartile range of pR, pL and PD

The striped boxes in Fig. 5 show the interquartile range (IQR) or the middle fifty of all measured values. The range of the volume of data is marked by vertical lines while one

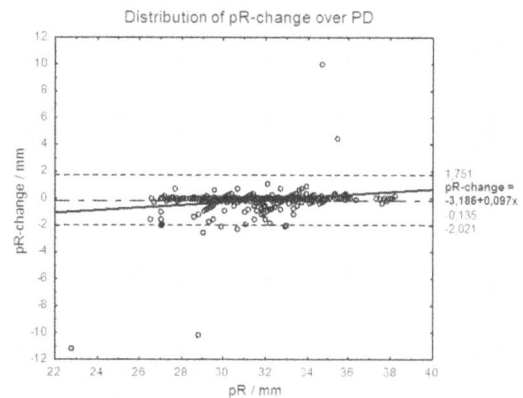

Figure 4: Display of pL-change in dependence on average pL. Mean value, regression line and 95 percent confidence interval (mean value ± 1.96*SD) are marked.

can read out the maximum as well as the minimum value (excluding extremes marked as circles).

The median IQR of pL is 0.7 mm while the median IQR of pR is 0.6 mm. The PD shows the smallest median IQR with 0.45 mm. The quartiles of pL are moved upwards, which also indicates the higher measured PD values on the left side. The scattering of pL is wider than the one of pR while the diversification of PD is the least.

3.4 Discussion

Measuring the exact pupil distance of the customers represents a major challenge. Since it is impossible to determine the absolute PD value by common measuring methods it is the goal of the study to establish a method measuring the PD value as exact as possible. Various error sources cause the difficulty of determining the real pupil distance. Because one cannot exclude all existing error sources at the same time there will always be a spreading. One error source is the medical device itself. Systematic errors often occur in devices. If known, they can sometimes be removed. The customer is the second error source because different customers appear with a varying degree of willingness. As

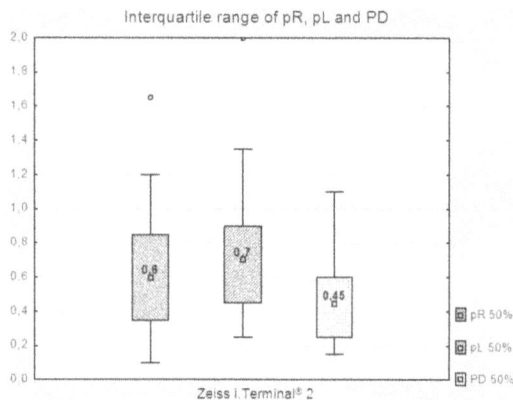

Figure 5: Display of the interquartile range (IQR) and of the median for pR values, pL values and PD values.

well as one customer may bring a strongly differing body tonus while measuring him a few times. This entails in deviating face values measured by video centration systems. Thirdly the user may also represent an error source. This error source only applies to portable video centration systems based on tablets and not to stationary systems like the Zeiss i.Terminal® 2. The last error sources are influences of all surroundings. Three measuring methods for determining the PD exist, as there are: manual methods, pupilometers and video centration systems. Compared with the first and second method video centration which is applied in this study provides the most accurate results with the highest reproducibility. Studies [2] – [6] show that manual measured values are achieved with a deviation up to sd = 0.71 mm, furthermore those values of pupilometers deviate from sd = 0.34 mm to sd = 0.55 mm and finally those values measured by video centration systems deviate from sd = 0.18 mm up to sd = 0.32 mm, all depending on device model and manufacturer. The standard deviation of the values obtained from video centration systems may be a result of the differing study designs [2] – [6]. It is conspicuous from all results measured with Zeiss i.Terminal® 2 that pL values are continuously larger than pR values. Assuming a perfectly symmetric face, pL and pR values would be identical. The shifting of 2 mm to the left side may indicate that the whole cohort of this study has a left asymmetric face. Though this unrealistic statement it cannot be disproved within this study. If this deviation is purely statistical or if there is a connection to the predominant right-handers in the population can also not be established within this study. It can either not be defined if this effect is resulting from the device itself, from the subject, from the user or from the surroundings. No distinction between male and female is made concerning the resulting values. Therefore it is left open if there may be a relation between gender and the proportionality of right and left pupil distance.

4   Conclusion

The aim of this study is to analyse the accuracy and reproducibility of pupil distance measuring methods. Deter-

mining the pupil distance for fitting glasses and lenses is performed by manual handling, by electronic devices or by video centred systems. As expected the manual alternative is the most imprecise method while video centration represents the most accurate method. The differing standard deviation values result in different study designs. Comparing all different measurement methods meaning all different measurement devices presented in studies [2] – [6] one further study with only one study design is required. This would gather credible and comparable results.

Furthermore the purpose of the study is to analyse the proportionality of the right and left pupil distance. Remarkable in this study is the left asymmetrically dominance of PD values. Possible reasons for the larger pupil distance on the left side have to be investigated in following studies. It is also necessary to examine the relevance and significance of this observation in further studies.

Acknowledgement

The work has been carried out within the scope of a 6-month intership at the Fielmann Akademie Schloss Plön.

5   References

[1] N. Dodgson, *Variation and extrema of human interpupillary distance* in: Proc. SPIE, San Jose, vol. 5291, pp. 36–46, 2004.

[2] J. Stollenwerk, *Videozentriergeräte. Als Präzisionsmessgeräte brauchbar?* in: Der Augenoptiker, Schömberg, vol. 2, pp. 2–6, 2015.

[3] W. Wesemann, *Messgenauigkeit und Reproduzierbarkeit von PD-Messgeräten und Unterschiede zwischen der Zentrierung auf Pupillenmitte bzw. auf Hornhautreflex.* in: DOZ-Verlag, Heidelberg, vol. 2, pp. 18–22, 1997.

[4] W. Wesemann, *Moderne Videozentriersysteme und Pupillometer im Vergleich, Teil 1 / Teil 2 Schluss.* in: DOZ-Verlag, Heidelberg, vol. 2 / vol. 7, pp. 18–22 / pp. 32–39, 2009.

[5] O. Brunner and F. Schlemmer, *Von der statischen zur dynamischen Messung: Das Dymeter-Verfahren zur vertikalen Zentrierung.* in: DOZ-Verlag, Heidelberg, vol. 2, pp. 32–38, 2010.

[6] R. Bodenstab, *Sportbrillenzentrierung – eine Fallstudie.* in: DOZ-Verlag, Heidelberg, vol. 07, pp. 48–54, 2010.

[7] Zeiss Homepage, *Call for Papers.* Available: http://www.zeiss.com/vision-care/en_de/products-services/dispensing-tools-and-instruments-by-zeiss/lens-fitting-and-consultation/iterminal-2.html [last accessed on 06.01.2016].

6

E-Health

Adaptation to a prototype system for the bidirectional communication in the MRI environment on clinical issues

M. Lewke[1], K. Rackebrandt[2], and H. Gehring[3]

[1] Medizinische Ingenieurwissenschaft, Universität zu Lübeck, mareike.lewke@student.uni-luebeck.de
[2] Institute of Medical Engineering, Universität zu Lübeck, rackebrandt@imt.uni-luebeck.de
[3] Department of Anesthesiology and Intensive Care, University Medical Center, Luebeck, hartmut.gehring@uksh.de

Abstract

The communication between the control room and the scanner room is limited during a MRI scan. The patient is observed through a window and is contacted via a speaker. To enhance the information transfer between the physician within the scanner room and the medical staff in the control room a bidirectional communication tool is developed via extending a video communication setup by an audio communication approach. The Wi-Fi based setup was tested for practicability and interferences by spurious noise scans, MRI scans of a water phantom and anatomical images of head and spine in MRI environment. The investigations revealed the benefits of the setup in terms of functionality and audio transmission quality. The interfering frequencies have no influence on the phantom scans and the anatomical images. Due to this a bidirectional communication between the MRI room and outside is possible in real time without interferences.

1 Introduction

Magnetic resonance imaging (MRI) is a non-invasive medical imaging modality, which creates sectional images in different regions of the body [1], [2]. MRI is used in medical diagnostics for visualisation of morphology due to its strong tissue contrast [1] - [3]. The MRI room is shielded to prevent an occurrence of interferences caused by magnetic or high-frequency radiation from outside. This shielding also separates magnetic and high-frequency fields of the MRI scanner from the surrounding rooms [2]. For an image recording with a tubular system the patient table is moved into the magnetic tube of the MRI scanner [2]. During a recording the physician is either in the scanner room or in the control room to provide medical surveillance. The communication between the patient in the examination room and the physician in the controlling space is limited due to spatial separation and in particular if the patient is sedated. The communication restriction is caused by a small window in the control room, noise level of the MRI scanner and present technical communication possibilities, like speaker and emergency button. In case that the physician remains within the scanner room, he is exposed to a high noise level and a strong magnetic field. Magnetic fields deployed in clinic application have magnetic field strengths of 0,2 up to 3,0 Tesla (T) [3]. The noise level of the MRI scanners are caused by frequently switched off and on gradient coils. These coils generate field gradients, which are necessary for the selection of a layer and section during medical imaging [2], [3]. Noise-cancelling headphones are worn for hearing protection.

The purpose of the study was to develop a bidirectional audio and video communication between the MRI scanner room and the control room. Hereby the medical staff would be enabled to monitor the patient or medical devices from the control room and to communicate with the attending physician in the MRI scanner room. Therefore a bidirectional, real-time capable communication tool is required, which does not restrict routine processes of the staff. The communication system, realized in [4], is based on an internal Wi-Fi network. It allows a video communication using back cameras and microphones of two iPads. This video communication setup was extended by an audio communication using Bluetooth headsets with built-in throat microphones. The audio signals of the throat microphones are generated by throat vibrations. The utilization of a throat microphones reduce the noise level of surrounding noises in a transmitted signal. External radiofrequency sources can generate artefacts in phase encoding direction on MRI scans (zipper artefacts) [3]. The interferences by the system are determined in the scanning room using spurious noise scans.

2 Material and Methods

The basic idea of the concept paper "Visualisation and Communication in MRI Environment ANVISCOM" was realised by [4] in a prototypical system for a bidirectional visual communication in MRI environment. The data exchange of the communication is implemented by the utilization of Wi-Fi (2.4 GHz frequency band). The Wi-Fi-network is generated by a router (300Mbps-Wi-Fi-N-Router TL-WR841ND, TP-LINK) inside the MRI scanner room. An access pointer (Wireless N WiFi Repeater, De-laycon) is set up in the control room and establishes the connection between the Wi-Fi device (iPad 2, Apple) and the network of the router. The wired connection between the router and the access pointer is constructed via the BNC interface. Both, Wi-Fi router and access pointer, feature RJ-45 ports, that is why the connection between the devices and the BNC interface is composed by midband coaxially network adaptors (TPA-311, TRENDnet). BNC cables with adaptors connect the network adaptors with the BNC sockets. The schematic structure is illustrated in Fig. 1. Two

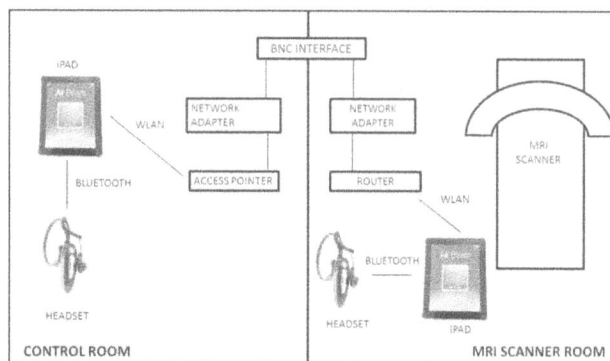

Figure 1: Setup of the bidirectional communication in the control room and the MRI room.

iPad applications regulate the data exchange. The application „Airbeam" (version: 2.1.1 (Build 1), Appologics UG) is deployed to send and receive the video signals, which are acquired with the back cameras. The image transmission rate of 15 pictures per second and a normal image resolution (50% of the camera resolution) allow a real-time video transmission. The audio signals are generated using Bluetooth headsets (BT sport, Midland) instead of the integrated iPad microphones. The headsets consist of in-ear headphones and throat microphones. The headsets transmit the audio signals to the iPads via Bluetooth (2.4 GHz frequency band). The audio signals are automatically transferred in real time between the iPads via the application „Walkie Talkie - Wifi & Bluetooth" (version: 1.5, Porchlight).

2.1   Measurement methods

The influences caused by the system and the functionality of the bidirectional communication in the MRI environment were determined by different measuring procedures in the scanner room. The functionality of the setup describes a stable audio and video transmission without time delays in the MRI room. The evaluation scale developed in [4] was employed for the assessment of the functionality of the communication system. The expanded evaluation scale, which uses time specification for delays, is described in Table 1. The continuous audio and video communication was evaluated based on the occurrence of time delays with the evaluation scale at different distances to the router. The minimum requirement for a video and audio signal usable in clinical applications is classified as a value of 3.

Table 1: Evaluation criteria for the the audio and video quality by [4]. An undisturbed audio or video signal is classified with the value 5. A 2-3 seconds time delay is acceptable in clinical application.

Impairment / Perception	Delay (s)	Value
Imperceptible	0	5
Perceptible, but not annoying	1-2	4
Slightly annoying	2-3	3
Annoying	from 3	2
Very annoying	from 30	1
No signal	∞	0

Interferences were determined in three positions of the scanner room either by spurious noise scans, MRI scans of a water phantom or anatomical images of the head and spine. The positions and measuring methods are shown in Fig. 2 and Table 2. The positions indicate areas in the scanner room where a physician provides medical care when the patient is placed on the patient table inside of the MRI scanner (position 1) or when the patient table is outside of the scanner (position 2). If the physician is not in the MRI room, the iPad can be attached to an installed holder on the wall to supervise the patient or medical devices (position 3).

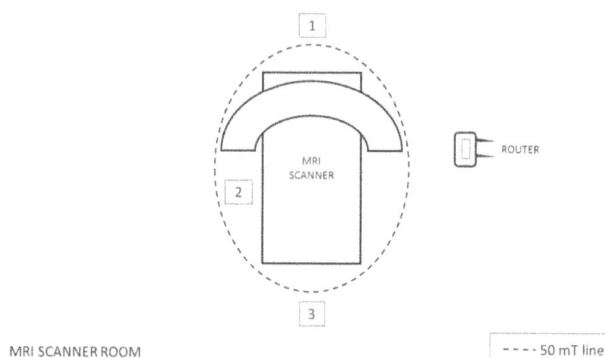

Figure 2: The positions of the setup in the MRI scanner room. Position 1 is near the top of the patient table. Position 2 is located to the left of the patient couch, if it is outside of the magnetic tube and within the 50 mT line. Position 3 is at the end of the MRI room.

Table 2: Measurements and positions in the MRI scanner room

measurements	position
spurious noise scan	1
water phantom	1, 2
anatomical images	3

2.2   Spurious noise scans

The noise scans were conducted to quantify the artefacts caused by Wi-Fi and Bluetooth. The spurious noise scans were measured with the following four center frequencies:

127.44 MHz, 127.61 MHz, 127.78 MHz, 127.95 MHz. The pixel bandwidth was 197 Hz/pixel and the aquisation matrix 1024x1024 pixel. The measurement covered a value range of approximately 0.8 MHz. A measurement without the communication system in the scanner room was produced to determine whether the MRI scanner room is interference-free. The interferences of the radiofrequency sources are determined by individually measuring Bluetooth and Wi-Fi. The interferences caused by the entire communication tool (audio and video transmission), were measured via spurious noise scans with and without communication.

2.3　MRI-sequences

The MRI scans were conducted in a Philips Ingenia 3.0 T MRI scanner. The head scans were measured with T1-TFE (TR/TE = 8.116/3.709 ms, FoV = 240x240 mm), T2-TSE (TR/TE = 3000/80 ms, FoV = 230x230 mm) and FLAIR (Fluid Attenuated Inversion Recovery, TR/TE/TI = 4800/302.374/1650 ms, FoV = 250x250) sequences. In addition, the MR images of the spine were recorded with sagital T1-TSE (TR/TE = 528.954/8 ms, FoV = 320x320 mm) and T2-TSE (TR/TE = 3500/120 ms, FoV = 320x320 mm), transversal T1-TSE (TR/TE = 606.377/8 ms, FoV = 180x180 mm) and T2-TSE (TR/TE = 2678.406/120 ms, FoV = 180x180 mm) data records and a STIR record (Short-Tau Inversion Recovery) using TR/TE/TI = 3371.895/60/210 ms and FoV = 320x320 mm. The water phantom scans were measured with T1-TFE (TR/TE = 8.119/3.712 ms, FoV = 240x240 mm), T2-TSE (TR/TE= 3000/80 ms, FoV = 230x230 mm) and FLAIR (TR/TE/TI= 4800/303.274/1650 ms, FoV = 250x250 mm) sequences.

3　Results and Discussion

In the following section the measurement results are presented and discussed.

3.1　Results

The integration of the headsets into the existing setup was realised and fully functional in the MRI environment. As shown in Fig. 3, a stable real-time audio and video communication was performed with the evaluation value 5 (Table 1) in the entire MRI room. This assessment was also accurate during the MRI recording. The throat microphones reduced the surrounding noises in particular those of the MRI scanner during the audio communication.

Interferences were not detectable in the spurious noise scans of the MRI room without the communication setup. Bluetooth and also Wi-Fi generated line artefacts in the noise scans, but the spurious noise scans of the Bluetooth included less line artefacts than the noise scans of the Wi-Fi. As soon as the system was involved in the information transfer, interferences were also measured. An example of these measured interferences in the spurious noise scans is illustrated in Fig. 4. The determined effects of the system proved no visible interferences on the MR images (Table 3).

The radio frequency waves caused by the system did not induce line artefacts in the measurement of the water phantom (Fig. 5) as well as in the images of the head and spine (Fig. 6).

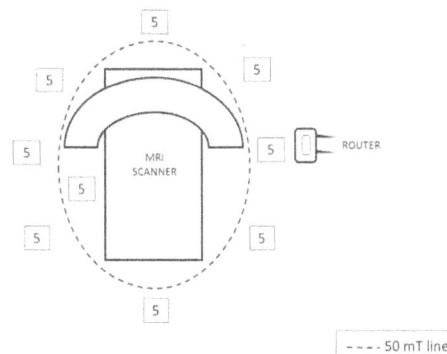

Figure 3: Valuation of the sound and video quality in the entire MRI scanner room with the value 5.

Figure 4: Spurious noise scan measured with a center frequency of 127.78 MHz for audio and video communication (Bluetooth and Wi-Fi). The artefact (white lines) are visible at multiple frequencies.

Table 3: Valuation of the measured MRI scans. A ✓ describes images without visible interferences and n.m. is the abbreviation for not measured MRI sequences.

MRI scans	water phantom	head	spine
T1-TFE	✓	✓	✓
T2-TSE	✓	✓	✓
FLAIR	✓	✓	n.m.
STIR	n.m.	n.m.	✓

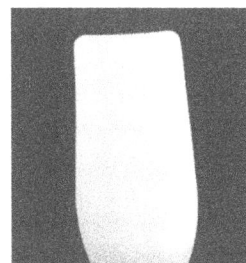

Figure 5: Undisturbed T1-TFE MRI scan of the water phantom with audio and video communication during the scan.

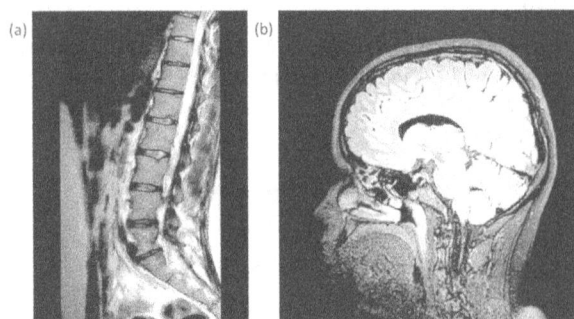

Figure 6: T2-TSE MRI scan of the spine (a) and FLAIR MR image of the brain (b) with audio and video communication during the scan.

3.2 Discussion

The bidirectional communication tool was established by the use of audio and video transference between scanner room and control room. This tool improved the communication between people in the MRI scanner room and the control room. The communication is capable of real-time and is undisturbed before and during the MRI scan. Therefore faster emergency responses are possible in the clinical application as well as the improvement of the medical care of the patient.

The realisation of the system via Wi-Fi and Bluetooth maintains the mobility of the medical employees. The in-ear earphone allows to carry the headset beneath the noise-cancelling headphone, which protects the hearing against the surrounding noises of MRI scanners. The throat microphones increase the sound quality by the reduction of the surrounding noises and provide an improved communication quality. The neckband of the headsets enables the medical staff to carry the headsets around the neck, therefore the hands of the medical employees are free and they are able to hold the iPad or supply the patient. After the establishment of the connection the audio and video communication does not have to be attended to further, if a stable Wi-Fi connection is provided. Therefore the medical staff is not distracted.

The spurious noise scans clarify that Wi-Fi as well as Bluetooth generate artefacts at different frequencies. These interferences were also visible during the communication via the system. Possibly electric cables like the power supplies of the routers or the network adaptors lead to interferences. Wi-Fi and also Bluetooth send in the 2.4 GHz frequency band. The frequency of the 3 Tesla MRI scanner is approximately 128 MHz. The receiving coils are focused on this frequency range. Due to the interference-free MRI scans of the spine and the head as well as the water phantom the caused interferences are substantially lower than the received benefit signals. Although body coils are more susceptible to interferences on account of the bigger diameter and measuring volume, the utilization of the head or the body coil makes no difference for the recordings. However, there were no recognisable interferences using the different standard MRI sequences. The interfering frequencies did not lead to visible artefacts on the MRI scans.

4 Conclusion

During the MRI scan the communication tool allows an undisturbed real-time communication (audio and video transmission) from the MRI scanner room to the control room without causing visible interferences in the MR image. The evolved enlargement of the setup can be used to improve the opportunity to monitor the patient and communicate with the physician from outside of the MRI scanning room in clinical application. Future work includes the testing of the system on utilization and suitability in clinical practice by medical staff. In addition an ethical application is necessary. Also the investigations whether the choice of the Wi-Fi channel influences the occurrences of line artefacts in the spurious noise scans and whether the disturbances affect not measured MRI sequences must be carried out. The applications of the communication system need to be simplified in one application in terms of easier handling.

Acknowledgement

The work has been carried out at the Department of Anesthesiology and Intensive Care, University Medical Center, Luebeck. The basic idea of the project is based on the concept paper "Visualisation and Communication in MRI Environment ANVISCOM" from Dr. med. Sebastian Brandt.

5 References

[1] M. Wrobel, M. Werth and T. Lahme, *Anästhesie-Fibel: Kompendium für den klinischen Alltag*. Urban& Fischer Verlag, München, pp. 80–81, 2011.

[2] A. Hendrix and J. Krempe, *Magnete, Spins und Resonanzen. Eine Einführung in die Grundlagen der Magnetresonanztomographie*. Siemens AG, Erlangen, 2003.

[3] D. Weishaupt, V. D. Köchli and B. Marincek, *How does MRI work? An Introduction to the Physics and Function of Magnetic Resonance Imaging*. Springer, Berlin/Heidelberg, 2006.

[4] J. Jayethevan, *Bidirectional communication via an internal Wi-Fi network in a MRI environment*. Department of Anesthesiology and Intensive Care, UKSH Lübeck, 2015.

[5] R. Schreiner, *Computernetzwerke: von den Grundlagen zur Funktion und Anwendung*. Carl Hanser Verlag GmbH Co KG, München, 2014.

Test Automation Framework for an HL7 integration engine

S. Heusel [1], S. Gisch [2], S. Mersmann [2], and J. Ingenerf [3]

[1] Medizinische Informatik, Universität zu Lübeck, sven.heusel@student.uni-luebeck.de
[2] Drägerwerk AG & Co. KGaA & Co, Lübeck, {Simon.Gisch, Stefan.Mersmann}@draeger.com
[3] Institute of Medical Informatics, Universität zu Lübeck, ingenerf@imi.uni-luebeck.de

Abstract

Functionality validation of software is driven by tests that evaluate the interaction of program modules, and determine whether the underlying scenario was handled correctly. Given a certain project size, writing tests by hand is inefficient compared to automated and continuous testing. This work presents a framework that processes abstract scenario information, runs corresponding tests on an existing program with minimal user interaction and therefore offers a more eligible way to software validation. The subject of testing is the *Health Level 7* message interface of the established *Integrated Care Manager©*, a patient data management system for hospitals developed by Drägerwerk AG & Co. KGaA. First, given requirements are analysed and according diagrams as well as overviews are shown. The subsequent implementation is showcased briefly due to the ongoing effort of fully implementing all features of the framework. Finally assets and drawbacks of continuous testing are presented and a conclusion is drawn.

1 Introduction

Functionality tests are a crucial part of any software project, especially if it has a clinical application area and therefore influences a patient's well-being. Damm et. al [1] state, that the effort of writing tests for software takes up to 50% of development time and therefore represents a considerable cost factor. The goal of this work is to reduce the amount of time required to establish test cases using automation. By conducting automated tests, a step towards test driven development [2] and continuous integration [5] is taken. The software to be tested is the *Health Level 7* (HL7) message interface of the hospital *Patient Data Management System* (PDMS) *Integrated Care Manager* (ICM) developed by Drägerwerk AG & Co. KGaA that is deployable in various hospital wards and is used to keep track of patient's vital signs, demographic data and location changes. ICM HL7 Test Tool uses test case scenario containing files as input that can be created and modified with little to no knowledge of computer-based test conducting. The framework then analyses the contents of the test file and starts the corresponding interaction with ICM to validate the consistency of the input-output sequence.

2 Material and Methods

This section highlights details about ICM, the tools used for creating ICM HL7 Test Tool and all imported data structures.

2.1 Software Platform and Programming Language

The framework titled "ICM-Hl7-Testtool" is a Microsoft Windows Forms application and uses the *application programming interface* (API) provided by Microsofts .NET framework. It is written in C-Sharp (C#), an object orientated programming language that was developed by Microsoft and submitted to standardization in 2000 in cooperation with Hewlett-Packard and Intel [3]. C# is, among others, influenced by the programming languages Java and C++. Concepts like pointers exist, albeit being only utilisable in code segments labelled as"unsafe" and therefore being more restricted than in the C++ implementation. On the other hand C# provides an implicit garbage collector, which is not part of the C++ implementation. It offers a compromise between the open mindset of C++ and the usability focus of Java. The framework was created using Microsoft's Visual Studio 2010 *integrated development environment* (IDE), since it natively supports C#. Windows Forms is a programming interface that allows the user to access Microsoft Windows user interfaces. These can be used to specify the look and behaviour of the *graphical user interface* (GUI) of the application. Fig. 1 shows a part of the framework's GUI where user input concerning scheduling and test periodicity is handled.

2.2 nHapi

nHapi is an open source software project that provides data structures for HL7 message handling. It realizes a port of the *HL7 application programming interface* (HAPI) sources and allows their usage in the .NET framework. HAPI is an

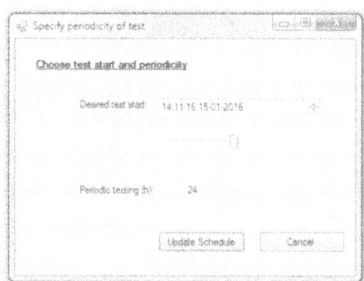

Figure 1: The scheduling GUI of ICM-HL7-Testtool.

open source, object orientated Java implementation that allows the instantiation and manipulation of HL7 messages as objects. It supports message encoding into strings of characters and vice versa which renders network communication with distant recipients possible. The provided data structures conform to the HL7 version 2.x specification of the HL7 organization and therefore serves as a means of communication between software systems.

2.3 Ranorex

Ranorex is a GUI based test automation framework developed and distributed by the Ranorex GmbH. It allows the user to run GUI based application tests by mapping GUI elements to a proprietary RanoreXPath expression. It is possible to track GUI elements on the users screen in real time in order to create an assembly of elements called a repository. A repository can store an arbitrary amount of GUI elements represented by their RanoreXPath. Furthermore, Ranorex provides a test automation library for .NET projects that enables the usage of Ranorex functionality in program code. It can be used to load a repository and use it to specify sequences of GUI interactions with the aim of navigating through an existing program and trigger a desired event, e.g. the admission of a patient in ICM.

2.4 Fluent nHibernate

nHibernate is a framework that is used to perform *object-relational mapping* (ORM) between database content and coded object instances. nHibernate is, analogous to nHapi (see section 2.2), a port of the Java implementation Hibernate. There are two ways to use Hibernate: Either to programmatically set up an object structure and to create the corresponding schema in the desired database or to read stored values and save them in the corresponding objects. Hibernate relies on an *Extensible Markup Language* (XML) file that contains the mapping information between database column and object variable. Using Fluent nHibernate, the usage of the XML-mapping file can be omitted. The mapping is specified inside dedicated class maps where each column is mapped on its own property. Subsequently each class map corresponds to a database instance and each corresponding object represents one row of the database.

2.5 Integrated Care Manager

Dräger's Integrated Care Manager is a customizable PDMS in use at ICUs in Germany, Austria as well as Switzerland. It is a medical product of class IIa certified in accordance with the *Medical Device Directive* (MDD) 93/42/*European Economic Community* (EEC) established by the Council of the European Union and the German "Medizinproduktegesetz" (law of medical products). It consists of four modules that are designed to operate in their respective area of clinical expertise and therefore function independently from each other: ICMi (intensive and intermediate care), ICM-neo (neonatology and paediatrics), ICMa (anaesthesia supervision and planning) and ICMe (emergency). Data compatibility in between different modules exists, hence information collected using e.g. an ICMe implementation used in the emergency room (ER) can be accessed in the ICMi instance of the ICU. ICM communicates with the *hospital information system* (HIS) using HL7 messages and implements drivers for medical devices in order to receive and display vital data. Demographic patient information can be viewed and updated, vital signs can be observed in real time via a daily chart. ICM is installed on every workstation that is subsequently called an ICM client. All clients communicate with a server that provides the necessary back end infrastructure such as a database instance and interfaces to other system components.

3 Results

This section explains the steps taken to create the interface testing framework in detail.

3.1 Analysis of Requirements

The first step was to analyse the given requirements in the context of programmatic realization:

- Scheduling and automated operation

- Generation of "random" demographical and clinical data

- HL7 template insertion

- Workflow verification (e.g. admission, transfer, discharge)

- Protocol support: *Network File System* (NFS), *Transmission Control Protocol* (TCP)/ *Internet Protocol* (IP), *File Transfer Protocol* (FTP)

- Sending and receiving messages to/ from ICM

- Database content verification

- Result logging

From these requirements an initial draft of the framework can be assessed. There are two kinds of test cases:
Inbound test cases:

1. Instantiate HL7 messages using nHapi and fill it with dynamically created data

2. Send message to ICM

3. Validate equality of input data and database content extracted using nHibernate

Outbound test cases:

1. Trigger event in ICM using the Ranorex UI testing library

2. Extract export data from the database using nHibernate and evaluate the subsequent HL7 message

Furthermore the program needs to support multiple data sources as input, execute at a specified time without user interaction and needs to log the results along with the reason for success or failure.

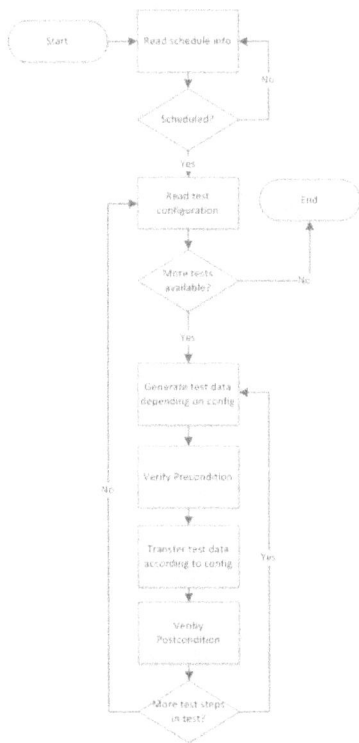

Figure 2: A flowchart depicting the desired workflow of the interface testing framework.

3.2 System Architecture

The typical first step in creating a new software project is to develop diagrams depicting the role of the new software in comparison to existing system components. The first created diagram is a specialization of the existent flow-chart (Fig. 2), which is shown in Fig. 3. It can be seen that inbound test cases use data created by the framework to validate the reaction of ICM. For outbound testing, ICM needs to be operated in order to trigger a specific scenario. The resulting messages are then validated to determine test success or failure.

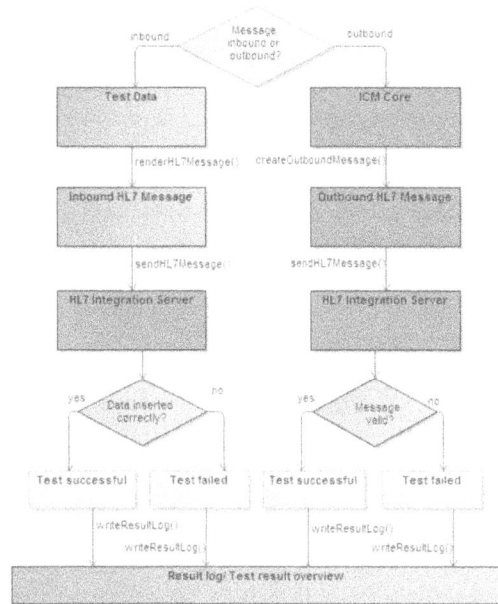

Figure 3: A specialization of the existent flow-chart (Fig. 2) specifying the diversification of processed bound to inbound and outbound tests.

Another aspect that needed to be considered was the number of possible user interactions, which was laid out using a use-case diagram shown in Fig. 4. It is apparent that the user does not have any influence on how the tests are conducted as it was specified in the requirement section (3.1).

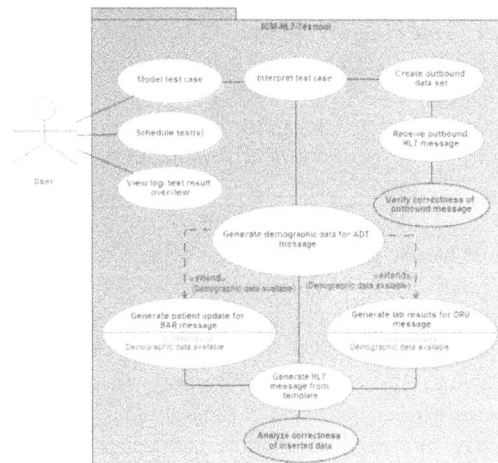

Figure 4: A use-case diagram showing possible user interactions and their desired handling.

3.3 Implementation

The framework is designed to follow the *Model View Controller* (MVC) pattern [4] and to implement an observer pattern. This effort was made to maintain flexibility and to ease extensibility of further features. The development was made using the principle of prototyping since the provided requirements allow to assess different approaches to implement a feature instead of forcing the development in a specific direction. Prototyping offers the possibility to flexibly implement change requests at a point in time, where changing parts of the program is still cost-efficient. Preset code conventions were to name variables in CamelCase notation while methods and properties were to be named in Pascal-Case. An overall level of code sophistication was maintained and surveyed. Special attention was paid to keeping the code clean of ambiguities and redundancy in naming and purpose of the code object.

The implementation of the framework is still ongoing and not all of the desired features have been realized yet. Scheduling and template insertion functionality as well as a certain set of inbound test cases can be executed, whereas the outbound direction of tests is still in construction.

Functionality validation of ICM HL7 Test Tool itself will be conducted following internal Dräger validation processes and unit testing once all features of the framework have been implemented.

4 Conclusion

This section provides a discussion of the advantages and disadvantages of the presented testing paradigms as well as a conclusion on ICM HL7 Test Tool.

4.1 Discussion and Outlook

The goal of this work was to establish a previously non existing framework for conducting tests automatically and continuously, in order to realize a paradigm shift in software testing. The two paradigms that were subject to this work are to write test manually and based on the functionality of singular modules or to continuously test the code at large. Using unit tests, the characteristics of the source code can be analysed precisely and since the complexity of a module by itself is bounded, achieving a good test coverage is relatively easy. Additionally, new functionality can be tested locally without releasing possibly erroneous software. A downside is that only the author of the code himself is able to write tests effectively. Any other person needs to become acquainted with the program code which is, depending on the size of the module, connected with expenditure of time and therefore not the optimal solution. This is of course not the case when using a framework that only requires abstract test case scenario information like ICM HL7 Test Tool. When conducting automated testing, it is possible to discern the effect of modules on the entirety of the program. Whenever a new module is inserted into the code base and tests subsequently fail, it is clear that new code corrupts existing code. Even if unit tests of the particular module have been passed successfully, the result of the system test indicates whether the feature has been implemented effectually. Automated testing is best run in an environment that bears resemblance to the production system [5]. Every new feature requires its own clone of the test environment

and therefore a fixed amount of dedicated resources. Summing up, it can be said that both testing paradigms are eligible for validating the functionality of software. When using automated testing however, while more resources need to be spent, an immediate and lasting evaluation of program functionality is provided. Ensured module validity does not guarantee program validity, since only the interaction of all modules makes the program function the way it should.

It has to be noted that the presented paradigms are not mutually exclusive and a combination of them both appears to be the most promising approach to functionality validation. This goal can be achieved by using ICM HL7 Test Tool. A formal evaluation cannot be conducted at this point since not all of the features have been implemented yet.

Functionality testing is an essential part of each software project and can be realized in different ways. It is possible to run tests after finishing a new module in order to validate its functionality, or set up an environment that enables automated testing. Both ways fulfil their purpose but in the end only the assertion of a working program as a whole defines the completion of a software project. A combination of both module and system testing appears to reflect the correct behaviour of a program best and should be the favoured way to functionality assertion.

Acknowledgement

The work has been carried out at Drägerwerk AG & Co. KGaA, Lübeck with the help of the members of the ICM development team. Special thanks to my supervisors Simon Gisch and Stefan Mersmann, who have been accompanying my endeavours each day with words and deeds.

5 References

[1] L. Damm, L. Lundberg and D.Olsson, *Introducing test automation and test-driven development: An experience report.* in: Electronic Notes in Theoretical Computer Science, Elsevier, vol. 116, pp. 3–15, 2005.

[2] E. Maximilien and L. Laurie, *Assesing test-driven development at IBM.* in: Software Engineering, 2003. Proceedings. 25th International Conference, IEEE, pp. 564–569, 2003.

[3] ECMA International, *ECMA-334: C# Language Specification.* Available: http://www.ecma-international.org/publications/files/ECMA-ST/Ecma-334.pdf [last accessed on 11.01.2016]

[4] Portland Pattern Repository, *Model View Controller.* Available: http://c2.com/cgi-bin/wiki?ModelViewController [last accessed on 12.01.2016]

[5] M. Fowler, *Continuous Integration.* ThoughtWorks, 2006. Available: http://www.martinfowler.com/articles/continuousIntegration.html [last accessed on 12.01.2016]

Converting HL7v2.6 to FHIR
– One method on how to perform it –

Mattes Rhein [1], Stefan Schlichting [2], Josef Ingenerf [3]

[1] Medizinische Informatik, Universität zu Lübeck, rhein@student.uni-luebeck.de
[2] Drägerwerk AG, stefan.schlichting@draeger.de
[3] Institute für Medizinische Informatik, Universität zu Lübeck, ingenerf@imi.uni-luebeck.de

Abstract

Because of the wide variety of medical messaging standards it is sometimes needed to convert one message with a particular standard to another standard. In this paper we describe a method to write Patient Care Device (PCD) Messages on a Fast Healthcare Interoperability Resources (FHIR) Server. FHIR is a draft standard created by the Health Level Seven International (HL7) while PCD is a domain standard defined by the Integrating the Healthcare Enterprise initiative (IHE). To write the information of the PCD-01 message we need to convert the information to FHIR conform resources. We describe the way to convert a PCD-01 message segment to the FHIR resource on the hand of the Patient Identification (PID) segment which represents the patient FHIR resource and write it on a FHIR-Server.

1 Introduction

A central problem in medical communication standards is the diversity of standards, which is caused by the variety of healthcare processes. Over the years the cost and complexity of implementing increased through the addition of more fields and optionality to the specification.

Fast Healthcare Interoperability Resources (FHIR) is trying to solve this problems by defining a framework for extending and adapting the existing resources. It tries to combine the best features from Healthlevel Seven International's (HL7) v2, HL7 v3 and Clinical and Administrative Domains (CDA) while using modern web standards and applying a thight focus on implementability.

HL7 v2 is a messaging standard which allows the exchange of clinical data between systems. Possibly it is the most widely implemented standard for healthcare in the world. It is designed to support a central patient care system as well as a more distributed environment where data resides in departmental systems [1].

HL7v3 represents a new approach to clinical information exchange based on a model driven methodology that produces messages and electronic documents expressed in XML syntax. The specification is built around subject domains which provide storyboard descriptions, trigger events, interaction designs, domain object models derive from the HL7 Reference Information Model, hierarchical message descriptors and a prose description of each element [2].

CDA is a document markup standard which specifies the structure and semantics of clinical documents with the purpose of exchange between healthcare providers and patients. By the definition of CDA a clinical document has six characteristics, Persistence, Stewardship, Potential for authentication, Context, Wholeness and Human readability [3].

To model healthcare data FHIR uses basic building blocks called resources. With this approach it is easier for healthcare providers to use and share the clinical data. Each of these resources can carry a human-readable text representation. For complex clinical information, where many systems using a simple text or document based approach, HTML is used as a fallback display option for clinical safety [4]. FHIR is a draft standard created by the Health Level Seven International (HL7) health care organization. The used resources of FHIR can be combined and used in working systems that solve real world clinical and administrative problems. FHIR is useable in a wide variety of context, like mobile phone apps, cloud communication or server communication in large institutional healthcare providers and many more [4]. In comparison to other HL7 standards, like HL7 v2 or HL7 v3, it offers many improvements like a strong focus on implementation, multiple implementation libraries, the support for RESTful architectures and a strong foundation in web-standards like XML, JSON or HTTP. Administrative concepts like patient, organization and device as well as clinical concepts like medications and diagnostics are covered by FHIR with its resources. In Fig. 1 we show an example FHIR resource written in XML, describing a patient resource and showing which part of the resource contains which information about the resource and the patient.

The Personal Care Device (PCD) domain is defined through the Integrating the Healthcare Enterprise initiative (IHE). PCD deals with use cases in which at least on participant is a controlled patient-centric point-of-care medical device that wants to communicate with, at least, one other participant

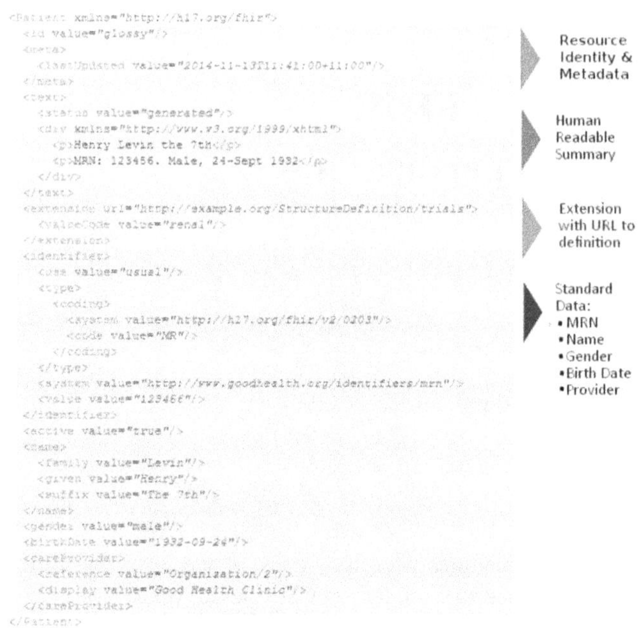

Figure 1: Example Resource of FHIR [4]

Segment	Meaning	Card	HL7 chapter
MSH	Message Header	[1..1]	2
[{SFT}]	Software Segment	[0..0]	2
[UAC]	User Authentication Credential	[0..1]	
{	--- PATIENT_RESULT begin		
[	--- PATIENT begin		
PID	Patient Identification	[1..1]	3
[PD1]	Additional Demographics	[0..0]	3
..[{NTE}]	Notes and Comments	[0..0]	2
..[{NK1}]	Next of Kin/Associated Parties	[0..3]	3
[	--- VISIT begin		
PV1	Patient Visit	[1..1]	3
[PV2]	Patient Visit – Additional Info	[0..0]	3
]	--- VISIT end		
]	--- PATIENT end		
{	---ORDER_OBSERVATION begin		
[ORC]	Order Common	[0..0]	4
OBR	Observation Request	[1..1]	7
[{NTE}]	Notes and Comments	[0..1]	2
[{	--- TIMING_QTY begin		
TQ1	Timing/Quantity	[1..1]	4
[{TQ2}]	Timing/Quantity Order Sequence	[0..0]	4
}]	--- TIMING_QTY end		
[CTD]	Contact Data	[0..0]	11
[{	--- OBSERVATION begin		
OBX	Observation Result	[1..1]	7
[{NTE}]	Notes and comments	[0..1]	2
}]	--- OBSERVATION end		
[{FT1}]	Financial Transaction	[0..0]	6
[{CTI}]	Clinical Trial Identification	[0..0]	7
[{	--- SPECIMEN begin		
SPM	Specimen	[0..0]	7
[{OBX}]	Observation related to Specimen	[0..0]	7
}]	--- SPECIMEN end		
}	--- ORDER_OBSERVATION end		
}	--- PATIENT_RESULT end		
[DSC]	Continuation Pointer	[0..0]	2

Figure 2: Content of PCD-01 [8]

like a medical device or an information system [5].
PCD coordinates with other IHE clinical specialty based domains like medical imaging, to ensure consistency of medical device integration solutions across all IHE technical frameworks [6]. It aims to raise the bar from the expensive integration projects to an easy out of the box interoperability solution [7].

In this paper we want to describe a way to convert incoming PCD-01 messages to FHIR conform resources to be written on a FHIR server.

2 Material and Methods

We are working with PCD-01 transactions of the IHE Patient Care Device Technical Framework. PCD-01 is used to transmit patient care device data between systems. The content of the PCD-01 message is written in HL7v2.6 syntax and it also contains the coding requirements related to observation reporting used for PCD data communication. In Fig. 2 we show the content of a PCD-01 message with the highlighted PID segment, which we want to use in this paper.
To convert the PCD-01 messages to a FHIR resource, a mapping of the content of the message is needed. This mapping is approximately shown in Table 1, while the Message Header (MSH) and Patient Identification (PID) are completely mapped to the corresponding FHIR resources, the information from Patient Visit (PV1), Observation Request (OBR) and Observation Result (OBX), depending on the content, are written in a diversity of resources.

PCD Segment	FHIR Resource
MSH	MessageHeader
PID	Patient
PV1	Encounter, Location (depending on the segmentpart)
OBR	Observation, Device, DeviceComponent, DeviceMetric (depending on the segmentpart)
OBX	Observation, Device, DeviceComponent, DeviceMetric (depending on the segmentpart)

Table 1: Short mapping overview of PCD-01 elements on FHIR Resources

As development environment we used Eclipse Mars and as programming language we used Java 1.8 with the FHIR API, called HAPI-FHIR. The server-application which was used for testing is a Spark Server from Furore and runs on a Microsoft Windows Server 2012 R2 system which is running in a virtual machine with an Intel Xeon Quadcore CPU E5-2680 with 2.5GHz and 16GB RAM. It is written in C# and is build for FHIR DSTU 2 [9]. The converter is running on the same system as the server-application. New messages are going to be received over the company intern network after that they are converted and written on the server.

To map the data of the message to FHIR resources we used the mapping which is given by the HL7 [10]. The resources on the FHIR server are saved under URL addresses which also represent the logical identifier of the resource. For example the resource for a test patient could be found under "baseserver\fhir\patient\id". This id is set by the server and is unique on the server system.

On the example of the PID segment of the PCD data we are showing how we established the converter. The PID segment contains all necessary information about the patient, like name, address, gender, birth-date, patient id and many others, see IHE PCD Technical Framework Volume 2 for the complete list of content [8].

In this paper we will exemplary concentrate on the mapping for the patient. The patient id will show how we got the information out of the message and afterwards we will show how we build the patient resource which will be written on the server.

The patient id is converted from the "Extended Composite ID" (CX[]) datatype to the identifier type of the FHIR resource. This identifier represents the business identifier which can be set by the software system.

The id we use as a the patient id is contained in the third part of the PID segment, the "Patient Identfier List". It is stored in an CX[] which itself is parted in ten parts, the ID Number, the Identifier Check Digit, Assigning Authority, Identifier Type Code, Assigning Facility, Effective Date, Expiration Date, Assigning Jurisdiction and Assigning Agency or Department.

The id which we want to use is within the first part of the CX[] datatype, the ID Number. The ID Number itself is stored as a String Data element, therefore the best way to use it further is to store it in a String. After receiving the message we read the patient identifier list and check if its not empty and not null. Then we read the first value of the patient identifier list and put it in a string.

After storing the ID in a String we need to get the "ID System" for the identifier, it describes the name space for the identifier. The ID System itself is build with the "Assigning Authority" and the "ID Type". Getting the other information's about the patient are implemented nearly the same way like the patient id.

Subsequently we get every information of the patient out of the message and we build the patient resource for the FHIR server. For that we fill the identifier, the address, the name, the birth-date and the gender with the values we got from the message. If there are no values for one of these information's the field in the resource is not going to be filled.

Once we set every received information of the resource we are checking if a resource with the same id has already been written on the server, if so the resource is going to be updated with the new values, if not the patient resource is written on the server. The update works like the create shown

in Fig. 3, the difference is that the update is getting the logical identifier for the patient resource to work with, while the create doesn't get any further information about the resource.

```
public MethodOutcome createPatient () throws Exception{

    // Create a DSTU2 context, which will use DSTU2 semantics
    FhirContext ctx = FhirContext.forDstu2();

    // This client supports DSTU2
    String ServerBase = "http://localhost:49734/fhir";
    client = ctx.newRestfulGenericClient(ServerBase);

    Patient patient = new CreatePatientFromMessage()
            .setup_patient();

    MethodOutcome outcome = null;

    List<IdentifierDt> matchingID = patient.getIdentifier();
    List<BaseIdentifierDt> matchingIdsList=
            new ArrayList<BaseIdentifierDt>(matchingID.size());
            matchingIdsList.addAll(matchingID);

    TestExistence existance = new TestExistence();
    boolean existancebool = existance.testExistens(matchingIdsList);

    if(existancebool == true) {
        try {
            MethodOutcome updatePatient = new UpdatePatient()
                    .updatePatient(patient);

            return updatePatient;
        } catch (InvalidRequestException invalidRequest){
            System.err.println
                    ("400_Bad_Request_for_Update");

            return outcome;
        }
    } else if(existancebool==false){
        System.out.println
                ("Entry_not_on_Server_creating_entry_"+patient);

        outcome = client.create()
                    .resource(patient)
                    .execute();
        return outcome;
    }
    return outcome;
}
```

Figure 3: Create Patient Method

Fig. 3 shows how a resource is created on the a server which is running on the localhost. If there are problems with the id, which is needed for the update request, the server will give back the HTTP "400 Bad Request" status code. The creation of other resources is in principle the nearly the same. The differences for the other resources are the content in these resources. For example a device metric, which represents measurements or changes in the status of a component, can be build with references to device components and a device. The components describe parts of a medical device, like sensors or pumps. Metrics and components are resources in FHIR which contain specific data and references to other metrics, components and devices.

3   Results and Discussion

```
PID|||867345^^^MR||Patient^Test^^^^L|^^^^^^U
```

Figure 4: The used PID-segment from the PCD-01 message

The PID-segment we used for the tests is shown in Fig. 4. We tested our converter on the same local machine as the server runs on. To test the receiving of a message we write the incoming messages in .txt files. For testing the writing on the server we read those .txt files and converted everything to the resources. In Fig. 5 we show the written resource with the converted values form the message. A complete message with 28 OBX segments needs 0.175 seconds to be read, converted and written on the designated server.

The main problem while implementing the converter was that FHIR is still in development and therefore not bug free and on some points the documentation is not complete. While working with FHIR we found a bug in the specifications, which makes it not possible to search on the server for Device Components by the logical identifier. But to check if a Device Component is already written on the server you need to search for it by the logical identifier. As a work-around we get every Device Component from the server and check the business identifier of every Device Component on the server with the new Device Component. If they are the same we assume that these two are the same Device Component. From the Device Component we then read the logical identifier, after that it is possible to update the Device Component with the new values.

```
- <Patient>
    <id value=" spark4" />
  - <meta>
      <versionId value=" spark6" />
      <lastUpdated value=" 2016-02-10T10:39:25.578+01:00" />
    </meta>
  - <identifier>
      <system value=" urn:fhir:id/PI/" />
      <value value=" 887345" />
    </identifier>
  - <name>
      <family value=" Patient" />
      <given value=" Test" />
    </name>
    <gender value=" other" />
  </Patient>
```

Figure 5: The written Patient resource on the server

4   Conclusion

Thanks to the implementation guides and documentation of FHIR, PCD and HL7v2.6, the mapping of the information given in the PCD messages to the FHIR resources are not that complicated. The bigger problem is that FHIR is still in under development therefore not everything went as we expected it to work. At some points we needed to build a work-around for problems which could not be solved with the methods given in the specifications.

For further development it would be necessary to optimize the converting process and to convert the FHIR resources back to a PCD-01 message and send them over a network.

Acknowledgement

The work has been carried out at Drägerwerk AG, Moislinger Allee 53, 23558 Lübeck. Under the supervision of Dr. Stefan Schlichting.

5   References

[1] HL7, "Hl7 version 2 product suite." http://www.hl7.org/implement/standards/product_brief.cfm?product_id=185.

[2] HL7, "Hl7 version 3 product suite." https://www.hl7.org/implement/standards/product_brief.cfm?product_id=186.

[3] HL7, "Cda release 2." http://www.hl7.org/implement/standards/product_brief.cfm?product_id=7.

[4] HL7, "Introducing fhir." http://hl7.org/fhir/summary.html.

[5] IHE, *IHE Patient Care Device(PCD) Technical Framework Volume 1 IHE PCD TF-1 Profiles*, 2014.

[6] IHE, "The webinar on patient care device," 2011.

[7] P. Paul Schluter, "Understanding interoperability with the ihe profiles," 2012.

[8] IHE, *IHE Patient Care Device(PCD) Technical Framework Volume 2 IHE PCD TF-2 Transactions*, 2014.

[9] Furore, "Spark fhir server." http://spark.furore.com/, Janurary 2016.

[10] HL7, "Resource patient - mappings." https://www.hl7.org/fhir/patient-mappings.html.

Device Metric processing with FHIR
– from device observation report to device metrics bundle –

D. Rehmann [1], G. Meinke [2], and J. Ingenerf [3]

[1] Medizinische Informatik, Universität zu Lübeck, daniel.rehmann@student.uni-luebeck.de
[2] Drägerwerk AG & Co. KGaA, Lübeck, gritje.meinke@draeger.com
[3] Institute of Medical Informatics, Universität zu Lübeck, ingenerf@imi.uni-luebeck.de

Abstract

Complying to standardized specifications requires the developer to update the specification for each new version. This work describes the steps taken to update an existing interface using Fast Healthcare Interoperability Resources (FHIR) highlighting the difficulties encountered through this update. It compares multiple representation alternatives as well as the different Reference Implementations (RIs) that are available to use. Further the difficulties to implement different versions of the same specification in one application are discussed. The paper concludes with an evaluation of the finished update and highlighting the most useful architectural designs and programming patterns encountered through the update. The conclusion further includes an evaluation of FHIR and the associated RIs in a productive environment. Finally an outlook to further improve the resource representation through profiles is provided.

1 Introduction

This paper deals with the update from Fast Healthcare Interoperability Resources (FHIR) with the Draft Standard for Trial Use 1 (DSTU-1) to the next version DSTU-2 with focus on device data exchange. One of the major differences regarding device data between these versions is that the device observation report was removed in the DSTU-2. The application that should be updated uses a REST interface to process device observations and is therefore dependent on this resource. This requires the evaluation of valid approaches to update the interface by choosing a different resource representation. Besides the decision on the correct resources to represent the information there was also an evaluation between the two reference implementations (RIs) that are mentioned on the official FHIR wiki [1] to determine which one should be used for this update.

2 Material and Methods

2.1 The REST interface to upgrade

The interface is responsible for receiving and processing data that contains vital signs measured by medical devices. It is embedded into a distributed client-server application that requires these measurements for further processing. This application uses representational state transfer (REST) over Hypertext Transfer Protocol (HTTP) to communicate. REST is a stateless, client-server communications protocol that can be used to communicate in distributed Systems [2]. The application receives data from a communication server invoking REST POST calls on the interface. This POST method contains a FHIR resource as JSON-code as payload, that has to be parsed for the application. Then it will be checked if all necessary information is contained in this message. Therefore a known medical code and a value with matching unit for this code needs to be present. Furthermore a device has to be contained in the request, representing the origin of these observations. Only observations that pass this check will be transformed into an domain object carrying only application relevant data over. The created domain objects will be saved to the database to be fully accessible for the whole application.

2.2 Programming languages and runtime environment

The production code for this project has been developed in Java 7. Java is an object orientated programming language and is also the most popular language in the enterprise back-end market [3]. Besides Java, Groovy has been used to write the Unit, integration and system tests. Groovy is a dynamic language that interoperates with Java-Code. Both languages get compiled into bytecode that is executable by the Java virtual machine like the Java Runtime Enviorment(JRE). The JRE enables multiplatform support for applications without the need to be recompiled for each different system [4].

2.3 Spring Framework

The Spring Framework provides comprehensive infrastructure support for developing Java applications. One of its core features is dependency injection and Inversion of Control (IoC): Decreasing the coupling between a class and a dependency [5]. It is an open source project often used in Java Enterprise Applications that contains many different modules that can be used on demand.

The Spring Modules used by the application that are connected to this update are:

- Core: providing the fundamental parts of the framework, including the IoC and Dependency Injection features

- Web: used to create the RESTful Web controller processing the FHIR resources

- Security: allowing a configuration for authorization and authentication allowing to process only data received from trusted sources

2.4   FHIR and reference implementations

FHIR is a standard developed by the HL7 organization for exchanging healthcare information electronically and especially developed to simplify implementation for the developers without sacrificing information integrity. It is meant to work with RESTful practices to enable interoperability between applications and organizations. At this time the current version of FHIR is the DSTU-2.

There are two different reference implementations of FHIR for Java that are published on the official downloads page on the wiki, which enable the use of FHIR-resources in the implementation [1]:

- The Native RI developed by the HL7 consortium consists of a model representation for FHIR resources and a parser to encode and read FHIR-JSON and XML representations.

- HAPI(HL7 application programming interface)-FHIR is a open source implementation maintained by the Universal Health Network (UHN). HAPI already existed as a Java parser for HL7 2.x message specification, that is used in healthcare information systems. When the FHIR standard was introduced HAPI-FHIR got developed, based on these specifications. Besides resource representations and parser the HAPI-FHIR framework also contains ready to use client-server implementations [6].

3   Results and Discussion

3.1   Analyzing alternate representations to substitute the DeviceObservationReport

One change between the FHIR versions DSTU-1 and DSTU-2 was the removal of the DeviceObservationReport resource. This was the base-resource used by the device data REST interface, containing all information about a device and the corresponding observations for this device. Therefore an adequate replacement was required.

There were four resources that seemed to be eligible serving as a container for the required device and observations resources:

1. MessageHeader

2. DiagnosticReport

3. List

4. Bundle

Further investigation showed the MessageHeader resource did not meet the requirements because its main purpose is to represent a message with optional payload and add metadata about the sending and receiving applications [7]. There is no need to process routing information for the controller and the resource seems to be inappropriate for the interface. The definition for the List resource states it to be a dynamic collection where items can be added and removed over time [7]. The use case for the interface does not require an alternation of the lists content.

The DiagnosticReport resource offers every feature needed by setting its subject to device and the results to the corresponding observations, but its definition to contain a clinical relevant report like a laboratory report clearly does not match the interface use case [7].

The common task of the Bundle to gather and send a collection of resources in a single instance is all that is required by the interface. Every information the application requires can be delivered by a Bundle Resource containing a Device and the corresponding observations. This construct seems to be appropriate for the application and is further referred to as device metric Bundle. In Fig. 1 the structural differences between the device metric representations can be seen. Because the application does not use the VirtualMedicalDevice, channel and metric information, these sections were not maintained in the device metric Bundle.

3.2   Decision on the Reference Implementations of FHIR

The original implementation used the native Java RI of FHIR. The updated interface should be built up in parallel to the existing one so that it allows the application to process data from both interfaces for a transitional period. This approach is costly to achieve by using the native Java RI considering that the different versions of the RI share exactly the same Classpath. Adding both as a dependency in the same project would result in an error. To resolve this issue it would be necessary to create and deliver an adapter for the different versions and connect them to an additional interface of the application. In addition the RI contained dependencies for test only packages which had to be actively excluded in the final application. Working with the resources often requires to write many similar blocks of code resulting in hardly readable sourcecode. With these problems in mind the use of another RI suggested by the official FHIR website HAPI FHIR (HAPI) was considered. HAPI offers to use multiple FHIR DSTUs in one application by including the versioning into the API's package

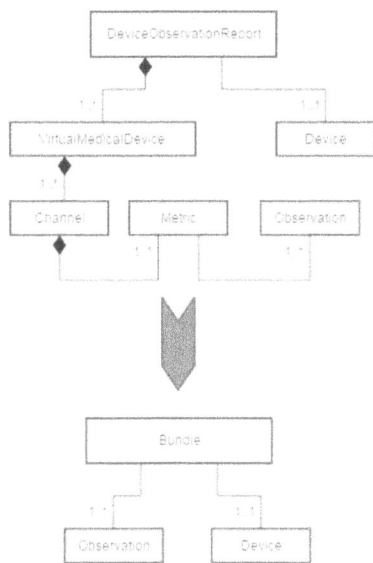

Figure 1: Structural representation of the relevant Observations and the Device in the original DeviceObservationReport and the new representation as Bundle.

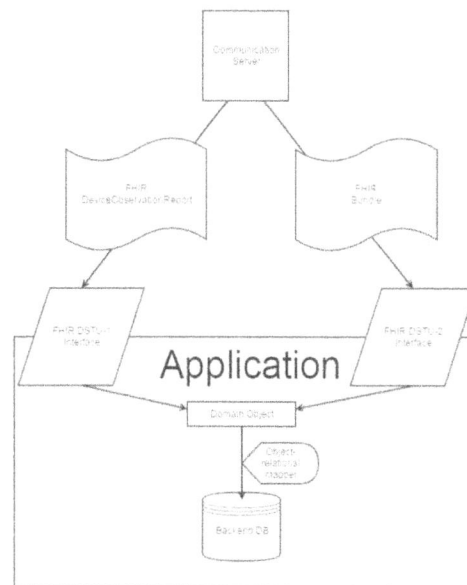

Figure 2: This figure shows the Application with two interfaces providing the same functionality for a FHIR DeviceObservationReport for DSTU-1 and a FHIR Bundle for DSTU-2, the device metrics Bundle. Resources containing identical information get processed in the same way resulting in an identical domain object that get saved to the database through an ORM.

structure allowing a parallel update approach. The construction and use of HAPI resources is much more developer friendly than the native RI because of its fluent interface, allowing the developer to chain multiple commands. The code becomes less cluttered and the use of temporary variables is minimized. The only drawback of HAPI is the use of an HAPIContext object, which is required to create a Parser for FHIR resources. Its instantiation is expensive. By just using one instance of this thread safe object this drawback can be easily canceled out. A comparison of these RIs favored the use of the HAPI FHIR API.

3.3 Implementation

There has been a decision that it would be desirable to have a transitional period where both either FHIR DSTU-1 resources or DSTU-2 resources should be able to be processed. This assures a timeframe for the sending application to be adjusted. Requests containing the same information should not deviate in the domain model used in the backend of the application, reducing the need for change to the interface like it can be seen in Fig. 2. The update followed the Test Driven Development (TDD) pattern. The already existing tests for the existing interface had to be copied for the updated version expecting the same outcome for input containing the same information.

First a Message Converter had to be implemented converting the JSON resource to a HAPI FHIR Object using the HAPI Json Parser. The spring framework automatically detects the FHIR resources and converts them. The architectural design of the application corresponds to the Clean Architecture Design [8] and consist of multiple layers. This

update needs to add a new controller and service to enable the processing of DSTU-2 Bundles.

The controller operates as the direct REST interface and delegates the processing of request to the service. The service serves as a connection between the controller and domain objects by transforming the received FHIR resources into domain objects. These objects can be saved to the database by using an object-relational mapper (ORM). The transformation is where most of the update happened. The structure of the Bundle that needs to be processed is equivalent to a collection containing a Device resource and at least observation resource. Each observation contains one measurement of the Device in the Bundle. The observations are checked for validity corresponding to type, unit and range of the measurement, mapped to the domain object containing all information needed by the application and finally saved to the database. If the received observations could be saved the controller returns a HTTP created status to the sending application, otherwise a HTTP error for bad request. Because there were no changes done to the domain object no other parts of the application needed to change and the update was complete.

4 Conclusion

Once a matching resource was found the replacement of the device observation report with the metric Bundle representation was possible without any problems. By using the Bundle as container object all necessary information were preserved while the structure of the resource was less complex. This resulted in a more understandable resource and less lines of code to represent it in a JSON-encoding. After the finished release of FHIR it would be necessary that adequate replacements for removed resources would be communicated, reducing the need to personal analyze possible replacements and to have a standardized approach to represent the data.

By having a layered architecture and loosely coupled dependencies the changes for this update only affected a small part of the whole application. This simplified a lot of the effort necessary to develop the new controller. The functionality delivered by the Spring Framework allowed an easy coupling of the controller, service and domain object allowing the developer to focus on the pure functionality. Because the update should not alter the interface's functionality most of the logical steps stayed the same. Therefore the update mainly altered the corresponding resources and the methods that are used to collect them. Using the TDD approach it was easy to recognize when the update has been completed successfully, since only all tests containing requirements for the old interface had also to pass for the new interface. TDD minimizes the possibility to write code delivering incorrect or unexpected results which do not get detected by the developer. Because it is much safer to change methods covered by testcases big chunks of code can be adapted without the risk to create an error elsewhere in the application.

The FHIR specification is a great way to ensure interoperability between different application without adding much complexity to the systems. Once a finished FHIR-version is released the necessity to update applications using this specification will drastically decrease, since there will be less frequent releases and probably minor change sets. Interfaces that only consume FHIR resources without using the resource representation internally do not have a lot of code working directly with FHIR resources resulting in less code that needs to be adapted. Once FHIR achieves a stable specification it might be a viable solution to use FHIR resources as domain objects. This reduces the need to convert the specification for complying interfaces.

The parallel development of the native RI shows that the developers are an important audience for the HL7 consortium since the success of a new specification greatly depends on its acceptance and use in newly developed applications. The RI contains all functionality necessary to use and parse FHIR resources but it requires an adapter per FHIR version used at the same time in an application. The HAPI RI offers a lot more functionality than the native RI. But most business application do not need the client and server features HAPI delivers. By including the FHIR version number into the package-names it is possible to use multiple FHIR versions in one application. While this allows to write multiple interfaces for different versions an update of the internally used FHIR version would require the developer to adapt all imports in these classes. The biggest improvement against the native RI is the fluent interface that allows the developer to write much more readable code with HAPI and therefore reduces the risk of errors.

5   Outlook

FHIR offers the use of profiles to further specify the resources used. In the device data context a suitable profile would be the DeviceMetricObservation that could be applied to the observation resource. This profile would explicitly describe the observations to be a direct or derived, qualitative or quantitative physiological measurement, setting or calculation by a medical device or a device component. Another possibility for a custom defined profile would be the Bundle resource restraining it to contain exactly one Device and at least one observation. These FHIR-profiles allow to add mandatory information to the resources while allowing to ignore the unnecessary information, communicating the desired fields directly to the sending application. This reduces overhead and reduces the risk to receive resources not containing the required data.

Acknowledgement

The work has been carried out at Drägerwerk AG & Co. KGaA, Lübeck.

6   References

[1] Health Level Seven International (HL7) health-care standards organization. *Downloads - FHIR v1.0.2.* Available: https://www.hl7.org/fhir/downloads.html [last accessed on 26.01.2016].

[2] E. Wilde. (2007). *Putting Things to REST.* School of Information. UC Berkeley: School of Information. Retrieved from: http://eprints.cdlib.org/uc/item/1786t1dm [last accessed on 09.02.2016].

[3] TIOBE Software, *TIOBE index.* Available: www.tiobe.com [last accessed on 26.01.2016].

[4] C. Ullenboom, *Java ist auch eine Insel.* Rheinwerk Computing, Bonn, 2010, ch. 1.2.

[5] R. Johnson et al., *Spring Framework Reference Documentation.* Available: http://docs.spring.io/spring/docs/current/spring-framework-reference/htmlsingle/ [last accessed on 26.01.2016].

[6] University Health Network, *Introduction to HAPI FHIR.* Available: http://jamesagnew.github.io/hapi-fhir/ [last accessed on 26.01.2016].

[7] Health Level Seven International (HL7) health-care standards organization, *FHIR Resource Index.* Available: https://www.hl7.org/fhir/resourcelist.html [last accessed on 26.01.2016].

[8] R. Martin, *The Clean Architecture.* Available: https://blog.8thlight.com/uncle-bob/2012/08/13/the-clean-architecture.html [last accessed on 26.01.2016].

7

Signal Processing

Memory Performance of Spiking Neural Networks
– How to Store and Retrieve Information –

S. Holtorf[1], A. Sinha [2], N. Davey [2] V. Steuber [2] and C. Metzner[2,3]

[1] Medizinische Ingenieurwissenschaft, Universität zu Lübeck, stella.holtorf@student.uni-luebeck.de
[2] University of Hertfordshire, Biocomputation Research Group, {a.sinha2, n.davey, v.steuber}@herts.ac.uk
[3] Institute of Robotics and Cognitive Systems, Universität zu Lübeck, metzner@rob.uni-luebeck.de

Abstract

Synaptic plasticity is the basis of memory formation in neural networks, and while networks undergo various changes as memory formation occurs, homeostatic plasticity mechanisms help these networks remain stable. In a recent paper, Vogels et al. use computational modeling to demonstrate the storage of associative memories in a network that uses a new symmetric spike time dependent plasticity (STDP) learning rule at inhibitory synapses to maintain stability. While it is accepted that memories form in excitatory neuron sets, we extend the model to investigate memory formation in inhibitory neurons and their effects on network stability and the memory storage capacity of excitatory neuron sets. We also investigate the interactions of various other types of synaptic plasticity with this stabilizing STDP and their effect on memory capacity.

1 Introduction

While it is widely established that plasticity at excitatory (E) synapses is necessary for the formation of memories in neural networks, the role of plasticity at inhibitory (I) synapses remains almost unexplored [1]. Recent work from Vogels et al. [2] highlights the importance of inhibitory synaptic plasticity in regulating excitation and inhibition in a neural network. They show that associative memories can be stored and recalled attractor-less, i.e. without sustained activity above background level. The following equation describes, how the evolution of synaptic weights is calculated:

$$W_{ij} \to W_{ij} + \eta(x_i - \alpha) : \text{presynaptic spike at time } t_j$$
$$W_{ij} \to W_{ij} + \eta x_j : \text{postsynaptic spike at time } t_i$$
$$(1)$$

This equation allows to determine the de- or increase of the weights of the connection $W_{i,j}$ depending on the time difference between pre (i) and postsynaptic (j) neuron spikes, by using the trace of synaptic activity x, a learning rate η and a depression constant α caused by presynaptic spikes. Figure 1 (taken from [2]) shows how synaptic weights will de- or increase as a function of the time difference between pre- and post-synaptic spikes and represents equation 1.

As you see in Figure 1, independent from which neuron (pre- or post-synaptic) spiking comes first, the weight of the connection will increase for near-coincident spikes. This rule forces postsynaptic neurons to keep their firing rates low and leads to an asynchronous irregular (AI) state. This AI state is necessary in order to effectively store patterns in

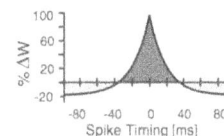

Figure 1: STDP which determines the hight of increase of synaptic weights as a function of pre-and postsynaptic spike-timing (Figure taken from [2])

a network. This homeostatic mechanism allows to create attractor-less memory [2]. An example of how the network is able to store information attractor-less can be seen in Figure 3.

We extend the simulation model used by Vogels et al. [2] to investigate how storing memory patterns in inhibitory neuron sets affects the stability of the network and in turn, the capacity of the network. Former studies have investigated the effects of different types of synaptic plasticity and different network configurations on the performance of associative memory networks with excitatory and inhibitory populations [3]. This work shows that restricting the flexibility of the network, by fixing the weights for certain types of connections, decreases the memory capacity up to the point where no patterns can be stored. Metaxas et al. [3], however, use simple binary neurons for their investigation. Here we wanted to follow up their work in the biologically more plausible model of Vogels et al. [2].

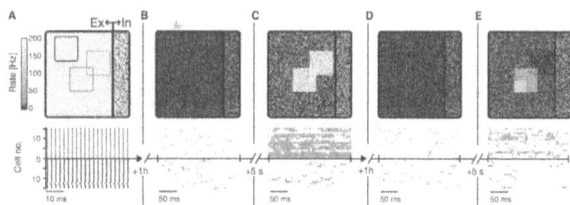

Figure 2: **Example of the five stages of the simulation,** **A** synchronous regular network dynamics, with high firing rates in the whole neuron set, **B** low firing rates in the whole neuron set in the asynchronous irregular state (AI) after switching on the I-to-E synaptic plasticity as a regulatory mechanism, **C** Presentation of two patterns increases the firing rates of pattern neurons, **D** Recovery of AI by plastic connections between I neurons which project to E neurons, **E** Recalling pattern by adding noise as a trigger to the network and getting high firing rates from patterns compared to background signal (patterns flash up). (Figure taken from [2])

2   Methods

We used the Auryn [6] simulator to replicate and then extend the model used by Vogels et al.. Our version is expected to store many (25 randomly generated) patterns like in the work of Metaxas et al. [3] and in contrast to the setup from Vogels et al. [2]. We used the signal to noise (SNR) metric proposed by Dayan and Willshaw [5] to measure the memory capacity of the network. We also tested various configurations of synaptic plasticity building on the work of Metaxas et al. [3].

2.1   Basic Network

The simulation consisted of 8000 excitatory and 2000 inhibitory spiking leaky integrate and fire (LIF) neurons. In the initial network configuration, all synapses other than the I to E synapses were static while the I to E synapses were plastic and followed the symmetric spike time dependent plasticity rule. The connections had a sparsity of 2 %, like it is found in the human brain [4]. Associative memories to be stored in the network consisted of randomly chosen subpopulations of neurons - 10 % of excitatory and 10 % of inhibitory neurons for patterns stored in excitatory and inhibitory populations, respectively. The learning of the patterns was realized by using one-shot Hebbian learning, i.e. an increase of the E-E and I-I weights by a multiplication factor. While investigating different combinations of plastic and static synapses, we only stored patterns in the excitatory subpopulation.

2.2   Simulations

Our model included four stages, which every simulation had to go through (summarized in table 1). Depending on the number of patterns, stage two to four had to be repeated for each case (presenting just one pattern up to presenting all of

them to the network) and the SNR was calculated for every recall.

Table 1: Stages of the Implementation

Stage	Task	Duration
First	Increase background noise, Switch on STDP	
	→ network goes to AI state	600 sec
Second	Stores patterns	1500 sec
Third	Switch off STDP, Stimulate recall subset	1 sec
Fourth	Record activity to calculate SNR	2 sec

2.3   Storing Patterns in Both Excitatory and Inhibitory Populations

. At first, we wanted to store patterns in the E and in the I population (so-called hybrid-patterns). Therefore, the first modification we did to the network, was to extend the pattern (originally only including E neurons) with a subset from the I neuron population. We chose a pattern size for the combined pattern of 1000, with 800 E neurons and 200 I neurons to keep a biologically plausible ratio of excitatory to inhibitory neurons [4]. We varied the multiplication factors of the weights of the synapses between neurons forming patterns from 0.01, 0.1, 0.5, 1.0, 2.5, 5.0, 7.5 up to 10.00 to investigate their effects on the memory performance of the network. In this test I-to-E connection weights were plastic, all the other connection weights were fixed.

2.4   Investigation of Different Synaptic Configurations

Next, we wanted to test different types of plasticity in the prescribed network setup. Similar to Metaxas et al. [3]. In our model, from all possible configurations, only the cases including I-to-E plasticity make sense, since we require a homeostatic mechanism to store patterns attractorless. Therefore we could rule out each of the 16 possible configurations, which did not include I-to-E plasticity. This resulted in the following five configurations

1. E-to-I (sym.) and I-to-E (sym.) plastic

2. E-to-I (asym.) and I-to-E (sym.) plastic

3. All plastic

4. I-to-I static

5. *-to-I static

Due to time constraints, we were so far only able to investigate the first two of the above five configurations. We also varied the constant depression factor α over the values 1, 3 and 9 for I neurons projecting to other neurons and 5, 15 and 45 for E neurons projecting to other neurons. To make sure the network had sufficient time to reach the AI state, we increased the stabilization time (second stage) from 1500 sec to 4000 sec. In this test series we tried to store only one pattern in the network and investigate which of the configurations from above were able to stabilize to the AI state and store a pattern.

3 Results and Discussion

3.1 Results of Storing Patterns in Both Excitatory and Inhibitory Populations

3.1.1 Varied Pattern Weights

The results for presenting 25 hybrid-patterns to the network which can be seen in Figure 3 shows, that, as we expected, high weights at E-to-E connections within patterns and low weights at I-to-I connections within patterns led to the highest storage capacity. This can be explained by the high excitation of neurons in the network while the inhibition is very low - this increases the SNR. Low synaptic weights in the inhibitory patterns resulted in higher SNRs. The lowest SNR was produced by a network with an inh weight of 10.0. The curves of lower synaptic weights between neurons in patterns stored in inhibitory neurons are slightly shifted to the right, meaning that the storage of patterns is better in those networks (e.g. maximal SNR at 4 patterns compared to the maximal SNR at 2 patterns a configuration with $pat_{exc} = 5.0$ and $pat_{inh} = 1.0$).

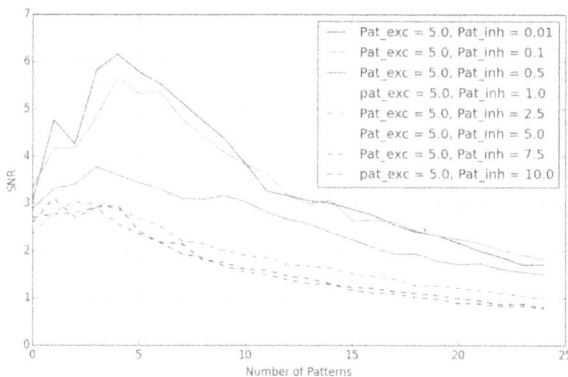

Figure 3: SNR graphs of network with I-to-E synaptic plasticity, network size of 8000 : 2000 (E : I) neurons. The excitatory weights were fixed to 5.0 and the inhibitory weights of the patterns were set to 0.01, 0.1, 0.5, 1, 2.5, 5, 7.5 and 10.0. All 25 pattern consist of 800 excitatory and 200 inhibitory neurons.

3.1.2 Varied Pattern and Network Sizes

The excitatory and inhibitory synaptic multipliers of pattern weights were fixed to 10 and 5, respectively. Figure 4 shows the SNR when we halve or quarter the initial size of the network of 8000 : 2000 (E : I) neurons. The pattern sizes corresponded to 10 % of the network size. The network with the highest number of neurons exhibited highest memory capacity, its SNR was almost three times the maximal SNR compared to the smallest network. However, the

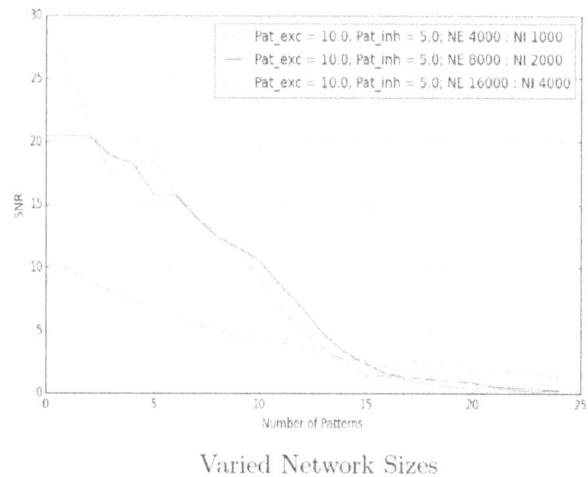

Varied Network Sizes

Figure 4: SNR graph of network with I-to-E synaptic plasticity. The excitatory weights were fixed to 10.0 and the inhibitory weights of the patterns were set to 5.00. All 25 pattern consist of 800 excitatory and 200 inhibitory neurons. The network size was varied with 4000 : 1000, 8000 : 2000 and 16000 : 4000 (E : I) neurons.

smallest network shows the smallest decrease of SNR with growing number of patterns, since the number of patterns where the SNR falls down under a threshold of two is almost the same for all networks.

In Figure 5 the network size was set to 16000 : 4000 (E : I) neurons and we varied the size of the patterns. The number of neurons in the pattern was set so 10 %, 5 % und 2.5 % of the particular neuron group. The simulation with the largest pattern size produced the largest maximal SNR, whereas a pattern size of 2.5 % of the network produced a SNR near zero. The pattern size with 5 % of the network size produced a maximal SNR of 5 but the decrease of the SNR was much slower than the decrease for the largest pattern size.

3.2 Results of Tests with Different Synaptic Configurations

3.2.1 Network with Symmetric E-to-I and I-to-E Plasticity

In our the tests with symmetric E-to-I and I-to-E synaptic plasticity, the firing rates over time for this configuration were able to reach the AI state. Initial high firing rates around 90 Hz decreased drastically within 60 sec to firing rates around 21 Hz (I neurons) and 4 Hz (E neurons) but there was another peak around 500 sec with firing rates of 55 Hz (I neurons) and 25 Hz (E neurons). After this initial phase, homeostatic plasticity forced the firing rates to decrease again and the activity level of both types of neurons was stable at around 5 and 15 Hz, respectively. Summarizing, it tok the network around 3000 sec to converge to the an AI state with low network activity.

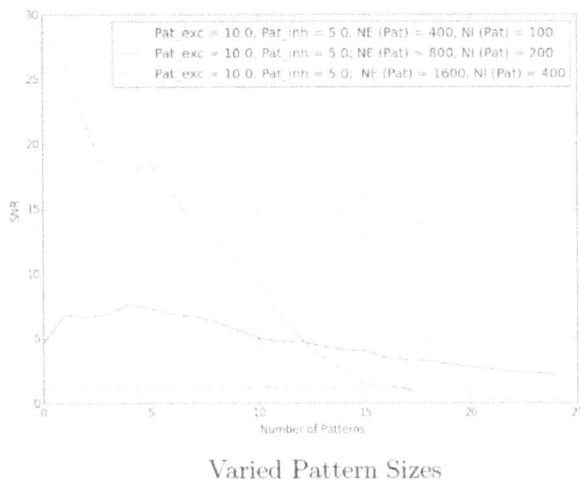

Varied Pattern Sizes

Figure 5: SNR graph of network with I-to-E plasticity, pattern size of 8000 : 2000 (E : I) neurons. The excitatory weights were fixed to 5.0 and the inhibitory weights of the patterns were set to 10.0. The pattern size of all 25 pattern was varied with 400 : 100, 800 : 200 and 1600 : 400 (E : I) neurons.

3.2.2 Network with Asymmetric E-to-I and Symmetric I-to-E Plasticity

Next, we introduced an asymmetric learning rule implemented in Auryn to the model [6]. Here we saw almost periodically appearing peaks of the firing rates of E neurons even after several thousands of seconds. This means that the network was not able to settle into the AI state and thus was not able to store any pattern.

4   Conclusion

In section 3.1, we show the effects of storage of patterns in the inhibitory neuron population on patterns stored in the excitatory set. A limitation of this approach is, that there was no correlation between patterns in the E and I neurons, and future work should explore what effect correlations between patterns in E and I neurons might have on memory capacity. In section 3.2, we report that the addition of asymmetric STDP plasticity at E-to-I synapses prohibited it from stabilizing to an AI state, while the addition of symmetric STDP to E-I synapses did not have this effect [6]. These results imply that there is a fundamental difference between these two configurations. It will be interesting to see whether this difference also holds when one includes even more plastic types of connections, as outlined in section 2.4.

Acknowledgement

The work has been carried out at the Biocomputation Research Group, University of Hertfordshire, Hatfield, United Kingdom.

5   References

[1] D. M. Kullmann, K. P. Lamsa *Long-term synaptic plasticity in hippocampal interneurons*. Nat. Rev. Neurosci., vol. 8, no. 9, pp. 687–699, 2007.

[2] T. P. Vogels, H. Sprekeler, F. Zenke, C. Clopath and W. Gerstner *Inhibitory plasticity balances excitation and inhibition in sensory pathways and memory networks*. Science, vol. 334, no. 6062, pp. 1569–1573, 2011.

[3] A. Metaxas, R. Maex, V. Steuber, R. Adams and N. Davey *The effect of different types of synaptic plasticity on the performance of associative memory networks with excitatory and inhibitory subpopulations*. in Information Processing in Cells and Tissues, Springer, pp. 136–142, 2012.

[4] V. Braitenberg, A. Schulz *Long-term synaptic plasticity in hippocampal interneurons*. Cortex: Statistics and Geometry of Neuronal Connectivity, 2nd edn. Springer, Heidelberg, 1998.

[5] D. Willshaw, P. Dayan *Optimal plasticity from matrix memories: What goes up must come down*. Neural Computation, MIT Press, vol. 2, no. 1, pp. 85–89, 1990.

[6] F. Zenke, W. Gerstner, *Limits to high-speed simulations of spiking neural networks using general purpose computers*. Front Neuroinform 8, 76. doi: 10.3389/fninf.2014.00076, 2014.

Identifying an Oxygenation Model from FiO2-SpO2 Relations in Ventilated Patients

J. Beuke [1], L. Kahl [2], D. Schädler [3], T. Becher [3], N. Weiler [3], and P. Rostalski [4]

[1] Medizinische Ingenieurwissenschaft, Universität zu Lübeck, jonas.beuke@student.uni-luebeck.de

[2] Drägerwerk AG & Co. KGaA, Lübeck, lorenz.kahl@draeger.com

[3] Department of Anesthesiology and Intensive Care Medicine, University Medical Center Schleswig-Holstein, Campus Kiel, {dirk.schaedler,tobias.becher,norbert.weiler}@uksh.de

[4] Institut für Medizinische Elektrotechnik, Universität zu Lübeck, philipp.rostalski@uni-luebeck.de

Abstract

Peripheral blood saturation measured by pulse oximetry (SpO_2) is a clinical standard to constantly monitor a patient's oxygenation. Under mechanical ventilation the fraction of inspired oxygen (FiO_2) is defined by the ventilator settings and not by ambient gas. An approach for relating changes in FiO_2 to the resulting SpO_2 by means of a dynamic model is presented and evaluated using measured data. For identification it is important to have several preoxigenation manoeuvres, where FiO_2 is temporarily increased to 100%, clinically used to ensure patient safety during interventions where mechanical ventilation is infeasible. These steps in the data are utilized to estimate the delay of changes in FiO_2 being distributed to the point of measurement. Afterwards, the shunt, a parameter describing the venous admixture, is estimated from the data. The procedure is sufficient to describe 89.4% of the measured data points within the 3% accuracy of the pulse oximeter.

1 Introduction

A thorough understanding of the human respiratory system is a crucial part in the development of better ventilation concepts and the design of medical devices. Using mathematical models is a common approach to characterize the complex interactions in the human body. These models can also be used for example in the training of medical staff without putting patients at risk.

This work aims to relate the *fraction of inspired oxygen* (FiO_2) to the *oxygen saturation in peripheral blood measured by pulse oximetry* (SpO_2). Both values allow for a continuous monitoring and can be easily measured in clinical routine. The obtained data will be used to estimate the parameters of a model.

Pulse oximetry is a noninvasive method for estimating the *arterial oxygen saturation of hemoglobin in blood* (SaO_2). It uses the fact, that the light absorption properties of oxigenated and deoxigenated blood differ at different wavelengths. By assuming that only the arterial component of the absorption is affected by the pulse, it is possible to distinguish between arterial and venous blood [1]. Pulse oximeters are empirically calibrated and their accuracy generally differs from ±1% to ±3% depending on probe type and measuring conditions. Motion artefacts or a bad perfusion can significantly lower the accuracy [2]. Measured values are averaged for stabilisation purposes and therefore slightly delayed [1].

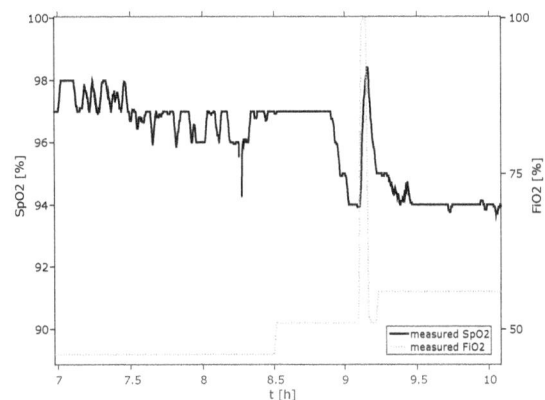

Figure 1: Section of the time series (smoothed with a moving average filter over 60s for visualisation purposes). The patients response to a preoxigenation peak can be seen at around $t = 9.2$h.

2 Material and Methods

The data used to test the presented approach was obtained monitoring one patient over a 20 hour period by a Masimo Radical 7 pulse oximeter and a Dräger Evita V500 ventilator. The resolution of the pulse oximeter is 1%. The pulse oximeter was placed on the patients finger and the measurement of FiO_2 was performed in the ventilator.

The time series from both devices were synchronized and

Symbol	Definition
$P_{H_2O} = 47mmHg$	Partial pressure of water vapour at $37°C$
$P_{bar} = 760mmHg$	Ambient pressure
$\lambda = 863mmHg$	Conversion coefficient
$V_A = 3.28l$	Alveolar volume
$V_T = 39.64l$	Tissue volume
$\dot{V} = 4.28\frac{l}{min}$	Air flow
$Q = 5.04\frac{l}{min}$	Blood flow
$CR_{O_2} = 0.242\frac{l}{min}$	Oxygen cosumption rate
$\alpha = 3.17 \cdot 10^{-5}\frac{ml}{ml \cdot mmHg}$	Solulability constant of oxygen
$Hb = 0.15\frac{kg}{l}$	Haemoglobin concentration
$k_H = 1.34\frac{kg}{l}$	Hüfner constant

Table 1: Constants used for simulation (based on [3]).

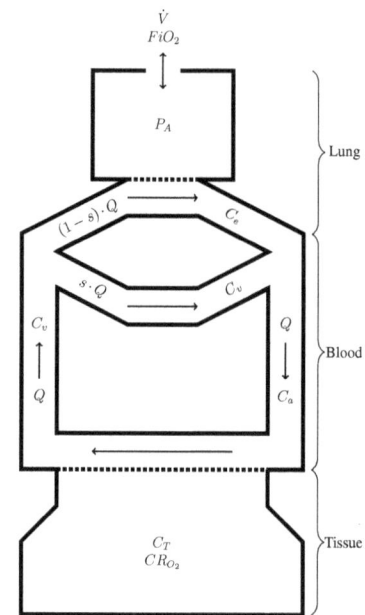

Figure 2: Schema of the different parts of the model (adapted from [3]). The dotted lines denote the diffusion into the blood and the tissue respectively.

the values were linearly interpolated to a sample rate of 1 Hz (on the basis of 1 Hz in the pulse oximeter data and 0.17 Hz in the ventilator data) using the mathematical software R (version 3.2.2). All SpO_2 measurements below 70% were considered to be erroneous and were therefore excluded from the calculations indicating a poor signal quality. Values of $SpO_2 = 100\%$ were set to 99.99% in order to avoid calculation problems with the oxygen dissociation curve.

The data was recorded under normal care and there was no protocol available of activities concerning the patient.

The peaks in FiO_2 are preoxigenation manoeuvres used to prepare the patient for a period without mechanical ventilation. This is standard clinical practice before eg. suctioning the patients airway. An example is given in Fig. 1.

The study was approved by the local ethics committee (D411/13) from the Christian-Albrechts-University Kiel with a waiver for informed patient consent due to the observational study design.

2.1 Model

A physiological model derived from [3] was chosen since it allows adjusting the model with just one parameter: the shunt. It describes the *venous admixture to the end-capillary blood* due to blood bypassing the gas exchanging part of the lung. The physiological shunt is normally around 2% of the blood flow [5] but since various types of lung diseases affect the gas exchange and therefore the *alveolar oxygen partial pressure* (PaO_2), the value s should better be considered *virtual shunt*. This simplifies the identification procedure and describes the patients medical condition at least partly (compare [2]). The model assumes a continuous flow in both lung and blood instead of simulating inhalation cycles and pulsation (see Fig. 2). The decision for a model type was based on the facts that the reaction to FiO_2 changes is rather slow, that pulse oximeter values are inherently inaccurate and that a simple physiological model will less likely overfit to the data.

The model was implemented in MATLAB (R2014b, The Mathworks, Natick, MA, USA) using the `ode15s` solver with a maximum step size of $30s$. Only the oxygenation part of the model was used, since the scope of this work

is primarily the oxygenation. This leads to the equations as described in the following. Unless stated otherwise, the equations are found in [3]. The constants used for the simulation are found in Table 1.

The *partial pressure of inspired oxygen* (P_i) is given by

$$P_i = (P_{bar} - P_{H_2O}) \cdot FiO_2. \tag{1}$$

The FiO_2 value refers to dry air and calculating P_i in (1) must correct for the humidification in the upper airways [4].

$$\frac{dP_A}{dt} = \frac{\dot{V} \cdot (P_i - P_A) - \lambda \cdot Q \cdot (1 - s) \cdot (C_e - C_v)}{V_A} \tag{2}$$

relates P_A to the change due to P_i and the flow into the blood given by the difference between *venous* (C_v) and *end-capillary* (C_e) *oxygen concentration*. The conversion from concentration to partial pressure is captured in the parameter λ. The symbols $\dot{V}$ and Q denote the *respiratory* and *heart minute volumes* respectively. V_A is the *volume of the lung's alveolar space* and thus the gas exchanging part of the lung.

$$C_a = (1 - s) \cdot C_e + s \cdot C_v \tag{3}$$

connects the different oxygen concentrations by means of the shunt fraction s.

$$\frac{dC_T}{dt} = \frac{Q \cdot (C_a - C_v) - CR_{O_2}}{V_T} \tag{4}$$

describes the concentration of oxygen in the body tissue C_T in relation to the volume flow in and out of the tissue. The *consumption rate of oxygen* CR_{O_2} due to the metabolism is considered to be constant.

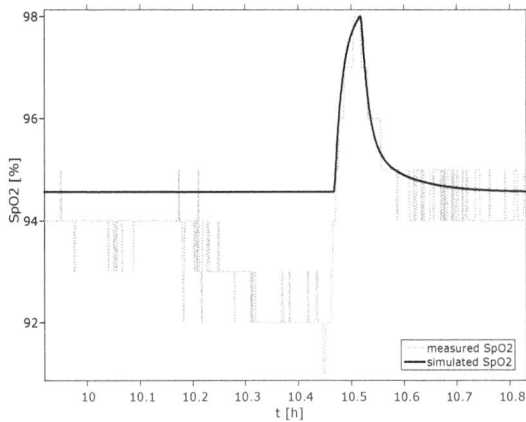

Figure 3: This example shows clearly how the model is able to accurately describe the SpO_2 for preoxigenation manoeuvres but is unable to depict the desaturation in the run-up. The simulation parameters are $s = 32.7\%$ and $\tau = 45s$.

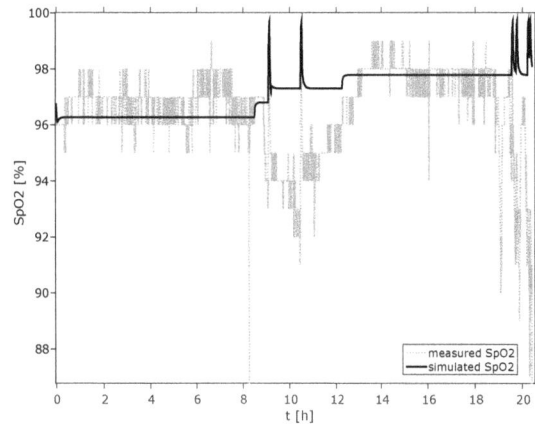

Figure 4: This example shows how the desaturation between the preoxigenation peaks is clearly out of the pulse oximeters error range whereas for the parts of constant FiO_2 the simulated values are mostly within the error range of $\pm 3\%$. The simulation parameters are $s = 26.5\%$ and $\tau = 45s$.

Assuming flow limitation and no diffusion impairment, the *partial pressures in end-capillary blood* (P_e) and *alveolar space* (P_A) leads to

$$P_e = P_A. \tag{5}$$

For the tissue part the equality of *partial pressures in venous blood* (P_v) and *tissue* (P_T) is defined by

$$P_v = P_T = \lambda \cdot C_T. \tag{6}$$

The conversion between PaO_2 and SaO_2 was implemented using the oxygen dissociation curve developed by Severinghaus [6]. A constant pH of 7.4 and a standard body temperature of $37°C$ were assumed since no blood gas samples or core temperature measurements were available.

Pulse oximeters measure the *functional saturation* (saturation of haemoglobin available for binding) whereas SaO_2 is the *fractional saturation* (saturation of the haemoglobin as a whole). To compensate for the effect of non-binding carboxihaemoglobin and methaemoglobin at normal levels a 2% surcharge is applied to SaO_2 [1]:

$$SpO_2 = 1.02 \cdot SaO_2. \tag{7}$$

For calculating the oxygen concentration both the haemoglobin and the plasma component in blood are considered in

$$CaO_2 = k_H \cdot Hb \cdot SaO_2/100 + PaO_2 \cdot \alpha \tag{8}$$

where applying Henry's law with a solubility factor of α accounts for the plasma part [1].

An iterative algorithm was used to convert CaO_2 to SaO_2 as (8) cannot be analytically inverted.

2.2 Identification

Since it takes some time for a change in FiO2 to be distributed through the lung, into the blood and finally to the position of the sensor, the reaction is delayed. This delay is mainly influenced by the heart's minute volume and therefore the speed of blood distribution, which is considered constant for this analysis. The delay was identified by simulating the model with the delay $\tau = 0s$ at a shunt value of $s = 20\%$ and aligning the preoxigenation peaks in the measured and the simulated time series. The shunt fraction is assumed to only change the peak size but not the delay time and thus permits the procedure above. The idea of using the simulated time series instead of directly aligning measured SpO_2 to FiO_2 is that the peak's shape should be similar between simulated and measured SpO_2 and the alignment might therefore be more accurate. The delay is implemented by shifting the time series by τ.

With this information the shunt fraction was estimated by comparing the model to the measurements using the sum of squared errors (SSE) criterion. The SSE is minimized by applying MATLAB's `fminbnd` with $0 < s < 50\%$. Values of $s > 50\%$ were not considered since according to [7] such values lead to "increases in FiO_2 having no effect on PaO_2". Instead of calculating the SpO_2 for every simulated data point, the comparison was based on the CaO_2 derived from the SpO_2 measurements. This is only done once before the minimization step and therefore significantly saves computation time.

3 Results and Discussion

An example for the exclusion of values can be seen in Fig. 1 at around $t = 8.3h$ due to zeros in the original data. It is clearly visible from the time series, that even though FiO_2 does not change for over an hour the values of SpO_2 mostly range between 96% and 98%. The FiO_2 change at $t = 8.5h$ does not seem to increase SpO_2 but to stabilise it. Another interesting observation about the SpO_2 is the drop beginning at $t = 8.9h$ which is not caused by an FiO_2 change. Medicaments, suctioning manoeuvres or patient movements might, among others, explain the unexpected course of SpO_2.

Fig. 3 shows the models performance for preoxigenation peaks. The shape and decay time are matching well, however the desaturation in the forefront is not reflected in the simulated time series. The patients SpO_2 in Fig. 4 was almost constant at a lower FiO_2 for hours. This might be, among other causes, explained by the accompanying drop in the PEEP (Positive End-Expiratory Pressure; not in the figure) resulting in a lower alveolar recruitment and thereby a lower diffusion into the blood [1]. But since no changes to the FiO_2 settings were made, the model does not depict this change.

The same seems to apply to the last hour of the time series, where the simulated value follows the course of FiO_2 but the measurements seem to be more or less independent from this. A closer look reveals, that there are peaks that are likely to correspond to the preoxigenation manoeuvres but the overall saturation is declining. Since there are several peaks in a short period of time and also a wide variation of measured SpO_2 values, there might have been some intervention maybe including movements or something of similar effect leading to a poor signal quality.

The calculated time delay τ is small with 45s compared to reported values of $130s$ at normothermia and in healthy subjects [8]. For the most part, the model does simulate the patient's SpO_2 rather well with 89.4% of the simulated values being within $\pm 3\%$ around the measured ones.

Another source of error apart from the pulse oximeter inaccuracies is the empiric oxygen dissociation curve used for converting SpO_2 and PaO_2. The dissociation curve was empirically fitted to data and does have a mean error of $0.16 \pm 1.12\%$ according to [9]. The curve is used in three places of the model: the diffusion of oxygen from alveolar space into blood, the partial pressure equilibrium between tissue and blood and when calculating the SpO_2 from CaO_2. This leads to an error propagation throughout the model. Since the curve is nonlinear and shifts to the right for higher body temperatures and more acidic conditions [5], small errors in the assumed values can have a high impact, depending on the slope of the curve for the given conditions.

4   Conclusion

The presented approach shows that even a simple model with only one parameter is able to describe the oxygenation of a patient accurately within the precision of the pulse oximeter for most of the time. The parameter estimation of the model is subject to several sources of error and the resulting value for the virtual shunt cannot be considered to directly reflect the patients real condition. The precision could be improved by adding more knowledge about the patient. Especially blood gas samples and values of expired CO_2 might be helpful to model the patients blood chemistry more accurately. Another way to improve the significance of the estimated patient condition could be to add diffusion impairment (see [7]). To depict the effect of

the PEEP a more precise model of the lung could be used. Some pulse oximeters provide information about the signal quality. This could be used to improve the robustness of the parameter estimation by weighing the prediction error with the signal quality of the measurement. Since the model calculates SpO_2 measurements from C_a, the sensor position and thus oxygen consumption up to this point are not accounted for. Comparing different sensor positions (finger, foot, earlobe) to blood gas samples might provide additional insights.

Acknowledgement

The work has been carried out at Drägerwerk AG & Co. KGaA in Lübeck and was partly supported by the German Federal Ministry for Education and Research under the project "WiM-Vent", Grant: 01IB10002.

5   References

[1] R. Kramme, ed., *Medizintechnik: Verfahren - Systeme - Informationsverarbeitung*. Springer Medizin, 4. ed., 2011.

[2] W. Karlen, C. L. Petersen, G. A. Dumont, and J. M. Ansermino, "Variability in estimating shunt from single pulse oximetry measurements.," *Physiological Measurement*, vol. 36, no. 5, pp. 967–981, 2015.

[3] L. Chiari, G. Avanzolini, and M. Ursino, "A comprehensive simulator of the human respiratory system: validation with experimental and simulated data," *Annals of Biomedical Engineering*, vol. 25, no. 6, pp. 985–999, 1997.

[4] J. Piiper and H. P. Koepchen, "Atmung," in *Physiologie des Menschen*, vol. 6, Urban & Schwarzenberg, 1972.

[5] K. Golenhofen, *Basislehrbuch Physiologie*. Urban & Fischer, 4. ed., 2006.

[6] J. W. Severinghaus, "Simple, accurate equations for human blood O2 dissociation computations," *Journal of Applied Physiology*, vol. 46, no. 3, pp. 599–602, 1979.

[7] A. Riedlinger, J. Kretschmer, and K. Möller, "On the practical identifiability of a two-parameter model of pulmonary gas exchange.," *Biomedical engineering online*, vol. 14, p. 82, 2015.

[8] D. B. MacLeod, L. I. Cortinez, J. C. Keifer, D. Cameron, D. R. Wright, W. D. White, E. W. Moretti, L. R. Radulescu, and J. Somma, "The desaturation response time of finger pulse oximeters during mild hypothermia.," *Anaesthesia*, vol. 60, no. 1, pp. 65–71, 2005.

[9] J.-A. Collins, A. Rudenski, J. Gibson, L. Howard, and R. O'Driscoll, "Relating oxygen partial pressure, saturation and content: the haemoglobin-oxygen dissociation curve.," *Breathe*, vol. 11, no. 3, pp. 194–201, 2015.

Detection of Joint Diseases in Horses through Classification of Vibroarthrography Signals using Mel Frequency Cepstral Coefficients

P. Bedei [1], A. Moeller [2], and A. Mertins [2]

[1] Medizinische Ingenieurwissenschaft, Universität zu Lübeck, patrick.bedei@student.uni-luebeck.de
[2] Institute for Signal Processing, Universität zu Lübeck, {moeller,mertins}@isip.uni-luebeck.de

Abstract

Vibroarthrography (VAG) presents an alternative to complicated procedures like computed tomography or magnetic resonance imaging for diagnosing joint diseases in horses. Therefore, two different classification systems with features based on Mel Frequency Cepstral Coefficients (MFCCs) were tested to classify signals into healthy or pathologic. One system uses features consisting of the mean of the MFCCs over all frames of the signal as well as the mean of the standard deviation over all coefficients. The other uses the coefficients themselves to classify every frame and then classify the whole signal based on the fraction of signal frames that were classified as pathologic. Both systems showed promising results on VAG signals from equine fetlocks with features based on standard deviation being the best with up to 0.96 correct classification rate. Classification with MFCC based features may provide an accurate diagnostic tool to help avoiding unnecessary further examinations.

1 Introduction

Vibroarthrography (VAG) is a non-invasive examination method which uses sounds produced by joint movement to diagnose diseases of the joint. Especially in veterinary medicine, this method could become significant. In contrast to VAG signal acquisition, the examination of horses via magnetic resonance imaging (MRI) or computed tomography (CT) is difficult and can lead to additional injury because the positioning of the horses requires anesthesia. A reliable classification into pathologic and healthy joints using the VAG signals could avoid unnecessary MRI and CT examinations and aid in the early detection of joint diseases. When using a classification system on horses with a suspected joint disease with the goal to avoid exposing healthy ones to the risks of MRI or CT, obviously a high accuracy, but especially a high specificity is desired. In addition, the system should be able to deal with slightly time shifted signals as it is common with the current recording procedure [?].

Mel frequency cepstral coefficients (MFCCs) for recognizing VAG signals of human knees was tested in the past with promising results [?]. The objective of this paper is to assess whether features based on MFCCs can be used for classification of VAG signals of equine fetlocks. Therefore, two different classification approaches are investigated. Also, the effect of different classifiers and frame lengths is considered. VAG signals are nonstationary [?], making a windowing into frames or short-term analysis necessary. The first approach uses features based on the standard deviation and mean of the MFCCs over all frames, the second one consists of a two-step classification system using the MFCCs to classify signal frames and then classify the whole signal based on the percentage of frames classified as pathologic and healthy, respectively.

2 Material and Methods

In this section, the recording and properties of the VAG signals used for this paper as well as MFCCs and the derived features used for the classification systems are described.

2.1 VAG Signals

In this paper, 20 VAG signals from equine fetlocks were used with 10 signals originating from ill and healthy horses, respectively. The signals have a recording frequency of 44,1 kHz and are about 10 seconds long. Each signal consists of two channels. The first channel contains sound information recorded by a sound sensor placed on the medial side of the metatarsus or metacarpus, the second by a sensor on the lateral side. During the measurement the joint of interest was bent and stretched by a veterinary in time with an optical metronome [?].

2.2 Mel Frequency Cepstral Coefficients

The unit mel represents the perceived pitch of a tone. The mapping between mel frequency (f_{mel}) and physical frequency (f_{Hz}) has been determined experimentally and reflects the nonlinearity of the auditory system between those two variables. It can be approximated as

$$f_{mel} = \frac{1000}{\log(2)} \left[1 + \frac{f_{Hz}}{1000} \right] , \qquad (1)$$

which means that the mapping is approximately linear below 1000 Hz and logarithmic above.

In the following, the computation of the so called mel cepstrum is described. The mel cepstrum is computed for every frame of the signal with each frame being of length $N < M$, with M being the signal length. In a first step, the short-term discrete Fourier transform (stDFT) is computed. Afterwards, the log total energy in a critical band around the mel frequencies is computed. Therefore, the magnitude of the short-term spectrum is filtered with critical-band filters. The bandwidth of those triangular filters varies with frequency, increasing logarithmically above 1 kHz. These filters have to be centered on frequencies sampled by the stDFT. That means for the i-th center frequency

$$f_{c,i} = k_i \frac{f_s}{N'} , \qquad (2)$$

with the sampling frequency f_s and k_i being some $k \in [0, N'-1]$. N' points were used to compute the stDFT. The logarithm of each filter output is taken resulting in the log total energy in each critical band. The output $Y(i)$ of the i-th filter H_i in frequency domain can be written as

$$Y(i) = log \left(\sum_{k=0}^{N/2} |S(k)| H_i \left(k \frac{2\pi}{N'} \right) \right) , \qquad (3)$$

where $S(k)$ denotes the short-term spectrum.

Finally, the discrete cosine transformation (DCT) is applied on the sequence of filter bank outputs $\hat{Y}(k)$ with

$$\hat{Y}(k) = \begin{cases} Y(i), & k = k_i \\ 0, & \text{other } k \in [0, N'-1] . \end{cases} \qquad (4)$$

This final step to compute the cepstrum $c(n)$ consisting of n cepstral coefficients can be written as

$$c(n) = \frac{2}{N'} \sum_{i=1,2,...,N_f} \hat{Y} \left(k_i \cos \left(k_i \frac{2\pi}{N'} n \right) \right) , \qquad (5)$$

where N_f is the number of critical band filters [?].

This work utilizes the implementation by Slaney [?]. The windowing to divide the signal into frames is realized using overlapping Hamming windows. It uses 40 normalized filters covering a frequency range from 133 to 6854 Hz with 13 of them spaced linearly and 27 logarithmically [?]. After the DCT, only the first 13 cepstral coefficients are kept. In the following, the term MFCCs is used to denote these 13 coefficients.

2.3 Features

This paper discusses two different classification systems using five different features in total.

2.3.1 One-step Classification System

In this approach, a single feature vector for the whole signal is computed out of, depending on the frame length, about 500 to 4000 vectors consisting of the 13 MFCCs. Four different feature vectors were tested for this system. One of those feature vectors is simply the arithmetic mean of the MFCCs over all frames of the medial and lateral channel of one signal. So, if $\overline{MFCC(i)}$ denotes the arithmetic mean of the i-th coefficients over all frames, the feature vector $\boldsymbol{MFCC}$ looks as follows

$$\boldsymbol{MFCC} = \begin{bmatrix} \overline{MFCC(1)} \\ \overline{MFCC(2)} \\ \vdots \\ \overline{MFCC(13)} \end{bmatrix} . \qquad (6)$$

This feature should work well if the MFCCs themselves are different in pathologic and healthy signals.

Another possibility might be that the MFCCs vary more between frames in signals from pathologic and healthy joints. Therefore the standard deviation of the MFCCs between frames is calculated. The result is a vector with 13 components representing the standard deviation of each coefficient. Then, the arithmetic mean over these components is taken yielding a single value for both the medial and lateral portion of each signal, say SD_{Med} and SD_{Lat}. One feature $\boldsymbol{SD_{MedLat}}$ is a vector consisting of those two values:

$$\boldsymbol{SD_{MedLat}} = \begin{bmatrix} SD_{Med} \\ SD_{Lat} \end{bmatrix} . \qquad (7)$$

Another one is the mean $\overline{SD}$ of those two values allowing for a decision based on a single value

$$\overline{SD} = \frac{SD_{Med} + SD_{Lat}}{2} . \qquad (8)$$

The mean of the MFCCs and the single value of the standard deviations are combined to a new feature vector $\boldsymbol{MFCC_{SD}}$ with

$$\boldsymbol{MFCC_{SD}} = \begin{bmatrix} \boldsymbol{MFCC} \\ \overline{SD} \end{bmatrix} . \qquad (9)$$

2.3.2 Two-step Classification System

The second classification system uses a single value which results from classifying each frame in a first step with the MFCCs as features. For training of the classifier in this first step, the assumption is made that each frame of a pathologic or healthy signal belongs to the class 'pathologic' or 'healthy', respectively. In a second step, the fraction p_{ill} of frames classified as pathologic is calculated for each signal:

$$p_{ill} = \frac{m_{ill}}{m_{total}} , \qquad (10)$$

with m_{ill} denoting the number of frames classified as pathologic and m_{total} the number of signal frames. p_{ill} is then used as feature to classify the whole signal.

3 Results and Discussion

In the following, the classification systems are evaluated using 20 VAG signals from the equine fetlock joint.

3.1 One-step Classification

The evaluation was performed by splitting the 20 signals into one data set used for training and one for test of the classifier with each set containing 5 signals from ill and healthy horses. All possible configurations to choose the training and test data were considered and the accuracy averaged to get the presented values.

Table 1 shows the results for the classification using the one-step classification system using MATLAB's own implementation of a Support Vector Machine (SVM) as a classifier. The accuracy is fairly high (above 0.85) for all feature configurations and frame durations. Nevertheless, a longer frame duration of 40 ms lead to higher accuracy for all features. So, the 20-40 ms frames usually used in speech processing seem to be an optimal choice for these VAG signals underlining their similarity to speech signals. Also, the features based on the standard deviation SD_{MedLat} and $\overline{SD}$ clearly outperform the ones containing the mean of the MFCCs. SD_{MedLat} and $\overline{SD}$ yield the highest value overall while $\overline{SD}$ performs slightly better for shorter frame durations. The combination of the standard deviation and mean $MFCC_{SD}$ did not improve the classification over the two features based on the standard deviation alone. Fig. 1 shows a plot of the SD_{MedLat} feature vector that gave the highest correct classification rate for the 20 signals. Pathologic signals always possess a higher value than healthy ones in at least one of the two channels (medial, lateral), making the two classes linearly separable.

Table 1: Correct classification rate for one-step classification via SVM for different frame lengths.

Feature	40 ms	20 ms	10 ms	5 ms
$MFCC$	0.88	0.88	0.87	0.86
SD_{MedLat}	0.96	0.95	0.93	0.88
$\overline{SD}$	0.96	0.96	0.94	0.90
$MFCC_{SD}$	0.90	0.88	0.89	0.86

Also, different classifiers were tested for the one-step classification system. Table 2 shows the results for MATLAB's SVM, Linear Discriminant Analysis (LDA), Naive Bayes and k-nearest-neighbor (KNN) classifiers. All classifiers showed similar behavior to the one shown for SVM in Table 1 in terms of accuracy depending on the frame duration. For the features based only on the standard deviation, all classifiers work well with an accuracy of about 0.95 while the accuracy of LDA and Bayes decreases for feature vectors containing the mean of the MFCCs. Overall, SVM seems to be the best choice for the features considered here.

Table 3 shows the rates for correct classification of signals from ill (sensitivity) and healthy (specificity) horses, respectively. For the features containing the mean of the

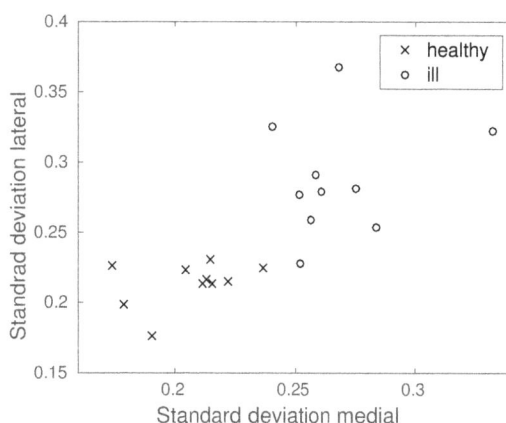

Figure 1: Average standard deviation of MFCCs of the medial and lateral channel of all 20 signals.

Table 2: Accuracy for the one-step Classification using different classifiers for 40 ms frame length.

Feature	SVM	LDA	Bayes	KNN
$MFCC$	0.89	0.77	0.80	0.86
SD_{MedLat}	0.96	0.95	0.94	0.94
$\overline{SD}$	0.96	0.95	0.95	0.95
$MFCC_{SD}$	0.90	0.80	0.86	0.89

MFCCs, both rates are around 0.9. The features based on the standard deviation lead to a slightly higher sensitivity, but stand out in terms of specificity with $\overline{SD}$ classifying every healthy test signal correctly. So, classification based on these features exhibits high sensitivity and very high specificity. These properties agree well with the ones desired for a test or classifier for reducing unnecessary and potentially dangerous further examination steps.

Table 3: Sensitivity and specificity for 40 ms frame length.

Feature	Sensitivity	Specificity
$MFCC$	0.89	0.89
SD_{MedLat}	0.93	0.99
$\overline{SD}$	0.92	1
$MFCC_{SD}$	0.89	0.92

To assess the generalization ability of the classification system, an extreme approach to split the data into training and test data was performed. For Table 4, only one healthy and one pathologic signal were used for training leaving 18 signals for testing. The mean over all possible configurations of training and test data as well as the standard deviation were computed. The two features derived from the standard deviation of the MFCCs still yield the highest accuracy although the SD_{MedLat} has the biggest loss with about 0.6 compared to when half of the signals are used for training (see Table 2). Overall, the accuracy is still very high with values around 0.9 despite the very small training data set consisting of only 10% of the data. These findings suggest that these features make for a high generalization ability.

Therefore, good results are expected when using a classifier based on the features considered here in practice.

Table 4: Accuracy and standard deviation using only 2 signals as training data.

Feature	Accuracy	Standard deviation
$MFCC$	0.88	0.09
SD_{MedLat}	0.90	0.05
$\overline{SD}$	0.92	0.05
$MFCC_{SD}$	0.88	0.08

3.2 Two-step Classification

For this system, a SVM was used as classifier for both classification steps. Fig. 2 shows the feature p_{ill} used for the two-step classification system. As expected, the signals from ill horses have a higher fraction of frames classified as ill and the two classes are perfectly separable. The percentage of frames classified as pathologic in signals from pathologic joints, however, is relatively low ranging from about 0.13 to 0.32. This might be because of the naive approach of simply assigning every frame of the signal to the class of the signal and training the classifier with those frames instead of only comparing frames from the same part of the signal. However, this approach makes the classification system more robust to time shifted signals and therefore able to deal with signals recorded as described in Section 2.1. With a test data set of ten signals for the first step and performing leave-one-out cross validation (using MATLAB's routine) the system classifies all test signals correctly. However, with a similar approach as in the one-step classification to reduce the training data to one signal from a pathologic and one from a healthy joint and leaving the rest for testing (see Table 4), the system only classified on average 0.73 of the data correctly. As opposed to the one-step classification, it seems like much more data for training is needed to achieve similar results.

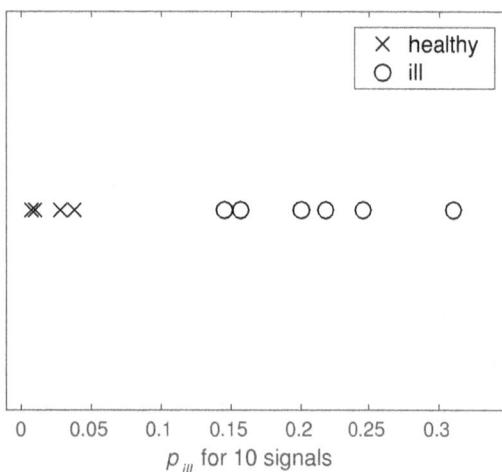

Figure 2: Share of frames classified as pathologic. Each mark represents one signal.

4 Conclusion

This paper investigated the classification of VAG signals from equine fetlocks into signals from pathologic and healthy joints. Therefore, features based on MFCCs were used in two different classification systems. Features based on the standard deviation of the MFCCs provide a nearly perfect overall accuracy and specificity. They also promise a high generalization ability. The proposed methods worked well on slightly time shifted VAG signals due to human error in the recording procedure. So, classifying VAG signals with these methods might provide an easy, non-invasive way to examine horses with a suspected fetlock joint disease and to avoid unnecessary MRI and CT examinations. In the future, the two-step classification system could be further evaluated to compare it directly to the one-step classification system. In addition, future works might include a more extensive testing with newly recorded signals since the work in this paper was conducted using only 20 signals.

Acknowledgement

The work has been carried out at Institute for Signal Processing, University of Luebeck.

5 References

[1] A. Moeller, "Classification of vibroarthrography signals from the equine fetlock joint by the local discriminant bases algorithm," Master's thesis, Institut für Mathematik, Universität zu Lübeck, 2013.

[2] G. H. Seng, T. T. Swee, and H. Chai, "Parametric study of vibroarthrographic signal characteristic for future osteoarthritis analysis," *Recent Advances in Electrical Engineering Series*, no. 10, 2013.

[3] S. Krishnan, R. M. Rangayyan, G. D. Bell, and C. B. Frank, "Adaptive time-frequency analysis of knee joint vibroarthrographic signals for noninvasive screening of articular cartilage pathology," *Biomedical Engineering, IEEE Transactions on*, vol. 47, no. 6, pp. 773–783, 2000.

[4] J. Deller, J. Hansen, and J. Proakis, *Discrete-Time Processing of Speech Signals*. IEEE Press, New York, pp. 380-384, 2000.

[5] M. Slaney, "Auditory toolbox," *Interval Research Corporation, Tech. Rep*, vol. 10, 1998.

[6] T. Ganchev, N. Fakotakis, and G. Kokkinakis, "Comparative evaluation of various MFCC implementations on the speaker verification task," in *Proceedings of the SPECOM*, vol. 1, pp. 191–194, 2005.

Movement identification based on exoskeleton sensor data for event marking of the electroencephalogram

N. Eckardt [1], M. Tabie [2], A. Seeland [2], E. Kirchner [2] and P. Rostalski [3]

[1] Medizinische Ingenieurwissenschaft, Universität zu Lübeck, nils.eckardt@student.uni-luebeck.de

[2] German Research Center for Artificial Intelligence - Robotics Innovation Center, Bremen,
{marc.tabie, anett.seeland, elsa.kirchner}@dfki.de

[3] Institute for Electrical Engineering in Medicine, Universität zu Lübeck, philipp.rostalski@uni-luebeck.de

Abstract

In the present work the development of an algorithm for movement identification based on exoskeleton sensor data is described. The exoskeleton is part of a project on post stroke rehabilitation. The algorithm shall be used to mark movement events in a simultaneously recorded electroencephalogram as a replacement for external motion tracking. The angular values for each joint of the exoskeleton are utilized by the algorithm to calculate a threshold and to decide, if a movement was executed or not. The quality of the algorithm is evaluated with an experiment, where four subjects have to do specific movements while wearing the exoskeleton. During the experiment, data from exoskeleton sensors, the electroencephalogram and motion tracking is recorded. The provisional results show, that the algorithm is able to detect the movements, but the threshold needs to be adapted to the status of the subject.

1 Introduction

Worldwide, stroke is the most frequent cause for middle and heavy acquired physical impairments [1]. In 30% to 66% of the cases, the paretic limb of the stroke patient, who suffers from paralysis, remains impaired 6 months after stroke, whereas only 5% to 20% of the patients recover to full functionality [2]. One concept for stroke rehabilitation is motor learning, which is based on repetitive, task-specific training. This makes robotic systems applicable to assist in therapy. The functionality and strength of the patient's impaired limb may improve through robot assisted therapy and training [3], so this approach has become an important technology applied in rehabilitation engineering [4].

The aim of the *Recupera-Reha* project at the German Research Center for Artificial Intelligence (Deutsches Forschungszentrum für Künstliche Intelligenz, DFKI) is the development of a mobile whole body exoskeleton, for robot-supported rehabilitation of neurological illnesses like stroke. During the development of the robot system most different aspects must be considered, as for example suitable kinematics, the choice of fitting joints and a correct actuator strength. Especially for the rehabilitation use case, appropriate, assistive control methods must be developed. The mechatronic attempts for actuation and control of the system get combined with methods of online evaluation of the electroencephalogram (EEG) and the electromyogram (EMG)[5].

The purpose of the algorithm proposed in this paper is to identify the movements of the wearer by using internal data of the exoskeleton. This offers the advantage of being independent of external systems and devices, like motion tracking systems or push buttons, which have to be setup before use and often comes with additional hardware. This limits the mobility and flexibility of experiments and application scenarios, which have to be adaptable, especially in the context of rehabilitation. A similar comparison with the same intention of more freedom and independency between a motion sensing suit and a motion tracking system was done in [6]. The information about the movements is on the one hand used to get a feedback, if the intended movement was absolved right or not and on the other hand it is needed for marking the EEG data of the user, to evaluate intention recognition methods and to adapt learning algorithms for context recognition. As a basis and test system of the algorithm described in this paper, the active exoskeleton of the project *CAPIO* is used, which was also developed at the DFKI before. This two-armed upper body exoskeleton is connected with the human body by eight contact points and the kinematic structure covers eight active degrees of freedom at the arms and four active degrees of freedom in the back [7].

2 Material and Methods

The algorithm and all additional scripts, which are related to the algorithm, were developed in the programming language Python 2.7. For later use, the algorithm will be embedded in the open source software *pySPACE* (Signal Processing And Classification Environment in Python) [8].

2.1 The detection algorithm

Before the evaluation of movements begins, the joint values are separated into two groups (left and right arm). While identifying movements it became apparent that voluntary movements needs to be distinguished from involuntary, as for example short twitching or smaller, passive co-movements of an arm while moving other body parts. At the same time, it is important to detect the correct movements precisely in time. To recognize whether a joint moves at the time point i, one can compare the current position x of the joint with its respective value t samples before, which represents a determined time span related to the sampling frequency. The difference of both values represent the average speed of change of the joint angle in the period of time. If this speed exceeds a certain threshold θ, a voluntary movement of the joint is assumed (the output of the algorithm $f_{t,\theta}(x_i)$ will be the value 1), otherwise, it is not counted as a movement (marked by the value -1):

$$f_{t,\theta}(x_i) = \begin{cases} 1, & \text{if } x_i - x_{i-t} \geq \theta \\ -1, & \text{otherwise} \end{cases} \quad (1)$$

As a initial value for the threshold the standard deviation of the activity in rest was chosen for each joint. In contrast to the difference of minimum and maximum value during rest, which could also be a possible threshold, the standard deviation was more robust against peaks caused by artifacts, but was too small, which made an additional multiplying factor for the threshold necessary. In this context it can be seen as a kind of noise, which is not assignable to any voluntary movement and can result from different causes. Now, using the outcome of the single joints, it had to be decided, whether the motions of the single joints were enough to be counted as a voluntary movement of the whole arm itself. The first attempt was to examine, which joint of each arm moved currently the most and then just focus on and compare this two for both arms. By this, the decision whether the respective arm moved was reduced to the value of this joint. Indeed, this version was able to lead to precise detection in time, with a delay minimum between the starting of the motion of one joint and movement detection of 0.1 seconds, but at other movements 0.7 seconds. Especially in case of more complex movements this delay fluctuated and it was difficult to find a threshold, which delivered good results for the whole sequence. The next idea was to take the sum of the differences of each joint j out of n joints for each arm and in accordance to this choose the sum of thresholds as the threshold for each arm to compare it with:

$$F_{t,\theta}(x_i) = \begin{cases} 1, & \text{if } \sum_{j=1}^{n}(x_{j,i} - x_{j,i-t}) \geq \sum_{j=1}^{n} \theta_n \\ -1, & \text{otherwise} \end{cases} \quad (2)$$

This resulted in a more robust detection of the decision algorithm $F_{t,\theta}(x_i)$, which also seemed to be easier to adjust, so preferably only relevant movements were detected.

2.2 Captured Data

As a basis for the development of the algorithm data from the exoskeleton were recorded, where the subject lifted and lowered the left and right arm alternately. The recording contained values of all joints of the exoskeleton with regard to the deflection of the respective joint to its ground position, the strength currently applied to the joint and values of the inverse dynamics. At the same time the movements of the user were recorded by a motion tracking system (Oqus, Qualisys AB, Gothenburg, Sweden) using six infrared cameras and two infrared markers fixed to the wrists of the operator. However, the data of both recordings had not been captured synchronously so a synchronization had to be carried out afterwards manually by using events, which could be identified in both records. For data preparation it was checked, which joints did not moved during the recording or were switched off. These joints (the four in the back of the exoskeleton) were sorted out, because they were not relevant for the recognition and differentiation of the arm movements. In addition, among all available parameters only the current joint angulars were used for the classification of the movements. To avoid unceratinties about the movements, the movement sequence for the evaluation data sets was planned concretely and was controlled during recording. In addition, the different data streams of the exoskeleton, the motion tracking system (with six markers fixed to the exoskeleton and three markers on each object (cup, bottle, box)) and, for further evaluation and testing for the actual use case of event marking, EEG were recorded with synchronized timestamps. The EEG was captured with a 32 electrodes system (actiCap, Brain Products GmbH, Munich, Germany).

2.3 Comparison

To examine if the movement detection performed by the algorithm is able to replace the identifying of movements by using the motion tracking data, both methods are compared. After manually synchronization the starting points of the movements and their duration are compared to the starting point of the respective detection and how long it stays positive. In addition, observing the motion tracking recording will help to check, how the movements were actually carried out or if there are any unexpected motions, which may be detected by the algorithm. Further, it was checked, how the algorithm is reacting to smaller movements, which are not counted as a real movement of one arm.

The comparison allowed to adapt the parameters of the algorithm (time window t, factor for threshold adjustment) to an optimum (500 ms, 15 times the standard deviation of the activity in rest) for the present data by increasing the values from 50 ms in 50 ms steps in the first case and from 1 to 15 in 0.5 steps in the case of the threshold factor. In both cases, a further raise would lead to an increased detection delay or even to more missed movements. Too small parameter values would lead to more detections, then expected, caused by wrong detection when no movement is actually done (*unexpected*) or by splitting up one actually

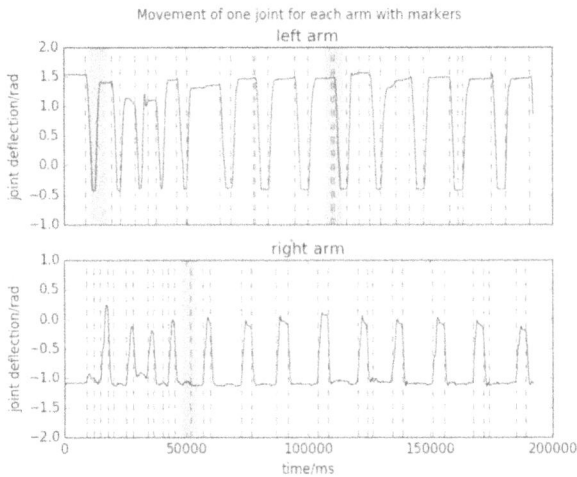

Figure 1: Illustration of the movement marking on the test data by the algorithm with adjusted threshold after comparison with the motion tracking data. The vertical, dashed lines mark the beginning of the detected movements, the solid lines the values of the current deflection of the first shoulder joint of the left (top) and right (bottom) arm over time. These joints were chosen as a optical reference because of their wide and clear shift during movements. A missed movement can be seen in top graphin the first grayed area, a splitin the second grayed area and an unexpected detection in the bottom graphin the third grayed area.

correct movement into two or more smaller ones, when the movement value is fluctuating around the threshold (*splits*). The results of the adapted algorithm and a differentiation of the different errors are given in Table 1. A visualization of the data with marks at the start of the movements is shown in Fig. 1 with the deflection values of one joint for each arm as a comparison.

Table 1: Differentiations of movement markings on the test data set

movement	left arm	right arm
expected	27	28
detected correctly	25	28
missed	2	0
splits	4	3
unexpected	2	2

2.4   Evaluation experiment

To evaluate the algorithm, adapted to the test data set, an experiment was set up, to obtain further data under controlled conditions. Four male subjects participated, the only restriction was a suitable body shape and strength to wear the exoskeleton.

After the motion tracking system is calibrated and the subject is prepared for EEG recording, he puts on the exoskeleton, which is additionally connected to a counterweight, which reduces the load for the subject by 20 kilogram. In

front of the subject three objects are located: a cup, a bottle and a box. These objects are used for some of the following movement tasks. The tasks can be separated into easy movements, like moving an arm forth or back from a relaxed position and more complex, more realistic movements like grasping a cup or bottle, filling the cup, raising it for drinking or lifting a box with both arms, like it probably will be done in a rehabilitation scenario. Behind the table, a screen is set up. On the screen, the subject receives commands, one by one and in an appropriate speed (5 seconds for easy movements, 8 for the more complex ones, with a 2 seconds break between each command). The commands tell the subject, which movement has to be performed, by showing stylized pictures, as shown in Fig. 2. Before the experiment starts, the subject gets an instruction on how to move with the exoskeleton and what movements are meant with the commands. The command sequence itself lasts ten minutes. It is assumed, that the subject will react to every command with just one movement of the respective arm. To support later analysis, it is documented during the sequence, how the movements are performed by the subject, to know, if the movements are done as expected.

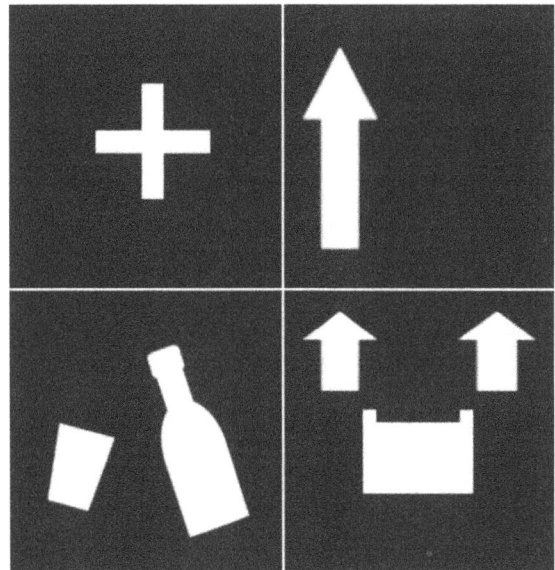

Figure 2: 4 examples out of 15 command pictures the subject gets to see during the experiment: Top left: relax, top right: left arm forth, bottom left: grasp bottle and cup, bottom right: pick box. These descriptions are depicted on the screen, too. The remaining command pictures are variations (reversion, change of sides, etc.) of the pictures shown here.

3   Results and Discussion

First results seem promising that the algorithm is capable of detecting movements. As a first test, the algorithm is applied to the new data sets with the parameters adapted on the test set. The results of these runs are shown in Table 2. It can be seen, that most of the missed movements were complex movements. On the other hand, most of the unexpected detections happened during the task times of easy movements. Partially, the unexpected detections can be explained by the performance of the subjects. For example, subject 4 received the command *left arm back*, but accidentily had moved his right arm back in three small steps, too. The algorithm detected one movement on the left and three shorter ones on the right, so an actual correct detection.

To figure out, if more errors can be explained just by the performance of the subjects, for example if some of the splits are actual separable movements during a complex task, a closer examination of the motion tracking recording is necessary. In addition, the movement marks will be applied to the EEG data to examine their accuracy for event marking. This and a more detailed analysis of the results and further steps to a more automatic threshold adjustment and embedding are currently in preparation.

Table 2: Differentiations of movement markings on the four obtained data sets

movement	left arm	right arm
type	easy / complex	easy / complex
expected	26 / 27	28 / 25
subject 1		
detected correctly	25 / 26	26 / 25
missed	1 / 1	2 / 0
splits	12 / 42	19 / 22
unexpected	9 / 1	9 / 8
subject 2		
detected correctly	26 / 19	28 / 25
missed	0 / 8	0 / 0
splits	15 / 9	13 / 28
unexpected	0 / 0	0 / 2
subject 3		
detected correctly	25 / 23	26 / 20
missed	0 / 4	2 / 5
splits	8 / 20	6 / 14
unexpected	3 / 1	0 / 1
subject 4		
detected correctly	26 / 20	28 / 19
missed	0 / 7	0 / 6
splits	10 / 13	4 / 7
unexpected	2 / 2	5 / 0

4   Conclusion

In this article, the development and evaluation of an algorithm for movement detection of an exoskeleton was presented. The algorithm, which was developed based on a test data set, was evaluated on four data sets recorded in a extra designed movement sequence experiment. The first results showed, that the parameters of the algorithm, which were optimized on the test data, are a good starting point, but have to be adjusted for every individual subject, to get the optimal outcome. Apart from a more thorough analysis and the embedding into a signal processing framework, there are a few possible enhancements, which could be added to

the algorithm. The use of kinematic information about the exoskeleton may allow to separate the detection to more differentiated categories, like which part of the arm is moving in which direction, which kind of movement is done etc. In addition, changes in the torque values of the joints could tell the algorithm, if the user is carrying something or in general, if something changed about the status of the subject. Furthermore, a closer look to the detections during the more realistic, complex movements could be a basis to draw up characteristic sequences for these kinds of movements.

Acknowledgement

The work has been carried out at the German Research Center for Artificial Intelligence - Robotics Innovation Center, Bremen.

5   References

[1] S. C. Johnston, S. Mendis, and C D. Mathers. *Global variation in stroke burden and mortality: estimates from monitoring, surveillance, and modelling.* The Lancet Neurology, vol. 8, no. 4, pp. 345–354, 2009.

[2] G. Kwakkel, B. J. Kollen, J. van der Grond and A. J. H. Prevo. *Probability of Regaining Dexterity in the Flaccid Upper Limb Impact of Severity of Paresis and Time Since Onset in Acute Stroke.* Stroke, vol. 34, pp. 2181–2186, 2003.

[3] H. Woldag, and H. Hummelsheim. *Evidence-based physiotherapeutic concepts for improving arm and hand function in stroke patients.* Journal of Neurology, vol. 249, no. 5, pp. 518–528, 2002.

[4] J. Mehrholz, A. Headrich, T. Platz, J. Kugler and M. Pohl. *Electromechanical and robot-assisted arm training for improving generic activities of daily living, arm function, and arm muscle strength after stroke.* Cochrane Database Syst Rev, vol. 6, 2012.

[5] DFKI GmbH - Robotics Innovation Center (RIC), *RECUPERA-Reha.* Available: http://robotik.dfki-bremen.de/de/forschung/projekte/recupera-reha.html [last accessed on 20.01.2016].

[6] Y. Menguc, et al. *Soft wearable motion sensing suit for lower limb biomechanics measurements.* Robotics and Automation (ICRA), 2013 IEEE International Conference on, pp. 5309–5316, 2013.

[7] M. Mallwitz, N. Will, J. Teiwes and E. A. Kirchner, *The capio active upper body exoskeleton and its application for teleoperation..* In: Proc. of the 13th Symp. on Advanced Space Technologies in Robotics and Automation (ASTRA-2015), 2015.

[8] M. M. Krell, et al. *pySPACE—a signal processing and classification environment in Python.* Frontiers in neuroinformatics 7, 2013.

Acquisition of respiratory signals and generation of visual realtime biofeedback using structured IR light surface scanning

E. Mücke [1], R. Werner [2], M. Wilms [3], H. Handels [3], and T. Gauer [4]

[1] Medizinische Ingenieurwissenschaft, Universität zu Lübeck, eike.muecke@student.uni-luebeck.de
[2] Department of Computational Neuroscience, University Medical Center Hamburg-Eppendorf, r.werner@uke.de
[3] Institute of Medical Informatics, Universität zu Lübeck, {wilms, handels}@imi.uni-luebeck.de
[4] Department of Radiotherapy and Radio-Oncology, University Medical Center Hamburg-Eppendorf, t.gauer@uke.de

Abstract

In radiotherapy of moving tumors, surrogate signals are used to track the patients breathing motion. To help the patient breath more consistently, one method of choice is to "biofeedback" the amplitude of the chest wall motion to the patient. The purpose of this work is to capture depth data of the upper body using the Microsoft Kinect with structured infrared light. An in-house software displays the depth data as visual biofeedback in a total of three different options: A direct representative of the acquired depth data, a bar and a circular visualization with changing height and radius. We evaluated the changes in respiratory motion influenced by those options. The evaluation is based on the data of four volunteers. All visualizations were presented at random order to each volunteer. As a result, the breathing period increased, and similarly does the standard deviation. The amplitude increased, depending on the biofeedback visualization.

1 Introduction

During radiotherapy, respiratory motion can cause unwanted geometric miss of the targeted tumor as well as hitting healthy structure with too much radiation dose. This problem arises especially for the treatment of thoracic and abdominal tumors, at which tumor movement is up to two centimetres [1]. Usually, surrogate signals are used to track breathing motion. These generated surrogate signals provide the possibility to shutdown or reduce radiation and to protect organs at risk in the case of breathing irregularities.

Motion impact can be especially critical in clinically applied treatment techniques where tumor dose escalation is attempted in a tight margin around tumors [2]. One way to compensate for tumor motion is to increase safety margins to cover the total motion range but this could expose high-risk organs to more radiation than justifiable. Technical approaches are tumor tracking [3] and gating [4]. Tumor tracking is an active process with the radiation beam following the tumor motion. Gating means that radiation is only active when the tumor is located at a predefined position window corresponding to a certain respiratory phase.

Biofeedback of respiratory motion, however, aims to stabilize the movement and to increase the awareness of the patient. These improvements can be understood as a minimization of standard deviation of the motion amplitude and period. Furthermore, applied biofeedback could benefit the treatment of radiotherapy in terms of reduced treatment times or to better control radiation dose handling in organs of risk. The input of the biofeedback is usually generated by respiratory belts or other surrogates for respiratory motion tracking.

In this work, we implemented a visual biofeedback system using the Microsoft Kinect. We implemented three different options of biofeedback and compared the options. The data was visualized by a Python-based software utilizing matplotlib, OpenCV and freenect libraries.

2 Material and Methods

2.1 Varian RPM System

For evaluation of the results of biofeedback influences, we used a clinically established system, which keeps track of all measured experiments. The Varian Real-time Position Management Respiratory Gating (Varian RPM) is a clinically used system to measure an external respiratory motion signal. For measurement, a marker block containing six infrared-reflective marker dots is placed on the patient's thorax or abdomen. This block is captured in all spatial directions by an infrared (IR) camera. The position of the marker block over time results in a motion signal that is used in clinical practice for 4D-CT reconstruction and gated tumor therapy.

2.2 Microsoft Kinect

The Microsoft Kinect is a low cost sensor to capture depth and RGB images. It was initially manufactured for gaming purposes and was aimed to be an additional input device. It consists of three different imaging devices: a color camera for the RGB images and an infrared projector and a camera to estimate depth values of the scene. Microsoft used the PrimeSense PS1080 sensor for depth measurements, which has a field of view of 58° horizontally and 45° vertically. The framerate of the Kinect is 30 frames per second and the principle of depth measurement is structured light [5]. The IR camera and the IR projector form a stereo pair. The IR projector sends out a fixed pattern of light and dark speckles. For a known baseline, depth can be calculated via triangulation against a known pattern from the projector. For a known depth, the pattern is stored and new depth values can be computed given in relation to this base value. For each pixel of the IR image, a small correlation window is used to compare local patterns around that pixel with the memorized patterns. The Kinect performs an interpolation of the best match of the correlation window to obtain sub-pixel accuracy [6]. Output values are delivered in 11-bit packages with a value of 2047 as a non-determineable depth at a pixel. The Kinect is able to capture multiple regions of the body simultaneously.

2.2.1 Calibration

To obtain correct depth values at desired regions of the image, the infrared camera was calibrated. This is necessary to estimate the inaccuracy of used lenses of the sensor. Afterwards, we undistort our video stream by applying a generic geometrical transformation. For calibration, the Mobile Robot Programming Toolkit (MRPT) Application which, amongst other things, provides an interactive calibration of the intrinsic parameters of the Kinect sensors using checkerboard [7]. OpenCV was used to apply the calibration of the infrared camera after intrinsic parameters were known. As stated in [8], the Kinect is capable to reliably capture respiratory surface motion with correlation coefficients up to 0.98 compared to the RPM system. Which, for our usage, is absolutely sufficient, because it is not necessary to measure that precisely to generate biofeedback.

2.3 Study Design

Four healthy volunteers were enrolled in this study. The group was composed of male students in the age of 21 to 27 years. No respiratory ailments were known.

2.4 Measurement Setup

In order to combine signals from the thoracic and abdominal ROIs measured by the Kinect and the marker block measured by the RPM, we blocked both views of the cameras to null the signal and achieve time synchronization. This needs to be done at the start and the end of each experiment. The Kinect camera was placed at a distance of about 1 meter to the volunteers under an angle of 30°. The markerblock was placed at the chests of the volunteers. Fig. 1 shows the setup. All measurements were performed in a clinical environment inside a CT gantry.

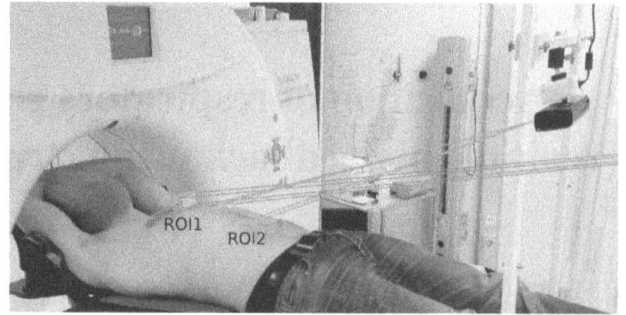

Figure 1: Setup of the Kinect (on the right) and the RPM (not in the image) onto the chest surface. ROI1 is placed directly below the marker block capturing the same region as the RPM system and ROI2 at the abdomen.

2.4.1 Experiments

At first, all volunteers were asked to breath freely to estimate a ground truth respiratory signal. We aimed to record our data for about four minutes to compensate for variations due to familiarization with the uncommon lying situation inside the gantry of the CT. After that, we presented our three different visualization approaches with a short demo of the to be expected plots. A laptop screen, was placed beside to CT table to present the visualization. ROI1 was set just below the markerblock, exactly over the block would cause too much noise, and ROI2 directly over the navel. Then, we went to the biofeedback in a randomized order.

2.4.2 Visual Biofeedback Options

The visual biofeedback options were implemented in Python via the matplotlib library and using the OpenKinect libfreenect library [9] bindings for a convenient access of the cameras of the Kinect. For visualization, ROI1 was used.

Figure 2: Overview of the tested visualization. From left to right: Direct plot of the depth data, bar plot and circle plot. The bar arises when the subject inhales as does the radius of the circle. Arrows indicate movement of the plot over time.

Fig. 2 illustrates the three options of biofeedback. The amplitude and height of the time resolved plot respectively bar plot were calculated by subtracting the current raw value of the ROI1 from the maximum value (which is 2047). So we basically inverted our signal to generate a more intuitive plot of the depth data. For the radii of the circle plot, we had to subtract an offset additionally to the inverting to obtain a small circle at maximum exhalation and a big circle at maximum inhalation.

Table 1: Mean amplitude from peak-to-peak (in mm) detected by the RPM with standard deviation.

Cases	w/o	Option 1	Option 2	Option 3
1	$14.2{\pm}0.8$	$18.1{\pm}1.1$	$17.3{\pm}0.9$	$16.6{\pm}0.9$
2	$1.8{\pm}0.4$	$2.7{\pm}0.6$	$3.0{\pm}0.6$	$2.2{\pm}0.4$
3	$5.1{\pm}0.3$	$8.6{\pm}1.9$	$6.1{\pm}1.6$	$13.1{\pm}4.9$
4	$3.1{\pm}0.8$	$3.4{\pm}1.0$	$3.1{\pm}0.9$	$4.4{\pm}0.9$

Table 2: Period (in seconds) detected by the RPM from maximum to maximum with standard deviation.

Cases	w/o	Option 1	Option 2	Option 3
1	$3.8{\pm}0.1$	$4.3{\pm}0.2$	$4.3{\pm}0.2$	$4.9{\pm}0.1$
2	$3.4{\pm}0.2$	$4.0{\pm}0.2$	$4.4{\pm}0.4$	$3.6{\pm}0.4$
3	$4.5{\pm}0.3$	$5.1{\pm}0.4$	$6.1{\pm}1.0$	$6.3{\pm}0.8$
4	$6.9{\pm}1.2$	$8.1{\pm}1.2$	$6.9{\pm}1.3$	$7.2{\pm}1.2$

3 Results and Discussion

Every volunteer successfully performed all visual biofeedback variations of our experiments.

3.1 Comparison "free breathing" to "biofeedback breathing"

Table 1 shows the captured amplitude of the RPM system for the different experiments. One can see that the standard deviation of normal breathing was lowest for all four volunteers, which is contrary to our expectations. However, the total amplitude increased for all volunteers in all biofeedback measurements.

Table 2 shows the captured period of the breathing cycles captured by the RPM. One can see that the free breathing yields the shortest period of all four types. Biofeedback results in an increase of up to 41% in case of the circle visualization of volunteer 3. No clear statement about the best or most reliable visualization option can be made; however, over all subjects, the bar plot yielded the least changes of the three and is also the least intuitive one, according to three of the volunteers. This is due to difficulties in translating an almost sinusoidal motion of the time resolved captured depth data to a straight bar plot, which does not give a hint of any previous respiratory motion. One can further see that the deviation of the period changed as the mean value increases, which is to be expected but not desired (subject 3 almost got a three times higher standard deviation compared to free breathing).

The results of circular and time resolved visualization show a slight increase of the standard deviation at a higher mean period. However, it is important to point out that we could not observe an increase in terms of smoothness of the breathing patterns of the volunteers. Contrary to the expectation of biofeedback helping the subjects breathing more evenly, which would result in a lower standard deviation of the period, we only observed an increase in total mean values of the period and a equal or slightly higher stan-

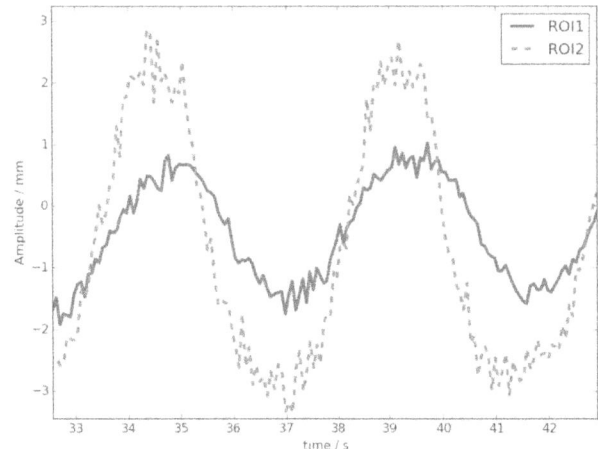

Figure 3: Differentiation between thoracic and abdominal breathing through two regions of interests. ROI1 placed at thorax and ROI2 placed at the abdomen. One can also see a period of around 5 seconds and an a higher amplitude for ROI2, which captured the abdomen. Amplitude in the matter of mm and time in seconds.

dard deviation. That means we did not achieve a significant improvement for the respiratory motion patterns with the current setup, but got promising results in terms of longer breathing phases in general.

These results are the reason that we will focus on the time resolved and circle visual biofeedback for future improvements.

3.2 Differentiation of thoracic and abdominal breathing

The Varian RPM only acquires a signal for one position which is determined by the marker block. With the Kinect, it is possible to track multiple regions of interest (ROIs) at a time. By analyzing both captured ROIs at the chest surface, we could decide which region had a higher amplitude. Fig. 3 shows a plot of the two ROIs corresponding to the setup shown in fig. 1. One can see that the signals amplitude of ROI2 is larger than of ROI1, which indicates abdominal breathing. That corresponds with the observation during the recording of the data. However, the volunteers were not able to perform a switch between abdominal and thoracic breathing, which would be desirable to further validate differentiation. It is imaginable that this knowledge could be used to train patients to thoracic or abdominal breathing, depending on the tumor location. For abdominal breathing, a tumor in the upper region moves less, and this could result in a more accurate treatment.

3.3 General aspects

Further improvements could be achieved by a similar method to [10], where a motion amplification based on a translation surface at the chest is used to accomplish less noisy images because of a flat surface compared to the uneven surfaces of the human body. However, these methods require additional tools at the patients chest surface, which might neither be suitable nor desirable for clinical use. The accuracy of distance measurements has also been shown to be increased with further processing [11], which is a promising addition for future development. According to preliminary results in [12], we expected reduced standard deviation of the respiratory signal compared to free breathing. Likewise, as also stated in [12], gated radiotherapy suffers from longer periods, the treatment time increases because the beam will be turned off for a longer period of time during the respiratory cycle. This will negatively effect dose delivery as well as patient throughput. It is imaginable that tumor tracking could benefit from a better predictability of the chest movement as irregularities would be less common for biofeedbacked motion. Additionally, with our implementation, we achieved a constant frame rate of 10 frames per second which is inferior compared to the RPM system, which delivers 26 frames per second. Another point to mention is that we only tested four subjects, which is not a convincing number of test subjects to reliably answer the final assertion that biofeedback reduces variations in respiratory motion.

4   Conclusion

In this paper, we introduced different options for visual biofeedback for respiratory signals, which have the potential to guide patients into a more regular and more stable breathing pattern. We captured depth data at two regions of interest and assessed influences compared to breathing without biofeedback. Future work will be facing the runtime of the algorithm to better suite our real-time visualization approach, which is currently limited. Besides the performance enhancements, other features, which are based on volunteer inputs, can further improve the software. For example, it would be helpful to see the last period (in case of the time resolved plot) or the last minimum and maximum values in the bar and circle plot. Thus, the patients would be able to know how deep he or she has to inhale and exhale. Another point that is important to address is to mention the compliance of patients or volunteers that use these kind of biofeedback systems. It is essential that the patients really make use of it and see a benefit of this visualization. Otherwise, there is the risk of ignoring the assistance and breath freely.

Acknowledgement

The work has been carried out at University Medical Center Hamburg-Eppendorf (UKE), Department of Radiotherapy and Radio-Oncology. We would like to thank participating volunteers for their assistance and feedback in regards of future improvements of the setup.

5   References

[1] Y. Seppenwoolde et al., *Precise and real-time measurement of 3D tumor motion in lung due to breathing and heartbeat, measured during radiotherapy.* International Journal of Radiation Oncology - Biology - Physics, vol. 53, pp. 822-834, 2002.

[2] M. Velec et al., *Effect of breathing motion on radiotherapy dose accumulation in the abdomen using deformable registration.* International Journal of Radiation Oncology - Biology - Physics, vol. 80, pp. 265-272, 2011.

[3] A. Schweikard, G. Glosser, M. Bodduluri, M. J. Murphy, and J. R. Adler, *Robotic motion compensation for respiratory movement during radiosurgery.* Computer Aided Surgery, vol. 5, pp. 263-277, 2000.

[4] H. D. Kubo and B. C. Hill, *Respiration gated radiotherapy treatment: a technical study.* Physics in Medicine and Biology, vol. 41, no. 1, p. 83, 1996.

[5] H. Sarbolandi, D. Lefloch, A. Kolb, *Kinect range sensing: Structured-light versus Time-of-Flight Kinect.* Computer Vision and Image Understanding, vol. 139, pp. 1-20, 2015.

[6] X. Zhou, *A study of Microsoft Kinect calibration.* Department of Computer Science, George Mason University, 2012.

[7] MRPT Mobile Robot Programming Toolkit. http://www.mrpt.org/, accessed 2016-01-27.

[8] J. Ortmüller, T. Gauer, M. Wilms, H. Handels, and R. Werner, *Respiratory surface motion measurement by Microsoft Kinect: implementation and evaluation of a clinical setup.* Current Directions in Biomedical Engineering, vol. 1, pp. 270–273, 2015.

[9] Openkinect. http://openkinect.org/wiki/Main_Page, accessed 2016-01-27.

[10] J. Xia and R. A. Siochi, *A real-time respiratory motion monitoring system using Kinect: Proof of concept.* Department of Radiation Oncology, University of Iowa, 2012.

[11] T. Wentz et al., *Accuracy of dynamic patient surface monitoring using time-of-flight camera and B-spline modeling for respiratory motion characterization.* Physics in Medicine and Biology, vol. 57, 13, pp. 4175-4193, 2012.

[12] P. H. Cossmann, *Video-coaching as biofeedback tool to improve gated treatments: Possibilities and limitations.* Zeitschrift für Medizinische Physik, vol. 22, pp. 224-230, 2012.

8

Medical Imaging

Impact of Different Breathing Analysis Techniques on Body-surface Indexed Respiratory-Related Stroke Volumes

M. Wattenberg[1], H. Körperich[2], P. Barth[2] and K.T. Laser[3]

[1] Medizinische Ingenieurwissenschaft, Universität zu Lübeck, maximilian.wattenberg@student.uni-luebeck.de

[2] Institut für Radiologie, Nuklearmedizin und Molekulare Bildgebung, Herz- und Diabeteszentrum Nordrhein-Westfalen, Ruhr-Universität Bochum, Bad Oeynhausen, {HKoerperich, PBarth}@hdz-nrw.de

[3] Klinik für Kinderkardiologie und angeborene Herzfehler, Herz- und Diabeteszentrum Nordrhein-Westfalen, Ruhr-Universität Bochum, Bad Oeynhausen, TLaser@hdz-nrw.de

Abstract

MRI can be used to measure flow in thoracic vessels. Respiratory-dependent stroke volumes normalized to body-surface area (SVs) can be extracted and can indicate preserved or impaired hemodynamics. A reliable tracking of the patients' respiration is mandatory because a temporal misregistration of the breathing curve and image generation is present. In order to examine the influence of the mismatch we compared a respiration belt signal with a simultaneously registered respiratory curve extracted from consecutively acquired MR images. SVs were measured in three vessels and tested for statistical significant differences based on mentioned respiration curves and four gates: expiration, end-expiration, inspiration and end-inspiration. The temporal delay between both curves was detected. The delay between image processed curves and respiration belt curves averaged out at 167 ms ($\pm$143 ms). A significant difference ($p < 0.05$) between SVs was observed in: end-expiration (vena-cava-Inferior), inspiration (ascending-aorta), end-inspiration (vena-cava-superior) and expiration (ascending-aorta, vena-cava-superior).

1 Motivation

Measurement of blood flow in the great vessels is mandatory to obtain hemodynamic information which is important for the clinical decision-making process. It has been shown that body-surface normalized respiratory-related stroke volumes (SVs) can differ in patients after comprehensive cardiac surgery resulting in preserved or impaired hemodynamics [1]. This might be used to evaluate the surgery outcome. It is well-known that SVs are dependent on the patients' breathing due to changes in intrathoracic pressures. The impact of respiratory-related SVs can be obtained by using real-time MRI. As shown in literature [1] the breathing curve is typically subdivided into four breathing states end-expiration, inspiration, end-inspiration and expiration. To assess respiratory-dependent SVs it is necessary to track the patients breathing curve reliably. Typically, this is achieved by placing an abdominal respiratory air-cushion belt, although, it has not been proven that image generation time corresponds to the data from the registered breathing curve, possibly resulting in a temporal mismatch. To address this problem the aim of this study was to develop a new software algorithm for recording and analyzing the breathing curve by tracking changes of the subcutaneous fatty tissue signal in the neighborhood of the examined vessel. The temporal mismatch of different breathing detecting techniques and the impact on respiratory-related SVs was analyzed.

2 Material and Methods

Normalized SVs were calculated from real-time MR image data recorded by a multi-transmit 3.0 T-MRI-system (Philips Healthcare, Best, The Netherlands). For real-time MRI a velocity-encoded echo-planar pulse sequence using parallel imaging and undersampling techniques were applied [2]. In this imaging method flow velocity is encoded by phase alterations by introducing bipolar flow-sensitive magnetic gradients [3]. To obtain accurate blood flow information the measuring planes were adjusted orthogonally to each examined vessel. For data analysis flow information was obtained from the ascending-aorta (AO), vena-cava-superior (VCS) and vena-cava-inferior (VCI), respectively. Quantitative evaluation of blood flow was achieved by defining region of Interests (ROI) in the according vessel above the total cross-sectional area in all temporal time frames for creating velocity-to-time, flux-to-time and stroke volume (integral of flow over time) curves [4].

2.1 Respiratory Belt

The standard method for breathing detection in MRI is placing a respiratory belt on the upper part of the abdomen [5]. This device consists of an air cushion facility. During inspiration the epigastrum lifts and compresses the cushion. Air volume changes were registered by a flow meter, which defines the breathing curve. Data was stored by the scanners' registration unit as scan physiological data (SP) and automatically marked with starting and stopping flags to define the beginning and completion of data acquisition.

2.2 Breathing Analysis through Thorax Tracking

Because of the spatial distance between the respiration belt and location of image data acquisition a temporal delay is likely leading to inaccurate flow information assignment. Therefore it is desirable to extract the respiration curve directly from real-time MR images which were recorded in the thoracic region. For this purpose a self-written program by the author called Breathing Analysis through Thorax Tracking (BATT) was used to extract this respiratory information (MATLAB R2014a). 500 magnitude images were recorded with a frequency of 24 Hz showing chest wall movements in a single slice over time. Based on the typically acquired axial slices in the present data set, the line between room air and fatty tissue above the sternum is characterized by a strong contrast (see Fig. 1). BATTT

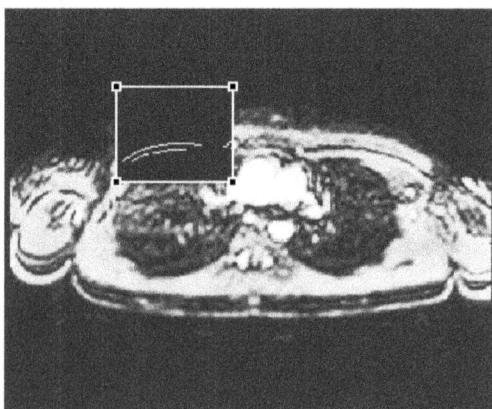

Figure 1: Single axial real-time MRI image. On top of the MRI thorax image a ROI is placed. Inside this ROI a Canny edge filter is applied characterizing the thoracic wall which will be tracked by BATTT.

semi-automatically detects this edge and traces its height in the image. First an image area with an intense contrast is selected and a canny edge filter is applied. This results in a binary edge image. Secondly, all edge representing pixels are set as starting points to search for the longest continuous line of edges, which most likely represents the room air/ fat tissue border. The used recursive algorithm takes the most left starting point as origin to search for the closest edge inside a given mask. In case of an empty mask the mask size increases and searches for an edge in the increased mask (see Fig. 2). In case of multiple edges inside the mask multiple pathways were saved and continuously completed as well. All these paths were scored. Shorter mask sizes and longer paths generate higher scores. The recursive algorithm stops when the mask size reaches its defined maximum size with no edge underneath, all saved paths are completed and all starting points were expired.

In the last step the path with the highest score is suggested as best representation for the thoracic wall. The extraction of the breathing signal uses this representation. All found edges where followed over time in height and additionally displayed as an average curve. The mean height of all edges in one image represents the breathing intensity at one specific time.

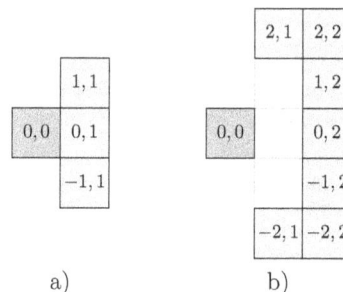

Figure 2: Mask for path detection with relative coordinate to starting point or current point $(0,0)$. Mask increases while the path detection algorithm searches for the closest edge pixel under the right mask domains. a) with mask size key = 1, b) with mask size key = 2.

2.3 Respiratory-Dependent Stroke Volumes

Post-processing was performed by using selfwritten flow quantification software developed by Peter Barth. To obtain respiratory-dependent SVs image DICOM data were accordingly assigned to the simultaneously but separately collected respiratory data saved as log files as described above. In a first step the respiration curve was subdivided into the four respiratory gates: expiration, end-expiration, inspiration and end-inspiration (see Fig. 3). The end-expiration interval was defined as the lower 15% of the difference between two adjacent extreme values on each side of the minimum whereas the end-inspiratory interval corresponds to the upper 15% of the difference between two adjacent extreme values on each side of the maximum. In a second step flow information was accordingly assigned to the different time intervals as illustrated in Fig. 3 to generate artificial respiratory-related stroke volumes.

3 Study Design

Data was available from 31 healthy subjects and was recorded between 06/2011 and 06/2012. One data set in the AO was excluded due to poor image quality. In the VCI ten data sets were acquired. The study was approved by the local Scientific Ethics Institutional Review Committee (RegNo. 09/2011) and informed, written consent was obtained from all study participants. Respiratory-related SVs were obtained in the AO, VCS and VCI. The volunteers were between eight and 40 years of age (mean $= 13.6 \pm 6.2$ years). The number of participants and the gender ratio are listed in Table 1.

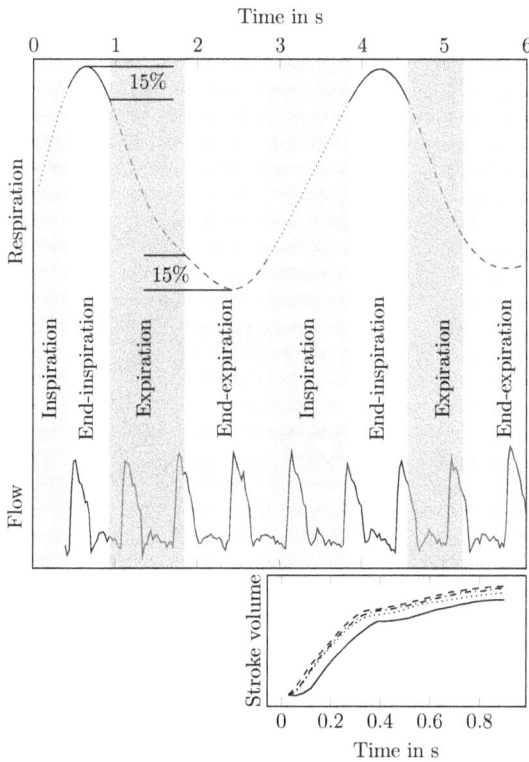

Figure 3: Generating breathing dependent stroke volumes (bottom) from flow (top). Dotted: inspiration, solid: end-inspiration, dashdotted: expiration, dashed: end-expiration.

	Total	Female	Male
AO	30	16	14
VCS	31	17	14
VCI	10	6	4
Delay	61	36	25

Table 1: Number of datasets used for examination and gender ratio of the subjects. Determining the delay ten datasets were excluded because no distinct maximum could be defined.

The patients' breathing pattern was obtained by applying the respiration belt or the BATT curve.

3.1 Respiratory Delay

First each pair of curves was checked on a measurable temporal delay, because a time shift is the most likely to influence SVs. The maximum of the end-inspiration was chosen as indicator for our test. End-inspiration shows a smaller peak with a mostly accurate shaped maximum. This lowers the error for exact evaluation. In contrast end-expiration forms a plateau which obviates to define a precise time stamp (see Fig. 4). After extracting the BATT curve all recorded maxima of end-inspiration were marked as well as the end-inspiration maxima of the corresponding SP curve. The time differences Δt of all maxima pairs were averaged

for each subject regardless of vessel.

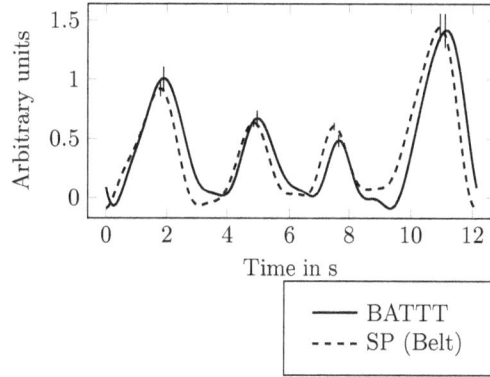

Figure 4: Comparison of the two breathing curves generated by BATT and air cushion belt. The curves were scaled and shifted in height to match the same interval. The temporal coordinates are unchanged.

3.2 Statistics

In order to verify a statistical difference in SVs between SP- and BATT-based SVs a paired student T-Test was applied. The Shapiro-Wilk-Test was used to test data on normal distribution in every curve/vessel/breathing-state set of data. The Levene-Test was used to validate homogeneity of variance in every vessel/breathing-state dataset paired by the two analysis techniques. In Shapiro-Wilk-Test and Levene-Test p-values > 0.05 confirm assumption of homogeneity or normality. In T-Test p-values < 0.05 were considered to assume statistically significant differences.

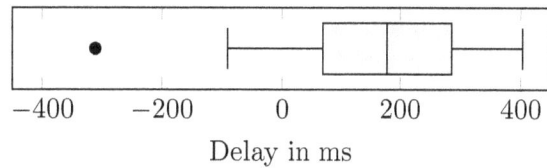

Figure 5: Delay of BATT curve against SP curve based on maximum of end-inspiration. The gray area describes the quartiles (25th and 75th percentiles). The Line inside the area indicates the median. The plotted whisker extend to the most extreme data value that is not an outlier (marked as dot). The shift of the boxplot from zero indicates the delay.

4 Results

The respiration curve recorded with the belt appeared in average 167 ms (standard deviation $= 143$ ms) earlier compared to the curve extracted with the BATT program (see Fig. 5).

Stroke volumes		AO	VCS	VCI
SP	end-expiration	82.31 ± 21.83	-29.59 ± 9.07	54.86 ± 18.20
	inspiration	77.72 ± 20.53	-30.92 ± 7.86	66.90 ± 28.67
	end-inspiration	73.49 ± 19.79	-27.55 ± 7.65	51.80 ± 25.63
	expiration	77.27 ± 21.28	-25.40 ± 8.40	44.64 ± 12.95
BATTT	end-expiration	81.77 ± 21.85	-30.11 ± 8.18	62.99 ± 22.75
	inspiration	76.70 ± 20.30	-30.20 ± 7.98	61.78 ± 29.59
	end-inspiration	74.52 ± 20.82	-26.19 ± 7.74	46.16 ± 16.48
	expiration	80.72 ± 21.81	-27.12 ± 9.71	46.89 ± 14.02

Table 2: Respiratory dependent stroke volumes of different vessels and breathing states including standard deviation.

As shown in Table 3 all data from the different vessels and breathing states appear normally distributed regarding to Shapiro-Wilk-Test. Referring to the Levene-Test all SP-BATTT-pairs satisfy homogeneity of variance. Therefore, it is allowed to use the T-Test as an instrument to test for statistical significant differences. The difference between SP-based SVs and BATTT-based SVs was significant in end-expiration for VCI, in inspiration for AO, in end-inspiration for VCS and, in expiration for AO and VCS (see Table 4).

Shapiro-Wilk-Test		AO	VCS	VCI
SP	end-expiration	0.769	0.743	0.570
	inspiration	0.640	0.622	0.487
	end-inspiration	0.663	0.873	0.648
	expiration	0.682	0.446	0.914
BATTT	end-expiration	0.780	0.262	0.930
	inspiration	0.782	0.713	0.567
	end-inspiration	0.498	0.590	0.740
	expiration	0.785	0.448	0.530

Table 3: Summary of p-values testing normal distribution using Shapiro-Wilk-Test. P-value > 0.05 confirms normal distribution.

5   Discussion and Conclusion

Applying the SP-based version: During inspiration diaphragm movement results in negative thoracic pressures. This movement appears earlier in the caudal position of the air-cushion belt. A temporal delay is present by using the BATTT evaluation program. This might be related to the more appropriate location of the applied ROI which matches the measuring position of the examined vessels. The ribcage movement might begin later and has got a bigger influence on the vessels.

This difference had a noteworthy impact on the respiratory-related SVs. A premature occurrence of the bulk blood flow was detected in the VCI with statistical significance (see Table 2, 4). These results illustrate the importance of respecting breathing analysis techniques while predicting surgical outcome based on respiration dependent SVs. In patients with cardiac disease the respiration is a major factor maintaining blood circulation [6]. Therefore repeating the cur-

rent tests with patients might help to strengthen the statement of this paper and expose the superior analysis technique.

T-Test	AO	VCS	VCI
end-expiration	**0.145**	**0.327**	0.023
inspiration	0.015	**0.780**	**0.144**
end-inspiration	**0.083**	0.017	**0.144**
expiration	0.000	0.030	**0.226**

Table 4: Summary of p-values for comparison SVs based on SP or BATTT curve tested with T-Test. P-value > 0.05 means the difference was not significant (bold).

Acknowledgement

The work has been carried out at *Herz- und Diabeteszentrum Nordrhein-Westfalen, Ruhr-Universität Bochum,* Bad Oeynhausen. I would like to thank the team of *Institut für Radiologie, Nuklearmedizin und molekulare Bildgebung* for the big support and great time.

6   References

[1] H. Körperich, et al., *Impact of respiration on stroke volumes in paediatric controls and in patients after Fontan procedure assessed by MR real-time phase-velocity mapping.* European Heart Journal, Cedex, 2015.

[2] H. Körperich, et al., *Flow volume and shunt quantification congenital heart disease by real-time magnetic resonance velocity mapping.* Circulation, Dellas, 2004.

[3] N. J. Pelc, *Signa advantage application guide.* 1994.

[4] F.A. Flachskampf, *Kursbuch Echokardiographie.* Thieme, Stuttgard, 2012.

[5] J. Bogaert, S. Dymarkowski, A.M. Taylor, *Clinical Cardiac MRI.* Springer, Heidelberg, 2006.

[6] V. Hjortdal, *Effects of exercise and respiration on blood flow in total cavopulmonary connection.* Circulation ,Aarhus, 2003.

Investigation of the suitability of a commercial 3D camera for observation of the surgical field

Albrecht Kleinfeld [1] and Norbert Linz [2]

[1] Medizinische Ingenieurwissenschaft, Universität zu Lübeck, albrecht.kleinfeld@student.uni-luebeck.de

[2] Institut für Biomedizinische Optik, Universität zu Lübeck, linz@bmo.uni-luebeck.de

Abstract

The development of 3D imaging cameras opens new perspectives for medical tasks. Several concepts using these devices are already made, like respiratory gating while tomographic imaging or robot-assisted tumor radiotherapies. In this paper, an investigation of a PMD ToF camera, regarding its ability to detect different materials under modified light conditions in the operating field, is presented. Surgical drapes and materials are tested with regard to their detectability, for which the reliability and precision of the measured distances and amplitude values are discussed. To analyze the performance of the ToF sensor and its sensitivity to operating light, measurements under dark and illuminated conditions are made and compared. In this paper, the tested CamBoard pico^S 71.19k (pmdtechnologies ag) shows a stable functionality and is able to localize most of the materials, however high reflective objects can't be reliably detected.

1 Introduction

In the last years, 3D imaging cameras became more important to gain spatial information of the environment in real time. They are mainly used for object and motion detection, collision prevention and gesture control. In future, threedimensional data will be more and more important for medical issues. First concepts using 3D cameras have been made, like autonomous robot systems, which are able to react to patients respiration during tumor radiotherapies [1]. Another example is gesture control of a performing surgeon, that enables him to work in a sterile interaction. To avoid malfunctions or failures of these systems, it is necessary to investigate the performance of the used 3D camera regarding the detectability of different materials in the surgical field under common operating (OR) illumination.

Today, many 3D sensing technologies are known. Time of Flight (ToF) sensors like the Photon Mixing Device (PMD) became the most important systems of common 3D cameras. They typically consist of a near-infrared (IR) light emitter and a CMOS receiver. The distance measurement is based on the principles of pulsed or continuous wave modulation [2].

The pulsed wave modulation is the most simple principle of the ToF method. By use of a single light pulse, the object position can be achieved by measuring the time the pulse needs to travel the distance between emitter, object and receiver. Using the speed of light, the distance can be determined from the time measurement. For obtaining a high spatial accuracy, the receiver has to measure the pulse delay within picoseconds. This level of precision is not easily achievable, so an alternative measurement method is commonly used [3].

The continuous wave method, also known as intensity modulation principle, is based on the correlation of the emitted (reference) signal $g(t)$ and the received optical signal $s(t)$, which is reflected by the object and detected by the camera. The correlation function can be written as

$$c(\tau) = s \otimes g = \lim_{T \to \infty} \int_{-T/2}^{T/2} s(t) \cdot g(t + \tau) \, dt, \qquad (1)$$

where t is the time and τ the phase offset. Assuming sinosoidal functions, e.g.

$$g(t) = cos(2\pi f_m t) \qquad (2)$$

and

$$s(t) = b + a \, cos(2\pi f_m t + \phi), \qquad (3)$$

the correlation function results in

$$c(\tau) = \frac{a}{2} cos(f_m \tau + \phi) + b, \qquad (4)$$

with amplitude a of the incident signal, the correlation bias or intensity b, the modulation frequency f_m and the phase shift ϕ between emitted and reflected light.

The correlation function $c(\tau)$ can be demodulated by using four discrete samples $A_i = c(i * \frac{\pi}{2})$, with $i = 0, ..., 3$. The absolute number of these recurring sample points is determined by the integration time $\triangle t$, which influences the signal to noise ratio, the depth accuracy and the motion blur. A schematic drawing of the correlation function, the integration time and the four samples are shown in Fig. 1. Using the discrete samples, amplitude, intensity and phase shift can be calculated by following equations:

$$a = \frac{1}{2}\sqrt{(A_1 - A_3)^2 + (A_2 - A_0)^2} \qquad (5)$$

$$b = \frac{A_0 + A_1 + A_2 + A_3}{4} \qquad (6)$$

$$\phi = arctan(\frac{A_1 - A_3}{A_2 - A_0}). \qquad (7)$$

With the help of the reconstructed phase shift ϕ, the object distance d can be easily determined by:

$$d = \frac{c}{4\pi f_m}\phi, \qquad (8)$$

where c is the speed of light [3], [4], [5]. Using d and knowing the direction of the received light, a threedimensional x,y,z-coordinate of an object point can be calculated. The z-component is correlated to the orthogonal distance between object and camera level. In this paper this value is called z-distance.

ToF camera systems allow to measure depth and amplitude images at the same time. Thereby the amplitude a is an indicator for the signal strength of the backscattered IR light. This parameter can be used to analyze the quality of the distance value, where a high amplitude is correlated to a good accuracy of the determined distance [6]. With the help of the amplitude value, a certain evidence about the detectability of different materials in the surgical field can be made. Especially low scattering materials or targets far from the camera may lead to low amplitude signals. In contrast, objects with a high reflectivity might saturate the sensor [4]. Beside the reflected light, unwanted ambient light can reach the camera. The influence of this extra signal can be analyzed by using the amplitude information. Bright ambient light may saturate the ToF sensor and reduces the signal to noise ratio [7]. Because of the reasons given above, an evaluation of different surgical materials and OR lighting conditions, concerning their influences on the suitability of a ToF camera, is performed.

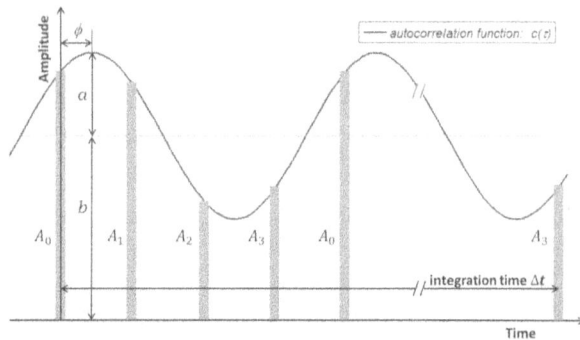

Figure 1: Schematic drawing of the continuous wave principle with the correlation function $c(\tau)$, the integration time $\triangle t$ and four discrete samples $A_0, ..., A_3$. a represents the amplitude, b the offset/intensity and ϕ the phase shift between the emitted and received signal.[3].

2   Materials and Methods

The following section presents the tested 3D camera and its output parameters. Additionally, the test bench is described and the materials are listed, which are used to investigate the detection capability of the camera.

2.1   CamBoard pico^S 71.19k

For the analysis of a commercial 3D ToF camera, the pmd[vision]$^{\circledR}$ CamBoard pico^S 71.19k, developed by pmdtechnologies ag (Siegen, Germany), is used. This device operates with the continuous wave principle, using IR light at 850 nm. The cameras working range is specified with $20 - 100$ cm and provides a resolution of 160×120 px. To obtain a stable and reliable measurement, the camera is used after a warm-up time of 20 minutes to an operating temperature of $41°C$. The integration time is set to one second.

For every measured point, the camera detects an amplitude value and calculates the z-distance in meter. The reliability and precision of the z-distances of an image are evaluated by the mean value and the standard deviation. In addition, the camera provides the option to analyze the measured values with regard to their validity. Therefore, three different identifiers are available. If the raw data values are inconsistent with each other, the point is marked as *inconsistent*. Inconsistency may be identified, if a fast changing scene is measured. An overexposed point is marked as *saturated* and an incomming signal that is too low triggers a *low-signal* identifier. If one of these three markers is set, the relevant point is considered as *invalid*. The changes of the total numbers of these identifiers are used to analyze the tested materials concerning their absorption, reflectivity and scattering. Due to a better comparison of the measured values, image sections including pure test material are manually selected.

2.2   Test bench and OR materials

For the experimental investigation, the CamBoard pico^S is placed 16 cm below a Polaris$^{\circledR}$ 600 OR light, which is designed by Drägerwerk AG & Co. KGaA in Lübeck (Germany). This OR light illuminates the simulated surgical field with 92 LEDs, emitting a maximum of 160.000 lux, because it can be assumed that saturation of the ToF sensor occurs with the highest illumination. The tested materials are placed on a height-adjustable measuring plate, positioned centrally under the ToF camera.

A realistic surgical environment is simulated by selecting four different drapes. These are colored in white, green, blue and black, to determine the influence of the color. All drapes are placed 30 cm, 60 cm and 90 cm below the camera. Furthermore, aluminium foil is analyzed to simulate the reflectivity of a warming blanket. A 18 cm $\times$ 20 cm large area of this foil is placed 10 cm above black drape and 30 cm below the camera (Fig. 2 (a)). All materials are tested under dark and maximal OR illuminated conditions. To simulate the reflectivity of OR instruments, a set of cutlery is used as a dummy. The instruments are placed on black drape at 30 cm distance below the ToF sensor. In addition, intentional saturation is introduced by adjusting the reflective side of a knife, so that the IR light of the camera LED is directly reflected to the ToF sensor.

To simulate realistic conditions, in which sterile latex gloves and human skin can be located in the OR field, a human hand with and without latex glove is tested. Black drape is placed 10 cm below the hand. The distance between camera and hand is set to 30 cm (Fig. 2 (b)). During the investigations a black wall and white paper are examined with regard to their reflectivity of IR light. Therefore the camera is placed 30 cm in front of the wall. The paper is imaged at 30 cm, 60 cm and 90 cm.

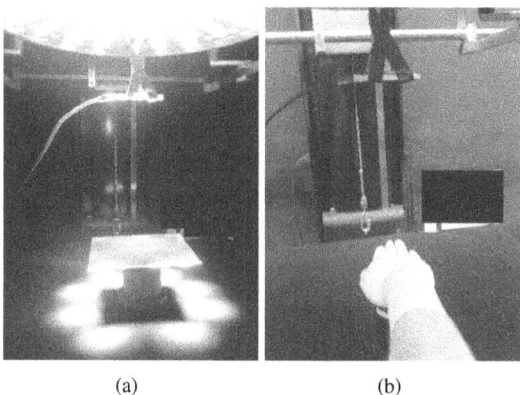

(a)   (b)

Figure 2: Test bench with the CamBoard pico^S, placed directly under the Polaris® 600. Aluminium foil (a) and a hand with latex glove (b) are placed 10 cm above black drape.

3 Results

Every dataset consists of about 19.000 points, where the selected and connected sections, including only drape or paper, have 6156 points. Due to the smaller size of the tested OR materials and instruments, their image sections contain less points. The average and standard deviation of the z-distances and amplitudes of the drapes and paper under different light conditions are shown in Fig. 3 and 4. No points in these datasets are marked as *saturated* and the inconsistency doesn't exceeds 3%. All measured points at a distance of 30 cm are valid and all points at 60 cm of white paper and white drape are marked as valid too. The number of *low-signaled* values correlates with the *invalid* point number and increases with a higher distance. At a distance of 60 cm, over 36% of the green and about 9% of the blue and black drape points are *invalid*. Almost no valid point of green drape is detected at 90 cm and the average number of *invalid* points of white, blue and black drape is 32%, 81% and 69%. The difference of these values, compared between dark and OR lighted condition, never exceeds 2%. The results for the OR materials and instruments are summarized in Table 1. The tested Aluminium foil causes 11.5% *saturated* points in dark and 14.5% with bright illu-

mination. Surgical instruments generate less than 1% over-exposed points, but more than 8% are *low-signaled*. An intentional saturation of the entire sensor by adjusting an instrument can be hardly realized and only for a short time.

Table 1: Results for materials, measured at a distance of 30 cm. The percentage of the *invalid* points, the average (∅) and standard deviation (σ) of the z-distances (z) and amplitudes (a) are listed.

Material	light	∅ inv.	∅ z /cm	σ z	∅ a /a.u.	σ a
alu. foil	off	11.5	63.2	8.3	772	686
alu. foil	on	15.1	62.7	9.3	809	778
hand	off	0	28.0	0.6	465	129
hand	on	0	29.1	0.8	461	141
latex glove	off	0	27.7	0.6	493	156
latex glove	on	0	28.3	1.0	511	152
surg. instr.	off	7.8	49.4	10.9	227	214
surg. instr.	on	8.8	50.0	11.1	229	223
black wall	off	100	37.1	20.4	17	5

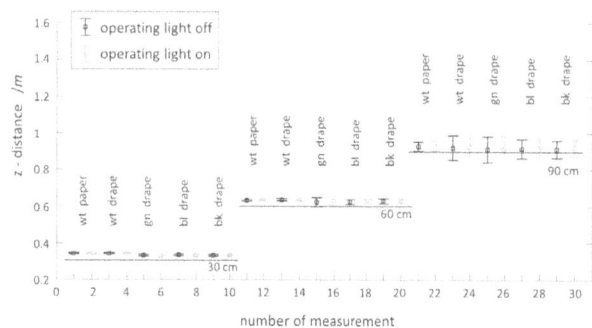

Figure 3: Average z-distances and their standard deviation for white paper and four different colored drapes.

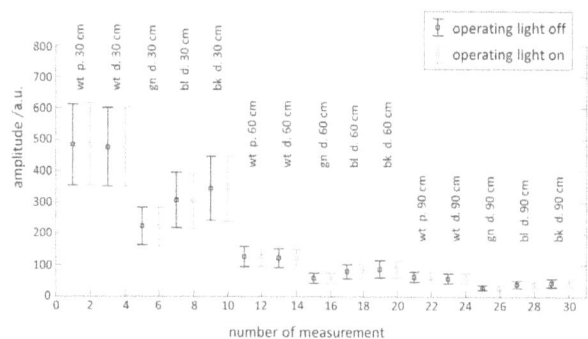

Figure 4: Average amplitudes and their standard deviation for white paper (p.) and four different colored drapes (d.).

4 Discussion

Due to its small size and low energy consumption, the CamBoard pico^S can be easily integrated into existing OR devices. Nevertheless, its ability of a robust object detection had to be tested and evaluated.

The four different colored drapes and the white paper are captured reliably at the investigated z-distances (Fig. 3).

The average values demonstrate a stable progression in relation to the preset distances. The low standard deviations at 30 cm and 60 cm indicate good precisions. But in Fig. 4 a significant decrease of the related amplitudes with higher distances can be perceived. This reduction of the detectability is consistent with the reduced number of valid points at 90 cm. Especially green drape could not be detected safely. Here, about 99% of the points are *invalid*, however 32% of the points of the white drape have a high enough signal. This difference might be explained by the higher absorption of IR light by green drape, which can already be seen by the reduced amplitude values at measurement number 5 and 6. The PMD camera shows good reliability concerning the z-distances and very high precision. However the measured distances are slightly longer than the real values. These are influenced by small ripples of the used drapes and an imperfect orientation of the camera and measuring plate.

From Fig. 3 and 4 it can be concluded that additional OR light doesn't influence the functionality of the ToF sensor. This fact might be explained by a small IR light sensitivity of the used Dräger Polaris® 600 and an included "Suppression of Background Intensity" (SBI)-filter inside the PMD camera. This technology is patented by pmdtechnologies ag and makes the sensor robust against ambient light. However, it can be seen in Table 1, that OR light, which is reflected directly into the sensor by aluminium foil, results in higher amplitudes. Here the increasing number of *invalid* value is based on the increment of *saturated* points. It can be observed that extremly high amplitudes cause overexposed camera pixels. Additionally, the measured average z-distance is twice as high as the space of 30 cm between camera level and aluminium foil. This can be explained by the fact, that the emitted light travels twice the way between camera, aluminium foil and the above positioned Polaris® 600. Due to this results, it should be avoided to use the PMD with highly reflective materials.

Small OR instruments, like scalpels, also cause false distance values and overexpose the sensor, but only in a small region. A manual and intended saturation of the camera by adjusting an instrument is hardly realizable and only for a short period of time. Due to this, the CamBoard can not detect small reflective instruments, but these objects do not influence the ability to locate other objects in the scenery.

The camera is able to capture human skin and latex gloves reliably. As presented in Table 1, no *invalid* value is generated. The deviation between adjusted and measured z-distances can be explained by the curvature of the hand.

In this study, a black wall is tested, whether it can be imaged by the camera. Due to the very high absorbtion of IR light, almost no valid point is generated. Therefore, imaging of highly absorbing material at 850 nm is not possible. However, it can be assumed that this type of objects doesn't prevail in the operating field.

In summary, the investigated ToF camera is able to capture most of the tested OR materials reliably at distances up to 60 cm, whereas high reflective objects can not be located with high accuracy. However these objects don't influence the detectability of other objects in the scenery.

5 Conclusion

A first investigation of the suitability of a PMD ToF camera in a surgical field is presented. This study demonstrates, that the used pmd[vision]® CamBoard pico^S 71.19k is able to detect most of the tested materials and that its functionality is not influenced by operating light radiation.

Four different colored drapes are tested concerning their detectability and all of them are reliably captured at distances up to 60 cm.

Materials in the OR field like latex gloves and human skin can be easily detected, but reflective objects cause false measured values. However, the camera sensor is hardly influenced by overexposure.

It is shown that detection of materials with high IR light absorption is not possible. To investigate this subject, further scattering tests and absorption spectroscopies should be made.

In summary, the tested camera can be used in most medical cases, for example in the direct operating field, where movement and gestures of the operating surgeon can be observed. The use of the camera in fast changing scenarios has to be investigated in further tests.

Acknowledgement

The work has been carried out at Drägerwerk AG & Co. KGaA, Lübeck (Germany).

6 References

[1] C. Ozhasoglu et al., *Synchrony - cyberknife respiratory compensation technology.* in: Medical Dosimetry, InTech, vol. 33, no. 2, pp. 117–123, 2008

[2] S. Hussmann, T. Edeler and A. Hermanski, *Real-Time Processing of 3D-TOF Data in Machine Vision Applications.* in: Machine Vision - Application and Systems, InTech, pp. 73–92, 2012.

[3] T. Ringbeck, T. Möller and B. Hagebeuke, *Multidimensional measurement by using 3-D PMD sensors.* Advances in Radio Science. 5, pp. 135–146. 2007.

[4] A. Kolb, E. Barth, R. Koch and R. Larsen, *Time-of-flight cameras in computer graphics.* Computer Graphics Forum, vol. 29, no. 1, pp. 141–159, 2010.

[5] C. Dal Mutto, P. Zanuttigh and G. M. Cortelazzo, *Time-of-Flight Cameras and Microsoft Kinect* Springer, New York, 2012.

[6] D. Lefloch et al., *Technical Foundation and Calibration Methods for Time-of-Flight Cameras.* in: Time-of-Flight and Depth Imaging, Springer, Berlin/Heidelberg, pp. 3–24, 2013.

[7] L. Li, *Time-of-Flight Camera – An Introduction.*, Texas Instruments, Dallas, 2014.

Phantom Design for a High Resolution Field Free Line MPI Scanner

F. Matysiak [1], M. Weber [2], S. Bruns [2], A. von Gladiß [2] and T. M. Buzug [2]

[1] Medizinische Ingenieurwissenschaft, Universität zu Lübeck, florian.matysiak@student.uni-luebeck.de
[2] Institute of Medical Engineering, Universität zu Lübeck, {weber, bruns, gladiss, buzug}@imt.uni-luebeck.de

Abstract

Magnetic Particle Imaging (MPI) is an imaging method with promising potential for clinical applications, since it features a very high temporal resolution. Furthermore, improving the spatial resolution of MPI might allow for new research in the area of real-time angiography or stem cell applications. In order to simulate biological processes, precise phantoms can serve as a model assumption to avoid animal experiments. This work describes the design and construction of a high-precision vessel phantom. In a first approach, a two-dimensional flow phantom is developed and evaluated in a 5 T/m MPI scanner featuring a field free line (FFL). A pumping system is implemented to imitate the natural blood flow with different flow rates. Theory predicts a spatial resolution of 1.1 mm concerning Resovist as a tracer. The presented results of a first measurement setup with a static phantom indicate great potential for further dynamic experiments.

1 Introduction

Magnetic Particle Imaging (MPI) is an imaging method with a high spatial and temporal resolution. The response of superparamagnetic nanoparticles in presence of an alternating external magnetic field is used to generate tomographic images of the particle distribution [1]. MPI shows a high potential for angiography, cancer imaging or cell tracking [2]. Due to the technical progress of MPI, real-time imaging and millimeter resolution are achieved [3]. There is an increasing demand for highly accurate static and flow phantoms.

A magnetic gradient is used to generate a field free point (FFP) or a field free line (FFL) to achieve spatial encoding. It is desirable to apply a high gradient, because the achievable resolution is directly proportional to the used gradient strength [4]. With an external magnetic field, called drive field, the FFP or FFL can be moved over the field of view (FOV). A projection image can be generated by using the FFL for MPI [5]. The two most common MPI reconstruction methods are frequency-domain reconstruction and time-domain reconstruction, also known as x-space reconstruction [1] [6]. In this work, the x-space reconstruction is used, as it is the most common for the FFL projection method.

Current phantoms do not consider the possible fields of application for MPI, e.g., dynamic processes as the blood flow of biological organism. First, a static phantom is designed providing the basis for further usage as a flow phantom. The pumping system contains a peristaltic pump to simulate blood flow. A FFL MPI scanner with gradient strength of 5 T/m is used for measurements. The reconstruction steps using x-space reconstruction and the evaluation of the designed phantom are described.

2 Material and Methods

In the following section the phantom design and pumping system is described. Furthermore, the measurement setup, the signal processing and reconstruction of the raw measured signal is explained.

2.1 Phantom Design

A flow phantom is designed in SolidWorks (Dassault Systems, Matham, MA). This phantom features a design which allows a static and dynamic imaging process. The inner tubes of the phantom are formed like a blood vessel structure. For later simulation of a blood flow within the phantom, a first test with a static model is conducted.

The design of the phantom must fulfill certain criteria. First, it is important, that the materials used for the phantom are non-magnetic to prevent any noise during the measurement. Second, it must have the correct dimensions in order to fit into the scanner. Furthermore, a connection for a liquid pump is needed. It is also important that the phantom is impermeable, to prevent liquid to flow into the scanner. Since an excitation of the particles with the FFL takes place in the horizontal plane, the vessels extend only in the vertical axis.

The two phantoms are printed with two different 3D printers. The first one with the Ultimaker 2 (Ultimaker B.V., Geldermalsen, The Netherlands) and the second one with the 3D Printer ProJet 3510 HD Plus (3D Systems GmbH, Darmstadt, Germany). Both printers have a different resolution. The Ultimaker prints with a layer thickness of 0.1 mm and a resolution in the xy-level of 0.4 mm. The ProJet prints with a layer thickness of 0.029 mm and a resolution in the xy-level of 0.034 mm. Phantom (I), printed with the Ultimaker, is made of transparent polyactide (PLA). Phantom (II), printed with the ProJet, is made of VisiJet M3-X (3D Systems GmbH, Darmstadt, Germany), a plastic material with acrylonitrile butadiene styrene properties. Both materials are plastics. Since plastic does not generate any disturbing signal, PLA and VisiJet M3-X are both suitable for the phantom design.

The phantoms have an outer diameter of 28 mm and a total length of 50 mm to fit through the bore of the scanner. The tubes of the vessel structure of phantom (I) have an inner diameter of 2 mm, phantom (II) of 1 mm. The vessels are aligned on a vertical plane with a size of 10×50 mm. Additional threaded nozzles are attached at the openings of the vessels of phantom (II). Phantom (I) has extended deepenings to use threaded nozzles. These are used to be able to connect the phantom with a pump. The designed 3D model of the phantom is shown in Fig. 1. The phantom is designed as a hollow cylinder with several openings so that it is possible to examine the internal structure, to avoid that parts of the inner structure of the phantom are not filled with liquid during the measurement. Fig. 2 shows the two printed phantoms.

A mounting is attached to the scanner to ensure that the phantom does not move in the scanner during the measurement. A threaded rod is screwed into the mounting, which can be attached to the phantom via an extension.

The pumping system simulates the blood flow of an organism. A peristaltic liquid pump with silicone tubing (Adafruit) is used. The pump is powered by a 12 V DC motor and has an indicated flow rate up to 100 ml/min. The motor is connected to and controlled by an Arduino UNO. The Arduino is powered by a 12 V DC power supply. An external switch allows to start and stop the pump. A second switch changes the predetermined flow rate.

The amount of pumped liquid for a period of one minute is measured to test the indicated flow rate of the pump. The desired values of the flow rate are calculated from the given maximum rate. The pumping system is inserted in a aluminum shielded box to reduce disturbing signals of the pump's magnetic fields.

2.2 MPI Measurement Setup

The imaging process is performed on a FFL MPI scanner featuring a gradient strength of 5 T/m. The gradient field is generated by opposing permanent magnets. A 25 kHz drive field is used to excite the particles. A receive coil measures the induced signal of the changing particle magnetization. The bore diameter of the scanner is about 30 mm. The phantom is mechanically moved with a threaded rod. A full rotation of the rod moves the phantom in a step size about 1.5 mm through the FOV of the scanner. The FOV is defined by the maximum shift of the FFL and the phantom movement through the scanner. This results in a FOV of 67.5×10.8 mm.

The phantom is filled with diluted Resovist (Bayer Scher-

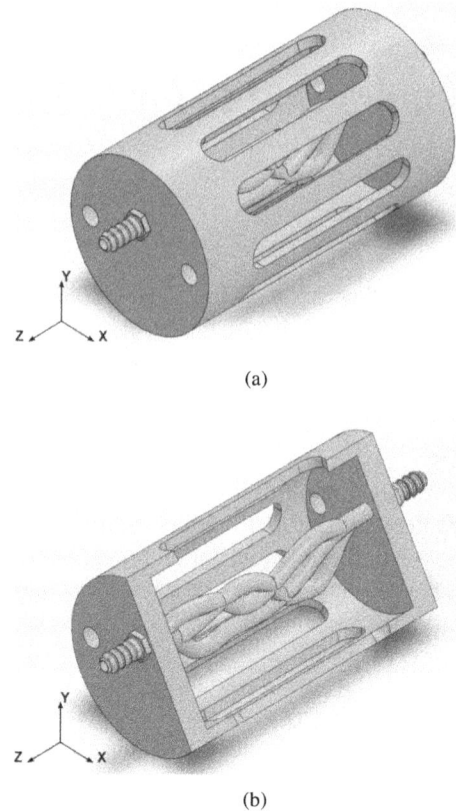

Figure 1: (a) shows the designed phantom in Solidworks. (b) shows the blood vessel-like structure inside the phantom.

Figure 2: (a) Phantom (I) printed with the Ultimaker. (b) Phantom (II) printed with the ProJet.

ing Pharma) with a ratio of 1:5. The particle diameter is assumed to be 23 nm [7].

2.3 Signal Processing and Reconstruction

The particle signal is measured with a sampling frequency of 25 MHz and an accuracy of 16 bits. The phantom is stepped through the scanner by the threaded rod. After each full rotation a 1D projection of the phantom image is taken. Altogether, 45 measurements are acquired to reconstruct a 2D image of the phantom.

X-space reconstruction method is used for the image reconstruction of the measured particle signal [6] [8]. An empty chamber measurement is recorded and subtracted from the measured phantom. The receiving chain causes a phase shift. Hence the measured data must be shifted by a constant phase. Due to a reinforcement of the frequencies from 800 Hz, the data is filtered with a low-pass filter in frequency space. Afterwards, the data is divided by the FFL speed. Since a division by zero induces an error at the FOVs edges, 2 % of the sample data is rejected at the FOVs edges. The actual spatial position is restored by regridding the time-dependent signal with a 1D-linear interpolation. Because a period of the measurement consists of two projections of the sample, the period is split into two parts, superimposed and normalized. As the final step, the signal is deconvolved with the theoretical PSF using the Wiener filter algorithm [9]. The algorithm for the reconstruction is implemented in MATLAB (Mathworks, Natick, MA, USA). With given diameter of the particles and gradient strength of the scanner, the theoretically achievable resolution Δx can be calculated [4]. It results a resolution of 1.1 mm.

3 Results and Discussion

In this section, the printed phantoms are evaluated and the calibration of the pumping system is performed. Then the measurements with the suitable phantom are shown and analyzed.

3.1 Evaluation of the Printed Phantoms and Flow Rate of the Pump

A CT image is taken to examine the printed quality of the two phantoms. Fig. 3 shows a sectional image of the vessel structure for each phantom. The vessel structure had to be printed for phantom (I) with a 2 mm diameter due to the low printer resolution of the Ultimaker 2. A previous test print with a 1 mm diameter could not be used, because the vessel structure was partly closed by the printer material. Although phantom (I) does not show any locally sealed structures, it is unsuitable for the measurement with the MPI scanner. As marked in Fig. 3(a), the wall of the structure clearly shows gaps. Phantom (I) is not suitable as a flow phantom since liquid could leak at a blood flow simulation through the gaps in the structure. However, phantom (II) in Fig. 3(b) shows a completely filled structure and no obstruction in the vessel wall. Hence it fulfills one of the predefined criteria. The other listed criteria are also fulfilled by the design and the material used for the phantom.

In order to test the pumping performance of the peristaltic pump, the amount of pumped liquid is measured for different pumping speeds and shown in Table 1.

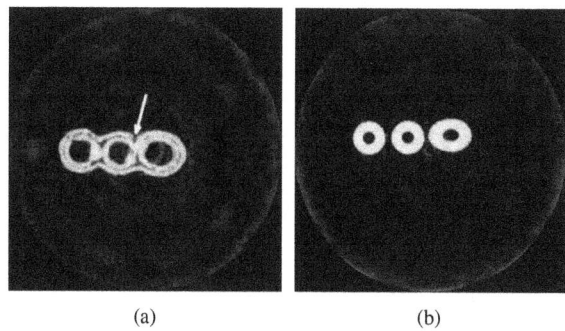

(a)   (b)

Figure 3: Comparison of two CT scans of the two printed phantoms. (a) shows the phantom printed with the Ultimaker. The arrow marks a gap in the structure. (b) shows the phantom printed with the ProJet.

Table 1: Results of measurements for the flow rate of the pump.

No.	Expected value [ml/min]	Actual value [ml/min]
1	100.0	59.3
2	95.0	55.6
3	90.0	52.2
4	85.0	49.6
5	80.0	45.0
6	75.0	40.7
7	70.0	36.9
8	65.0	33.3
9	60.0	29.2
10	55.0	25.1
11	50.0	20.2
12	45.0	17.0

The comparison between expected and measured values shows considerable differences. A deviation from up to 59 % can be found. Furthermore, short scattered breakdowns of the pump can be noted starting from the tenth measurement. Rollers are arranged in a clover pattern inside the pump. The three rollers press the fluid through the tube when the motor of the pump turns. If the torque of the DC motor is too low, it does not have enough power to move the rollers. As a consequence, the engine spins without the rollers. The pump does not continue to pump for a flow rate smaller than 17 ml/min. Although the maximum indicated flow rate is not reached, the pump can be nevertheless used as the pumping system of the flow phantom.

3.2 Phantom Measurements

Fig. 4 shows the imaged phantom filled with Resovist. The measured and deconvolved signal of the complex phantom is shown in Fig. 5(b). Additionally Fig. 5(a) shows a sectional image of the CT dataset of the flow phantom. The structure of the flow phantom filled with Resovist can be seen clearly. Also parts of the structure which are not filled because of air bubbles can be recognized. Only parts of

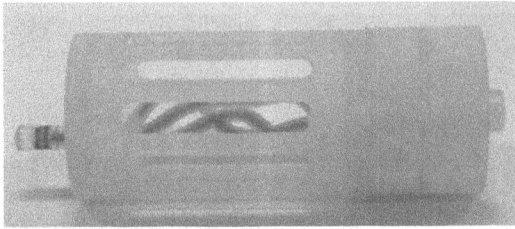

Figure 4: Constructed flow phantom. The inner structure is filled with diluted Resovist.

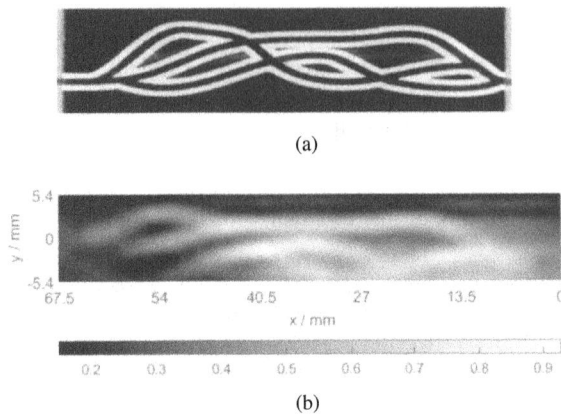

(a)

(b)

Figure 5: (a) shows an sectional image of the vessel system. (b) shows the deconvolved reconstructed image of the flow phantom.

the structure at the edge of the FOV are slightly blurred. It is possible that the vessel structure was not aligned perpendicular to the FFL after the full rotation during the measurement. The full-width at half-maximum (FWHM) is calculated to determine the achieved resolution. The FWHM of the deconvolved image of the flow phantom is about 2.0 mm. Due to the high variance from the expected resolution of 1.1 mm, further measurements with a delta sample or special resolution phantoms are necessary to receive the precise achieved resolution.

The high resolution and high contrast of the blood vessel-like structure in MPI is shown in Fig. 5(b). Therefore, it could be well suitable for real-time angiography to represent the blood flow in vessels. Furthermore, a combination of MPI and CT could generate accurate functional and anatomic correct images of the brain or lungs.

4 Conclusion

The first flow phantom for a high resolution FFL MPI scanner was presented in this work. The main criteria for the phantom were described along with their implementation with selected materials, size, usage as a static as well as flow phantom and design. The measurement setup and the FFL reconstruction with the x-space method are listed. Further, a static 2D measurement of the flow phantom with a 5 T/m MPI scanner is shown. By using the Wiener decon-

volution method the resulting FWHM value is 2.0 mm.

The next step will be the initial startup of the pumping system during a MPI measurement and the real-time imaging. Furthermore, a automatic step-by-step movement to replace the threaded rod is planned to perform smaller steps through the FOV the scanner. In the future a rotating FFL could be implemented to receive a high resolution 3D image.

Acknowledgement

The work has been carried out at the Institute of Medical Engineering at the Universität zu Lübeck, Germany.

5 References

[1] B. Gleich and J. Weizenecker, "Tomographic imaging using the nonlinear response of magnetic particles," *Nature*, vol. 435, pp. 1214–1217, June 2005.

[2] K. Lüdtke-Buzug, "From synthesis to clinical application. magnetic nanoparticles," *Chemie in unserer Zeit*, vol. 46, no. 1, pp. 32–39, 2012.

[3] J. Weizenecker, B. Gleich, J. Rahmer, H. Dahnke, and J. Borgert, "Three-dimensional real-time in vivo magnetic particle imaging," *Physics in Medicine and Biology*, vol. 54, no. 5, pp. L1–L10, 2009.

[4] J. Rahmer, J. Weizenecker, B. Gleich, and J. Borgert, "Signal encoding in magnetic particle imaging: properties of the system function," *BMC Medical Imaging*, vol. 9, no. 1, pp. 1–21, 2009.

[5] J. Weizenecker, B. Gleich, and J. Borgert, "Magnetic particle imaging using a field free line," *Journal of Physics D: Applied Physics*, vol. 41, no. 10, p. 105009, 2008.

[6] P. W. Goodwill and S. M. Conolly, "The x-space formulation of the magnetic particle imaging process: 1-d signal, resolution, bandwidth, SNR, SAR, and magnetostimulation," *IEEE Transactions on Medical Imaging*, vol. 29, pp. 1851–1859, Nov 2010.

[7] J. J. Konkle, P. W. Goodwill, O. M. Carrasco-Zevallos, and S. M. Conolly, "Projection reconstruction magnetic particle imaging," *Medical Imaging, IEEE Transactions on*, vol. 32, pp. 338–347, Feb 2013.

[8] K. Bente, M. Weber, M. Graeser, T. F. Sattel, M. Erbe, and T. M. Buzug, "Electronic field free line rotation and relaxation deconvolution in magnetic particle imaging," *IEEE Transactions on Medical Imaging*, vol. 34, pp. 644–651, Feb 2015.

[9] T. Knopp, M. Erbe, T. F. Sattel, S. Biederer, and T. M. Buzug, "A fourier slice theorem for magnetic particle imaging using a field-free line," *Inverse Problems*, vol. 27, no. 9, p. 095004, 2011.

Sensitivity of Magnetic Particle Imaging Compared with 19F Magnetic Resonance Imaging
– Preliminary MPI Results –

S. Bruns [1], F. G. Heslinga [2], D. W. Hensley [3], P. Keselman [3], M. Weber [4], E. Yu [3], B. Zheng [3], X. Y. Zhou [3], P. W. Goodwill [3], S. M. Conolly [3,5], and T. M. Buzug [4]

[1] Medizinische Ingenieurwissenschaft, Universität zu Lübeck, steffen.bruns@student.uni-luebeck.de
[2] Department of Neuroimaging, University of Twente, f.g.heslinga@student.utwente.nl
[3] Department of Bioengineering, University of California at Berkeley,
{dwhensley,pkeselma,elaineyu,bozheng,xyzhou,goodwill}@berkeley.edu
[4] Institute of Medical Engineering, Universität zu Lübeck, {weber,buzug}@imt.uni-luebeck.de
[5] Department of Electrical Engineering & Computer Sciences, University of California at Berkeley,
sconolly@berkeley.edu

Abstract

Cell tracking methods are very promising for both the diagnosis and the treatment of tumors. Besides achieving a high sensitivity, it is highly desired to avoid ionizing radiation and toxic tracer material. Thus, we evaluated the cell tracking potential of magnetic particle imaging (MPI), a novel non-invasive medical imaging technique based on superparamagnetic iron oxide nanoparticles (SPIONs) as a tracer. Therefore, we measured six point sources of SPIONs with different iron contents in an MPI system. We were able to show that MPI can already detect 188 ng of iron at a signal-to-noise ratio (SNR) of 3 within a clinically reasonable scan time of 9.5 minutes. This implies that MPI may have the potential to be a sensitive, non-invasive alternative to 19F magnetic resonance imaging (MRI) for cell tracking.

1 Introduction

Magnetic particle imaging (MPI) is a novel non-invasive tracer imaging modality providing high resolution and sensitivity as well as positive contrast [1]. It utilizes the nonlinear magnetization response of superparamagnetic iron oxide nanoparticles (SPIONs) to alternating magnetic fields. Because of their low frequency, these fields do not experience any attenuation in tissue. Furthermore, no ionizing radiation needs to be applied to the patient [2]. In tandem with the fact that only the SPIONs contribute to the signal and not tissue, MPI has great potential to become the leading imaging modality for angiography and *in vivo* stem cell tracking [3], [4].

In this work, we aim to determine the minimum amount of iron that can clearly be distinguished from the noise floor in a clinically relevant setting for MPI. Scanning parameters such as scan time, field of view (FOV), and spatial resolution need to be reasonable and also comparable to a future equivalent setup with 19F magnetic resonance imaging (MRI).

1.1 MPI Theory

MPI images the spatial distribution of SPIONs [1]. These ferromagnetic iron oxide tracer particles do not have any

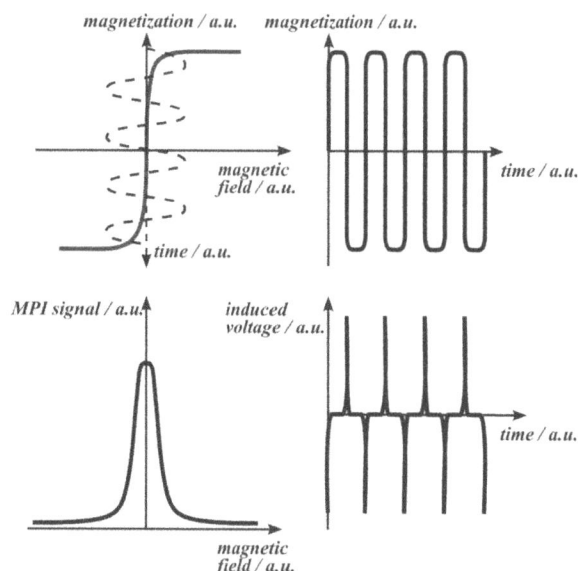

Figure 1: The magnetization M of SPIONs is dependent on the applied magnetic field H and can be modeled with the Langevin function (top left, bold line). A time-varying magnetic field (top left, dotted line) causes a time-dependent magnetization of the SPIONs (top right). Changes in magnetization can be measured as induced voltages (bottom right). The point spread function (PSF) in MPI is the derivative of the Langevin function (bottom left).

hysteresis once their diameter becomes sufficiently small [5]. The nonlinear magnetization behavior of the SPIONs can be modeled using the Langevin theory of superparamagnetism [6]. The bold line in the top left corner of Fig. 1 shows the Langevin function which is only valid when relaxation effects are neglectable [2]. An oscillating magnetic field with $f_0 = 25$ kHz, called drive field, is utilized to excite the SPIONs (Fig. 1, top left, dotted line). The resulting change in magnetization (Fig. 1, top right) can be measured inductively using receive coils — the induced voltage can be seen in Fig. 1 in the bottom right. Both the fundamental frequency f_0 and its harmonics are received, where the particle signal at 25 kHz is superimposed by the direct coupling of the drive field signal [7]. A temporary stationary gradient field enables spatial encoding. Apart from a small region termed field-free point (FFP), all particles are saturated and cannot follow the drive field and therefore, cannot change the magnitude of their magnetization anymore. Thus, only tracer material from the FFP contributes to the signal at a certain time. Shifting the FFP across the whole FOV, gives a spatial representation of the scanned object [1]. A second, sophisticated MPI model with a field-free line (FFL) has recently been introduced [8] and both theoretically [9] and experimentally been evaluated [5], [10]. In comparison to FFP MPI, this technique provides an estimated improvement in signal-to-noise ratio (SNR) of a factor 10 due to the fact that more particles at a time contribute to the signal while keeping the spatial resolution constant [8], [10]. To date, two main reconstruction models are used. The originally introduced harmonic reconstruction technique measures the whole system response to consecutively shifted point sources as a reference [1]. Secondly, Goodwill and Conolly [2] have shown that one can also model MPI as a scanning process in space, which is termed x-space MPI.

1.2 Signal-to-Noise Ratio in MPI

Regardless of the reconstruction technique, the derivative of the Langevin function (Fig. 1, bottom left) constitutes the point spread function (PSF) in MPI. Thus, it is crucial for calculating important imaging features such as sensitivity or resolution. It is specific to the tracer material which also significantly affects the SNR. When describing the MPI imaging process as a convolution of the particle distribution with the derivative of the Langevin function, one can derive the expression [2]

$$\mathrm{SNR}_{1D} = M_0 \frac{B_1 kRG}{3NF} \sqrt{4k_B \Delta f (T_{\mathrm{coil}} R_{\mathrm{coil}})}$$
$$\approx M_0 \frac{B_1}{3NF4k_B T_{\mathrm{coil}}} \sqrt{\frac{\pi \mu_0 mRG}{2\ln(2) R_{\mathrm{coil}}}} \tag{1}$$

where M_0 is the total magnetization, B_1 is the receive coil sensitivity, RG is the slew rate, NF is the noise figure of the preamplifier, Δf is the bandwith, T_{coil} is the temperature of the receive coil, R_{coil} is its resistance, $m = \pi/6\, M_{\mathrm{sat}}\, d^3$ is the magnetic moment with saturation magnetization

$M_{\mathrm{sat}} = 0.6\, T/\mu_0$ (for magnetite particles) and particle diameter d, and both k_B and μ_0 are natural constants with Boltzmann's constant $k_B = 1.38 \cdot 10^{-23}\,\mathrm{m}^2\,\mathrm{kg}\,\mathrm{s}^{-2}\,\mathrm{K}^{-1}$ and vacuum permeability $\mu_0 = 4\pi \cdot 10^{-7}\,\mathrm{N}\,\mathrm{A}^{-2}$. Most important for our sensitivity study is the proportionality of SNR and the number of SPIONs that directly influences the total magnetization. Equation (1) also illustrates that the SNR in MPI is inversely proportional to the square root of the bandwith. The more harmonics one uses for reconstruction, the better the spatial resolution becomes, but at the same time, SNR drops. The SNR increases with scanning rate due to a higher signal and more averages possible during the same scan time. With increasing size of the FOV, less averaging can be achieved. This is important with regard to the reduction of noise which is governed by coil noise in MPI [2]. Because of positivity constraints, the noise is considered to be Rician-distributed. This means that in regions with very low or zero signal, noise is distributed around its mean value $\mu_{\mathrm{noise}} \neq 0$ with the standard deviation σ_{noise} [11].

2 Material and Methods

To be able to make a fair comparison between MPI and 19F MRI, we need the imaging processes and parameters to be comparable. In particular, this relates to the size of the FOV, total scan time, and spatial resolution. When finally comparing an SNR value of MPI to MRI, one needs to normalize for these factors.

2.1 Preparing Point Sources

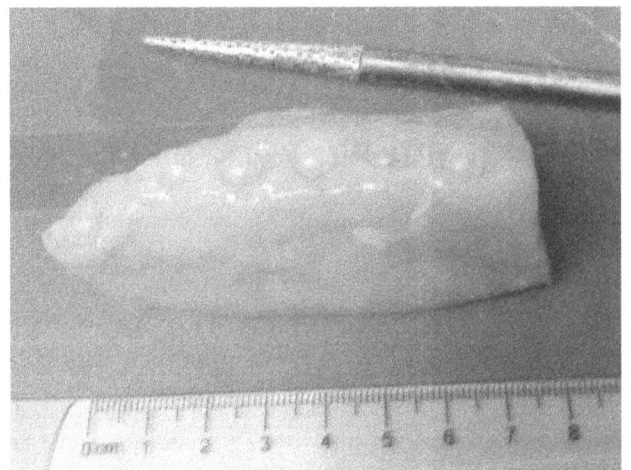

Figure 2: Tubes with six point sources filled with UW (University of Washington, USA) SPIONs inserted into chicken breast. Iron content: 1000, 500, 250, 125, 62.5, and 31.25 ng.

To achieve a good MPI performance in terms of resolution, UW (University of Washington, USA) SPIONs are used. They have experimentally shown to provide both the highest spatial resolution and the highest sensitivity so far. In

MPI, the SPIONs from within one voxel also affect the signal in neighboring voxels due to the shape of the PSF. Therefore, point sources are measured to make sure the induced signal for one voxel stems from only this voxel. Table 1 shows the mass concentrations of the stock dilutions and the total mass of iron in the six point sources. They are prepared by pipetting 500 nL from the particular stock, cutting off the end of the pipette tip, and wrapping it into impermeable plastic foil. These point sources are put into 200 μL PCR tubes which are stuck into 50 g of raw chicken breast (Fig. 2). This is done to get as close as possible to the clinical situation where SPIONs will be inside a human body. Since tissue constitutes the most relevant noise source in MRI, this has to be taken into account to ensure a fair comparison. However, it should not affect the MPI signal significantly since MPI is still coil-noise dominated.

Table 1: Mass concentration and mass of iron in the six 500 nL point sources.

No.	mass concentration / $g\,L^{-1}$	mass / ng
1	2	1000
2	1	500
3	0.5	250
4	0.25	125
5	0.125	62.5
6	0.0625	31.25

2.2 Scanning Parameters

The piece of chicken breast with the tubes and the point sources inside is placed into the UC Berkeley FFP Scanner *Gertrude* (Fig. 3) with gradient strengths of 7 T m^{-1} in x and 3.5 T m^{-1} in y and z direction, respectively. The ob-

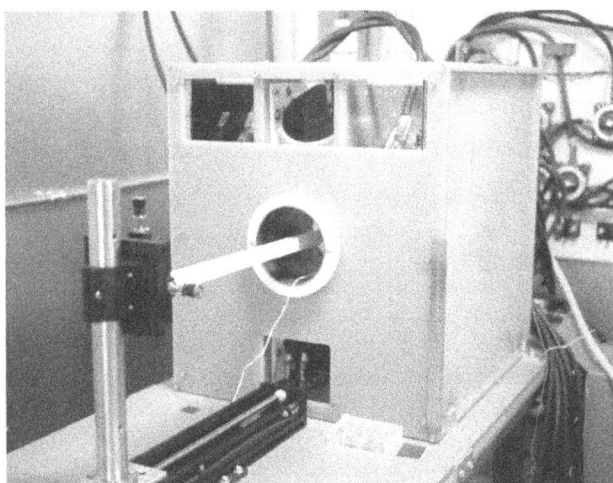

Figure 3: UC Berkeley FFP MPI scanner *Gertrude*.

ject is pushed through the scanner in z direction in 1 mm steps. For each z slice, the FFP is shifted across the FOV on a Cartesian trajectory with 64 lines in y direction. A $3 \times 3 \times 8$ cm^3 FOV is acquired within 9.5 minutes and reconstructed using x-space theory.

2.3 SNR Calculation

The reconstructed 3D volume is used to compute the SNR, and therefore, the sensitivity of this particular MPI experiment. The mean μ_{noise} and the standard deviation σ_{noise} of the noise are determined from the image background ($10 \times 64 \times 100$ voxels) using custom MATLAB (Math-Works, Natick, USA) scripts. Due to the Rician distribution of noise in image regions with low signal, the SNR for the point sources is calculated as follows:

$$ \mathrm{SNR} = \frac{S_{max}}{\mu_{noise} + \sigma_{noise}}. \qquad (2) $$

Here, S_{max} is the maximum signal intensity of each point source in the image which can easily be found in a maximum intensity projection (MIP). All found SNR values are used for a linear fit to determine the sensitivity of MPI. We define this to be at an SNR of 3. Note that this SNR calculation is rather conservative.

3 Results and Discussion

For better visual clarity, the MPI image of the chicken with the point sources is shown as a MIP in Fig. 4. Even the fourth point source containing 125 ng of iron is distinguishable from the background, although the overall noise level is higher in the MIP than in each slice of the 3D volume itself. The linear fit of the calculated SNR values for the four

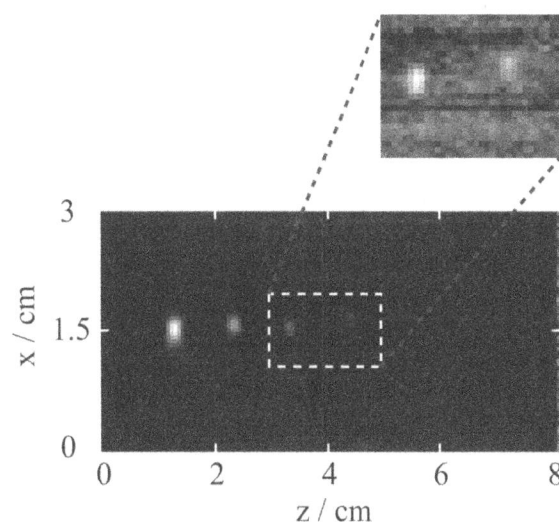

Figure 4: MPI maximum intensity projection (MIP) in y direction of six UW (University of Washington, USA) point sources containing 1000, 500, 250, 125, 62.5, and 31.25 ng of iron.

identifiable point sources shown in Fig. 5 indicates an MPI sensitivity (SNR = 3) of 188 ng of iron. Together with a proven uptake of 78 pg iron per human mesenchymal stem cell [4], this would imply a possible detection of about only 2400 stem cells in this MPI setup. Note that this number of 78 pg per cell has been found for Resovist (Bayer Schering AG, Leverkusen, Germany), a different SPION tracer.

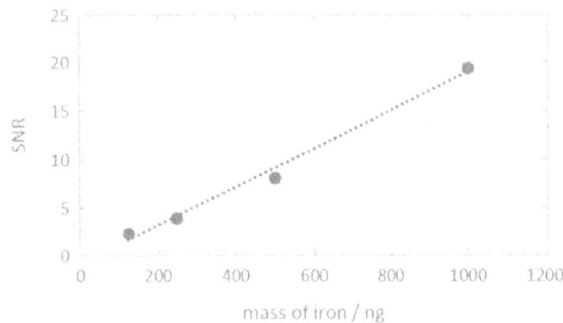

Figure 5: SNR of the four detectable UW (University of Washington, USA) point sources in an FFP MPI system assuming Rician noise.

Resovist was chosen despite its lower sensitivity and spatial resolution because it is already FDA-approved. However, recent studies have shown that even Resovist MPI measurements may provide an improvement of at least one order of magnitude in sensitivity compared to 19F MRI [12]. Combining that with the findings of this work, MPI can be considered a real competitor of 19F MRI. In addition, the MPI performance regarding SNR can still be enhanced. Note that boosting the SNR by either using less harmonics for reconstruction or by increasing the number of averages always comes with drawbacks. Decreasing the bandwith causes a significantly worse spatial resolution. Although one can trade SNR at the cost of scan time, FOV size, or resolution, these important imaging features must be within a reasonable range. Furthermore, a normalization for these three properties has to be done when comparing the SNR of MPI and MRI. True SNR enhancements can, however, be achieved e.g. by using an FFL instead of an FFP, by improvements in hardware design (receive coils), or by the development of tailored SPIONs.

4   Conclusion

Within a 9.5 minutes MPI scan covering a FOV of 3 cm by 3 cm by 8 cm with a spatial resolution of roughly 1.25 mm in each direction, we were able to detect 188 ng of iron at an SNR of 3. This corresponds to a number of 2400 stem cells detectable in this setup. By using an FFL instead of an FFP, we will be able to further decrease this number to only 250 and even more as there are still a lot of possibilities to improve the MPI hardware. In contrast to PET, MPI does not need ionizing radiation. In contrast to optical imaging, MPI does not suffer from attenuation through tissue. When MPI is now able to detect significantly smaller amounts of cells than MRI, it has great potential to become the leading imaging modality for medical applications like stem cell tracking.

Acknowledgement

The work has been carried out at the Department of Bioengineering of the University of California at Berkeley.

5   References

[1] B. Gleich and J. Weizenecker, *Tomographic imaging using the nonlinear response of magnetic particles*. Nature, vol. 435, no. 7046, pp. 1214–1217, 2005.

[2] P. W. Goodwill and S. M. Conolly, *The x-space formulation of the magnetic particle imaging process: One-dimensional signal, resolution, bandwith, SNR, SAR, and magnetostimulation*. IEEE Trans. Med. Imag., vol. 29, no. 11, pp. 1851–1859, 2010.

[3] B. Zheng et. al., *Magnetic particle imaging tracks the long-term fate of in vivo neural cell implants with high image contrast*. Sci. Rep., vol. 5, pp. 14055, 2015.

[4] B. Zheng et. al., *Quantitative magnetic particle imaging monitors the transplantation, biodistribution, and clearance of stem cells in vivo*. Theranostics, vol. 6, no. 3, pp. 291–301, 2016.

[5] K. Bente, M. Weber, M. Gräser, T. F. Sattel, M. Erbe, and T. M. Buzug, *Electronic field free line rotation and relaxation deconvolution in magnetic particle imaging*. IEEE Trans. Med. Imag., vol. 34, no. 2, pp. 644–651, 2015.

[6] S. Chikazumi and S. H. Charap, *Physics of magnetism*. John Wiley & Sons, 1964.

[7] K. Lu, P. W. Goodwill, E. U. Saritas, B. Zheng, and S. M. Conolly, *Linearity and shift invariance for quantitative magnetic particle imaging*. IEEE Trans. Med. Imag., vol. 32, no. 9, pp. 1565–1575, 2013.

[8] J. Weizenecker, B. Gleich, and J. Borgert, *Magnetic particle imaging using a field free line*. J. Phys. D: Appl. Phys., vol. 41, pp. 105009, 2008.

[9] T. Knopp, T. F. Sattel, S. Biederer, and T. M. Buzug, *Field-free line formation in a magnetic field*. J. Phys. A: Math. Theor., vol. 43, pp. 012002, 2010.

[10] J. J. Konkle, P. W. Goodwill, O. M. Carrasco-Zevallos, and S. M. Conolly, *Projection reconstruction magnetic particle imaging*. IEEE Trans. Med. Imag., vol. 32, no. 2, pp. 338–347, 2013.

[11] H. Gudbjartsson and S. Patz, *The Rician distribution of noisy MRI data*. Magn. Reson. Med., vol. 34, no. 6 pp. 910–914, 1995.

[12] F. G. Heslinga et. al., *Stem cell tracking potential of magnetic particle imaging compared with 19F magnetic resonance imaging*. in: International Workshop on Magnetic Particle Imaging 2016, Infinite Science Publishing, Lübeck, accepted.

GATE based Monte Carlo Simulation of Planar Scintigraphy using the XCAT Phantom

R. Schäfer [1], M. Schaar [2,3], and M. Rafecas [2]

[1] Medizinische Ingenieurwissenschaft, Universität zu Lübeck, richard.schaefer@student.uni-luebeck.de

[2] Institute of Medical Engineering, Universität zu Lübeck, {schaar, rafecas}@imt.uni-luebeck.de

[3] Graduate School for Computing in Medicine and Life Sciences, Universität zu Lübeck

Abstract

Monte Carlo simulations are an essential tool in nuclear imaging. In this work, we have used the toolkit GATE to simulate acquisitions of bone scintigraphy. GATE provides accurate physical models of photon transport and can handle voxelized phantoms. However, this high accuracy results in long simulation times. To accelerate the simulations, the emission angle of the sources can be constrained, given that an ideal parallel collimator should only allow the detection of photons impinging perpendicular to the camera. In doing so, photon interactions in the collimator are ignored and the results thus become less accurate. To this aim, different emission angles were considered and the resulting projection data were compared. For analysis, various metrics were proposed. The results show that small angles allow for fast simulation but mainly at the cost of artificially reducing background noise. To obtain realistic data, emission angles should not be smaller than $120°$.

1 Introduction

Planar scintigraphy is a method commonly used in clinical routine for diagnostic and treatment monitoring, i.a. malignant tumor, osteomyelitis and bone infarction. Bone scans using the radioactive tracer Technetium-99m (^{99m}Tc) are regarded as not-very-specific, but highly sensitive [1]. Planar scintigrams are acquired using gamma cameras, which mainly consist of a scintillation crystal and a collimator. The goal of the gamma camara is to detect those gamma rays emitted as a result of the radioactive decay. The collimator selects the photons which impinge along the same direction as the collimator holes. Monte Carlo Simulations (MCS) are commonly used in nuclear imaging to investigate novel imaging setups, new algorithms for image reconstruction or correction techniques. Accurate and versatile simulation packages like GATE include well-validated physical models for nuclear imaging [2]. A drawback of MCS is the long computation time. For example, an accurate GATE simulation of a scintigraphic exploration might take days on a conventional computer. To reduce the simulation time, the emission angle of the emitting sources can be constrained. The rationale is that an ideal parallel collimator should only allow the detection of photons impinging perpendicular to the camera. On the other hand, if the angle is constrained, some physical effects, such as scatter within the septa or septa penetration, are ignored and the resulting image becomes biased. In this work, we have investigated the effect of constraining the solid angle of the emission sources on the image accuracy and simulation time.

For this purpose, we have simulated a gamma camera and implemented a voxelized human phantom for bone scintigraphy.

2 Material and Methods

The evaluation described in this work is based on computer simulations of planar scintigraphy. Therefore, we start introducing these concepts in order to better understand the methods.

2.1 Monte Carlo Simulations

MCS is a statistical method to evaluate various topics in nuclear imaging, especially to determine scatter contributions, collimator design and to investigate effects of various parameters upon image quality [2], [3]. These methods can simulate detailed radiation transport from the phantom into the imaging system by utilizing random numbers. Monte Carlo methods are based on probability density functions combined with physical models to simulate various particle-tissue interactions that can occur in the human body and in the radiation detector. Simulations, based on Monte Carlo methods, generate clinically realistic images if appropriate models for imaging system, phantom and activity distribution are used. Various packages to compute MCS exist e.g. SimSET, SIMIND, MCNP and GATE [4]. This work focusses on GATE.

2.2 GATE

GATE stands for *Geant4 Application for Emission Tomography* and provides a script-based environment including voxelized phantoms [2]. This toolkit uses accurate physical models of the Geant4 database for calculating the photon transport in precise way. These models have been validated several times [5]. The first key component in GATE is the definition of the scanner geometry. It follows the process of particle propagation, which consists in three main steps. In the first step a particle is emitted with predefined properties like particle type, time, momentum and energy. The second step includes calculation of an elementary step size, which correspond to a trajectory between discrete interactions. In this step, changes of particle's energy and momentum on the trajectory are calculated. In the third step, if the particle interacts within the detector, the information about the interaction is stored and termed hit. Step 2 and 3 are repeated until one of the following situation occur: (a) the particle is registered by the detector; (b) the energy of the particle reaches a threshold, or (c) the particle exits a predefined volume.

2.3 General Setup

In this work GATE 7.1 and Geant4 10.1 patch 2 were used to simulate particle interactions in an anthropoid phantom to generate scintigraphic data. The simulated projection data (the planar image) are stored in Interfile format. The image analysis was performed on MATLAB® 2015b. Simulations were computed on an Intel® i3 CPU 550, 2 x 3.2 GHz with 8 GB DDR 3 RAM at 1333 MHz.

2.4 Voxelized Phantom and Source

Digital phantoms are used to provide a computerized model of human's anatomy and physiology [6]. MCS using such phantoms allow realistic data to be simulated, including the behaviour of absorption and scattering found in human tissue [6]. One of such synthetic phantoms is the adult male XCAT phantom which allows user defined anatomies and breathing cycles. In this work, we implemented a dataset consisted of 479 transaxial slices containing the torso and the limbs with a matrix size of 256 x 256 pixels with a length of 3.125 mm. The slice thickness is 3.125 mm. The simulated phantom consisted of a 3D distribution of the radionuclide Technetium, uniformly distributed in the bone tissue. In this work no attenuation in the phantom was considered. Fig. 1 shows the source phantom.

2.5 Detector Setup

We have built a planar gamma camera using the *SPECT-Head* scanner model provided by GATE.

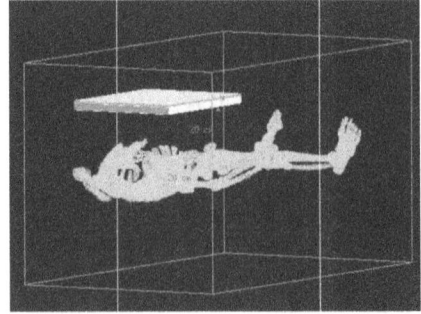

Figure 1: Simulation setup using the adult male XCAT phantom and *PHILIPS BrightView XCT* detector. The detector covers the torso and is placed approximately 10 cm above the phantom.

The specifications are inspired by the *Philips BrightView XCT*.

- Camera dimensions: $54.1 \times 40.1 \times 3.5$ cm^3,

- Parallel hole low energy general purpose collimator made of lead,

- Septa: length 1.4 mm, depth 24.7 mm and thickness 0.18 mm,

- Scintillation crystal: Natrium iodide (NaI), thickness 9.5 mm,

- Energy window: 0 - 140 keV.

In GATE there is no equivalent to photomultiplier of real gamma camera, so the signal was directly read from crystals with a resolution of 150×200 square pixels and a size of 2.705 mm.

2.6 Data Acquisition

Similar to real scintigraphy, the number of decays to be simulated is specified by the activity in Becquerel (Bq) and duration of simulated image acquisition. In a realistic scenario, the gamma rays are emitted isotropically and many of these photons do not intersect the physical volume of the camera. Additionally, most photons are stopped within the collimator. Tracking the later within the simulation cost time but produce no signal.

In this work we have constrained the emission angle of the photons, so that photons are emitted towards the detector. To reduce simulation time and validate the effect of constraining the angle on the projection images, we have considered polar angles between 1° and 180° and the corresponding solid angles Ω are shown in Table 1. We have simulated source emission for each Ω with a photon flux density of 95 $\frac{MBq}{s \cdot sr}$. The data were added up to one image per angle.

Table 1: Acquisition parameters with associated properties to generate simulation data.

Data set	α / °	Activity / MBq	Ω / sr
D_1	1	$23 \cdot 10^{-3}$	0.002
D_{30}	30	20	0.214
D_{60}	60	80	0.842
D_{90}	90	176	1.840
D_{120}	120	300	3.140
D_{150}	150	445	4.660
D_{180}	180	600	6.280

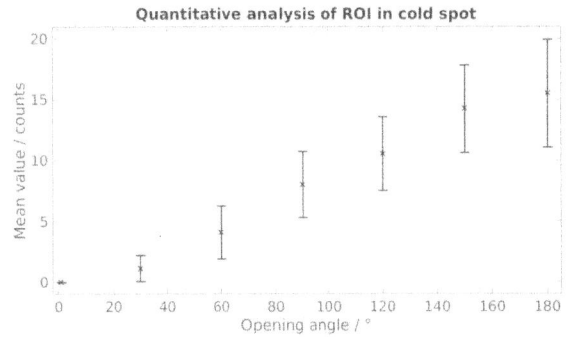

Figure 2: Total photon counts in the ROI containing background signal. The error bars correspond to one standard deviation.

2.7 Data Analysis

To compare the results and estimate the error from constraining the source, we developed the following relative error metric:

$$\delta_{\alpha,i} = \frac{|D_{\alpha,i} - D_{180,i}|}{D_{180,i}}, \ i \in \{1, ..., N\} \qquad (1)$$

$$\alpha \in \{1°, \ 30°, \ 60°, \ 90°, \ 120°, \ 150°, \ 180°\}$$

Equation (1) defines the normalized difference of data set $D_{\alpha,i}$ to the reference D_{180} for pixel i. To have a common measure to quantify of the relative error $\delta_{\alpha,i}$ and simulation time t_α we introduce χ_α:

$$\chi_\alpha = \frac{\sum_i \delta_{\alpha,i}}{t_\alpha}, \ i \in \{1, ..., N\} \qquad (2)$$

The numerator of (2) is the sum of δ_α over all N pixels in an image. Here N is 30000. t_α is the computation time in seconds without initialization of the simulation geometry. Hence, χ_α allows estimation of a trade-off between accuracy and computation time. Additionally, we compared the projection data as follows: Regions of Interest (ROI) were defined for a cold spot containing background signal, as shown in Fig. 3. The mean value and standard deviation of the ROI were calculated.

We used the sum of $\delta_{\alpha,i}$ divided by the number of pixels N to exploit another error metric ψ_α as

$$\psi_\alpha = \frac{\sum_i \delta_{\alpha,i}}{N}, \ i \in \{1, ..., N\}. \qquad (3)$$

This can be seen as the relative error of the data set D_α normalized by the number of pixels.

3 Results and Discussion

To investigate and compare the impact of constraining the solid angles, the mean values and the standard deviation for all data sets in the ROI were determined. The results are shown in Fig. 2. It can be seen that both the mean value of the cold ROI and the standard deviation increases with larger solid angles. This is to be expected since larger solid angles mean that the emitted photons reach a larger area. Some of those photons will penetrate the septa and be detected in the crystal. Some others might deviate because of scatter within the septa. Additionally, crystal scatter might

also occur. All effects contribute to the higher background signal. The larger the angle, the larger their contribution. It should be noted that our reference image is also affected by statistical noise due to the nature of MCS.

Difference images using D_{180} as a reference were generated and normalized to D_{180} as introduced in (2). The results are shown in Fig. 3. As a result of the aforementioned normalization difference image values are distributed between zero (black) and one (white). Light areas represent larger differences and darker points close to 0. Visual inspection reveals notable differences between the selected data set D_1, D_{60} and D_{120}. Fig. 3 shows $\delta_{\alpha,i}$ is increasing up to D_{120} for background pixels. $\delta_{\alpha,i}$ can be interpreted as a relative error caused by synthetically limiting the solid angle. As aforementioned, the larger the angle, the more photons contribute to image blurring, since a higher fraction of photons do not impinge perpendicularly to the camera. On the contrary, for small angles a higher fraction of photons reaches the detector without being deviated. As a result, constraining the emission angle leads to an artificial image improvement.

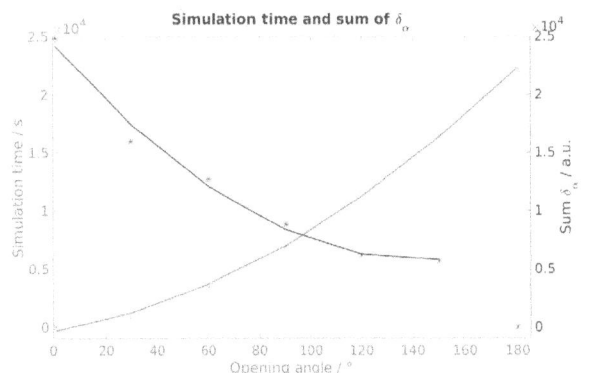

Figure 4: Relative errors for $\sum_i \delta_{\alpha,i}$ and the corresponding simulation time t_α. The grey line represents the simulation time and the black line the sum of $\delta_{\alpha,i}$. Due to the fact that δ_{180} is the reference image, it is not consider for the fit of the sum of $\delta_{\alpha,i}$.

To have a measure for quantification of the relative er-

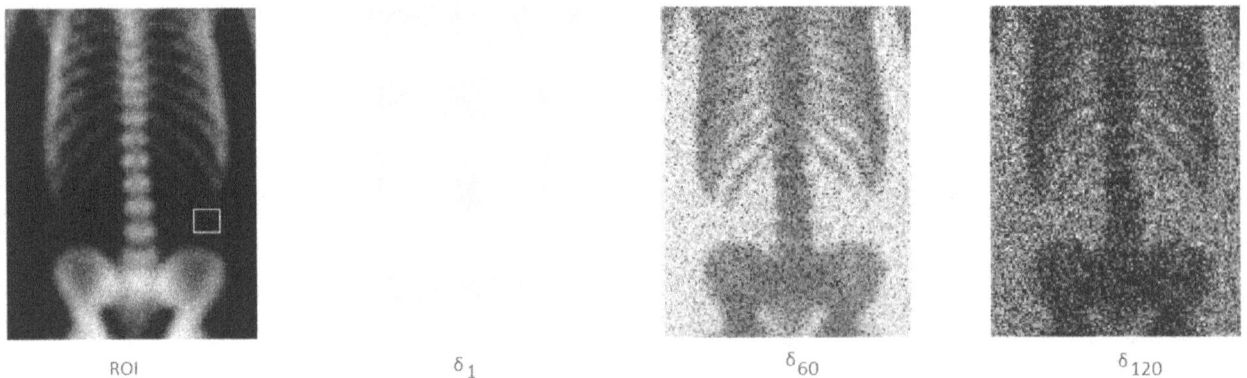

Figure 3: Images of the relative difference of three exemplary projections compared to D_{180}. Images were normalized to δ_{180}. The calculated values are displayed as grey values between zero (black) and one (white).

ror $\delta_{\alpha,i}$ and simulation time t_α we introduced χ_α in (2). This estimate a trade-off between accuracy and computation time. The results are displayed in Table 2. Comparing D_{120} and D_1, χ_α is increased approximately 2000 times. For D_{120} and D_{90} the increase is only 2 and for D_{120} D_{60} 6.5. Summarising Fig. 4 and Table 2, emission angles down to $120°$ lead to only few errors in the projection data while simulation time can be shorten.

To investigate the relationship between t_α and $\delta_{\alpha,i}$ for different α we compare the sum of $\delta_{\alpha,i}$ for each data set. Fig. 4 shows, that the $\sum \delta_{\alpha,i}$ follows a quadratic function, which decreases for larger angles. ψ_1 is approximately 2.5 times smaller than for ψ_{120} where t_α is increased by a factor of 1000. This is due to the fact that D_{120} had a higher activity per voxel to keep a constant photon flux density, which in turn lead to higher simulation time. Also the probability of being detected in the neighbouring pixel increased. As a result larger angles lead to longer simulation time but less error in the projection data.

Table 2: χ_α for all data sets δ_α

Data set	ψ_α	t_α / 10^3 s	χ_α
D_1	0.83	$15\cdot10^{-3}$	1213
D_{30}	0.53	0.85	20.2
D_{60}	0.46	3.43	4.1
D_{90}	0.52	6.92	1.3
D_{120}	0.33	11.7	0.6
D_{150}	0.23	16.4	0.4
D_{180}	0	22.1	0

4   Conclusion

We have used GATE to simulate scintigrams for varying directional characteristic of source emission. Finally, one can conclude that limiting the emission angle of the source shortens simulation time but artificial improves the image quality. Angles down to $120°$ results in a good agreement between the computational effort and the accuracy of simu-

lation performance. This will be useful when more complex simulation models or phantoms are used.

Acknowledgement

We would like to thank Dr. Gernot Ebel (Scivis wissenschaftliche Bildverarbeitung GmbH) for providing insightful help with the simulation setup.

5   References

[1] T. Grüning, C. Brogsitter, I. W. Jones and J. C. Heales, *Resolution recovery in planar bone scans: diagnostic value in metastatic disease.* In: Nuclear Medicine Communications, Vol. 33, No. 12, pp. 1307–1310, 2012.

[2] M. Ljungberg, S.-E. Strand and M. A. King, *Monte Carlo calculations in nuclear medicine.* CRC Press, Boca Raton, 2013.

[3] E. E. Bolch, *The Monte Carlo method in nuclear medicine: current uses and future potential.* In: The Journal of Nuclear Medicine, doi: 10.2967/jnumed.109.067835, 2010.

[4] D. Sarrut, M. Bardiès, N. Boussion, N. Freud, S. Jan et al., *A review of the use and potential of the GATE Monte Carlo simulation code for radiation therapy and dosimetry applications.* In: Medical Physics, Vol. 41, No. 6, pp. 064301-1–064301-13, 2014

[5] G. Mok, Y. Du, Y. Wang, E. Frey and B. Tsui, *Development and validation of a Monte Carlo simulation tool for multi-pinhole SPECT.* in: Molecular Imaging and Biology, Vol. 12, Issue 3, pp. 295–304, 2010.

[6] W. P. Segars, G. Sturgeon, S. Mendonca, J. Grimes and B. M. W. Tsui, *4D XCAT phantom for multimodality imaging research.* Medical Physics, Vol. 37, No. 9, pp. 4902–4915, 2010.

3D printed NEMA NU 4-2008 phantom for small animal PET: a comparison study using different modalities

M. Grehn [1], J. Kurth [2], J. Stenzel [2] and M. Rafecas [3]

[1] Medizinische Ingenieurwissenschaft, Universität zu Lübeck, melanie.grehn@student.uni-luebeck.de

[2] Multimodal Small Animal Imaging, Universitätsmedizin Rostock, {jens.kurth, jan.stenzel}@med.uni-rostock.de

[3] Institute of Medical Engineering, Universität zu Lübeck, rafecas@imt.uni-luebeck.de

Abstract

In the field of imaging, 3D printing could be a cost-effective and fast way for developing and testing new phantoms for device and application evaluation. In this work two different 3D printed versions of the NEMA NU 4-2008 image quality phantom for small animal Positron Emission Tomography were tested using different imaging modalities. For comparison, a commercial phantom and a manufactured copy were also tested. Our results show that relevant properties of 3D printed phantoms differ from the commercially produced one. The support material of the 3D printer affects the later phantom quality. A generalization of the suitability of 3D printed phantoms for all imaging modalities does not seem to be permissible.

1 Introduction

Three-dimensional (3D) printing is a rapidly evolving technology with many fields of application in industry and medicine. In the field of imaging, 3D printers offer a cost-effective and fast way to develop and test new imaging phantoms towards device and application evaluation.

The main goal of this work was to study the suitability of 3D printed phantoms for small animal Positron Emission Tomography (PET). For this purpose, two 3D printed copies from the well-established image quality phantom for small animal PET, the NEMA NU 4-2008 [1], were produced.

Nowadays PET systems are always combined with other image modalities, e.g. Magnetic Resonance Imaging (MRI) or Computed Tomography (CT). Therefore the suitability of the PET phantoms were evaluated using various imaging modalities: a clinical MRI, a clinical CT, a preclinical PET/CT and a Micro-CT.

To investigate possible leakage and the effects of the material porosity, measurements were carried out directly after filling the phantoms and one or two days later. The attenuation properties of the materials could be obtained through CT measurements. For comparison, a manufactured copy and a commercially available phantom were also measured.

2 Material and Methods

In the following, the phantoms and image modalities are described.

2.1 Phantoms

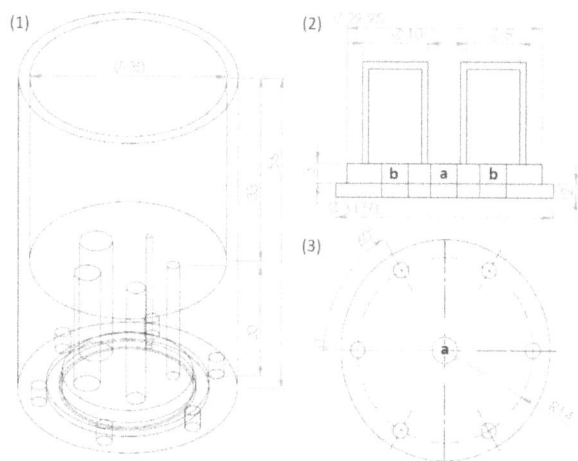

Figure 1: The parts of the NEMA NU 4-2008 IQ phantom: (1) "Body", (2) "Top Cover", (3) "Bottom Cover". The boreholes a and b in (2) and (3) are for filling and draining the phantom (a: activity, b: water or air) and get locked with plastic screws.

Four NEMA NU 4-2008 phantoms were used in this work: two 3D printed copies, one copy manufactured at the Scientific Workshop of the University of Lübeck and a commercially available phantom (QRM, Germany [2]). The latter was used as reference. Following the specifications of the NEMA protocol [1] the construction plans were drawn up in SolidWorks (2015 Edition SP04, SolidWorks Corporation, USA). The NEMA NU 4-2008 phantom is 63 mm long, with a diameter Ø of 33.5 mm (Fig. 1). The phantom consists of three parts: "Top Cover" (two cold chambers filled with air or water, 15 mm long, Ø 8 mm), "Bottom Cover" (removable) and "Body" (homogeneous part: 33 mm long, hot rod part: 20 mm long, Ø 1-5 mm, both filled with ^{18}F-Fluordesoxyglucose). These parts were printed individually.

A first 3D printed copy (henceforth termed *IMT Phantom*) was made by a ProJet® 3510 HDPlus (3D Systems, USA) with the available material VisiJet® X. Another copy was printed by a Perfactory® 3 Mini Multi Lens (Envision-TEC, USA) with the material E-Shell 600 clear (*Fraunhofer Phantom*).

2.2 Imaging Modalities

MRI: The MRI measurements were carried out on a Philips Ingenia 3.0 T (Philips, Germany). The used sequence was a T1 Spin Echo sequence with a pixel size of 0.6 mm^2, a slice thickness of 3 mm and a matrix size of 150 x 150. The reference and the IMT phantoms were measured, both filled with water. To avoid susceptibility artifacts during the measurement the phantoms were arranged in a larger case, also filled with water.

Micro-CT: Data of one "Top Cover" chamber of the two 3D printed copies were acquired on a SkyScan 1172 100 kV (SkyScan, Belgium). The measured chambers were filled with water and were mounted in the center of rotation of the scanner. The measurements were made at 40 keV and 250 μA. The HD images of the IMT phantom were recorded immediately after filling and 24 hours later (pixel size: 9.84 μm^2, matrix: 1280 x 1280). In contrast, the Fraunhofer phantom was measured after filling and 48 hours later with a lower resolution because of time constraints (pixel size: 33.96 μm^2, matrix: 640 x 640). Additionally, one HD measurement of the Fraunhofer phantom was performed for comparison of the print slice orientation (same parameter as IMT phantom).

CT: The data were acquired at a Siemens Somatom Balance (Siemens Healthcare, Germany). The measurement protocol included a slice thickness of 1 mm, a tube voltage of 110 keV, an image matrix of 512 x 512, a pixel size of 0.32 mm^2 and the U90 ultra sharp reconstruction kernel. The four phantoms were attached to a self-built styrofoam bracket (dimensions: 120 x 120 mm, with cutouts at the bracket corners for the phantoms). The CT data allowed the calculation of the mean attenuation coefficients in the material of the four phantoms. The attenuation coefficients μ were calculated under the assumption that $\mu_{water} \approx 0,1625cm^{-1}$ at 110 keV and with the Hounsfield equation, [3]:

$$HU(tissue) = \frac{\mu_{tissue} - \mu_{water}}{\mu_{water}} \cdot 1000 HU. \quad (1)$$

PET: The scans were acquired on a Siemens Inveon (Siemens Healthcare, Germany). The scan time was 20 minutes at an energy window of 350-650 keV and a coincidence time window of 3.438 ns (slice thickness: 0.8 mm, pixel size: 0.78 mm^2, matrix: 128 x 128). The acquired data were attenuation corrected and reconstructed using the filtered back projection (Fourier Rebinning, Ramp-Filter, Cut-off: 0.5 ν). Immediately before filling the specific activity amounted to 0.257 MBq/ml. First, all four phantoms were measured individually: one hour after filling (reference), after 1.5 h (Fraunhofer), after 2 h (IMT), after 2.5 h (manufactured) and 4 h after filling (all four at once, directly attached to each other).

2.3 Image Quality in small animal PET

To evaluate the achieved PET image quality the homogeneity, the spillover ratio (SOR) and the recovery coefficients (RC) were determined. The SOR and RC are measures of the undesired image background, as well as contrast and resolution, respectively.

To determine the homogeneity, a volume of interest (VOI) was drawn in the central part of the phantom with a diameter of 22.5 mm and 10 mm height. Subsequently, the average activity concentration, the maximum and minimum value within the VOI and the standard deviation were calculated. The determination of the SOR is carried out drawing VOIs (4 mm diameter, 7.5 mm length) in the non-active chambers (one filled with air, the other with water) . For this purpose, the mean activity is determined and the ratio to the average value of the homogeneous region is calculated. A small SOR stands for well-working correction and reconstruction methods. For calculation of the RCs, the central 10 mm image slices of the 1-5 mm rods were averaged to get one slice with a better signal-to-noise ratio. Around the rods, Region of Interests (ROIs) were drawn with a diameter twice the physical rod diameters. The image pixel coordinates of the maximum ROI values are used to generate line profiles along the rods. Then, the measured pixel values are divided by the average activity concentration of the homogeneous phantom part. The RCs are factors that indicate how much the measured activity of structures of different size differ from the true activity. [1], [4]

3 Results and Discussion

In the next sections the results for the different measurements are given.

3.1 Phantoms

Table 1 provides an overview of the phantom properties. The density and the filling volume have been determined mathematically. The dimensions of the Fraunhofer phantom differ largely from the planning ones (up to 0.4 mm in length, height and diameter). According to the NEMA protocol only deviations below 0.1 mm are valid. The IMT phantom complies with the requirements with slight variations below 0.1 mm. The manufactured phantom corresponds exactly to the planning dimensions.

Table 1: Overview of the phantom properties.

Properties	Reference	IMT	Fraunhofer	Manufactured
Material	PMMA (Basis)	Synthetic with ABS Characteristics (Urethane acrylate: 30-40 %, Phenylbis (2,4,6-trimethylbenzoyl) phosphine Oxide: 0.1-0.5 %)	7,7,9-Trimethyl-4,13-dioxo-3,14-dioxa-5,12-diaza-hexadecan-1,16-diol dimethacrylate: 5-20 %, Tetrahydrofurfuryl methacrylate: 10-25 %, Diphenyl (2,4,6-trimethylbenzoyl)	PMMA
Density	1.42 g/cm³	1.18 g/cm³	1.20 g/cm³	1.19 g/cm³
Empty Weight	43.9 g	40.91 g	41.55 g	40.75 g
Filling Weight	64.65 g	61.5 g	62.61 g	61.45 g
Filling Volume	20.79 ml	20.6 ml	21.11 ml	20.74 ml
Total Weight	65.36 g	62.31 g	63.35 g	62.2 g
Color	translucent	nature	translucent	translucent
Production Time	N/A	ca. 31 h	N/A	7 h
Cost	ca. 1000 €	ca. 100 €	ca. 50 €	10 €

3.2 Imaging Modalities

MRI: The MRI images show irregularities on the inner edge of the IMT phantom, as well as on the outer edge, which point out to small air bubbles (Fig. 2, top). In the region of the chambers of the IMT phantom emerged a ring artifact that indicates a "chemical shift" (Fig. 2, top right). The reason for that could be the waxy support material of the 3D printer, which could not be removed completely. The evaluation of the image histograms showed an average gray value in the material of 43.3 in the IMT phantom and 37.84 in the reference one (gray scale: 0-255).

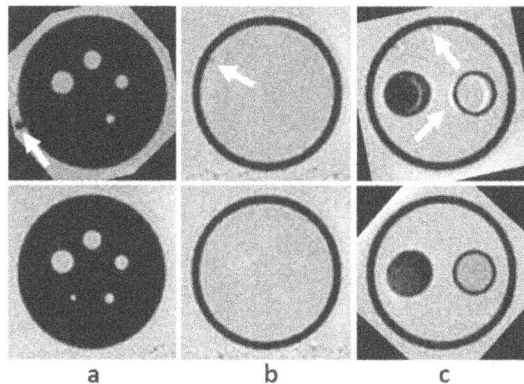

Figure 2: MRI Images of the IMT (top) and the reference phantom (bottom) of the different phantom parts (**a**: hot rods within "Body", **b**: uniform part within "Body", **c**: cold chambers within "Top Cover"). The visible air bubbles and the ring artifact are marked with arrows.

Micro-CT:

The IMT phantom shows many air bubbles on the inner walls, which are not freely movable (despite avoiding air bubbles during filling e.g. by tapping). The measurement after 24 h showed an enlargement of the air bubbles in terms of number and size (Fig. 3). Between day 1 and day 2 the histogram distribution changed. The average gray level in the chamber wall increases slightly (+2, gray scale: 0 -255). At the Fraunhofer phantom, no air bubbles are visible on the inner walls. Only at the upper end a floating bubble can be detected, which is due to filling. Measurement after 48 hours shows an increase in the average gray value in the

chamber wall of +8 (gray scale: 0 - 255) and the formation of another bubble at the chamber ground. The reason for the bubble on the ground of the chamber seems to be an air lock in the ground material with connection to the inner chamber. Furthermore, various printing errors are clearly visible (e.g. not exactly round chamber). In both cases the individual printing layers are clearly visible and show the reason for trapped air bubbles in the IMT phantom (Fig. 4).

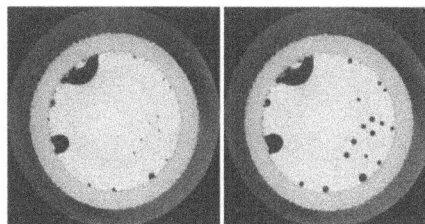

Figure 3: Micro-CT Images of the IMT Phantom directly after filling (left) and 24 h later (right).

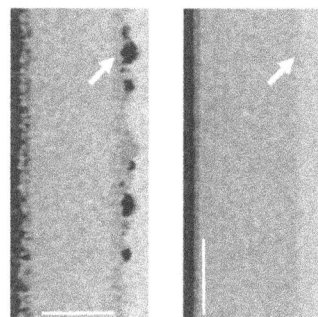

Figure 4: Micro-CT Images of the chamber wall of the IMT (left) and the Fraunhofer Phantom (right). The print directions are marked with lines. The arrows show the inner chamber wall.

CT:

The Fraunhofer, IMT and the manufactured phantoms showed similar mean linear attenuation coefficients: 0.1808 cm^{-1}, 0.1802 cm^{-1}, 0.1803 cm^{-1}, respectively. The reference phantom has a higher coefficient (0.182 cm^{-1}). These results are in agreement with the calculated density values of the material (Table 1). In the CT images various material defects of the Fraunhofer phantom become visible, e.g. air locks in material. The IMT Phantom is the only one which shows a big air bubble in the homogeneous part and it features, like before in the Micro-CT, air inclusions at the chamber and inner phantom walls. In the manufactured phantom only a few little air bubbles concentrate at the bounding surface. The surface of the commercial phantom seems to be least prone to the formation of air bubbles.

PET: The images of the individual phantoms show no significant differences in terms of homogeneity, SOR and RCs (Table 2, Fig. 5), possibly because of VOI position: the described problems, e.g. the air bubbles on the inner wall of the IMT phantom, lie outside the VOIs.

The Field of View of the Siemens Inveon was not large enough to fully image the four phantoms simultaneously (Fig. 6). Still, the respective values for the homogeneity, SOR and contrast resolution could be determined for each phantom. Compared to the results of the individual measurements, the values in terms of homogeneity and spillover ratio deteriorated considerably (Table 2).

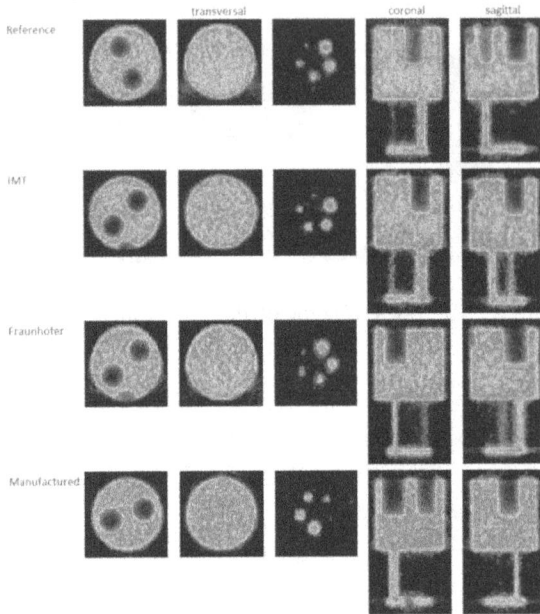

Figure 5: PET images of the individually measurements.

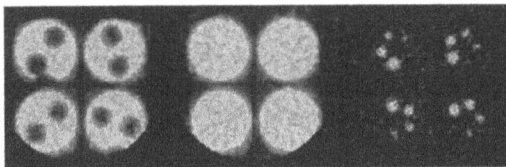

Figure 6: PET images of the "all four at once" measurement: Fraunhofer (top left), Reference (top right), IMT (bottom right) and Manufactured (bottom left).

Table 2: Calculated values from the PET-data.

Phantoms individually	Homogeneity			Recovery Coefficients					Spillover Ratio	
	% SD	Max/Mean	Min/Mean	1 mm	2 mm	3 mm	4 mm	5 mm	Water	Air
Reference	9.1	1,39	0,69	0,40	0,66	0,75	0,79	0,86	5,5%	5,7%
Fraunhofer	10	1,35	0,63	0,24	0,57	0,75	0,83	0,87	5,1%	6,1%
IMT	10	1,47	0,65	0,21	0,60	0,84	0,86	0,92	4,2%	4,7%
Manufactured	11	1,47	0,62	0,29	0,68	0,77	0,80	0,88	4,6%	4,8%
all four at once										
Reference	17	2,02	0,50	0,23	0,57	0,65	0,85	0,95	10,5%	8,9%
Fraunhofer	17	1,73	0,46	0,26	0,52	0,65	0,90	0,95	8,4%	6,2%
IMT	17	2,00	0,47	0,14	0,41	0,61	0,82	0,87	8,9%	8,3%
Manufactured	17	1,92	0,44	0,17	0,39	0,67	0,84	0,92	9,3%	10,2%

4   Conclusion

Bieniosek et al. showed in a previous study that measured differences between a commercially produced phantom and

a 3D printed replica are due to measurement inaccuracies and otherwise the measurement characteristics seem to be equivalent [5]. Their results are attributable to a single measurement at a preclinical PET/CT. Our findings partly differ from the ones presented in the aforementioned work. We show that not every 3D printed phantom possesses the same properties as a commercially produced phantom. Before using a 3D printed phantom, a close examination of the suitability should be carried out. It is thus not possible to generalize the suitability of 3D printed phantoms for all imaging modalities. However, for PET-only measurements the 3D printed alternative seems to fulfill the requirements.

In planning and implementation of a 3D print, attention should be paid to structures where the printer uses support material. Not completely removed support material can, for example in MRI scans, result in artifacts which hinder image evaluation. In turn, avoiding the support material might cause significant deviations in the printing accuracy, because in this case most dimensions were not conform to the planning dimensions. Likewise the direction of the printing layers should find attention.

Acknowledgement

The work has been carried out at the Institute of Medical Engineering, Universität zu Lübeck (IMT), the Core Facility Multimodal Small Animal Imaging, University Hospital Rostock and the Department of Radiology and Nuclear medicine, UKSH Campus Lübeck. The authors would like to thank Dr. C. Joram (CERN), D. Steinhagen (IMT), A. Timmermeyer (IMT), M. Bobek (IMT), M. Stille (IMT), R. Schulz (Scientific Workshop, Universität zu Lübeck) and D. Wendt (Fraunhofer Institution for Marine Biotechnology).

5   References

[1] National Electrical Manufacturers Association, *NEMA Standards Publication NU 4-2008, Performance Measurements of Small Animal Positron Emission Tomographs*, USA 2008

[2] http://www.qrm.de/content/products/microct/micro-pet-iq.htm, last access: 11/27/2015, 1:19 pm

[3] Krieger H.: *Grundlagen der Strahlungsphysik und des Strahlenschutzes*, Springer, 2012, DOI 10.1007/978-3-8348-2238-3

[4] DiFilippo F. P., Patel S., Asosingh K. and Erzurum S. C.: *Small-Animal Imaging Using Clinical Positron Emission Tomography/Computed Tomography and Super-Resolution*, Molecular Imaging, Vol. 11, No. 3, 2012, 210:219

[5] Bieniosek M. F., Lee B. J. and Levin C. S.: *Technical Note: Characterization of custom 3D printed multimodality imaging phantoms*, Medical Physics 42, 5913 (2015), DOI 10.1118/1.4930803

A PET/MR Rat Brain Atlas for
Automated Time Activity Curve Extraction

K. Kläser [1,2], G. Cowin [3,4], A. Janke [3,4], G. Angelis [2], K. Mardon [3,4], A. Kyme [2], J. Gillam [2], R. Fulton [2], S. Meikle [2,4], and W. Ryder [2,4]

[1] Medizinische Ingenieurwissenschaft, Universität zu Lübeck, Germany, kerstin.klaeser@student.uni-luebeck.de

[2] Brain and Mind Centre, University of Sydney, Australia, {georgios.angelis, andre.kyme, john.gillam, roger.fulton, steven.meikle}@sydney.edu.au, wjryder@ieee.org

[3] Centre for Advanced Imaging, University of Queensland, Australia, {gary.cowin, andrew.janke}@cai.uq.edu.au, k.mardon@uq.edu.au

[4] National Imaging Facility, Australia

Abstract

A rat brain atlas can be used for automated extraction of regions of interest (ROIs) from positron emission tomography (PET) images. When the atlas is spatially co-registered with the reconstructed PET volumetric data, it can be used to accurately delineate anatomical structures and extract corresponding regional kinetic parameters without operator bias. To develop the rat brain atlas, we performed rat brain studies and acquired PET and MR data simultaneously. Post-image reconstruction, the Johnson atlas was spatially registered to the MR images resulting in a PET/MR atlas in a common space. Further we 3D printed a physical rat brain phantom that was derived from the animal PET/MR data to acquire experimental MR/PET data. Future work includes further improvement of our atlas, automated extraction of ROIs and regional kinetics like time-activity curves, simulated data studies, testing of image registration algorithms and the development of a 3D printed phantom that includes structure.

1 Introduction

In PET studies one of the biggest interests is the quantification of biochemical parameters and processes, such as metabolism, receptor density and receptor occupancy. For this reason it is necessary to identify key anatomical or functional regions in the brain on each and every subject that are often reconstructed in completely different orientations. Drawing those regions is associated with several problems. First of all it is a very tedious and time consuming procedure that requires a trained expert. Another aspect that states a common problem is that PET images contain functional information that makes them inappropriate for drawing anatomical regions. Further the fact that different tracers have different uptake distributions in the brain makes it difficult to draw regions in some cases. In addition, a user-dependent bias may be introduced to the analysis that leads to variations in the dimensions of the regions for each subjected analysis. These problems make an automated extraction of regional kinetics important. Therefore a brain atlas is very beneficial since each PET study can be transformed directly to the atlas space and regional values can be extracted automatically. Additionally the comparison of regional parameters from one scan to another in the same subject (intra-subject or longitudinal studies) or between subjects at the same time point (inter-subject or cross-sectional studies) is often necessary, where-fore requiring reliable, reproducible ROI selection is important. The atlas contains delineated anatomical regions in a binary format. In some cases the registration of PET data to the atlas might not be straightforward due to missing anatomical information. Thus it is useful to have a PET and a MR template that are built out of averaged PET and MR data to register the acquired data to, so that one can extract regional values using the atlas that is in the same space. We further created the atlas from the rat strain that is normally used for imaging studies in our laboratory which will potentially simplify future registrations.

For those reasons the aim of this work was to generate a rat brain atlas based on simultaneously acquired PET/MR data. The main advantage of this approach is that PET data and MR data are by default in a common space so one can use either or both for the registration. Further the atlas facilitates a straightforwardly automatic region extraction. In addition we aimed to create a physical phantom that can be used to acquire test PET and/or MR data to validate the developed atlas.

2 Material and Methods

In this section the whole development of the atlas and the physical phantom is described: animal studies, image pre-processing, the workflow of the registration, averaging the simultaneous acquired MR images, co-registration with the Johnson MR template [1], creation of physical phantom and PET data acquisition with the phantom.

2.1 Animal Studies

We used four postnatal-week 10 Sprague Dawley rats for the image acquisition which were injected with 70-80 MBq of ^{18}F-FDG in 300 μl via the tail vein. PET and MR data were acquired simultaneously for 90 minutes on a Bruker Clinscan 7T PET/MR system and the data were reconstructed using the default scanner software.

2.2 Image Pre-processing

The PET datasets ($128 \times 128 \times 89$) are reconstructed directly to the default volume size (0.731842 mm $\times$ 0.731842 mm $\times$ 0.731842 mm) wherefore no image pre-processing is necessary. However, the dimensions of the acquired MR datasets are in complementary anisotropic sampling orientations ($59 \times 256 \times 256, 256 \times 59 \times 256, 256 \times 256 \times 59$) that leads to Nyquist limited isotropic voxels (0.6 mm $\times$ 0.136719 mm $\times$ 0.136719 mm). Therefore the combination of these datasets is necessary to produce one isotropic volume.

To produce an isotropic volume we combined the datasets using the Minc Toolbox [2]. The Minc Toolbox contains features for advanced image processing, statistical analysis and visualization. *Volregrid* is a Minc tool that uses k-space regridding to create an isotropic volume. In the first step this feature defines an output sampling. Afterwards each point in the input dataset is interpolated to the output sampling. This value can be weighted with different functions like Kaiser-Bessel, Gaussian, linear or by a nearest function and is then written into the current output volume. Furthermore the weight is added into a "count" volume. The last step is to sum up all weighted values and divide them by the "count" volume on a voxel by voxel basis. By changing the input parameter "regrid_radius" one sets a limited window around the sampling point in which the weighted values will be calculated.

2.3 Registration Flow

The goal is to automatically extract the mean ROI values from a PET image with less registration steps. The conventional way is to register either the PET or the MR image directly to the anatomical Johnson atlas. Especially for PET images this is not always straightforward since the information in an atlas is very different to the information in a PET image. The idea is that it is much easier to register the PET image to a PET template that is already in the same space as the anatomical atlas. In this case the registration requires only one registration step. The acquired image is registered to the PET template using the affine transformation matrix M_{PET} that includes 3 rotations, 3 shears, 3 translations and 3 scales. The resulting image can easily be overlaid with

the Johnson atlas since they are in the same space. This is demonstrated in Fig.1. Therefore we hypothesize that intra-modality registration gives more accurate results than inter-modality registration.

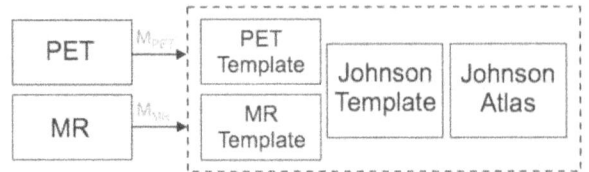

Figure 1: The idea is to register the acquired PET or MR images to either a PET or MR template that is already in the same space (dashed box) as the Johnson template and the Johnson atlas using the corresponding transformation matrix M to achieve more accurate results.

2.4 Averaging MR and Registration

In order to achieve a more uniform brain image and to reduce the noise level we averaged the T2 weighted MR images from the four rat studies using the Minc Toolbox. Prior to averaging, the images were aligned to each other using *MNI_AutoReg*, an automated method that uses multi-scale, three dimensional (3D) cross-correlation for the registration of two datasets. The averaged MR template is shown in the top left of Fig. 2.

The first step of the linear registration includes the usage of *minctracc*, a tool that aligns the image of an individual to a model and results in an output transform file that includes a full affine transformation matrix with 12 parameters. In the following step the original input image is resampled using the *mincresample* tool. This tool applies the transformation file to the input data and gives an output file that contains of the same sampling as the original model [3].

MNI_AutoReg was also used to spatially co-register the averaged MR images to the Johnson space, a magnetic resonance histology template, resulting in the aimed MR template in a common space (Fig. 2 bottom left). The PET images were processed in a similar way: they were aligned to each other, averaged and transformed to the Johnsonn atlas using the same transformation as for the MR images. To aid the automatic extraction of the regional values from the acquired data we also co-registered the 27 brain labels of the Johnson atlas [4] to the Johnson template using this Minc tool. Since the Johnson template is in the same space as the PET/MR templates, the Johnson labels, which are already in the same space, can be fused without an additional registration step.

2.5 3D Printed Phantom

We 3D printed a physical phantom which would allow us to acquire experimental MR and PET data to evaluate the newly created atlas. The physical phantom was created from the averaged MR images, which makes it easier to evaluate the registration of the developed atlas. At first the

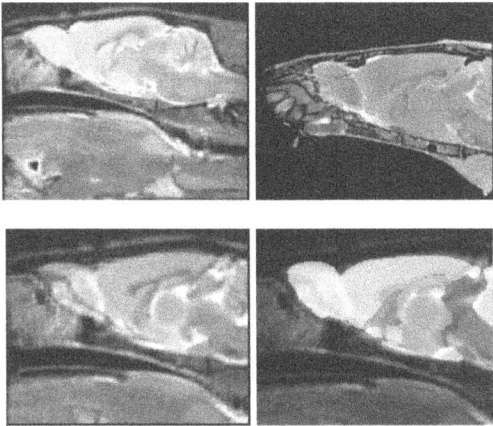

Figure 2: Top left: Averaged MR template created by averaging four MR images. Top right: Original Johnson template. Bottom left: Averaged MR template fused with Johnson template in the same space. Bottom right: Segmented Johnson atlas overlaid with MR template.

brain was segmented using Seg3D [5] by applying a threshold to do a pre-segmentation followed by a final manual segmentation to delete residual noise.

The 3D brain was build with the same software and the isosurface was exported and loaded into Rhino3D [6], where the actual phantom was created. We subtracted the isosurface from a solid block and added a filling hole to be able to fill the phantom with contrast agent and/or isotopes. Finally the physical phantom was loaded into the printer software and printed on a CreatBot 3D printer using Acrylonitrile butadiene styrene (ABS) as print material.

The printed phantom is shown in Fig. 3 before it was manually adjusted by sandpapering excess material off and before the cut surfaces of both parts were greased with silicon and stuck together. Four screws on the edges ensure that both parts are hold together. The final production step was the sealing of the phantom with varnish to prevent leaking after filling it with MR contrast agents or PET isotopes.

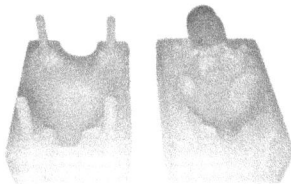

Figure 3: A 3D-printed rat brain phantom split in the transverse plane. The phantom consists of two parts, dorsal (left) and ventral (right).

2.6 PET Data Acquisition

For the evaluation of the atlas we acquired PET data for 30 minutes on a Siemens microPET Focus 220 with the physical phantom. It was filled with 2 ml ^{18}F with an initial activity of 38.6 MBq. The PET scan started 18 minutes af-

ter filling the isotope into the phantom. Following the law of radioactive decay

$$A(t) = A(0) \cdot 2^{(t/T_{1/2})} \qquad (1)$$

the activity A at the beginning of the measurement t is 34.5 MBq with the initial activity $A(0)$ and the half-life $T_{1/2}$. In order to reduce image artefacts in form of capping artefacts an attenuation correction of the emission scan is important. Those artefacts become noticeable through reduced activity closer to the centre. Therefore we did a transmission scan with an external ^{57}Co point source (4.25 MBq) using 32 passes (equivalent to ~140 minutes). The number of passes should be high, because more acquired data leads to less noisy transmission data. Therefore it is less likely that reconstructed images are affected by noise that is propagated by the transmission data. The PET data was reconstructed using the 3D Filtered Back-Projection by Re-Projection reconstruction algorithm (3DFBPRP) [7].

3 Results and Discussion

In this section the results of this work are presented. Up to this point of the project we created the atlas and were able to extract ROIs in both PET and MR images.

3.1 Co-registered Atlas

Fig. 4 shows a reconstructed PET image that was acquired using the physical phantom and the corresponding registration with the new MR template. Due to the direct registration with the new template it is possible to achieve more accurate registrations. The same template can be used for MR images as it is shown in Fig. 5.

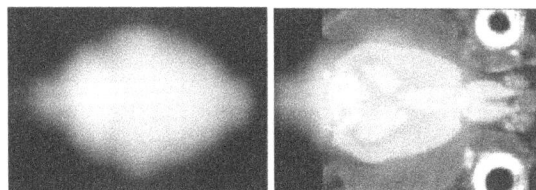

Figure 4: Top: PET image that was acquired using the physical phantom filled with ^{18}F. Bottom: Resulting image after registration of the acquired data with the new PET/MR template.

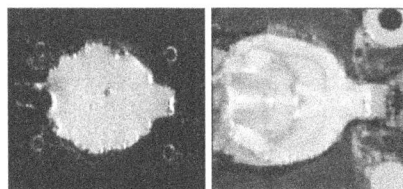

Figure 5: Left: MR image of the physical phantom filled with Vitamin E oil as a contrast agent. Right: The MR data was registered to the PET/MR template.

3.2 ROI Extraction

The next idea was to extend the prior registration by overlaying the Johnson atlas with the PET images to delineate important brain regions. Fig. 6 shows the acquired phantom PET images that were registered to the new PET template fused with the 27 labels from the Johnson atlas. It is now possible to extract the activity of a specific region of the static image that shows the overall tracer distribution once the tracer settled in the tissues.

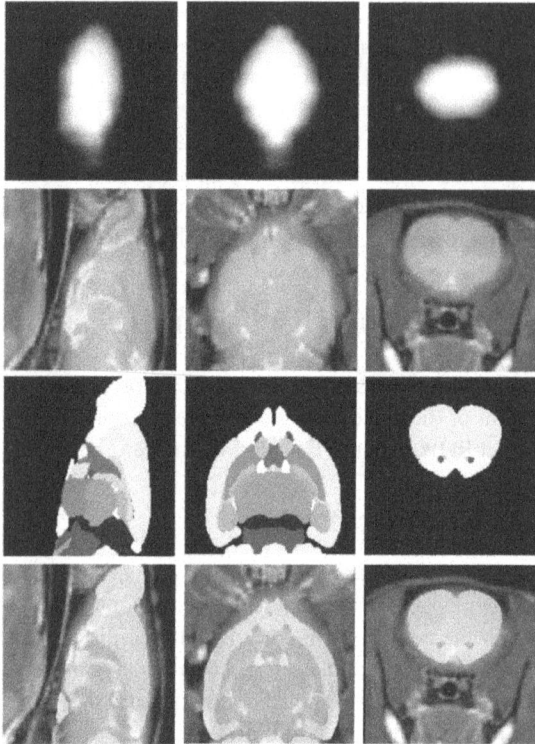

Figure 6: The first row shows PET images that were acquired with a ^{18}F filled rat brain phantom. The images were registered and overlaid with the PET/MR template, which is presented in the second row. The third row demonstrates the labels of the Johnson atlas and in the last row the fusion of the reconstructed PET images of the physical phantom, the PET/MR template and the Johnson atlas can be seen. The sagittal plane is shown in the first column, the coronal plane in the second and the transverse plane in the last column.

4 Conclusion

A novel PET/MR atlas for atlas-based image registration has been presented in this work. We demonstrated the capability of the template for the registration of both MR and PET images and the automated extraction of ROIs.

The work on this project will continue, therefore the future work includes the comparison of the proposed method to manual ROI delineation, in terms of speed, reproducibility, variability and error/bias, studies with simulated data and testing of image registration algorithms. The next step will be the transfer of the registration strategy to dynamic images. This facilitates the automatic extraction of regional kinetics, e.g. time-activity curves. Dynamic images are a series of 3D images, where each image represents the distribution of the radiotracer at a certain time. The time-activity curve for a specific ROI shows a plot of the mean radioactivity value in this region across a sequence of PET images. Each data point corresponds to the mean pixel value in a common ROI in a given time frame . Further we want to compare the registration with the original Johnson MR atlas with the registration using our new created PET/MR atlas. Another idea is to develop a new physical phantom that includes some structure to fill it with different contrast agents and/or isotopes at the same time.

Acknowledgement

The work has been carried out at the Brain and Mind Centre, University of Sydney. The authors would like to thank Gary Perkins and the staff at the cyclotron facility in Camperdown for providing the isotope for the PET imaging experiments presented in this paper.

5 References

[1] G. A. Johnson, E. Calabrese, A. Badea, G. Paxinos and C. Watson, *A Multidimensional Magnetic Resonance Histology Atlas of the Wistar Rat Brain*. NeuroImage, 62, pp. 1848–1856, 2012.

[2] The McConnell Brain Imaging Centre , *Software*. Available: http://www.bic.mni.mcgill.ca/ ServicesSoftware/ServicesSoftwareMincToolKit [last accessed on 21.01.2016].

[3] D. L. Collins, P. Neelin, T. M. Peters and A. C. Evans, *Automatic 3D Inter-Subject Registration of MR Volumetric Data in Standardized Talairach Space*. Journal of Computer Assisted Tomography, 18(2), pp. 192–205, 1994.

[4] E. Calabrese, A. Badea, C. Watson and G. A. Johnson, *A quantitative magnetic resonance histology atlas of postnatal rat brain development with regional estimates of growth and variability*. Neuroimage, 71, pp. 196–206, 2013.

[5] Center for Integrative Biomedical Computing, *Seg3D: Volumetric Image Segmentation and Visualization*. Available: http://www.seg3d.org [last accessed on 21.01.2016].

[6] Robert McNeel & Associates, *Rhinoceros3D*. Available: https://www.rhino3d.com/ [last accessed on 21.01.2016].

[7] P.E. Kinahan and J.G. Rogers, *Analytic 3D Image Reconstruction Using All Detected Events*. IEEE Trans. Nucl. Sci., 36(1), pp. 964–968, 1989.

9

Image Processing I

Evaluation of medical image registration algorithms on pulmonary inhale and exhale CT scans

M. Schumacher [1], T. Klinder [2], A. Schmidt-Richberg [2], S. Kabus [2] and M. Heinrich [3]

[1] Medizinische Ingenieurwissenschaft, Universität zu Lübeck, mona.schumacher@student.uni-luebeck.de
[2] Philips Research, Europe, Hamburg, {tobias.klinder, alexander.schmidt-richberg, sven.kabus}@philips.com
[3] Institut für medizinische Informatik, Universität zu Lübeck, heinrich@imi.uni-luebeck.de

Abstract

Medical image registration is a fundamental task in medical image processing. It establishes spatial correspondences between two or more images, possibly from different modalities or different time points. Until now, a large number of methods for image registration have been described in the literature. In this paper, we compare two methods: *elastix*, a widely-used toolbox for parametric registration with a collection of algorithms and the more recent technique *deeds* (dense displacement sampling), a symmetric and diffeomorphic registration approach based on discrete optimisation. The resulting deformation fields are evaluated with an implemented validation framework using the Jacobian determinant, the entropy using mutual information and landmark transformations using the target registration error.

1 Introduction

Since computer aided diagnosis is becoming increasingly important, medical image registration is a frequently used technique in medical image analysis. Areas of application for matching images include follow-up analysis, motion correction in radiotherapy or image-guided neurosurgery, where magnetic resonance (used to locate the target structure) and computed tomography (for guidance) are combined. In general, one image, the so called moving image I_M, is deformed to fit the second image, the fixed image I_F. Accordingly, the aim of this process is to find a displacement u, which is applied to the moving image. As a result of this deformation, the fixed and the moving images are aligned spatially in order to fuse and compare information.

To obtain a non-rigid displacement, an optimisation strategy has to be specified, which is formulated as a minimisation (or maximisation) problem:

$$u = arg\ min\ C(u; I_F, I_M), \qquad (1)$$

with the aim to optimise the following cost function:

$$C(u; I_F, I_M) = S(u; I_F, I_M) + \omega R(u), \qquad (2)$$

which defines the quality of alignment and depends on a similarity measure S, a regularisation term R and a weighting factor ω. Well-known examples for the similarity term S are mean squared difference, normalised correlation coefficient or mutual information. The similarity measurement consists of an energy landscape with multiple local minima. Depending on the choice of the optimisation method, different local minima and therefore different displacements can

be obtained, which results in a different alignment of the images. In order to constrain the search space and reduce the number of local minima the regulariser is added to the energy functional. In addition to the purpose of smoothing, it penalizes physically inconsistent deformations and avoids singularities in the displacements, wherein the magnitude of the influence depends on the weighting factor ω. Examples for R are the diffusive or total variation regulariser [1].

In the last years, many algorithms have been proposed based on different concepts of the components described above. The goal of this paper is to compare and evaluate two of those registration techniques on the basis of the publicy available CT data of patients from the COPDgene study [2]. However, evaluating the accuracy of registration results is a difficult task since there is no gold standard for non-rigid methods and, to make matters worse, there is no ground truth for clinical data [3].

2 Material and Methods

In this section, the design and the various components of the two registration models will be described and the evaluation strategy will be formulated.

2.1 Registration algorithms

The first registration approach is called *elastix*. It is an established and widely-used framework. It will be compared with a more recent technique called *deeds* (dense displacement sampling). In the following, the methods will be described in more detail.

2.1.1 Elastix

Elastix is a publicly available toolbox[1] for intensity-based medical image registration. It is a collection of components including several optimisation methods (e.g. Robbins-Monro or gradient descent), multiresolution schemes (e.g. Gaussian or shrinking pyramid), interpolators (e.g. nearest neighbour or N-th order B-spline), transformation models (affine or B-splines), and metrics (e.g. mean squared difference or normalized correlation coefficient). The framework is based on a parametric approach, which means that the number of possible transformations is limited. This leads to the optimisation problem

$$\hat{\mu} = \underset{\mu}{argmin} C(T_{\mu}; I_F, I_M), \tag{3}$$

where μ respresents a parameterisation of u. To find the optimal set of parameters $\hat{\mu}$, an iterative optimisation method is adapted in a multi-resolution scheme. For more details see [4], [5].

2.1.2 Deeds

Deeds is a symmetric and diffeomorphic registration approach. The optimisation strategy is formulated as a discrete model. Discrete optimisation is usually formulated on a Markov random field, an undirected graphical model, instead of a continuous vector field. The graph consists of a set of nodes $p \in P$ corresponding to groups of voxels (or control points) and a set of (hidden) labels f_p for each node, corresponding to discrete displacements. The energy functional

$$E(f) = \sum_{p \in P} D(f_p) + \alpha \sum_{(p,q) \in \mathcal{N}} R(f_p, f_q) \tag{4}$$

consists of the data cost D, which measures the similarity of a node in the fixed image related to a displaced node in the moving image and is independent of the displacement of its neighbours. The metric used for deeds is self-similarity context (see [6] for details). The pair-wise regularisation term R for each node q that is connected with a node p is defined as the quadratic norm of their displacement difference.

Deeds aims to find the global minimum in just two passes of dynamic programming using a reduced neighbourhood interaction (a minimum-spanning tree), so the grid is not fully connected. The edge weights are defined as the sum of absolute difference between the intensities of all voxels in node p and the respective voxels in node q.

To calculate the global optimum and thereby the best displacement, the cost function at each node p can be found, given the displacement f_q of its parent q:

$$C_p(f_p) = \underset{f_p}{min}(D(f_p) + R(f_p, f_q) + C_c(f_p)), \tag{5}$$

where c are the children of p. The tree is passed from the leaf nodes to the root node, because the cost for the leaves

---

[1] http://elastix.isi.nl/download.php

can be calculated directly.

To avoid local minima a multi-level scheme is added, in which only the highest resolution is needed. For more details see [7].

2.2 Evaluation strategy

We used pulmonary inhale and exhale CT scan pairs for evaluating the registration accuracies. Pulmonary registration is a challenging task since the lungs are highly deformable for small structures like lung vessels and the contrast is changing because of higher tissue density in consequence of lung compression. For these reasons, a non-rigid method is required. To compare and evaluate the results of the registration methods, no metric alone is sufficient since there is no ground truth for clinical data. In order to obtain a good indication of the performances, we implemented a validation framework (a graphical user interface) and use information from different values and metrics:

- **Jacobian determinant**:
  Using the Jacobian determinant

$$J(u) = det(\nabla u + \mathbb{1})$$

$$= \begin{vmatrix} \frac{\partial u_1}{\partial x_1} + 1 & \frac{\partial u_1}{\partial x_2} & \frac{\partial u_1}{\partial x_3} \\ \frac{\partial u_2}{\partial x_1} & \frac{\partial u_2}{\partial x_2} + 1 & \frac{\partial u_2}{\partial x_3} \\ \frac{\partial u_3}{\partial x_1} & \frac{\partial u_3}{\partial x_2} & \frac{\partial u_3}{\partial x_3} + 1 \end{vmatrix}, \tag{6}$$

the (physical) plausibility of the deformation fields can be analyzed. It is an indicator of volume changes at a given point: $J(u) > 1$ corresponds to an expansion, $J(u) = 1$ to volume preservation and $J(u) < 1$ to contraction. $J(u) = 0$ describes the limiting case of the expansion and different points are mapped on the same position after transformation. For the case $J(u) < 0$, there is a mathematically change of orientation. Thus the last two cases are physiologically inconsistent and the resulting deformation grid is not diffeomorph. This can be seen from the fact that the grid lines intersect or, for the case $J(u) = 0$, the grid lines become tangent to each other.

- **Target Registration Error (TRE)**:
  Given a set of landmarks for the fixed and the moving images, the TRE can be calculated by:

$$TRE = \sum_{i=1}^{n} ||L_i(I_F(x)) - L_i(I_M(x + u(x)))||, \tag{7}$$

where $L(I_F(x))$ is the landmark-position in the fixed image, $L(I_M(x + u(x)))$ the landmark-position in the deformed moving image, n the total number of landmarks and $|| * ||$ the Euclidian norm. The smaller the value, the better the registration result.

- **Mutual Information (MI)**:
  MI indicates how much information the images share, by comparing the entropy values:

$$H(x) = -\sum_i p(g_i) ln(p(g_i)), \tag{8}$$

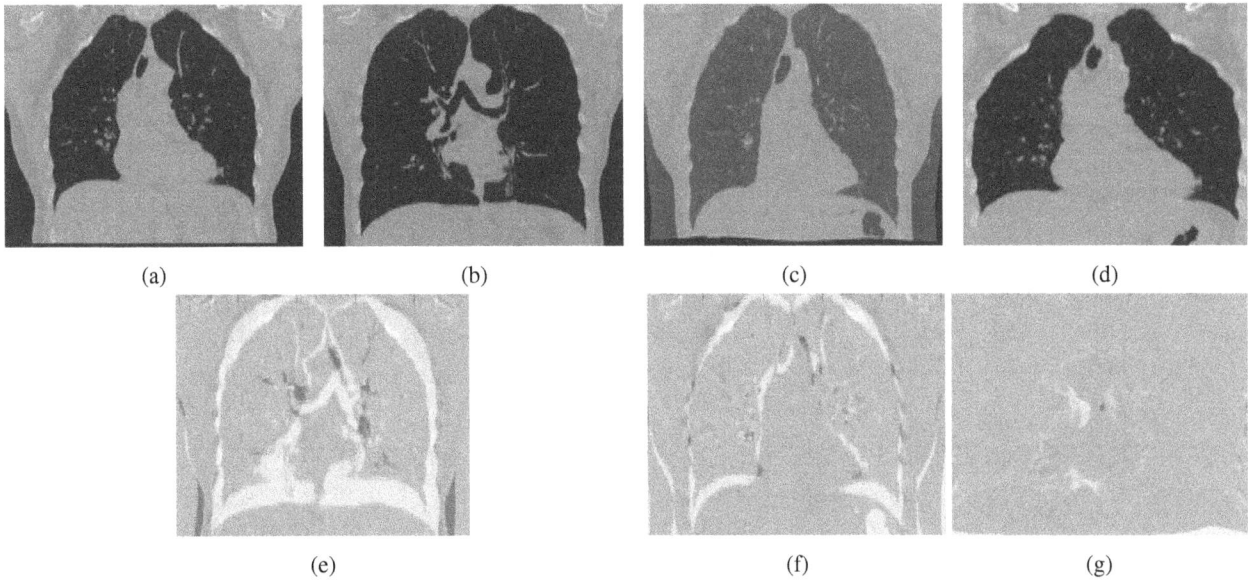

Figure 1: (a) an example slice of the fixed image (exhale), (b) same slice of moving image (inhale), (c) result elastix, (d) result deeds, (e) difference between (a) and (b), (f) difference between (a) and (c), difference between (a) and (d).

where $p(g_i)$ being the intensity of the images gray values g_i. The distance metric is calculated as follows:

$$MI(R,T) = H(I_F) + H(I_M(x + u(x)) \\ -H(I_F, I_M(x + u(x)), \quad (9)$$

where $H(I_F, I_M(x+u(x))$ is the combined entropy of the moving and the fixed images. The higher the value, the better the correspondence between the images.

3   Results and Discussion

To evaluate the described registration methods, we used the publicy available CT dataset of [2]. It consists of ten inhale and exhale scan pairs of patients from the COPDgene study with a spatial resolution ranging from 0.586 to 0.742 mm in xy-direction and 2.5 mm in z-direction. An expert identified 3000 anatomic landmark features (300 for each pair), with an observer error of 1.5 ± 0.7 mm. Fig. 1 (a), (b) and (e) shows an example slice of an inhale and exhale scan pair and the difference between them.

We tested different parameter settings for both algorithms. Until now, the following parameter settings performed best:

elastix: All scan pairs were first registered using an affine transform with six levels and 2000 iterations. For the main registration, we chose a non-rigid B-spline registration with normalized correlation as similarity measure and gradient descent as optimisation method. The B-spline registration was performed using the following parameters:

- maximum number of iterations: 5000

- resolution levels: 5

- grid spacing in physical units: $8.0 \times 8.0 \times 8.0$

- pyramid schedule in mm: $16, 8, 4, 2, 1$

- $a_k = a/(A+k+1)^\alpha$, with $\alpha = 0.6$, $A = 50$, $a = 8000$ for first, $a = 2000$ for other levels

- number of histogram bins: 32

- number of spatial samples: 3000

deeds: To perform a registration with deeds, the volumes were resampled to an isotropic voxel-size of $1 \times 1 \times 1$ mm. The registration was performed using the following parameters:

- levels: 5

- grid spacing: $G = 6, 5, 4, 3, 2$

- maximum search radius: $L = 6, 5, 4, 3, 2$

- quantisation of search step size: $Q = 5, 4, 3, 2, 1$

- regularisation parameter: $a = 2.0$

- number of random samples per node: 50

Discussion of results:
Fig. 1 (c), (d), (f) and (g) displays the same example slice as Fig. 1 (a) after registration. First visual examination of the resulting images reveals that deeds provides better registration results in comparison to elastix. The difference images (Fig. 1 (f) and (g)) show that the rib cage as well as the lung vessels match better by registration with deeds and until now, registration fails locally with elastix. Table 1 listed the quantitative registration results. The average TRE is 1 mm with a deviation of 0.2 mm and therefore decreased by about 22.6 mm. On the contrary, elastix shows an average TRE of 15.2 mm with a deviation of 4.3 mm, which is much higher than the value of deeds (Table 1 (b)). This can be confirmed by calculating MI. The value of resulting

Table 1: Registration results

Case #	initial	elastix	deeds
1	0.90	1.23	1.89
2	0.96	1.17	1.63
3	1.26	1.40	2.22
4	0.94	1.27	1.67
5	1.16	1.36	1.82
6	0.99	1.37	2.09
7	1.19	1.60	2.03
8	1.09	1.44	1.08
9	0.08	1.12	1.73
10	1.02	1.86	1.80
avg	1.05	1.38	1.80

(a) MI

Case #	initial	elastix	deeds
1	26.3	18.1	0.9
2	21.9	10.4	1.7
3	12.6	10.9	1.0
4	29.6	23.1	1.0
5	30.1	19.9	1.1
6	28.5	18.7	0.8
7	21.6	8.3	1.3
8	26.5	17.5	1.3
9	14.9	14.2	0.9
10	21.8	10.8	1.0
avg	23.4	15.2	1.0
std	10.1	4.3	0.2

(b) TRE in mm

Case #	elastix avg	elastix std	deeds avg	deeds std
1	0.92	0.20	1.02	0.36
2	0.93	0.17	1.01	0.35
3	0.99	0.10	1.06	0.37
4	0.88	0.28	1.04	0.31
5	0.94	0.25	1.11	0.38
6	0.89	0.22	1.04	0.46
7	0.99	0.12	1.04	0.34
8	0.93	0.13	1.04	0.42
9	0.96	0.18	1.04	0.39
10	0.91	0.16	1.00	0.42
avg	0.93	0.18	1.04	0.38

(c) Jacobian determinant

images of elastix increases from 1.05 up to 1.40, but deeds gains the advantage with an increase of about 0.75 (Table 1 (a)). The Jacobian determinants of the deformation fields can be found in Table 1 (c). The values are close to one for both algorithms, which equates to small changes in volume. Compared to the initial images in which the difference in volume is large because of the inhale and exhale phases, we expect values that have a greater variance of one. This will be discussed in future work.

4 Conclusion

We have compared two registration approaches: the wideley-used registration toolbox elastix and the more recent technique deeds. The experiments indicates that deeds provides better registration results. However, the correct parameter setting for elastix have not been found yet. Future research will include this search to improve the comparison and evaluation of these two methods. Additionally, we want to include another registration algorithm to the evaluation, which is called FEIR (Fast Elastic Image Registration).

Acknowledgement

The work has been carried out at Department Digital Imaging, Philips GmbH Innovative Technologies, Research Laboratories Hamburg.

5 References

[1] A. Sotiras, C. Davatzikos, and N. Paragios, "Deformable medical image registration: A survey," *Medical Imaging, IEEE Transactions on*, vol. 32, pp. 1153–1190, July 2013.

[2] R. Castillo, E. Castillo, D. Fuentes, M. Ahmad, A. M. Wood, M. S. Ludwig, and T. Guerrero, "A reference dataset for deformable image registration spatial accuracy evaluation using the copdgene study archive," *Physics in Medicine and Biology*, vol. 58, no. 9, p. 2861, 2013.

[3] G. E. Christensen, X. Geng, J. G. Kuhl, J. Bruss, T. J. Grabowski, I. A. Pirwani, M. W. Vannier, J. S. Allen, and H. Damasio, "Introduction to the non-rigid image registration evaluation projct (nirep)," *Lecture Notes in Computer Science*, vol. 4057, pp. 128–135.

[4] S. Klein, M. Staring, K. Murphy, M. A. Viergever, and J. P. W. Pluim, "elastix: A toolbox for intensity-based medical image registration.," *IEEE Trans. Med. Imaging*, vol. 29, no. 1, pp. 196–205, 2010.

[5] D. P. Shamonin, E. E. Bron, B. P. Lelieveldt, M. Smits, S. Klein, and M. Staring, "Fast parallel image registration on cpu and gpu for diagnostic classification of alzheimer's disease," *Frontiers in Neuroinformatics*, vol. 7, no. 00050, 2014.

[6] M. P. Heinrich, M. Jenkinson, B. W. Papiez, M. Brady, and J. A. Schnabel, "Towards realtime multimodal fusion for image-guided interventions using self-similarities.," in *MICCAI (1)* (K. Mori, I. Sakuma, Y. Sato, C. Barillot, and N. Navab, eds.), vol. 8149 of *Lecture Notes in Computer Science*, pp. 187–194, Springer, 2013.

[7] M. P. Heinrich, M. Jenkinson, M. Brady, and J. A. Schnabel, "Globally optimal deformable registration on a minimum spanning tree using dense displacement sampling.," in *MICCAI (3)* (N. Ayache, H. Delingette, P. Golland, and K. Mori, eds.), vol. 7512 of *Lecture Notes in Computer Science*, pp. 115–122, Springer, 2012.

Myocardial Strain Measurement
– A Registrational Method with Block Matching Approach–

L. Landwehr [1], C. Bouwman [2], J.P. Aben [2] F.W. Prinzen [3] and M.P. Heinrich [4]

[1] Medizinische Ingenieurwissenschaft, Universität zu Lübeck, lena.landwehr@student.uni-luebeck.de
[2]Pie Medical Imaging, Maastricht, The Netherlands, {chris.bouwman, jeanpaul.aben}@pie.nl
[3]Cardiovascular Research Institute Maastricht, Maastricht University, Maastricht, The Netherlands
[4] Institute of Medical Informatics, Universität zu Lübeck, heinrich@imi.uni-luebeck.de

Abstract

Strain measurement is the key to an early diagnosis of contractile dysfunction within the myocardium. Methods like Myocardial Tissue Tagging in Magnetic Resonance Imaging or Speckle Tracking in Echocardiography are available to measure myocardial strain, but incorporate extra image acquisition or poor image quality. The following approach introduces a way to analyse cardiac Cine-Magnetic Resonance images of the heart. The algorithm uses a Block Matching approach to analyse the myocardial motion and creates a dense displacement field with a Full-Search algorithm. This displacement field is used to calculate the strain on short-axis images. Right-sided pre-excitation, as seen during left bundle branch block, was artificially applied on canine heart and compared with Pre-LBBB datasets. The strain validation was done qualitatively by comparison with earlier findings. Overall it is a promising approach to estimate myocardial strain without the requirement of additional acquisition time.

1 Introduction

Changes in heart motion can shed light on several cardiac diseases. An early manifestation of contractile dysfunction is the key for a successful therapy in several diseases like acute myocardial dysfunction [1]. Strain and strain rate examination of the left ventricle (LV) can help to determine an early diagnosis and improve the prognosis in contrast to ejection fraction (EF) or visual estimation [2], because strain measurement offers a greater accuracy.

Measuring myocardial strain can be done by Echocardiography. One of the major drawbacks is the limitation by inadequate acoustic windows and poor image quality [3]. In contrast to that, the Cardiac Magnetic Resonance (CMR) imaging is a useful method to measure accurate, reproducible and high-contrast images of the heart. The strain measurement though is a difficult problem, due to the uniformity of the myocardial tissue in Magnetic Resonance Imaging (MRI). In contrast to Echocardiography no features within the myocardium can be seen, which complicates the motion estimation within the tissue. There are techniques like MRI Tagging, which generate artificial features to track. In practice a second imaging sequence in addition to the standard method Cine-CMR imaging, which describes a collection of images over the cardiac cycle, must be acquired. That has the disadvantage of additional acquisition time and signal fading [4]. So an approach for calculating the strain on usual Cine–CMR images is introduced to overcome those problems and qualitatively tested.

We propose a Block Matching (BM) approach with the

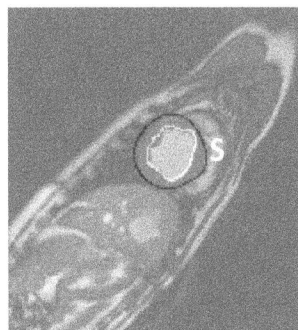

Figure 1: A short axis CMR image with highlighted endocardial (light) and epicardial (dark) border. The septal part of the myocardium is labeled (S). The contours are detected with the CAAS MRV 3.4 system (Pie Medical Imaging, Maastricht, NL).

Cross Correlation (CC) as similarity measure. The myocardium itself and the endocardial and epicardial walls are tracked in time. The borders are features, which can be well estimated, see Fig. 1. The contours of the myocardium are also used to estimate the maximal motion beforehand and to introduce regularization. This ensures a smooth displacement field and avoids wrong matching due to noise. The final strain tensor is derived from the displacement field [5] calculated by the BM algorithm.

For strain evaluation canine datasets with and without induced dyssynchrony due to ablation of the left bundle branch block (LBBB) [6] are used. Therefore change in myocardial motion occurs, which can potentially be detected with our algorithm.

2   Material and Methods

For the evaluation of cardiac parameters Cine-Magnetic Resonance (MR) images were used. They were acquired with the Single State Free Precision Sequence. This sequence establishes a steady state of magnetization and therefore images with a high tissue contrast and a relatively high resolution are created. Another advantage of Cine-MR images is, that it is already a standard method for cardiac diagnosis, so the acquisition time is not extended [4]. We use Short Axis (SA) images for calculation. They show the cross section of the left and right ventricle, as seen in Fig. 1.

2.1   Registrational Strain Calculation

At first we try to estimate the displacement of each image pixel, so a dense displacement field is calculated. For this purpose we use a non–rigid registration algorithm between two images. A Block Matching approach is used. In general, Block Matching means that we have a patch P_F in the fixed image F and we are looking in a certain region of interest (ROI) for the most similar patch P_M in the moving image M. This can be mathematically expressed as

$$\max_u S(P_F(x), P_M(x+u)) \qquad (1)$$

with the displacement u between the pixels of the two images and the similarity S. The patch size is estimated through the myocardial thickness. So not only myocardial intensity information is considered, but also border information.

We have a monomodal problem between two MR images. The best similarity measure for this problem is the CC

$$CC_{FM} = \frac{\sum\limits_{i=1}^{n}(P_{F,i}-\bar{P}_F)(P_{M,i}-\bar{P}_M)}{\sqrt{\sum\limits_{i=1}^{n}(P_{F,i}-\bar{P}_F)^2}\sqrt{\sum\limits_{i=1}^{n}(P_{M,i}-\bar{P}_M)^2}} \qquad (2)$$

between the moving image and the fixed image. The amount of pixels per patch equals n. We are looking for the highest absolute value, which ranges from 0 to 1 [7]. When the patch with the highest correlation is found, the displacement between those two patches is saved at the corresponding pixel position. After repetition for every pixel in the image, a dense displacement field is created.

In general the BM approach provides good estimations of displacement fields, but for our homogeneous myocardium the intensity changes are affected by noise. Therefore an regularization is used to avoid erroneous estimations affected by noise. The BM approach is completed by information about the contour and its movement. The contours are extracted by the CAAS MRV 3.4 software package (Pie

Medical Imaging, Maastricht, NL). Integral images are applied to reduce the computation time. After the integral image computation only a constant number of computations is needed. Especially for bigger patches, evaluating the whole width of the myocardium, integral images are useful, because the computation time is independent of the size [8]. To be able to calculate relatively big displacements a pyramidal BM approach is utilized. A computation of n scaled down images is performed. On each Level L_n, the displacement is calculated. Afterwards the calculated displacement u is scaled up due to

$$u^{L_n} = \frac{u}{2^{L_n}} \qquad (3)$$

[9]. A cubic spline interpolation step follows and this interpolated image is used for the next level. The displacement is summed up over n steps and gets more accurate in each stage. A special step is the one with the smallest image. It has to be guaranteed that the image is not too small, otherwise the estimation of the biggest motion in the level L_0 is wrong and the whole displacement field would be erroneous.

To achieve a smooth displacement field, an anisotropic filter is used between each level in the pyramidal approach. This guarantees that the motion is smooth like real myocardial motion, but preserves the edges. After calculating the displacement field, the strain calculation follows. Strain is defined as deformation relative to its length. Hence, it is a dimensionless value, which is usually measured in percent. It does not give information about the motion itself. If all parts of the object move in the same direction no deformation is expected and due to that no strain will be measured. We will use the Lagrangian strain tensor

$$[\mathbf{E}] = \frac{1}{2}(\mathbf{C} - \mathbf{I}) \qquad (4)$$

with $\mathbf{C} = \mathbf{F}^T\mathbf{F}$, $\mathbf{F} = \mathbf{I} + \nabla u$ and the identity matrix I to evaluate the strain for the myocardium. The displacement gradient ∇u is defined by

$$[\nabla \mathbf{u}] = \begin{bmatrix} \frac{\delta u_x}{\delta x} & \frac{\delta u_x}{\delta y} \\ \frac{\delta u_y}{\delta x} & \frac{\delta u_y}{\delta y} \end{bmatrix}. \qquad (5)$$

Consequently the strain is the spatial derivative of the displacement, as seen in (5) [10].

Furthermore, special definitions are used for cardiac strain. In general the strain is measured in circumferential, transmural and longitudinal direction. This can be seen in Fig.2. Locally the circumferential direction can be described as tangential vector along the myocardial contour. Because we only use SA images we concentrate on the circumferential and transmural direction.

Thus, the previously defined strain tensor must be transformed into a local Cartesian Coordinate system [11] corresponding with Fig. 2. The transformation is defined as

$$[\mathbf{T}]' = [\mathbf{Q}]^T[\mathbf{T}][\mathbf{Q}] \qquad (6)$$

with $Q_{ij} = cos(\mathbf{e}_i, \mathbf{e}'_j)$. The unit vectors e of each Coordinate System are related by the orthogonal tensor Q [10].

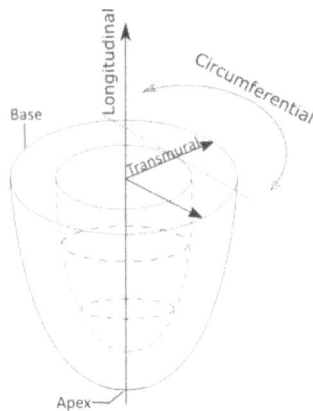

Figure 2: The main directions of the principal strain of the left ventricle. The three directions are locally perpendicular to each other and can therefore build a coordinate system.

All in all we get the local strain in circumferential and transmural direction.

2.2 Validation Methods

To validate the algorithm a two stage process of validation was used. At first the displacement field, calculated by the BM algorithm, was investigated. This was mainly done with registration. The calculated displacement field was used to deform the initial image. This image was afterwards visually compared with the original image and 30 comparisons were done. Another approach for the validation of the displacement field, was the usage of another registration algorithm. With the help of the Elastix package [12] the transformation between two images was calculated. Afterwards the provided displacement field was qualitatively compared with our approached displacement field.

In the second step the transmural strain calculation was evaluated. We considered a dataset of previously acquired MR images in dogs before (n=5, pre-LBBB) and after inducing LBBB (n=5, post-LBBB), as described before [6]. These datasets were analysed with the help of our Cine-MR image approach. The in-plane resolution was 1,5625 mm x 1,5625 mm and the slice thickness 8 mm. The resulting strain was qualitatively validated against the findings in [6]. A bull's eye plot for multiple points in time was created, see Fig.4. Each circle matches with one slice from apical to basal height. The regional strain was parted in 32 segments and averaged.

3 Results and Discussion

In the following the results of both, the validation of the displacement and the strain evaluation are carried out.

3.1 Validation of the Displacement

With the help of the above mentioned algorithm a dense displacement field is created. The example of a displacement field between two consecutive frames is visualized in

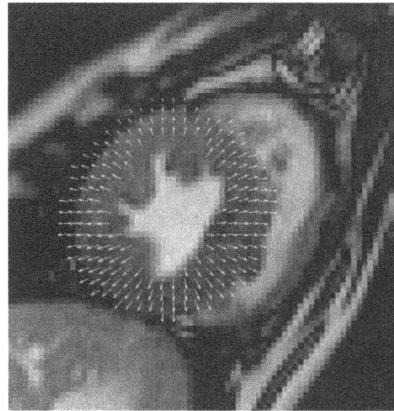

Figure 3: The displacement field indicated as vectors overlaid above the appertaining image frame.

Fig. 3. The analysis was done frame to frame to ensure that brightness consistency is given. For bigger displacements, the frame to frame displacement fields were summed up. As a result the validation of small motion between two consecutive frames is positive. In every case the correct motion could be calculated. The verification with the Elastix software showed not only similarities in direction, but also in magnitude. Small variations were seen, but due to the good interpolation result they were not further pursued.

3.2 Strain Validation

The above mentioned datasets pre-LBBB and post-LBBB differ due to ablation of the left bundle branch. This causes changes in the myocardial shortening of the septal wall [6], which can be observed in the transmural strain analysis. In all five datasets this change in strain can be seen at the septal side of the myocardium, see Fig. 4.

3.3 Discussion

The introduced algorithm performs registration on myocardial images and estimates the frame to frame displacement accurately. A true displacement field is the first step to calculate the strain. But the algorithm incorporates parameters which need to be well estimated. The amount of pyramidal levels should be adequate, due to the image size of the zeroth level. Also the estimated contours are important to asses the myocardial thickness and therefore calculate the block size. Besides it is used for regularization. So a validation of the impact of the contours should be done to be sure about the right parameters and its influences on errors. Due to a lack of quantitative analysis no further conclusions can be drawn. For example how the algorithm performs in contrast to proven methods like MRI-Tagging or echocardiographic Speckle Tracking. Therefore more evaluation needs to be done to prove the benefit achieved with the Cine-MRI algorithm.

Another point is the displacement field comparison only took place with other medical imaging methods, but also a comparison with the ground truth of the myocardial motion

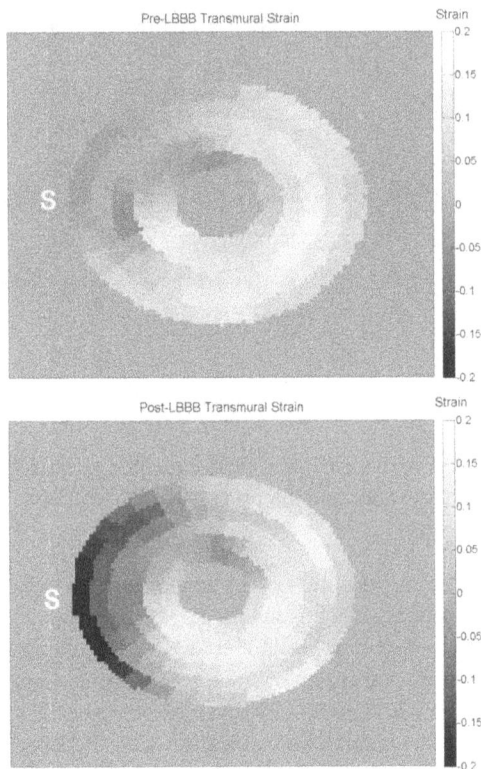

Figure 4: The transmural strain is illustrated as bull's eye plot. The inner circle shows the apical slice and the the outer circle the basal slice. At the top the strain pre-LBBB can be seen. The differences to the post - LBBB (bottom) are clearly to recognize. Especially the septal part (S) differs.

should be taken into account. Depended on the myocardial motion further regularization needs to be introduced, so the displacement field gets more accurate. An alternative approach might only consider the endocardial and epicardial motion.

Despite all this concerns the algorithm seems to perform well and is able to distinguish between stiff and contractile myocardial tissue. Further validation needs to be done to proof its capacity.

4 Conclusion

All in all we proposed an algorithm, which is be able to calculate the circumferential and transmural strain of Cine-MR images and therefore distinguish myocardial contractility. After further evaluation and extension an easier analysis of myocardial strain might be possible without additional acquisition times and with the possibility to apply it on Cine-CMRI datasets in retrospect.

Acknowledgement

The work has been carried out at Pie Medical Imaging, Maastricht, The Netherlands.

5 References

[1] E. Maret, T. Todt, L. Brudin, E. Nylander, E. Swahn *et al.*, "Functional measurements based on feature tracking of cine magnetic resonance images identify left ventricular segments with myocardial scar," *Cardiovasc Ultrasound*, vol. 7, no. 1, pp. 1–14, 2009.

[2] T. Butz, C. N. Lang, M. van Bracht, M. W. Prull, H. Yeni *et al.*, "Segment-orientated analysis of two-dimensional strain and strain rate as assessed by velocity vector imaging in patients with acute myocardial infarction," *Int J Med Sci*, vol. 8, no. 2, pp. 106–113, 2011.

[3] S. Orwat, A. Kempny, G. P. Diller, P. Bauerschmitz, A. C. Bunck *et al.*, "Cardiac magnetic resonance feature tracking: a novel method to assess myocardial strain," *Kardiol Pol*, vol. 72, no. 4, pp. 363–371, 2014.

[4] D. M. Harrild, Y. Han, T. Geva, J. Zhou, E. Marcus *et al.*, "Comparison of cardiac MRI tissue tracking and myocardial tagging for assessment of regional ventricular strain," *Int Journal Cardiovas Imaging*, vol. 28, no. 8, pp. 2009–2018, 2012.

[5] H. Wang and A. Amini, "Cardiac motion and deformation recovery from MRI: a review," *IEEE Trans Med Imaging*, vol. 31, no. 2, pp. 487– 503, 2012.

[6] M. Strik, L. B. van Middendorp, and K. Vernooy, "Animal models of dyssynchrony," *J Cardiovas Transl Res*, vol. 5, no. 2, pp. 135 –145, 2012.

[7] L. G. Brown, "A survey of image registration techniques," *ACM Comput Surv*, vol. 24, pp. 325–376, 1992.

[8] G. Facciolo, N. Limare, and E. Meinhardt-Llopis, "Integral images for block matching," *Image Processing On Line*, vol. 4, pp. 344–369, 2014.

[9] J.-Y. Bouguet, "Pyramidal implementation of the lucas kanade feature tracker," *Intel Corporation, Microprocessor Research Labs*, 2000.

[10] W. M. Lai, D. Rubin, and E. Krempl, *Introduction to Continuum Mechanics (Fourth Edition)*. Boston: Butterworth- Heinemann, 2010.

[11] C. Tobon-Gomez, M. De Craene, K. McLeod, L. Tautz, W. Shi *et al.*, "Benchmarking framework for myocardial tracking and deformation algorithms: an open access database," *Med Image Anal*, vol. 17, no. 6, pp. 632–648, 2013.

[12] S. Klein, M. Staring, K. Murphy, M. A. Viergever, and J. P. Pluim, "Elastix: a toolbox for intensity-based medical image registration," *IEEE Trans Med Imaging*, vol. 29, no. 1, pp. 196 – 205, 2010.

Image processing of pulmonary radiography of pediatric patients to assess tuberculosis or pneumonia for the application in diagnosis in low- and middle-income countries

I. Nehrhoff [1], S. Rodríguez López [2], J.E. Ortuño [2], O. Maier [3], M.J. Ledesma-Carbayo [2], and H. Handels [3]

[1] Medizinische Ingenieurwissenschaft, Universität zu Lübeck, imke.nehrhoff@student.uni-luebeck.de

[2] Biomedical Image Technologies, ETSI Telecomunicación, Universidad Politécnica de Madrid, {sararodrigue-zlopez,juanen,mledesma}@die.upm.es

[3] Institute of Medical Informatics, Universität zu Lübeck, {maier,handels}@imi.uni-luebeck.de

Abstract

Tuberculosis (TB) is an infectious disease with high prevalence in low- and middle-income countries, where the often modest clinical equipment renders the diagnosis challenging. This holds particularly true for children, for whom the conventional sputum analysis is not feasible. This work presents the architecture of an semi-automatic algorithm to detect TB in sets of frontal and lateral children's radiographs consisting of four parts: 1) lung segmentation with an adapted level-set approach 2) anatomically motivated cropping to the hilum region 3) patch extraction and feature computation 4) patch-wise classification and condensation into the diagnosis. In a preliminary evaluation, we investigate the robustness of the lung segmentation, confirm the reliability of the hila detection and determine the suitability of the selected kernel density estimation based feature.

1 Introduction

The infectious disease pulmonary tuberculosis (TB) is transmitted by the bacterium Mycobacterium tuberculosi through droplet infection and considered one of the deadliest infectious diseases in the world. In 2013 approximately 9 million people suffered from a TB-infection, 1.5 million of which died [?] despite available antibiotic-based treatment options. To stop the spread of TB and to cure the infected persons, a fast and secure diagnosis of TB is required. In adults, the gold standard is the microbiological sputum analysis. But from the high risk group of children and infants, sputum often cannot be obtained and the available tests show poor reliability [?]. An alternative approach for TB in children is the detection of hilar lymphadenopathy, an enlargement of the pulmonary lymph nodes, around the hila. Advances in medical imaging, such as computed tomography and magnetic resonance imaging, enable a reliable and fast assessment, but are expensive and not widely available in low- and middle-income countries. In hospitals with modest equipment, the diagnosis therefore still depends on the anamnesis and the findings in the radiographs of the thorax alone [?], which are challenging to interpret: The poor contrast and the superposition of structures of interest with anatomical structures renders it difficult for even experienced radiologists to make a safe diagnosis. Therefore, there is a need to develop automated tools that could help radiologists to make a safer interpretation and also relieve the radiologists of the large number of cases that need

to be diagnosed every day. There already exists a software named CAD4TB [?], which was developed at the Radboud University Medical Center in the Netherlands to detect TB in adults. But these methods are not readily transferable to the case of children due to the large differences in anatomy as well as TB manifestation. The Biomedical Image Technologies group of the Universidad Politécnica in Madrid collaborates with physicians in Mozambique, where a software to detect TB in the radiographs of children would improve the situation in the hospitals significantly. The idea of this work is to create a trained model with radiographs of infants and children from different age groups acquired by the doctors in Mozambique and apply it to classify formerly unseen radiographs of infant patients who are suspected to have a TB-infection. Our proposed method starts with a semi-automatic segmentation of the lungs in the frontal view, followed by an anatomically motivated detection of the hila area in both radiographs and finally a hilar lymphadenopathy diagnosis based on patch-wise classification.

2 Material and Methods

According to the tutorial of Savvas Andronikou [?] and other radiologists, the diagnosis of TB in children and infants requires detecting hilar lymphadenopathy in both, lateral and frontal radiographs [?], where it shows up as an pathologically enlarged hyperintense area (see Fig. ??). We received a total number of fours cases with frontal and lateral thorax radiographs for training and testing purposes, two of which were diagnosed with TB and the other two as healthy. In this section, we describe the steps of our algo-

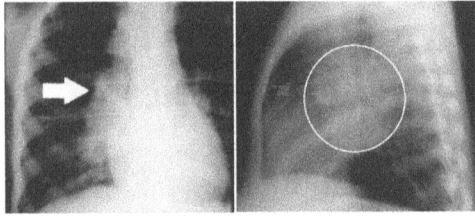

Figure 1: Left image: Lymphadenopathy in the region of the left hilum in a frontal radiograph. Right image: Lymphadenopathy in the region of the hila in a lateral radiograph forms a circular spot.

rithm subdivided into the segmentation of the lungs in the frontal radiographs; the determination of the region around the hila in both, frontal and lateral radiographs; and the patch-wise classification. A flow-chart of the involved steps is given in Fig. ??.

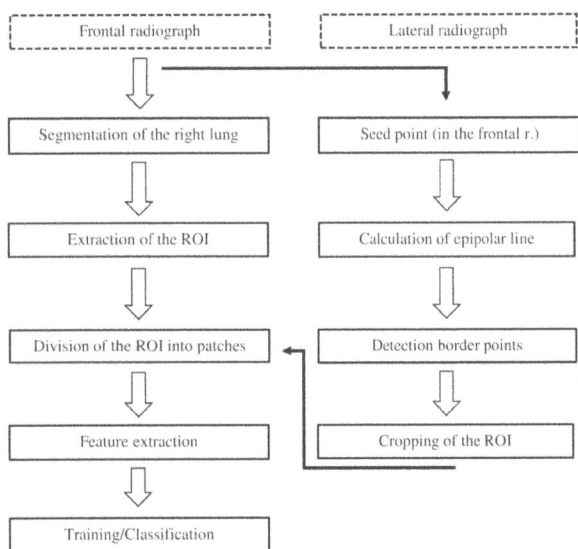

Figure 2: Structure of the TB-detection algorithm.

2.1   Lung Segmentation

The first step in the processing of the frontal radiographs is to segment the organ of interest i.e. the lungs. For adults, various well evaluated methods exist [?], including active shape models, active appearance models and pixel-wise classifiers. But our data situation did not allow us to train any meaningful model nor could the already trained models be applied to our data due to anatomically differences between adult and infant lungs. We instead decided on a semi-automatic level-sets method, which evolves a curve constrained by internal forces and guided by external forces based on image features through the minimization of an energy function associated [?]. Typically, these models are based on edge detection using gradient information,

which is known to fail in the case of poorly defined borders [?] as we are facing in our radiographs. Chan and Vese [?] proposed to consider the values outside the curve as well and thus improving the robustness. But this approach only works well for the segmentation of homogeneous areas. To solve this problem, Fernandez-de-Manuel et al. [?] substituted the term that computes the intensity difference between a point and the average inside and outside of the contour by a term that computes the absolute difference between these averages and included gradient information to achieve a more robust algorithm [?]. We adapted their algorithm to our needs and applied it to segment the lungs in the frontal radiographs. The user first places a number of seed points (4-8 seed points) inside the lung area to initiate the contour and then obtains the lung segmentation mask. Since TB manifest itself mainly in the right lung and the hilum in the left lung is often superimposed by the heart, we restrict our method to the right lung.

2.2   Determining the region around the hila

In both radiographs we automatically determine a region of interest (ROI) that reliably includes the hila while being as small as possible to render the subsequent classification more robust. The only user interaction is the setting of a single seed point in the frontal radiograph at the end of the trachea, where it splits into the left and right main bronchus (Fig. ?? (right), red dot), which can be performed during the lung segmentation. In the frontal view of the lungs,

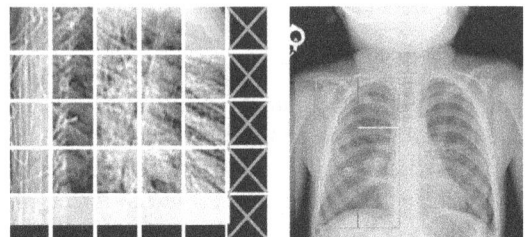

Figure 3: Left: Patches extracted from the lateral view's ROI. Crossed out patches are excluded due to unsufficient image information. Right: Seed point at the end of the trachea in the frontal radiograph (red dot) and determination of the region of the hilum (bottom right rectangle). Colors in the online version denote the ROI detection procedure's steps.

the hila is situated medial-central (Fig. ??). Since we already obtained a lung segmentation for this radiograph, we simply take the medial half of the segmentation mask and crop it by the treachea seed point in height, obtaining the frontal ROI (Fig. ??, (right)). Lung segmentation in lateral radiographs is particularly challenging. We therefore chose a rule based approach motivated by anatomical constraints and supported by the already detected anatomical locations in the front view to determine the ROI. First, the trachea seed point is projected onto the lateral image resulting in a line. This is implemented with the epipolar geometry of stereo vision by using information about scanner and patient positions from the meta-data of the radiographs: first both images are placed in a 3D common space, then the epipolar line associated with the point in the frontal view is determined and finally its projection on the lateral view is computed. The epipolar line in the lateral radiograph gives us a good upper border of the ROI because lymphadnopathies manifest themselves where the trachea splits into the left and right main bronchus, and this part is seen better in the frontal radiograph (Fig. **??**, horizontal line). Next the body

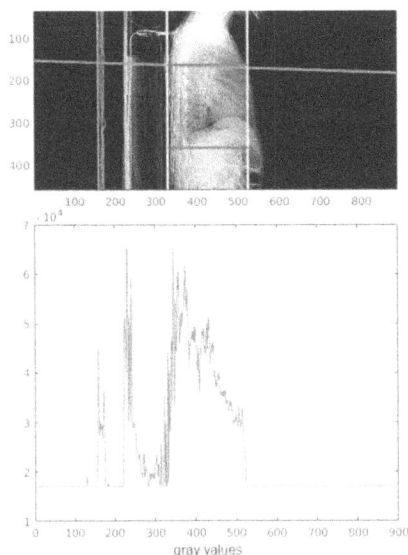

Figure 4: The horizontal epipolar line (red) in the lateral radiograph, the gray values plotted along the horizontal epipolar line (plot below), the two intersection points of the line with the body (yellow vertical lines) and the determination of the region of the hila in the lateral radiograph (square region between blue and yellow lines).

limits are determined via a intensity peak analysis along the epipolar line. Therefore, we first normalize the image's gray values and apply a window where fat and tissue vanishes from the thorax radiograph to get more adequate intersection points. After this, we collect the gray values along the epipolar line (Fig. **??**, plot at the bottom) and automatically select the widest peak as the confinements of the body. The ROI is then a square region with the sidelength of the body width, anchored below the epipolar line (Fig. **??**).

2.3 Patchwise classification

The extracted ROIs contain beside the hilum additionally background. We therefore decided to take a patch-wise classification approach with non-overlapping rectangular patches extracted from the intensity normalized ROIs. Each patch is passed through a previously trained classifier (one for frontal and lateral respectively) resulting in a classification as pathological or healthy / background. If more than a predetermined number of patches is assigned the pathology label, the whole case is diagnosed as TB-positive. The chosen patch size is 20 mm to ensure that at least some of them contain the hila. Patches with less than 50% image

pixels are discarded. As feature the patch histogram is employed, computed via the kernel density estimation (KDE). The KDE is an estimation of the probability of the density of each patch, by which effectively a smoothed representation of the histogram is obtained, thereby avoiding the known problem with histograms that the choice of binning can have a disproportionate effect on the resulting interpretation of the data. Since our training patches contain a number of false positives (background patches from TB-positive cases), the k-nearest neighbor (K-NN) algorithm (K = 5) with votes weighted by the inverse Euclidean distances is employed for classification. Assuming that the classes are separable using the chosen feature, we estimate that the real TB-negative patches will outvote the false TB-positive candidates.

3 Results and Discussion

Since only four cases were available, the following evaluation has to be considered preliminary.

3.1 Lung Segmentation

The level-set based lung segmentation algorithm was applied to segment the right lungs in the frontal radiographs with four to eight seed-points as manual input. In the absence of ground truth masks, the results were considered as success if the lung was successfully outlined and the hila included. Out of the four cases, the segmentation failed in one case, where the hila border was to pronounced for the contour to cross over (see Fig. **??**). For the further exper-

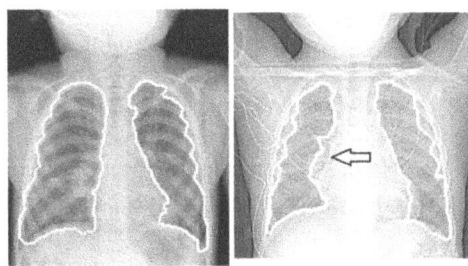

Figure 5: Left: A case where the level-sets algorithm achieved a good segmentation of the right and left lung. Right: The region of the hilum of the right lung is excluded in the segmentation and needed to be post-processed.

iments, we applied a morphological operator to the failed case's segmentation to include the hila. The results are not satisfying, but the poor data situation prevented us from employing methods known to be suitable for the task [**?**], as explained in the Methods section above.

3.2 Determining the region around the hila

After setting the trachea seed-point in the frontal radiograph, the anatomically motivated ROI determination procedure successfully returned ROIs which had the hila and a minimum of background included. Fig. **??** and Fig. **??**

(right) show exemplary results for one of the four cases. While a larger evaluation is clearly necessary to draw any definite conclusions, the implemented approach robustly detected the border of the body and provided meaningful ROIs.

3.3 Determining the region around the hila

Averaging the probability density extracted from all patches with the KDE, we obtain the distributions depicted in Fig. ??. While overlapping, the distribution of gray values

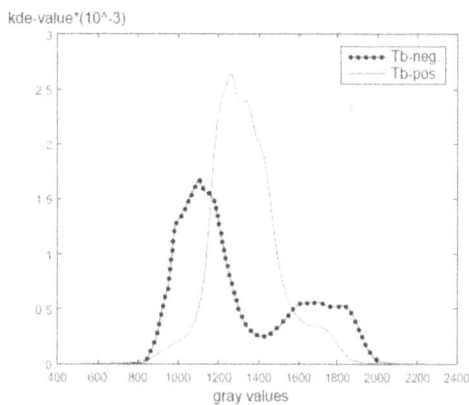

Figure 6: Mean of all estimated density values for TB-positive and TB-negative patches on the y-axis plotted against the gray values on the x-axis of the images.

in the TB-positive patches clearly differs from the one of the TB-negative patches, hinting that the classes might be separable. We conducted an classification experiment with the patches using always two cases for training (one pathological and one healthy) and the other two for testing, leading to a total of four experiments. The obtained mean accuracy for correctly classified patches was 0.3. Additional experiments employing support vector machines and decision trees as classifiers did not lead to higher values. It is difficult to determine whether the low accuracy obtained is caused by systematic errors in the method itself or an effect of the very poor data situation. A training with only two cases is simply impossible to train a classifier with robust generalization abilities. It might become necessary to change the training labels of the non-hila patches in the TB-positive cases for a more robust classification.

4 Conclusion

We have presented an architecture for a TB-classification from infant radiographs consisting of a lung segmentation, an anatomically motivated hila extraction and a patch-wise classification approach. The extraction of the ROIs seems to be robust, the segmentation of the lungs acceptable and the classes might be separable with the chosen feature. A lack of suitable cases seriously impeded the evaluation of the developed method: Drawn conclusions can only be considered preliminary and further evaluation is required. We nevertheless believe that we developed a suitable starting-point

for further investigation when new data becomes available. For the future we would like to implement and evaluate one of the tested automatic lung segmentation algorithms [?], evaluate the classification approach on a sufficient large dataset and introduce further features if required.

Acknowledgement

The work has been carried out at the Biomedical Image Technologies Group of the Universidad Politécnica of Madrid.

5 References

[1] *Global tuberculosis report.* WHO/HTM/TB/2014.08, Geneva, 2014.

[2] A. L. García-Basteiro and E. López-Varela, *Radiological Findings in Young Children Investigated for Tuberculosis in Mozambique.* PLoS ONE, vol.10, no.5, 2015.

[3] L. Hogeweg, C.I. Sanchez, P. Maduskar, R. Philipsen, A. Story, R. Dawson, G. Theron, K. Dheda et al. *Automatic detection of tuberculosis in chest radiographs using a combination of textural, focal, and shape abnormality analysis.* IEEE Transactions on Medical Imaging, vol. 34, no.12, pp. 2429-2442, 2015.

[4] S. Andronikou, *Diagnosis of Pulmonary TB in Children.* at <http://e.itg.be/dxtb/index.html>.

[5] M. Carlos, M.D. Perez-Velez, and B.J. Marais, *Tuberculosis in Children.* The new England journal of medicine, vol. 367, pp. 61-348, 2012.

[6] L. Fernandez-de-Manuel, J.L. Rubio, M.J. Ledesma-Carbayo,J. Pascau, J.M. Tellado, E. Ramon et al. *3D Liver Segmentation in Preoperative CT Images using a Level-Sets Active Surface Method.* Engineering in Medicine and Biology Society, EMBC. Annual International Conference of the IEEE, 2009.

[7] T.F. Chan and L.A. Vese *Active contours without edges.* Image Processing, IEEE Transactions on Image Processing, vol. 10, no.2, pp. 266 - 277, 2001.

[8] B. van Ginneken, M.B. Stegmann, M. Loog *Segmentation of anatomical structures in chest radiographs using supervised methods: a comparative study on a public database.* Medical Image Analysis, vol. 10, pp. 19-40, 2006.

Automatic Orientation Detection of Hand Structures in Digital X-Ray Images

I. M. Baltruschat [1], M. Hensel [2], and M. P. Heinrich [3]

[1] Medizinische Ingenieurwissenschaft, Universität zu Lübeck, ivo.baltruschat@student.uni-luebeck.de

[2] Philips Medical Systems DMC GmbH, Hamburg, marc.hensel@philips.de

[3] Institute of Medical Informatics, Universität zu Lübeck, heinrich@imi.uni-luebeck.de

Abstract

In the last decade portable flat panel detectors have gained increased usage in medical X-ray imaging. Since portable detectors can be used in different orientation, one cannot assume the anatomical structure of interest is upright. This means clinical staff has to manually rotate images. The aim of this paper is to improve the clinical workflow by using automatic orientation detection for hand structures and apply an automatic rotation. We include knowledge of the given anatomy by searching for intersection points of the arm with any image border. The algorithm extracts all border intersections. After analyzing those based on their histogram the intersection of the arm is found. Orientation can be calculated with a perpendicular line to the intersection line and the center of gravity. The results demonstrate that our algorithm is very fast and accurate for anteroposterior and posteroanterior projection images and can be employed to ensure an upright position.

1 Introduction

Image processing is a necessary tool for digital X-ray imaging with flat panel detectors. It is used in two ways. First and the most important one is to improve image quality. The second is to perform certain analysis tasks automatically. Automatization helps to improve clinical workflow and to save precious time at a digital radiography system.

Most modern X-ray systems use a fixed flat panel detector [1]. This ensures a predetermine orientation of the anatomical structure. However, over the last years portable detectors were invented. They can be used for new tasks e.g. acquiring an X-ray image of bedridden patients. Those portable detectors have the drawback that image orientation depends on how the detector was used. If the image orientation is not upright, the medical staff will have to manually rotate the image. The rotation is restricted to 90 degrees, 180 degrees and 270 degrees, because one does not want to loose image informations. Fig. 1 illustrates a rotation of a falsely oriented image (Fig. 1a) to the upright position (Fig. 1b). Those manual rotation disturb the clinical workflow and cost precious time.

A fully automatic orientation detection of anatomical structures can be used to rotate automatically the image. This work extends a previous Master thesis on automatic rotation for images at Philips [2]. Their approach was to employ a Canny edge detection and a Hough transformation for orientation calculation. The downside of this approach was its long calculation time of 210 ms. The requirement of Philips was a maximum of 100 ms for this processing step.

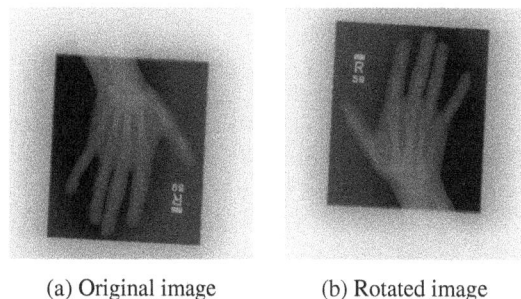

(a) Original image   (b) Rotated image

Figure 1: The hand structure in image (a) is falsely orientated. Either the medical staff will manually rotate the image or an automatic orientation detection will automatically rotate the image. The result of the rotation is shown in image (b).

We developed a new algorithm by defining specific features of hand images. The most important feature is the arm which has to intersect at least one image or shutter border. Also the line profile of the arm is used as a second feature. Those features should be invariant to rotation, translation and gray value changes. The objective of this work is to improve the clinical workflow for hand structures of anteroposterior (AP), posteroanterior (PA) and oblique projections by our automatic orientation detection.

2 Material and Methods

This section starts by explaining the automatic orientation detection which is divided into feature extraction and orientation calculation. Afterwards the created data set and the experiment is described.

2.1 Feature Extraction

In order to do the orientation calculation, the edge intersections and their features need to be extracted from an input image.

The intersections can be found in the binary sub image $B_{sub}(x,y)$ of the input image $f(x,y)$. To obtain the binary sub image B_{sub}, several processing steps are necessary.

The input image $f(x,y)$ is a preprocessed X-ray image and it is necessary to extracted the region of interest (ROI). The Philips shutter detection was used to obtain information about the corners of the shutter area as shown in Fig. 2a. Those points are required to extract the ROI $f_{sub}(x,y)$ of the input image $f(x,y)$.

Afterwards the histogram of the $f_{sub}(x,y)$ is calculated to determine a threshold t_{min}. The value of t_{min} is defined as the 70th percentile of the histogram and the binary image $B_{sub}(x,y)$ is created by a thresholding [5].

$$B_{sub}(x,y) = \begin{cases} 1, & \text{if } t_{min} < f_{sub}(x,y) \\ 0, & \text{else} \end{cases} \tag{1}$$

In the last step the binary sub image $B_{sub}(x,y)$ is analyzed to find every image border intersection. Fig. 2 shows an example of the processing steps.

Each intersection has two features. The length of the intersection and the median of the line profile. To reduce the number of intersections a minimal length of 2 cm is required and intersections with a connection point are merged. This merging is used for intersections as in Fig. 3b where the top and right intersections are connected.

2.2 Orientation Calculation

The arm is found by analyzing the extracted intersections. It is trivial if only one intersection was found. This occurs for images like Fig. 3c. Finding the arm in more than one intersection is done by comparing their median, because the intersection with the minimum median is most likely the arm.

After the arm was found, the hand orientation is determined. In the first step the center of gravity (COG) $\vec{cog} = (\bar{x}; \bar{y})$ [5] from the sub image B_{sub} is calculated with (2),(3) and (4).

$$A = \sum_{i=1}^{N} \sum_{j=1}^{M} B_{sub}[i,j] \tag{2}$$

For the given binary image, A is the number of pixels which are 1.

$$\bar{x} = \frac{1}{A} \sum_{i=1}^{N} \sum_{j=1}^{M} j \cdot B_{sub}[i,j] \tag{3}$$

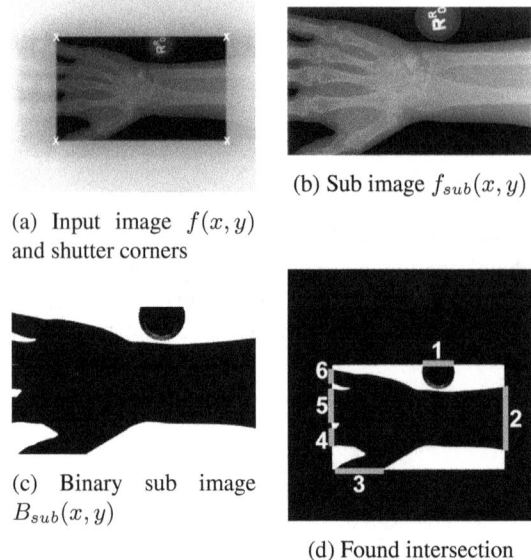

(a) Input image $f(x,y)$ and shutter corners

(b) Sub image $f_{sub}(x,y)$

(c) Binary sub image $B_{sub}(x,y)$

(d) Found intersection

Figure 2: The images (a)-(d) are demonstrating the intersection extraction. (a) Full size input image with 3001×3001 pixels. The shutter corners are marked with a white cross. (b) Sub image with 1457×854 pixels. The histogram is calculated and the 70th percentile is determined for a thresholding (c) Result of the thresholding. This image is analyzed to find all intersection. (d) Illustration of the found intersection which are defined by a line.

$$\bar{y} = \frac{1}{A} \sum_{i=1}^{N} \sum_{j=1}^{M} i \cdot B_{sub}[i,j] \tag{4}$$

The origin of the image coordinate system is translate to the cog. Thus the start point $\vec{P}_s = (x_s, y_s)$ and end point $\vec{P}_e = (x_e, y_e)$ of the arm intersection are defined as followed:

$$\vec{P}'_s = \vec{P}_s - \vec{cog} \qquad \vec{P}'_e = \vec{P}_e - \vec{cog} \tag{5}$$

Those two points define a line which is used to calculate the angle to the x-axis. If $x_s \neq x_e$, the slope-intercept form is defined as:

$$\begin{aligned} y &= (x - x'_s)\frac{y'_e - y'_s}{x'_e - x'_s} + y'_s \\ &= \frac{y'_e - y'_s}{x'_e - x'_s}x + y'_s - x'_s\frac{y'_e - y'_s}{x'_e - x'_s} \\ &= m\,x + b \end{aligned} \tag{6}$$

A normalization is done to transform this slope-intercept form to the Hesse normal form.

$$\begin{aligned} \frac{m}{\sqrt{m^2+1}}x - \frac{1}{\sqrt{m^2+1}}y + \frac{b}{\sqrt{m^2+1}} &= 0 \\ x \cdot \cos\alpha + y \cdot \sin\alpha - p &= 0 \end{aligned} \tag{7}$$

Comparing this result to the Hesse normal form we calculated the angle α_{half} to the x-axis.

$$\alpha = \begin{cases} \arccos\left(\dfrac{m}{\sqrt{m^2+1}}\right) & \text{, if } b < 0 \\ \arccos\left(-\dfrac{m}{\sqrt{m^2+1}}\right) & \text{, else} \end{cases} \quad (8)$$

The range of inverse cosine function $arccos$ is $[0,\pi]$. To get an α_{full} with a range of $[0,2\pi]$ a last calculation is done. One of the shutter corner point $S_i = (x_i, y_i)$ is used to check if it lies on the line defined by α_{half}.

$$\alpha_{full} = \begin{cases} \alpha & \text{, if } x_i \cos(\alpha) + y_i \sin(\alpha) + p = 0 \\ \alpha + \pi & \text{, else} \end{cases}$$
$$(9)$$

Previously, we stated that $x_s \neq x_e$ needs to be true. If that is not the case α_{full} will be calculated as followed:

$$\alpha_{full} = \begin{cases} \pi & \text{, if } x_i > 0 \\ 0 & \text{, else} \end{cases} \quad (10)$$

2.3 Data Set

In order to quantitatively evaluate the performance of the algorithm, a reference data set was created containing hand images of three different projection types AP, PA and oblique. The data set consists of 426 AP/PA images and 396 oblique images. The images were labeled with a specific angle which was defined as angle between fingertips and x-axis. An example of some images and the corresponding angles is shown in Fig. 3. The images were randomly taken without considering their orientation.

2.4 Experiment

The question of the experiment is whether or not the images need to be rotated and if we can find the right rotation angle. Furthermore we measured the runtime of our automatic orientation detection. The images are only rotated with multipliers of 90 degree otherwise we would loss image informations. The angles were divided into four intervals as shown in Fig. 4.

Using this definition all images were labeled with one of the interval names. To measure the performance of our automatic orientation detection we compared its results to the reference classification of the data set.

As an evaluation tool we used a standard statistics contingency table [3] as illustrated in Table 1. On the horizontal axis are the reference results and on the vertical axis are the test results. The test is positive if the interval labels of our automatic orientation detection and the reference results are identical. It is negative if the labels are different. For example, if the reference angle α_{ref} is 226° and the calculated angle α_{auto} is 224°, the angles will be assigned to different regions and the test will be failed.

All calculation were performed on a computer with four cores and a frequency of 3.4 GHz (Intel Xeon Processor E3–1231 v3), 16 GB RAM and Windows 7 64-Bit. The

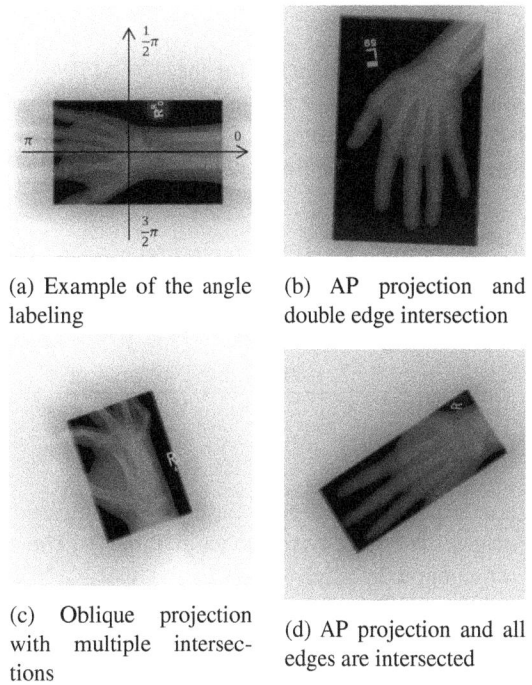

(a) Example of the angle labeling

(b) AP projection and double edge intersection

(c) Oblique projection with multiple intersections

(d) AP projection and all edges are intersected

Figure 3: (a) Visualization of the angle labeling. (b)-(d) X-ray hand images of different projections, orientations and shutter intersection counts. The corresponding angels are 266 degree, 260 degree and 212 degree in alphabetic order.

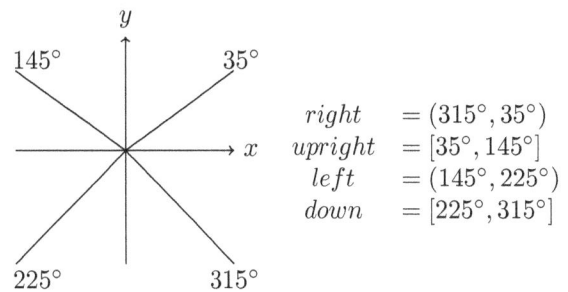

$$right = (315°, 35°)$$
$$upright = [35°, 145°]$$
$$left = (145°, 225°)$$
$$down = [225°, 315°]$$

Figure 4: Definition of the four angle regions which were used to decide if a rotation is necessary or not.

Table 1: Standard statistics contingency table which is used for evaluation.

	Rotation	No rotation
Test positive	True positive (tp)	False positive (fp)
Test negative	False negative (fn)	True negative (tn)

implementation was done in C++.

3 Results and Discussion

The experiment was carried out on the data set explained in section 2.3. In the first section the results are presented and the discussion is done afterwards.

3.1 Results

The experiment was evaluated for the 426 AP/PA projection images. We obtained results as presented in Table 2. For those images the accuracy of the automatic orientation detection was 100% and the average calculation time was 2 ms.

Table 2: Test results for AP/PA projection images. The total count is 426 images and the accuracy of our automatic orientation detection is 100%

	Rotation	No rotation
Test positive	250	0
Test negative	0	176

The same experiment was done for the 396 oblique projection images. The results are shown in Table 3. The automatic orientation detection had an accuracy of 95.71% and an average calculation time of 2 ms. Furthermore we calculated the specificity and recall of our algorithm.

$$Specificity = \frac{tn}{tn + fn} = 95.35\% \qquad (11)$$

$$Recall = \frac{tp}{tp + fn} = 95.98\% \qquad (12)$$

Table 3: Results for oblique projection images. The total count is 396 images and the accuracy is 95.71%

	Rotation	No rotation
Test positive	205	8
Test negative	19	164

3.2 Discussion

We tested our automatic orientation detection successfully for AP/PA projection images. The accuracy is 100% and the calculation time are promising. Our simple algorithm helped to reduce calculation times compared to the previous approach [2]. The data set had 250 images out of 426 which had to be rotated manually. Our automatic rotation reduces this to 0. This helps to improve the clinical workflow for hand images.

The results for oblique images are not as good as for AP/PA images. The accuracy of 95.71% is still good but the specificity of 95.35% is a remaining challenge. It can happen that the medical staff expects a upright image and our automatic detection rotates it in a wrong orientation. The problem with our approach is the assumption that the arm has the minimum median of all intersections. Furthermore we do not have an confidence index, which would help to reduce such a problem. Nevertheless our automatic detection reduces the manual rotation from 224 to 17. This is still a very good workflow improvement. But from a customer's point of view a false positive result is something they would not accept. The recall of 95.98% is not as bad as the specificity because either way the image

needs to be rotated manually.

4 Conclusion

Our findings demonstrate that the automatic detection works for hand images of AP/PA and oblique projections. A total accuracy of 97.93% for our data set is a good result. However, up to now the algorithm exhibits too many false positive. Nevertheless, the algorithm can improve the clinical workflow. In total we reduced the manual rotation from 474 to 17. This can save a lot of time.

The calculation time of 2 ms compared to 210 ms is a tremendous improvement. This was a necessary step for using an automatic orientation detection.

Our approach is based on the assumption that a hand structure is in the image. Extending the automatic orientation detection with anatomy detection would be a good improvement. Also a confidence index would be required for a medical application.

Acknowledgement

The work has been carried out at Philips Medical Systems DMC GmbH, Hamburg. I would like to thank C. Hamann and S. Wulfgramm for providing me the opportunity to do the internship and to work on such interesting topic.

5 References

[1] MarketsandMarkets, *X-Ray Detectors Market by Detector Type (Flat Panel, Computed Radiography, CCD, Line-Scan), Applications (Mammography, Cardiovascular Imaging, Dental, Security, Veterinary, Industrial) & by FPD Panel Size (Large Area, Small Area) – Global Forecast to 2019.* MarketsandMarkets, 2014.

[2] F. Werner, *Bestimmung der Orientierung anatomischer Strukturen in digitalen Röntgenaufnahmen.* Hamburg, 2014.

[3] D. M. Powers, *Evaluation: from precision, recall and F-measure to ROC, informedness, markedness and correlation.* Bioinfo Publications, 2011.

[4] L. Papula, *Mathematische Formelsammlung: Für Ingenieure und Naturwissenschaftler.* Springer Fachmedien Wiesbaden, 2014.

[5] H. Handels, *Medizinische Bildverarbeitung: Bildanalyse, Mustererkennung und Visualisierung für die computergestützte ärztliche Diagnostik und Therapie.* Vieweg+Teubner Verlag, 2009

Denoising spectral CT images using parallel level sets

J. Krooß [1], B. Brendel [2], T. Köhler [2], and M. Heinrich [3]

[1] Medizinische Ingenieurwissenschaft , Universität zu Lübeck, jakob.krooss@student.uni-luebeck.de
[2] Philips GmbH Innovative Technologies, Research Laboratories, D-22335 Hamburg
[3] Institute for Medical Informatics, Universität zu Lübeck, heinrich@imi.uni-luebeck.de

Abstract

Spectral CT is a new imaging method that measures, in addition to the photon flux, the energy of the transmitted photons. This has the advantage that not only the X-ray attenuation of a scanned object can be reconstructed, but also the material composition. However, reconstructed material images suffer from strongly amplified noise that is highly correlated between the images. Hence, denoising of the material images after reconstruction is required. The use of regularization terms like *Joint Total Variation* in image denoising leads to undesired smoothing and overshooting on edges. Thus, novel regularization terms are considered. This paper investigates the use of *Parallel Level Sets* to denoise aforementioned images, aiming for a better preservation of edge information. While being significantly slower regarding the convergence behavior of the tested denoising algorithms, image quality can be further improved in some cases.

1 Introduction

Conventional computed tomography (CT) is only capable of measuring the flux of photons. Recently, new methods for measuring, in addition, the energy of photons have been introduced.

One way to measure photon energy is the use of a dual layer detector with two detector arrays behind each other. Low energy photons are likely to be absorbed in the first detector layer, high energy photons in the second. A differentiation into only two energy levels makes sense, since the X-ray attenuation in human tissue is mainly due to two physical effects with different photon energy-dependency, namely photo electric effect and Compton scatter. Fig. 1 shows the energy-dependency due to the attenuation of Compton scatter and the photo electric effect.

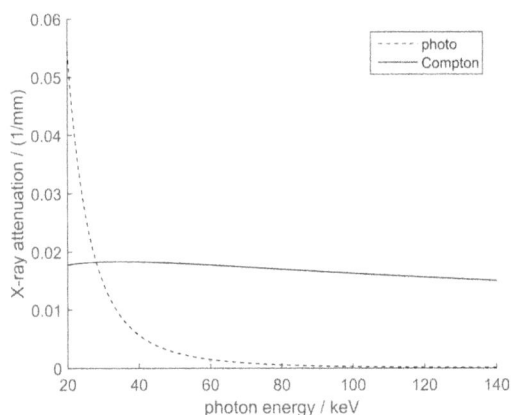

Figure 1: Energy dependency of Compton scatter and photo effect.

The obtained high and low energy data can be decomposed into attenuation portions of these two physical effects. Subsequently a photo electric effect image p and a Compton scatter image c are reconstructed from the decomposed data (called photo image and Compton image for brevity in the following). However, the decomposition is ill-posed and nonlinear, and results in amplified, but also correlated noise in both images. Thus, a denoising of the images has to be performed to increase the diagnostic value of the images.

One common method for joint denoising is statistical iterative denoising, which is performed as a minimization problem using, for example, the *Joint Total Variation* Regularization (JTV, 2.2.1) [1]. JTV is easy to implement and is converging in reasonable time but leads to undesired artifacts, mainly over- and undershooting on strong edges.

This paper evaluates the use of *Parallel Level Sets* (PLS) [1], [2] instead of JTV in order to improve the quality of denoising. Instead of minimizing the difference to neighboring pixels, PLS penalize nonparallel egdes in the photo and Compton images.

2 Material and Methods

As mentioned above, statistical iterative denoising is performed as a minimization problem. This requires two components: A cost function Φ that quantifies the quality for each denoised pair of photo and Compton images (lower value = better quality), and an iterative optimization method that changes the images in a way that the cost function reduces in every iteration. After convergence, the cost function should reach a minimum.

The cost function consists of a sum of a data term D, measuring the similarity of the denoised images p and c to the input images $\hat{p}$ and $\hat{c}$ (e.g., from filtered back projection reconstruction), and a regularization term R that enforces the denoising, by e.g. favoring piecewise constant images.

R can be weighted using a parameter β. A high β results in better smoothing but also less preservation of small detail.

2.1 Data Term

Considering that the noise in photo and Compton images is correlated, the data term uses a covariance matrix C of a multivariate Gaussian noise distribution:

$$D_{i,j} = \frac{1}{2} \begin{pmatrix} p_{i,j} - \hat{p}_{i,j} \\ c_{i,j} - \hat{c}_{i,j} \end{pmatrix}^T C_{i,j}^{-1} \begin{pmatrix} p_{i,j} - \hat{p}_{i,j} \\ c_{i,j} - \hat{c}_{i,j} \end{pmatrix} \quad (1)$$

with

$$C = \begin{pmatrix} \sigma_{i,j}^p & \sigma_{i,j}^{pc} \\ \sigma_{i,j}^{pc} & \sigma_{i,j}^c \end{pmatrix} \quad (2)$$

and i, j being the 2D coordinates in p and c. σ^p and σ^c denote the standard deviation of p and c, and σ^{pc} the covariance between p and c.

Due to the strong negative noise correlation in photo and Compton images, a larger difference between e.g., $\hat{p}$ and p is comparably cheap only if the difference between $\hat{c}$ and c has an opposite sign and a certain amplitude.

This can lead to undesired over- and undershooting artifacts: As the algorithm tries to denoise the images by e.g. minimizing differences between neighbouring pixels, and since the Compton image values are about ten times higher than those of the photo image, its cheaper for the regularization to attenuate edges in the Compton image while strengthening the edges in the photo image (Fig. 4 (a-b)).

2.2 Regularization Term

Minimizing the cost function without a regularization term would not change the image, since differences from input images are penalized by the data term. Regularization terms are used to improve image quality while still being aligned to the input images.

2.2.1 Joint Total Variation

A well known and often used example for regularization is *Joint Total Variation* [1]. Minimizing the differences between neighbouring pixels, it generally smoothes the whole image. As a drawback, edges are smoothed too.

$$R_{i,j}^{JTV} = \sqrt{\nabla_i p^2 + \nabla_j p^2 + \nabla_i c^2 + \nabla_j c^2 + \epsilon^2} \quad (3)$$

$\nabla_k x$ denotes the gradient in direction k in the image x. To avoid stripe artifacts, gradients are calculated asymmetrically, e.g. $\nabla_i p = p_{i+1,j} - p_{i,j}$. $\epsilon > 0$ ensures differentiability of the regularization term for the case that all gradients are zero.

2.2.2 Parallel Level Sets

It is common knowledge, that for all vectors $\nabla p = (\nabla_i p, \nabla_j p)^T$ and $\nabla c = (\nabla_i c, \nabla_j c)^T$

$$|\langle \nabla p, \nabla c \rangle| = |cos(\theta)| \cdot ||\nabla p|| \cdot ||\nabla c||, \quad (4)$$

θ denotes the angle between ∇p and ∇c.

Because of (4), for all vectors ∇p and ∇c,

$$R_{i,j}^{PLS} = ||\nabla p|| \cdot ||\nabla c|| - |\langle \nabla p, \nabla c \rangle| \geq 0 \quad (5)$$

with equality only if ∇p and ∇c are parallel or zero. Hence, (5) can be used as a measure for parallelism.

This does not change when using a more generalized approach [1], [2]:

$$R_{PLS} = \psi(\phi(||\nabla p|| \, ||\nabla c||) - \phi(|\langle \nabla p, \nabla c \rangle|)) \quad (6)$$

with ψ and ϕ being strictly increasing functions $[0, \infty) \to [0, \infty)$. In the following, $\psi(s) = \sqrt{s}$ and $\phi(s) = s^2$ are used. There are other possibilities, but this is out of the scope of this paper.

As in 2.2.1, ϵ is used to ensure global differentiability of the regularization term:

$$|\langle \nabla p, \nabla c \rangle| = \sqrt{(\nabla_i p \nabla_i c + \nabla_j p \nabla_j c)^2 + \epsilon^4} \quad (7)$$

$$||\nabla p|| = \sqrt{\nabla_i p^2 + \nabla_j p^2 + \epsilon^2}. \quad (8)$$

PLS as cost function is not convex, thus local optima might occur.

To get a global measure, the cost function is integrated over the whole image:

$$\Phi = \sum_i \sum_j D_{i,j} + \beta R_{i,j} \quad (9)$$

2.3 Optimization

Numerous optimization algorithms have been implemented and investigated in this work. All of them need the first derivative for each pixel and each image. So called Newton methods can provide better convergence speed but need additionally the second derivative. L-BFGS estimates the second derivative with first derivative information.

Detailed information and formulas for all algorithms can be found in the referenced literature.

2.3.1 Nonlinear Conjugated Gradient

The *Nonlinear Conjugated Gradient* method (NCG) [3] uses the fact that the negative first derivative $-\nabla f(x)$ is always pointing towards a descent direction which will lead to a decrease of the cost function.

Since the negative gradient is not pointing directly to the minimum, it can be advantageous to use nonorthogonal (*conjugated*) search directions in each iteration. It might be useful at some point to reset and restart with a new, unconjugated gradient descent.

As $-\nabla f(x)$ just gives the direction of descent, an additional line search has to be performed to find the right stepsize α. For this, the *Golden Section Search* (2.3.4) is used.

2.3.2 Iterative Coordinate Descent

The main problem with NCG is that all pixels are updated with the same stepsize. Thus, it is very difficult for the algorithm to perform big changes for single pixels.

The *Iterative Coordinate Descent* (ICD) [4] is updating every pixel individually. Consequently, for each pixel only a two-dimensional optimization problem has to be solved (one dimension for the change in each image).

2.3.3 Limited-Memory-BFGS

BFGS [5] is a quasi-Newton algorithm that approximates second derivative information by also evaluating updates of previous iterations. From these, the local curvature (Hessian) of the cost function can be approximated without knowing the exact second derivative. Similar to NCG, all pixels are updated simultaneously in one iteration.

Since a fully populated Hessian would require a lot of memory ($(2k)^2$ entries, with k being the number of pixels), the "Limited Memory"-version of BFGS is used. L-BFGS does not approximate the Hessian H itself, but the product $H\nabla f$, which is of the size $2k$.

2.3.4 Golden Section Search

All discussed optimization methods rely at a certain point on a line search algorithm to determine the correct stepsize α. The *Golden Section Search* [6] calculates the cost function for different stepsizes α and is constantly converging to the minimum by setting new search values in the golden section ratio $(0.39 : 0.61)$ until the optimal stepsize α is found.

2.4 Phantoms

For testing purposes, a numerical phantom is used. It is shown in Fig. 2. The signal (and thus SNR) in the Compton image is about ten times higher than in the photo image. Strongly correlated noise (correlation coefficient of -0.95) is added to the noise free phantom images to evaluate the denoising properties of the different algorithms.

It is worth mentioning that the used phantoms cannot represent the complexity of real clinical data. While giving good information about general behaviour of algorithms, convergence speed and image quality may change for real data.

3   Results and Discussion

To test convergence behaviour and to compare the different optimization methods, multiple tests have been performed. The maximum number of iterations was set to 10000. For

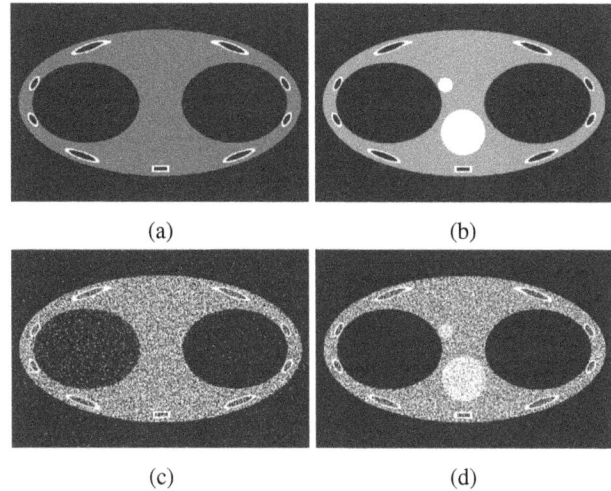

(a)   (b)

(c)   (d)

Figure 2: (a) Phantom of the photo image p. (b) Phantom of the Compton image c. (c) Phantom of the photo image p with added noise. (d) Phantom of the Compton image c with added noise. Images are windowed between -910 and -890 HU for p and -130 and -55 HU for c.

PLS, β was set to 0.3 and ϵ to 0.3 HU. To achieve the same noise level with JTV, β was set to 0.1 and ϵ to 0.1 HU.

Commonly, the noisy material images are used as input data $\hat{p}$ and $\hat{c}$ and start images p and c. In some cases however, using gaussian smoothed or empty (zeros) images as start images p and c can speed up and stabilize the convergence process of the optimization. Thus, different start images have been tested. Since ground truth is known, it can be used as a start image to evaluate the convergence behavior. L-BFGS has been used with 25 saved iterations, NCG with resets after 1000 iterations. While JTV converges quite fast with all tested algorithms, PLS did not fully converge with any method within 10000 iterations.

3.1   Convergence Behaviour

Table 1 shows the cost function after 10000 iterations for a selection of the tests performed with PLS. Fig. 3 shows the progress of the cost function over the iterations.

Since the cost function values for the different start images (performed by L-BFGS) are very similar, it is likely that they all converge to the same image. For complex images, this is not guaranteed. Indeed, images with multiple minima have been observed in tests with other data.

Table 1: Final cost function after 10000 iterations for all tested algorithms using PLS.

Method	starting image	cost function
NCG	noisy	460,095
ICD	noisy	508,523
L-BFGS	noisy	459,510
L-BFGS	smooth	459,533
L-BFGS	ground truth	459,465
L-BFGS	zeros	462,543

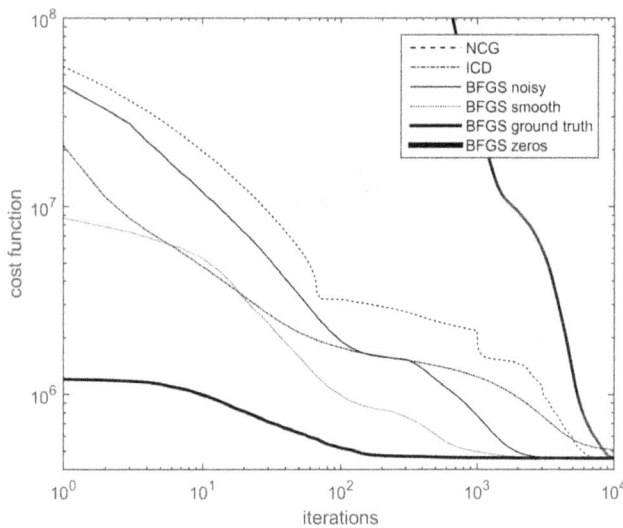

Figure 3: Behavior of the cost function for different algorithms using *PLS*. The cost function decreases with different speed.

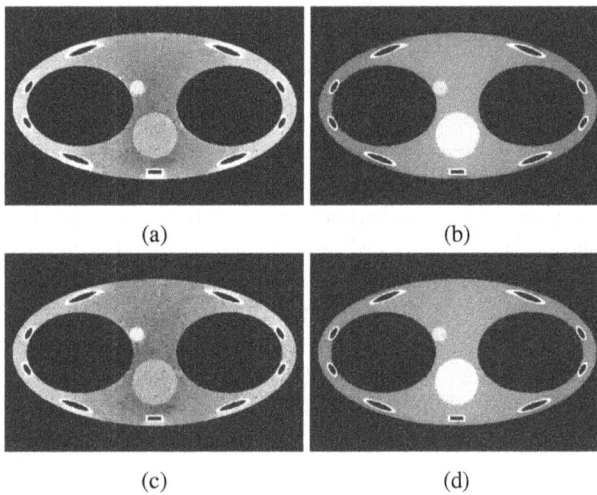

(a)   (b)

(c)   (d)

Figure 4: (a) p after denoising with *JTV*. (b) c after denoising with *JTV*. (c) p after denoising with *PLS*. (d) c after denoising with *PLS*. All images are windowed as in Fig. 2.

While performing much better at the start, the final outcome of the algorithm performs worse when initialised with smoothed images. This is probably due to the loss of high frequencies in the smoothing process.

Obviously, for a zero-filled start image, the cost function is very high at the beginning. Nevertheless, after enough iterations, minimization seems to yield the same results.

All tested mimization methods converge slowly, with L-BFGS being slightly faster than NCG. ICD performs good in the first iterations but is worse than both L-BFGS and NCG after a certain amount of iterations.

3.2 Comparison to Joint-TV

Fig. 4 shows the denoised images for both JTV and PLS. While providing approximately the same denoising properties, PLS images show less artifacts. The enhancement

of thin structures in the JTV images (Fig. 4a, left and right side) is clearly reduced in the PLS images (Fig. 4c). Crosstalk can be seen in both cases, but is slightly less noticeable in the PLS images. Over- and undershooting is clearly visible in the JTV images but barely present in the PLS images.

On the other hand, while JTV images appear to be rather smooth, the noise structure of PLS is patchy and thus unfavourable for clinical applications.

4   Conclusion

Considering processing time, convergence speed has to be further improved to yield results in acceptable time.

Apart from that, PLS might be an alternative to conventional JTV regularization.

The main argument for PLS is the reduction of artifacts caused by the covariance in the dataterm. While minimizing differences between pixels smoothes borders and thus produces overshooting artifacts (JTV), PLS tries to align edges. This leads to small, irregular patches withuniform value, which greatly reduces overshooting artifacts, but leads on the other hand to a patchy noise structure.

Furthermore, the possibility of multiple minima is potentially disadvantageous for clinical applications.

Thus, further testing with real clinical data has to be done.

Acknowledgement

The work has been carried out at Philips GmbH Innovative Technologies, Hamburg.

5   References

[1] M. J. Ehrhardt, K. Thielemans, L. Pizarro, D. Atkinson, S. Ourselin, B. F. Hutton, and S. Arridge, "Joint reconstruction of pet-mri by exploiting structural similarity," *Inverse Problems*, vol. 31, no. 1, p. 015001, 2015.

[2] M. J. Ehrhardt and S. Arridge, "Vector-valued image processing by parallel level sets," *Image Processing, IEEE Transactions on*, vol. 23, no. 1, pp. 9–18, 2014.

[3] J. R. Shewchuk, "An introduction to the conjugate gradient method without the agonizing pain," tech. rep., 1994.

[4] J.-B. Thibault, K. D. Sauer, C. A. Bouman, and J. Hsieh, "A three-dimensional statistical approach to improved image quality for multislice helical ct," *Med. Phys.*, vol. 34, 2007.

[5] J. Nocedal and S. J. Wright, *Numerical Optimization*. Springer, 2006.

[6] W. H. Press, S. A. Teukolsky, W. T. Vetterling, and B. P. Flannery, *Numerical Recipes*. Cambridge, 2007.

Using Convolutional Neural Networks
for Cut Line Detection in Medical Images

T. Langner [1] and T. Käster [2]

[1] Medizinische Informatik, Universität zu Lübeck, taro.langner@student.uni-luebeck.de

[2] Pattern Recognition Company GmbH, Lübeck, tk@prcmail.de

Abstract

A Convolutional Neural Network was used to design a detection algorithm which is able to locate features of a grid-like structure in images. Developed at the *Pattern Recognition Company GmbH* (PRC) in Lübeck, it allows for the automation of a production step used by an industry partner. For the design of the algorithm, multiple Convolutional Neural Networks were trained to perform a recognition task on image patches containing the desired features. Using the most successful resulting network, a detection algorithm with a search strategy was then developed to extract these patches from the base image and supply them to the network for classification. By comparing the results of the detection to manually placed annotations, an evaluation was carried out on a set of test data.

1 Introduction

By applying concepts of machine learning to the field of pattern recognition, it has become possible to simplify the process of designing solutions and powerful systems capable of recognizing abstract image features [1] . In contrast to traditional approaches with a hand-crafted structure of feature extraction and classification, *Deep Learning* allows for the training of models that unite these steps automatically. Modern *Convolutional Neural Networks* are able to extract hierarchical features from images, allowing for ever more abstract properties to be algorithmically processed. Today, networks of this kind are for example applied to the recognition of hand-written digits [2] and traffic signs [3] and are leading in competitions like the ImageNet Recognition Challenge (ILSVRC)[4].

Among the solutions offered by the *Pattern Recognition Company GmbH* (PRC) for computer vision and data analysis, approaches based on machine learning play a prominent role. The algorithm developed at PRC in the context of this project aims to automatize a processing step for an industry partner. In this production step, slides covered with cell cultures for medical diagnostics are first marked with a grid pattern and then cut by a machine along the markings. After the cutting procedure, the original image areas need to be re-identified in the resulting piece, as the mechanical stress usually leads to deformations. This goal is achieved by finding the grid points where the cut lines meet, which can then be compared to the original markings. For this task, a Convolutional Neural Network was trained to recognize image patches that are centred around a grid point. This network was then utilized in a detection algorithm to allow for the search of grid points and locating their posi-

tions within the image. In this paper, a description of the training process for the recognition task as well as a presentation of the detection algorithm and its evaluation on test data is conducted.

2 Material and Methods

For the design of the algorithm three grey value images were provided by the industry partner. Each of them was obtained as a scan from production, showing a slide covered with cell cultures that has been cut by a machine along a rectangular grid pattern. Due to the high resolution of around 9000×6000 pixels there are a high number of positions where the lines of the grid cross, yielding about 1500 grid points in each image. The positions of these grid points vary between the images, so that they can not directly be copied from one to another. Additionally, the grid lines are not perfectly aligned with the image edges but may be slightly rotated. Due to being composed of different scanned segments, there are occasional discontinuities in the lines of the grid where they are shifted by multiple pixels. In some areas, the seams of the scan also appear as streak artefacts, forming unusually bright lines with seemingly arbitrary grey values. Further artefacts are caused by other grid-like structures that are not part of the cutting process as well as air inclusions. The heterogeneous distribution of the cell cultures also leads to a high variance of brightness and contrast within and between the images. Strong edges and various structures are visible in the cell cultures that form the background. A segment of one of the images is seen in Fig. 1.

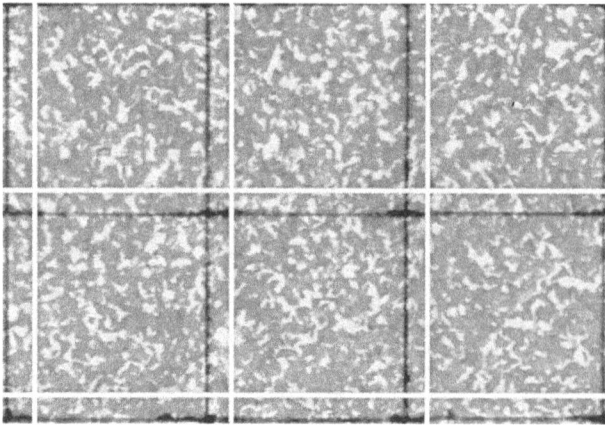

Figure 1: Goal of the detection is to find the correlation between the white grid denoting the original markings and the black grid that resulted from the cutting process. Based on the results, a corresponding transformation between both grids can be calculated in a later processing step.

A classical solution to finding the grid points would utilize the gradients of the image. Using an edge-detection filter, the grid lines can be extracted. To isolate the actual lines from the background structures, the *Hough transform* can be used [8]. The detected edge pixels are thereby transformed into a *Hough space*, where they are represented as sinusoids parametrized by their coordinates in the image. In this space, their alignment on lines in the original image space can be expressed as intersections of the sinusoid representations.

An alternative approach consists in defining a template that describes either the actual grey values expected in the area around a grid point or a representation of its features like gradients. By using the technique of *template matching* [10], a given image area can be compared to this template. A high degree of similarity is considered to indicate a successful detection.

A technique similar to feature-based template matching is based on a *Histogram of Oriented Gradients* (HOG). Among other applications, this representation of local shape has been used for human detection [9]. By defining a descriptor based on the expected gradients and their directions around the desired features, their occurrence in the image is found. As the arbitrary gradient directions of the background are rejected, this method is potentially more robust with regard to false positives.

All of the listed approaches require a formal description of the relevant features as well as a classifier of some form to be defined during their design. The search for reliable features would furthermore be challenged by the high amount of clutter observed in the images. By using a Convolutional Neural Network for the recognition task, the approach presented in this paper aims to automatize these steps. The data-driven training process is instead expected to allow the network to determine the relevant features and the best parameters for the classification by learning them.

2.1 Used Frameworks

Caffe is a deep learning framework developed by the Vision and Learning Center of the University of California, Berkeley [7]. It allows for the training of neural networks and offers pre-trained models and example cases for the classification of benchmark data sets.

The browser-based interface *DIGITS* developed by Nvidia integrates Caffe and offers visualizations for the assembly of datasets as well as the configuration and training of networks. The included cuDNN library developed by Nvidia accelerates the training process by optimizing the required calculations for GPU processing. Although DIGITS allows for existing datasets to be converted to a format suitable for Caffe, it does not provide the functionality required to extract training data patches from an image. For this purpose, a custom tool was implemented in C++ using the application framework *Qt*. This tool enables the user to place markers as annotations on a loaded image. Based on the selected positions, image patches with desired parameters like size, rotation and shift can be exported as individual images.

2.2 Recognition

In order to automatically find the grid points contained in the input images, a Convolutional Neural Network was trained to identify the intersection of two grid lines in the centre of a given image patch. The model used for this recognition task is based on the network architecture by *Krizhevsky et al.* [5] with five convolutional and three fully-connected layers using dropout [6].

For the input data, a training set of patches was manually extracted from the images, forming two classes. The first class consists of image patches with a central grid point. Any patch containing either an off-centre grid point or none at all was assigned to the second class. Examples of patches for both classes are seen in Fig. 2. The validation set was assembled in the same way, without using those source images that provided the patches of the training set.

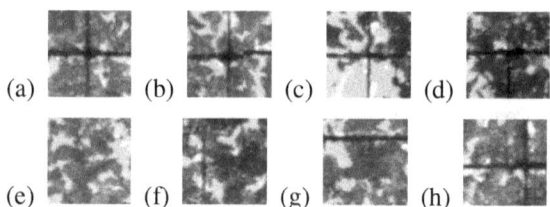

Figure 2: Image patches of 65x65 pixels used for the training of the networks with grid points (a-d) and counter examples (e-h)

2.3 Detection

With a classifier for grid points being thus obtained, an algorithm was developed to supply it with patches from the input image. By classifying a patch from a certain position as being centred on crossing grid lines, the network predicts a grid point to be located there. However, a naive scan across the entire image, with all possible patches being classified, proved to be too time consuming to be viable. Instead, the number of considered patches had to be reduced by introducing a searching strategy using assumptions about the grid structure.

The resulting algorithm attempts to locate grid points in search spaces around their estimated positions. For the first point, a large search space is used in the upper left image corner, as only few assumptions can be made about its position. Next, a neighbouring point is searched at the horizontal distance of an estimated edge length of a grid cell. As more points are found, the estimate for this distance is further refined based on the previously found positions. For each inspected position all the patches contained in the search space around it are supplied to the network. Of the patches being classified as grid point the one with the highest confidence rating is chosen as the result. In this way the algorithm proceeds from row to row until the entire image is processed.

2.4 Evaluation

An evaluation of the accuracy of the network in the recognition task was easily possible in Caffe by examining the ratio of correctly classified image patches in a given data set.

To evaluate the results of the detection task, the locations of the grid points predicted by the algorithm can be compared to a ground truth representing the correct placements. The conformity between the pairs of corresponding placements can be rated with an overlap measurement based on the *PASCAL Visual Object Classes Challenge* [11]. With a rectangular box for the detection result B_p and the ground truth B_{gt} the overlap a_0 can be defined according to equation (1).

$$a_0 = \frac{area(B_p \cap B_{gt})}{area(B_p \cup B_{gt})} \qquad (1)$$

A detection is considered successful if the value of a_0 for a pair of both a predicted and a ground truth annotation exceeds a given threshold.

3 Results and Discussion

During the training process of the Convolutional Neural Network, a high accuracy was reached for the predictions of the classes on the validation set. Multiple variations of the network with different training sets and parameters were created, with the best classifying $99,6\%$ of the validation set correctly. Due to the relatively small amount of training samples, different methods of data augmentation were attempted. By using variations created by mirroring and rotating the patches, the size of the training sets was vastly increased. The effect on the accuracy in recognition and eventual performance on the detection task was negligible, however, and the best-performing network was trained without any form of augmentation.

To evaluate the performance of the detection algorithm, it was run on one of the provided images. As no further data was available, this test was executed on the same image that had previously provided the validation data. The ground truth data set used for the evaluation is therefore composed of the same patches that were used for the validation of the network before. The ground truth data set that the resulting placements were compared to does accordingly contain the same patches that the accuracy of the network was tested against during its training. With these two sets of placements the evaluation according to (1) was possible, so that the overlap between the generated and manually provided placements was inspected. Rather than just using a single threshold for a_0, a range of values was used. Fig. 3 shows the ratio of annotation pairs exceeding the respective amount of overlap plotted against values for a_0.

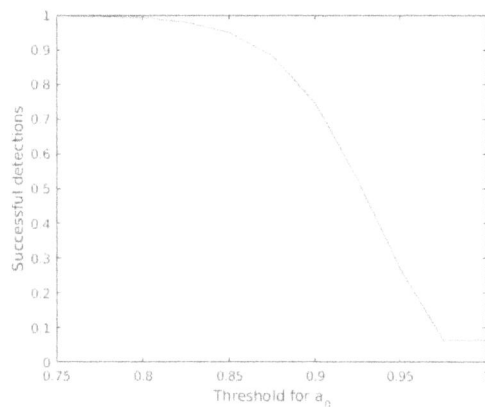

Figure 3: The percentage of successful detections by the algorithm plotted against a range of thresholds for the required minimum ratio of overlap a_0 according to (1)

Due to the low resolution of the patches, an overlap of $97,5\%$ and greater corresponds to every pixel of both annotations being at the same position, thus achieving complete coverage. Although even for the best network only about 7% of the generated annotations reach this value, it can not be directly concluded that their placements are inaccurate. In the majority of cases the grid lines are multiple pixels wide. Despite being used as ground truth, the manual placements therefore only represent one of multiple equally valid solutions. In some of the cases that failed the threshold test for lower values of a_0 it was observed that the manual placement was indeed visually less plausible than the generated one.

When comparing the differently trained networks, large differences in the detection results were observed. Despite the high accuracies during the training process, some networks predicted the wrong class when used within the detection algorithm, producing false-negatives and false-positives. Examples of these failed classifications are seen in Fig. 4. Since the search strategy depends on the previously found grid points, these errors had a strong effect on the result.

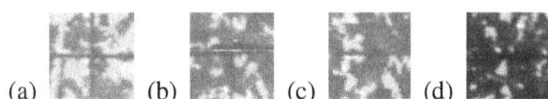

(a)   (b)   (c)   (d)

Figure 4: Image patches being recognized as false negatives (a, b) and false positives (c, d) by variations of the network

For false-negative patches, where a grid point was not recognized as such and could not be found near its estimated position, the algorithm therefore had to be adjusted. After an unsuccessful search in increasingly large search spaces, the grid point is eventually assumed to be at the centre of the search space. False-positive patches, where a patch is falsely classified as a grid point, likewise proved to be a hazard for the searching strategy. After a false-positive patch is mistakenly classified as a grid point, the following points are often searched at positions off of the grid. This problem is aggravated by the tendency of the network training process to favour classifications with high confidences. An improvement was achieved by reducing the size of the search space and relying more on the empirical edge lengths previously encountered during the searching process.

When using the best network, no patches were classified as false-positives or false-negatives and almost 95% of the placements match the ground truth with an overlap of at least 85%. It can therefore be concluded that the algorithm, while highly dependent on the quality of the network, is able to fulfil the detection task with high precision on the test image.

4   Conclusion

The Convolutional Neural Network trained for the recognition task proved to be highly potent in the classification of the image patches, reaching accuracies of up to $99, 6\%$ on the validation set. Even on visually ambiguous and degenerate cases the best-performing network yielded correct results. The original goal of solving the detection task with the help of the learning ability of the network alone was only partially achieved, as an algorithm with a searching strategy had to be developed to allow for an acceptable computation time.

The resulting detection algorithm as a whole was able to achieve high quality results. The evaluation according to (1) shows a high degree of overlap with the results of a manual detection. Based on a closer inspection of the mismatching cases it can be assessed as being on par with the original human performance on the test data. However, the results of the searching algorithm are highly dependent on the performance of the network.

A larger training set would potentially allow for even more accurate predictions by the network as well as more extensive testing of the detection algorithm. Given further images, the detection approach as a whole could therefore be made more robust and generalized and its precision in the placements further improved.

Acknowledgement

This work has been carried out at the *Pattern Recognition Company GmbH* (PRC) in Lübeck.

5   References

[1] K. Fukushima. *Neocognitron: A self-organizing neural network model for a mechanism of pattern recognition unaffected by shift in position* in: Biological cybernetics, 36(4), 193-202, 1980

[2] Y. LeCun, L. Bottou, Y. Bengio, P. Haffner. *Gradient-based learning applied to document recognition* in: Proceedings of the IEEE, 86(11), 2278-2324, 1998.

[3] D. Cireşan, U. Meier, J. Masci, J. Schmidhuber. *Multi-column deep neural network for traffic sign classification* in: Neural Networks, 32, 333-338, 2012.

[4] O. Russakovsky, J. Deng, H. Su, J. Krause, S. Satheesh, S. Ma, Z. Huang, A. Karpathy, A. Khosla, M. Bernstein, A. C. Berg, L. Fei-Fei. *Imagenet large scale visual recognition challenge* in: International Journal of Computer Vision, 1-42, 2014.

[5] A. Krizhevsky, I. Sutskever, G. E. Hinton. *Imagenet classification with deep Convolutional Neural Networks* in: Advances in neural information processing systems, 2012, pp. 1097-1105.

[6] G. E. Hinton, N. Srivastava, A. Krizhevsky, I. Sutskever, R. R. Salakhutdinov, *Improving neural networks by preventing co-adaptation of feature detectors* in: arXiv preprint arXiv:1207.0580, 2012.

[7] Y. Yangqing, E. Shelhamer, J. Donahue, S. Karayev, J. Long, R. Girshick, S. Guadarrama, T. Darrell. *Caffe: Convolutional Architecture for Fast Feature Embedding* in: arXiv preprint arXiv:1408. 5093, 2014.

[8] P. V. Hough. *Method and means for recognizing complex patterns* (No. US 3069654), 1962.

[9] N. Dalal, and B. Triggs. *Histograms of oriented gradients for human detection.* in: Computer Vision and Pattern Recognition, 2005. CVPR 2005. IEEE Computer Society Conference on. Vol. 1. IEEE, 2005.

[10] R. Brunelli. *Template matching techniques in computer vision: theory and practice.* John Wiley & Sons, 2009.

[11] M. Everingham, L. Van Gool, C. K. Williams, J. Winn, A. Zisserman. *The pascal visual object classes (voc) challenge* in: International journal of computer vision, 88(2), 303-338, 2010.

10

Image Processing II

Detection of Non-rigid Motion, while Conducting a Static 3D Scene Reconstruction in Realtime

S. Löw [1], S. Klement [2], D. Münster [2], and E. Barth [3]

[1] Medizinische Ingenieurwissenschaft, Universität zu Lübeck, Sebastian.Loew@student.uni-luebeck.de
[2] gestigon GmbH, Lübeck, {Sascha.Klement,Dennis.Muenster}@gestigon.de
[3] Institut für Neuro- und Bioinformatik, Universität zu Lübeck, barth@inb.uni-luebeck.de

Abstract

Using depth data alongside conventional amplitude images to detect and track motion is an established approach, since it provides more robust features than 2D data [1]. One of the constraints of using only momentary depth data is the effect of occlusion, which could lead to severe loss of information when tracking motion (i.e. movement of a person). A possible way to prevent this would be to reconstruct the scene in a volumetric manner. Conventional 3D-Reconstruction algorithms like KinectFusion [2] are able to reconstruct a scene in realtime using only depth data. The main constraint of this process however is the requirement of an entirely static scene, which prevents the detection of non-rigid motion. In this paper we present our approach to detect non-rigid motion while reconstructing the scene, leading to a volumetric representation of the scene or person retaining the information independent of occlusions, which is crucial for motion tracking.

1 Introduction

Microsoft's KinectFusion is an algorithm used for reconstructing static 3D scenes in realtime using only depth data and the capabilities of modern GPUs. It offers the possibility to reconstruct an arbitrary 3D object simply by scanning it from different directions with the depth sensor. The *kinfu implementation* of the open source PCL (Point Cloud Library) was used as a basis for this study, since it already provides the volume integration, camera tracking and raycasting of the scene, which are the main algorithms needed for the static scene reconstruction. We integrated the algorithm into a processing pipeline, providing a visualization of the reconstructed scene in form of a point cloud and rendered images from arbitrary directions. Further, we modified the algorithms to allow changes in various parameters regarding either detail and position of the reconstruction volume or the peformance and accuracy of the camera pose estimation, without having to restart the application.

Lastly, we engaged the idea of [3] to use ICP outliers, which are obtained in the camera tracking process to detect drastically changing elements in the scene. These are used to estimate regions of non-rigid motion, which are afterwards separated from the reconstruction process.

2 Material and Methods

In this section, we describe the main algorithms to achieve the 3D reconstruction of a scene. The PCL implementation is based on [2] and makes use of the NVidia Cuda Toolkit for parallel programming to improve the performance. In our experiments we used an NVidia GeForce 650 TI with 2 GB VRAM.

2.1 Camera tracking

The first problem with this reconstruction approach is the absence of a motion sensor, which means there is no other source of information besides depth data about the relative movement and orientation between the sensor and the global coordinate system. The knowledge of the exact position in relation to its environment is essential for the integration of incoming depth data into the estimated surface saved in the reconstruction volume. This requires a transformation from the camera coordinate system into the global one.

The position and orientation of the sensor can be represented by a transformation matrix T_k^g containing a rotation matrix $R_k \in \mathbb{R}^{3\times3}$ and a translation vector $t_k \in \mathbb{R}^3$.

$$T_k^g = \begin{bmatrix} R_k & t_k \\ \vec{0}^T & 1 \end{bmatrix} \tag{1}$$

In every frame an incremental transform $T_k^{g,inc}$ that is composed onto the transform of the previous frame T_{k-1}^g needs to be estimated to get the current transformation matrix T_k^g. To achieve this a vertex- and normal map are calculated given the incoming bilateral filtered depth map $D_k(\vec{u})$ [4], where $\vec{u}$ indicates the coordinates $(u,v)^T$ on the sensor plane. The vertex map V_k can be obtained through back-projection of the depth values using the the homogeneous vector $\vec{u}^h = (\vec{u}^T|1)^T$ and the camera calibration matrix K holding the intrinsic parameters of the sensor,

$$V_k(\vec{u}) = D_k(\vec{u})K^{-1}\vec{u}^h. \tag{2}$$

The surface normals, which are stored in the normal map N_k, are then calculated for each vertex by a principal component analysis using the covariance matrix created from the nearest neighbors of that vertex.

Additionally, the already existing reconstruction volume is used to obtain the global vertex- and normal map $\hat{V}^g_{k-1}$ and $\hat{N}^g_{k-1}$ with the same resolution as the other maps (further discussed in 2.3). Note that these correspond to the surface from the previous frame, because the camera has moved since the last volume update.

Given these maps, an *iterative closest point algorithm* (ICP) can be performed to fit the measured data into the surface estimation, yielding the transformation matrix T^g_k that minimizes

$$ E = \sum_{\vec{u}} \left\| \left(T^g_k V^h(\vec{u}) - \hat{V}^g_{k-1}(\hat{u}) \right)^T \hat{N}^g_{k-1}(\hat{u}) \right\|_2, \quad (3) $$

where $V^h(\vec{u}) = (V(\vec{u})^T | 1)^T$ and $\hat{u}$ represents the corresponding coordinates of $\vec{u}$. The sum of the energy function $E(T^g_k)$ only contains summands of correspondency pairs $\{V(\vec{u}), \hat{V}^g_{k-1}(\hat{u})\}$, which are found via a distance- and angle threshold and is then minimized in an iterative manner, which will not be explained in further detail.

2.2 Volume integration

Once the transformation matrix T^g_k is known, the connection between newly observed data and the reconstruction volume is given. To understand the process of volume integration it is necessary to know how the surface is actually represented.

The algorithm makes use of a volumetric *truncated signed distance function* (TSDF), which basically consists of values in the range $[-1, 1]$, indicating the surface at the zero crossing, whereas negative values indicate voxels behind the surface and positive values indicate voxels in front of the surface (see Fig. 1). Each voxel center $\vec{p} \in \mathbb{R}^3$ of a predefined volume is assigned a value $F(\vec{p})$ in the given range, which is stored on global GPU memory. To perform an update of the volume, the incoming data has to be of the same data structure as the stored TSDF volume. Therefore the so called *projective TSDF* is calculated as follows:

$$
\begin{aligned}
F^p_k(\vec{p}) &= \Psi \left(\lambda^{-1} \|t_k - \vec{p}\|_2 - D_k(\vec{u}) \right), \\
\lambda &= \|K^{-1}\vec{u}^h\|_2, \\
\vec{u} &= \lfloor \pi \left(K(T^g_k)^{-1}\vec{p} \right) \rfloor, \\
\Psi(x) &= \begin{cases} \min(1, \frac{x}{\mu})sgn(x) & x \geq -\mu \\ 0 & \text{otherwise} \end{cases}
\end{aligned}
\quad (4)
$$

The projected pixel coordinates $\vec{u}$ are determined with the nearest neighbour operator $\lfloor \cdot \rfloor$, while $\pi(\cdot)$ is used for dehomogenization after the transformation and λ for converting the distance between $\vec{p}$ and the sensor to a depth value, that is then compared with the measured depth $D_k(\vec{u})$.

In addition to $F(\vec{p})$ another value $W(\vec{p})$ is taken into account for the volume update, which denotes a weight for

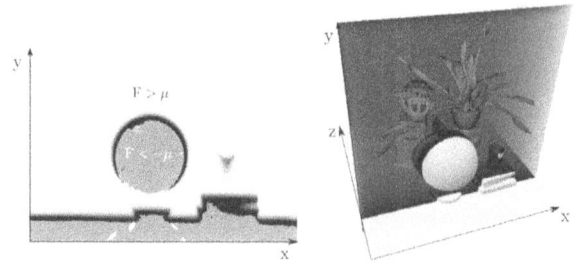

Figure 1: This figure shows a slice through the *truncated signed distance function* (TSDF), with voxels that have not yet had a valid measurement or are shrouded by a surface ($F < -\mu$) and voxels in front of the surface that had a valid measurement ($F > \mu$). The dark grey values show the area of the zero crossing and indicate the uncertainty of the surface estimation [2].

each point $\vec{p}$ in the volume. In each frame a tupel $F^p_k(\vec{p})$ is calculated,

$$
\begin{aligned}
F_k(\vec{p}) &= \frac{W_{k-1}(\vec{p})F_{k-1}(\vec{p}) + F^p_k(\vec{p})}{W_{k-1}(\vec{p}) + 1} \\
W_k(\vec{p}) &= W_{k-1}(\vec{p}) + 1
\end{aligned}
\quad (5)
$$

while $W(\vec{p})$ is initialized with 1 and incremented by 1 each frame. By doing this the update of the global TSDF becomes the arithmetic mean of all previous measurements. This procedure results in a denoised surface representation.

2.3 Raycasting

The raycast used in this algorithm serves multiple purposes. First, it is used for estimating the surface, which has been reconstructed so far in form of a vertex- and normal map $\hat{V}^g_k$ and $\hat{N}^g_k$ with the given sensor resolution and perspective. As described in section 2.1 these are important for the camera pose estimation step. Secondly, these maps can be used for visualizing the surface afterwards.

For each pixel coordinate $\vec{u}$ of the sensor plane there is a set of voxels, which lie on the corresponding ray $T^g_k K^{-1}\vec{u}^h$. Assuming an already reconstructed surface is present, for some $\vec{u}$ this set of voxels contain both negative and positive values $F(\vec{p})$ as described in (4). Starting at a certain minimal distance from the sensor the algorithm marches along the ray until either the reconstruction volume is exited or the sign of $F(\vec{p})$ changes, which means the surface has been reached. Once the position $\vec{p}$ of the zero crossing is found, it can be transformed into the camera frame to obtain the vertex $\hat{V}^g_k(\vec{u})$. The normal map is estimated by calculating the gradient $\nabla F_k(\vec{p})$ of the TSDF at $\vec{p}$, since the normal vector at that point is assumed to be orthogonal to the gradient.

2.4 Detecting non-rigid motion

Our approach of detecting non-rigid motion is using *ICP outliers*. These are represented by a set of pixel coordinates of the sensor plane, which are not included in the correspondency pairs found in the pose estimation step and thus saved in an *outlier map*. On the one hand this is caused by noise and uncertain measurements, especially noticeable on the edges of depth segments, where the surface is nearly orthogonal to the sensor planes normal vector. On the other hand ICP outliers will be visible when non-rigid motion occurs in the scene. The challenge is the distinction of the two scenarios. That means for the segmentation of the pixel coordinates related to non-rigid motion, some preprocessing steps have to be executed on the outlier map. Our approach was to use morphological operators on the map to erode smaller patches and eliminate connections inbetween them. Afterwards, a connected component algorithm [5] counts and labels all patches that are isolated from one another to distinguish between larger and smaller ones. Removing the smaller patches from the outlier map results in a mask holding only the coordinates of the pixels corresponding to nonrigid motion. A dilation operation in the end reverses the loss of information on the relevant patches, caused by the erosion step.

3   Results and Discussion

Given the surface estimates, only a few steps are necessary to render an image. With an artificial light source, which is ideally set to the viewers/sensors position, a weight is computed as a scalar product of the normals from the normal map $\hat{N}_k^g$ and the normalized position vector of the light source.

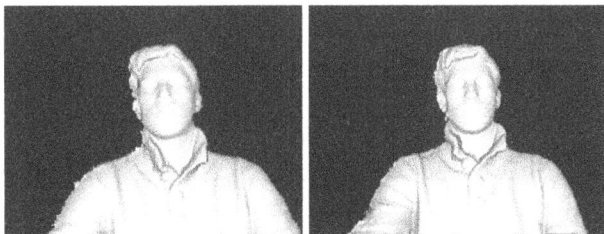

Figure 2: Raycast normal map from the cameras point of view, rendered as a greyscale image using a reconstruction volume consisting of 320^3 voxels and a volume size of $(1\,\text{m})^3$ (left) and $(60\,\text{cm})^3$ (right). Due to a higher resolution, the model in the right image is more detailed.

The weight scales the greyscale values in the rendered image in a range of $[50, 255]$. In Fig. 2 an example of a reconstruction with two different volume sizes and the same number of voxels is shown. For our purposes we extended the algorithm, so an arbitrary camera position and orientation around the volume can be set via config providing a qualitatively better view of the scenery during the reconstruction process. There is no need for the vertex- and normal maps in this scenario to have the same resolution as the input data, since they are not used in the camera estimation step. This way a more detailed view of the scene becomes possible, but at a loss of performance. The grey

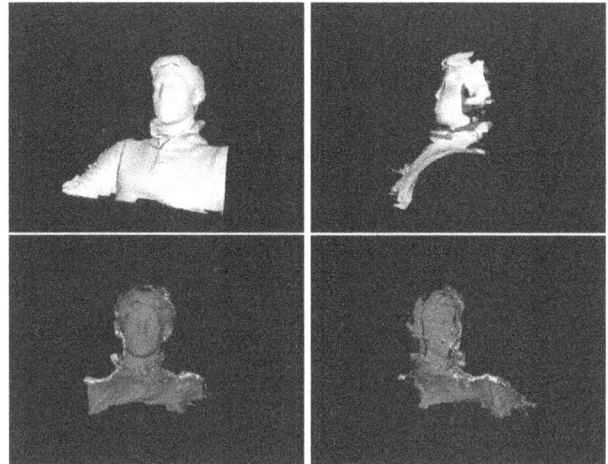

Figure 3: Raycast normal maps from an independent viewpoint, rendered as a greyscale image using a reconstruction volume consisting of 320^3 voxels and a volume size of $(50\,\text{cm})^3$. The images are screenshots of a real time visualization from different angles during one scanning scenario. The grey values of the back face pixels are scaled by 0.4 for better destinction (bottom).

values of the back face are scaled by 0.4 overall for better distinction (Fig. 3).

In addition to the rendered images we implemented a visualization of the point cloud representing the reconstructed model, which can be extracted in real time (Fig. 4). Next to the point cloud we display the boundaries of the reconstruction volume so the user can assess the dimensions of the space available for scanning. The size of the volume is set by the user and thus can be individually adjusted, according to the size of the object that is to be scanned. Reducing the volume size, while maintaining the same number of voxels results in an improved resolution. The camera angle can also be switched between an independent viewpoint and the view from the camera (similar to the rendered image in Fig. 3). The independent viewpoint is crucial for the current workflow, because it is necessary for planning a

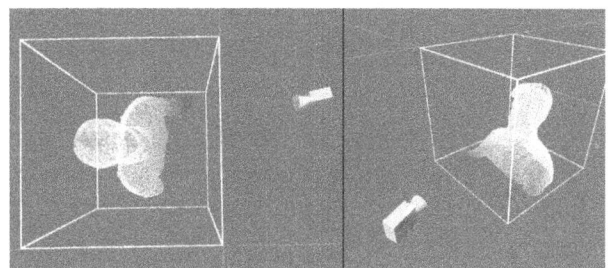

Figure 4: Point cloud of the reconstructed scene from two different independent viewpoints using a reconstruction volume consisting of 320^3 voxels. The camera model in both images indicates the position of the sensor. The white box indicates the boundaries of the available space for scanning, which was set to the dimension of $(60\,\text{cm})^3$.

scan of an object that occupies a certain amount of space. For all visualizations we use the 3D gaming engine Unity. To visualize the results of the ICP outlier detection and processing, we reconstructed a static scene during some non-rigid motion with and without the processing steps. Fig. 5 shows how the invalid ICP outliers would effect a segmentation attempt. In Fig. 5(b) the outliermap of the static scene is shown, which under ideal circumstances should not show anything. Fig. 5(c) and 5(d) show the outlier maps during non-rigid motion, before the processing steps, while in Fig. 5(e) and 5(f) the results after all processing steps have been executed can be seen. These maps can now be used as masks to exclude the pixels from the reconstruction step, as well as the tracking in the next frame. This offers the possibility of interacting in the scene without disturbing the reconstruction process.

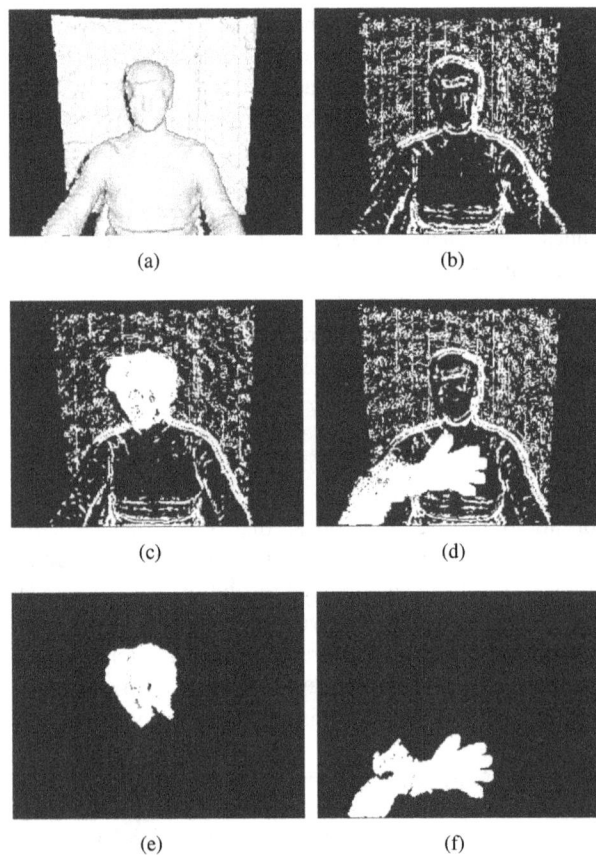

(a)   (b)

(c)   (d)

(e)   (f)

Figure 5: These figures show, how the invalid ICP outliers would effect a segmentation attempt for the patches indicating non-rigid motion. (a) shows a raycast normal map from the sensors point of view. In subfigures (b) - (d) the ICP outlier map without any processing steps is shown, whereas in (c) headmovement and in (d) hand movement is detected. (e) and (f) are the resulting ICP outlier maps after applying 2 iterations of erosion, a threshold of 2000 pixels for the connected component analysis and then 2 iterations of dilation on (c) and (d).

4 Conclusion

We used PCL as a basis to implement a 3D-Scanning application and included it in our processing pipeline. To improve the workflow, we introduced configurable parameters regarding volume size and position as well as the ICP algorithm that can now all be changed on runtime. Additionally, we are able to use any calibrated sensor that provides depth data, since the algorithm works on the depth values alone. This improves the adaptability of the application and thus broadens the spectrum of use cases. Further, we implemented a visualization based on the gaming engine Unity to improve computation and visualization workflow. The addition of the processing pipeline introduced in [3] for ICP outlier separation serves as an approach to detect and mask non rigid motion, while reconstructing the scenery. This approach offers very promising results regarding body scanning and retaining information about occluded body parts.

Acknowledgement

The work has been carried out at gestigon, Lübeck. First of all I want to thank Prof. Erhardt Barth from the *Institute of Neuro- and Bioinformatics* at *Universität zu Lübeck* for being the supervisor of my work. Of course I thank Sascha Klement and Moritz von Grotthus for giving me the opportunity to work at gestigon and making this project possible. Lastly, I want to sincerely thank my direct supervisor Dennis Muenster, who spent a lot of time integrating me into the company and the daily workflow as well as helping me with the project.

5 References

[1] L. Xia, C. C. Chen, J. K. Aggarwal, *Human Detection Using Depth Information by Kinect.* in: Computer Vision and Pattern Recognition Workshops, pp. 15-22, 2011.

[2] R. A. Newcombe, S. Izadi, O. Hilliges, *KinectFusion: Real-time dense surface mapping and tracking.* in: International Symposium on Mixed and Augmented Reality, pp. 127–136, 2011.

[3] S. Izadi, D. Kim, O. Hilliges, *KinectFusion: Real-time 3D Reconstruction and Interaction Using a Moving Depth Camera.* in: Proceedings of the 24th Annual ACM Symposium on User Interface Software and Technology, pp. 559–568, 2011.

[4] C. Tomasi, R. Manduchi, *Bilateral Filtering for Gray and Color Images.* in: Computer Vision, pp. 839–846, 1998.

[5] O. Kalentev, A. Rai, S. Kemnitz, R. Schneider, *Connected component labeling on a 2D grid using CUDA.* in: Journal of Parallel and Distributed Computing, vol. 71, no. 4, pp. 615–620, 2011.

Comparison of Frequency Selection Methods for Image Reconstruction in Magnetic Particle Imaging
– Improving Image Quality –

Y. Blancke Soares [1], A. v. Gladiss [2], M. Ahlborg [2], K. Gräfe [2] and T. M. Buzug [2]

[1] Medizinische Ingenieurwissenschaft, Universität zu Lübeck, yvonne.blanckesoares@student.uni-luebeck.de
[2] Institute of Medical Engineering, Universität zu Lübeck, {gladiss, ahlborg, graefe, buzug}@imt.uni-luebeck.de

Abstract

Magnetic Particle Imaging provides real time functional imaging information of injected superparamagnetic iron oxides. Its advantages over other medical imaging modalities lie in the speed of imaging, sensitivity as well as resolution and the absence of side effects like radiation. One common approach in order to reconstruct the acquired data is using a system function. The data that is stored in the system function can be represented in the frequency domain. This is useful to apply spectral filters. In this work, single frequency components are selected for the reconstruction process. The tested methods achieve a major improvement of the reconstruction quality that could be achieved so far. Single-sided MPI is unique for its good flexibility due to patient accessibility, its high resolution, sensitivity and the possibility of 3D imaging, which no other medical imaging modality can achieve today in this context without any side effects.

1 Introduction

In 2005 Weizenecker and Gleich introduced the novel imaging modality Magnetic Particle Imaging (MPI) [1]. This imaging technique facilitates real time functional imaging of superparamagnetic iron oxide nanoparticles (SPIONs). Electromagnets with opposing currents or permanent magnets generate magnetic fields that create a field free point (FFP). Another magnetic field, called drive field, moves the FFP over a field of view (FOV). SPIONs are used as tracer material. Their core exhibits superparamagnetic properties and therefore enables imaging and their coating provides biocompatibility [2]. The measured signal originates from the region of the FFP, where a change of the particle magnetisation takes place.

Whereas the size of the measurement range of the conventional scanner is limited due to the coil configuration, the single-sided MPI topology requires the coils to be arranged on a single side and therefore, it has no object size limitations. This allows more flexibility and patient accessibility, facilitating imaging of difficult regions, e.g. the axilla. Therefore, single-sided MPI has a high potential e.g. for finding the sentinel lymph nodes (SLN) in patients suffering from breast cancer. It has a high sensitivity and no side effects caused by radiation or toxicity, which is a major advantage over the current used SLN biopsy method that uses radioactive tracer material [3].

This work concentrates on improving the quality of image reconstruction of single-sided MPI phantom data. In theory, the Chebyshev polynomial of the second grade can be used as a transformation to reconstruct image data without any

quality loss, but this model only works for simple trajectories [4]. For complex trajectories, a system function (SF) is used to reconstruct image data. The SF exhibits the correlation of spatial distribution of injected SPIONs and the acquired signal. It contains a set of linear equations, which has to be solved to get information about the tracer distribution. The data of the SF can be represented in frequency domain, which is a precondition to apply spectral filters. This is useful to select those frequency components for the reconstruction process, which contain most of the signal.

In this work, eight different frequency selection approaches are tested. The quality of reconstructed images that can be achieved so far is delivered by a simple signal-to-noise ratio (SNR) approach, which has been included into the comparison study. The acquired frequency components can be arranged with respect to their mixing order [6], which has been used to generate approaches for a frequency selection.

2 Material and Methods

MPI uses the nonlinear magnetisation properties of the tracer material to obtain a signal in the receive coils [1]. The drive field amplitude and its waveform as well as the magnetisation curve of the particles determine the intensity and weight of higher harmonics in the signal spectrum. A calibration measurement has to be done to acquire scanner properties and particle characteristics [5]. Therefore, a sample is measured on each position of the FOV and the received signal is stored in the system matrix (SM) [7].

2.1 Experimental Setup

The SM contains the signal of 225 measured discretized positions. An empty measurement has been made to cope with the constant noise caused by the scanner. Over the base frequency of 2.5 MHz the excitation frequencies can be calculated, which have been 25.252 kHz for the x-channel and 26.042 kHz for the y-channel [7]. A frequency ratio of 33/32 has been used for the drive field to create a 2D Lissajous trajectory. Two receive channels have been used for the x- and y-direction. The signal has been sampled with a frequency of 17.5 MHz for each channel. 11089 frequency components have been received on each channel for a respective measurement, which results in a 22178×225 SM. Frequency components up to 1 MHz have been used.

One phantom has been measured with a 2D single-sided MPI scanning device [7]. The phantom consists of four holes (Fig. 1) that are filled with the tracer material Resovist (Bayer Schering Pharma). The holes have a diameter of 2 mm and are arranged differently with respect to the scanner center (Fig. 4), so that there have been six different phantom arrangements.

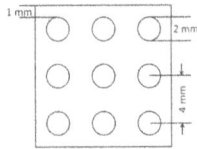

Figure 1: This image shows the composition of the phantom. This phantom has been arranged differently with respect to the scanner center (after [7]).

2.2 Frequency Selection

This section describes the selection of frequency components that will be used for reconstruction. Some frequency components contain more signal than others, therefore eight different methods have been elaborated and used to select only the frequency components, which contain the most signal and small noise.

2.2.1 SNR Selection (SNRS)

The first approach is the measurement of the SNR, which delivers the status quo in reconstruction quality. The following equation has been used for the SNR calculation

$$\mathrm{SNR} = \frac{\|S - E\|}{\mathrm{std}(S_{\mathrm{empty}})}, \tag{1}$$

where S is the SM, S_{empty} represents an SM of empty measurements, E is the empty measurement and std is the standard deviation [7]. Subsequently, only frequency components with an SNR above a certain threshold have been used for reconstruction. Different thresholds have been tested for each receive channel.

2.2.2 Mixing Order (MO)

Because of the nonlinear magnetisation characteristics of the tracer material, the excitation frequencies and the particle response get mixed nonlinear. For the discrete frequency component k holds

$$f_k = \sum_i |n_{i,k} f_i|, \quad f_i = f_x, f_y. \tag{2}$$

The mixing order of a frequency component is defined as the sum of the multiples $n_{i,k}$ [6],

$$m_k = \sum_i |n_{i,k}|, \quad n_i \in \mathbb{Z}. \tag{3}$$

A high mixing order implies a low SNR as it can be seen in Fig. 2. The method MO allows a maximum for the mixing order of the frequency component, so that every frequency component with a mixing order higher than the threshold gets filtered.

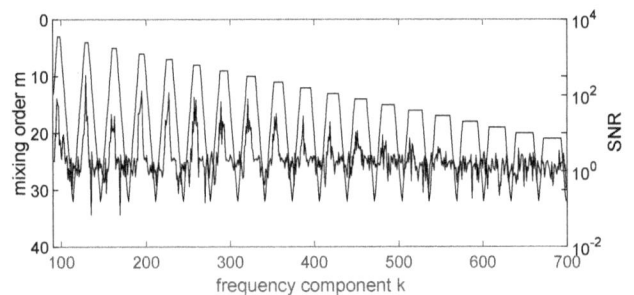

Figure 2: This figure shows the mixing order plotted against the frequency component as well as the SNR plotted against the frequency component. The y-axis of the mixing order is reverse so that there is a better way to compare mixing order and SNR. It is conspicuous that the mixing order is low when there is a high SNR and vice versa.

2.2.3 Weighted Mixing Order (WMO)

The method MO has been upgraded by additional terms that get added to (3) to receive a weighted mixing order, which has been used as a third method for frequency selection. Using WMO the mixing order depends on the receive channel, so that the mixing order curves differ among each other unlike in MO, where the mixing order of each receive channel is the same. As it can be read in [6], there is a correlation between the acquired signal and the closeness of the multiples of the excitation frequency, which is implemented in the following equation that got firstly introduced in [8]

$$
\begin{aligned}
m_{j,k} =\ & \sum_i |n_{i,k}| \\
& + \ \alpha \cdot \sum_i (|n_{j,k}| - |n_{i,k}|) \\
& + \ \beta \cdot \sum_i ||n_{j,k}| - |n_{i,k}||,
\end{aligned}
\tag{4}
$$

where i, j are the different receive channels. The α-term ensures a lower multiple of the frequency for a certain receive channel when the multiple of the other receive channel is higher. Additionally, the β-term calculates a lower mixing order when the multiples of the two receive channels are similar. Different parameters for α and β have been tested. A maximum threshold is used to filter the frequency components lying above.

2.2.4 Local Minimum Methods

As it has been shown that a low mixing order implies a high SNR (Fig. 2), a next approach has been, to select frequency components that lie near a local minimum of the mixing order curve. There were three ways to truncate the selection made by the local minimum method. The first was to choose an upper frequency index to sort every component that lies above (MINFREQ). The second one was to calculate the local minimum of the conventional mixing order curve and choose a threshold for the mixing order (MINMO) and the third was to use the weighted mixing order curve with a certain threshold for the mixing order (MINWMO).

2.2.5 SNR Combination Methods

The last two methods are a combination of SNRS and the mixing order approaches. Therefore SNRS and MO (SNRMO), and SNRS and WMO (SNRWMO) have been applied one after another.

2.3 Image Reconstruction

For the image reconstruction the linear equation

$$Sc = u, \qquad (5)$$

has to be solved. In (5), S is the SM, c the concentration of the particles and u the measured signal. Because of (5) being an ill-posed problem, it has been transferred to the least squares problem, which has been extendet by a Tikhonov regularisation

$$\|Sc - u\|_W^2 + \lambda\|c\|^2, \qquad (6)$$

where λ is the regularisation parameter [6]. The subscripted W stands for a weighting that has been applied to the selected frequency components. It uses the reciprocal of the energy respectively to normalise each frequency component [5]

$$W = \mathrm{diag}\left(\frac{1}{\tilde{w}_k^2}\right), \qquad (7)$$

where

$$\tilde{w}_k = \|g_k\| \qquad (8)$$

is the Euclidean norm of the kth row g_k of the SM S. The regularisation parameter λ has been chosen manually after an evaluation of images that have been reconstructed using different values for λ.

To compare each reconstruction result, the NRMSE has been calculated as follows

$$\mathrm{NRMSE}(x_1, x_2) = \frac{\sqrt{\frac{1}{n}\sum_{k=1}^n (x_{1,k} - x_{2,k})^2}}{\max(x_1) - \min(x_1)}, \qquad (9)$$

where x_1 denotes the phantom and x_2 the reconstructed image. n represents the number of picture elements. The reconstructed image had to be normalised. In this case, the maximum value of the image was set to 1, respectively.

3 Results and Discussion

To interpret the results of every frequency selection approach, the NRMSE of each reconstructed image has been compared. It can be shown that seven tested frequency selection methods could improve image quality of the reconstructed images in comparison to the gold standard achieved with SNRS (Fig. 3). This holds for all six phantom arrangements.

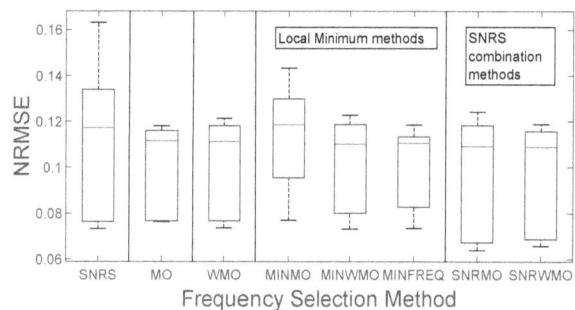

Figure 3: This boxplot graphic shows the NRMSE of each frequency selection approach of reconstructed images of every phantom arrangement. The best results in terms of NRMSE are achieved with MINFREQ, which has a comparatively low median value as well as a small distance between the 25th and 75th percentile. Best image results have been achieved with SNRMO.

3.1 SNRS

The results of SNRS represent the status quo in view of reconstructed image quality. In this comparison study it delivered the worst results in terms of the NRMSE (Fig. 3) as well as the in subjective visual evaluation (Fig. 4). Note, the chosen parameters λ, SNR thresholds and number of used frequency components for image reconstruction differ in comparison with [7].

3.2 Mixing Order

The mixing order approaches performed better than the results of SNRS. In comparison, SNRWMO achieved the lowest median for the NRMSE followed by SNRMO, MINWMO and MINFREQ, whereas SNRWMO exhibits a huge distance between the 25th and 75th percentile (Fig. 3). This

(a) phantom   (b) SNRS   (c) WMO   (d) SNRMO   (e) MINFREQ

(f) phantom   (g) SNRS   (h) WMO   (i) SNRWMO   (j) MINFREQ

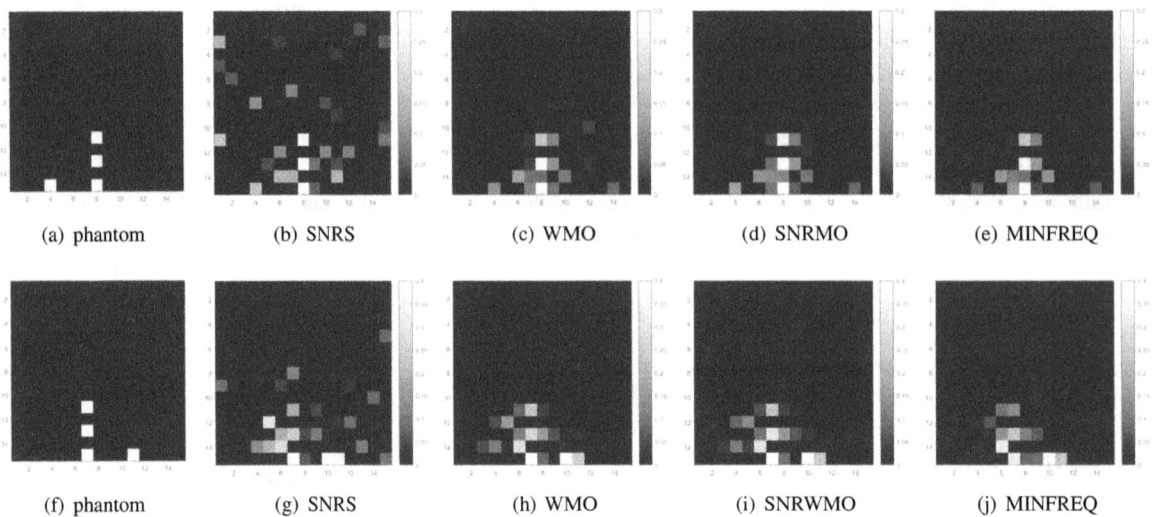

Figure 4: This figure shows a comparison of two phantom arrangement reconstructions. Those have been the most representative results. The SNRS results (b), (g) show the quality that could be achieved so far. (a) and (f) show two optimal reconstructions, (c) and (d) show the results of the methods that achieved the best image quality for phantom (a). (h) and (i) show the results of the methods that achieved the best image quality for phantom (f). (e) and (j) show the MINFREQ approach which has turned out as the best method tested in this work in terms of the combination of image quality, that can be achieved and the reliability of the method over all six phantom arrangements.

means that the results of SNRWMO differ for each phantom arrangement. The best results of all approaches have been achieved with MINFREQ, which has a small distance between the 25th and 75th percentile and a comparatively low median value. The best results of every method can be seen in Fig. 4.

4   Conclusion

It could be shown that there are various options on improving image quality with different frequency selection approaches. For the presented comparison study the MINFREQ approach delivered the best reconstruction results in terms of image quality and reliability. To provide a higher significance it would be necessary to repeat this study with more than six data sets and a different arrangement, e.g. more complex phantoms. It would also be interesting to test, if the described methods work for data measured with another MPI scanning device.

Additionally, it is possible to use the proposed methods as weighting of frequency components.

Acknowledgement

The work has been carried out at Institute of Medical Engineering, Universität zu Lübeck.

5   References

[1] B. Gleich and J. Weizenecker, *Tomographic imaging using the nonlinear response of magnetic particles*. in: Nature, vol.435, no. 7046, pp. 1214–1217, 2005.

[2] N. Panagiotopoulos et al., *Magnetic Particle Imaging: Current Developments and Future Directions*. in: International Journal of Nanomedicine, vol. 10, pp. 3097–3114, 2015.

[3] D. Finas et al., *Distribution of Superparamagnetic Nanoparticles in Lymph Tissue for Sentinel Lymph Node Detection in Breast Cancer by Magnetic Particle Imaging*. in: Springer Proceedings in Physics, vol. 140 ,pp. 187–191, 2012.

[4] M. Grüttner et al., *On the Formulation of the Image Reconstruction Problem in Magnetic Particle Imaging*. in: Biomedical Engineering, vol.58, no. 6, pp. 583–591, 2013.

[5] T. Knopp et al., *Weighted iterative reconstruction for magnetic particle imaging*. in: Physics in Medicine and Biology, vol.55, pp. 1577–1589, 2010.

[6] J. Rahmer et al., *Analysis of a 3-D System Function Measured for Magnetic Particle Imaging*. in: IEEE Transactions on Medical Imaging, vol. 31, no. 6, pp. 1289–1299, 2012.

[7] K. Gräfe et al. *2D Images Recorded with a Single-Sided Magnetic Particle Imaging Scanner*. in: IEEE Transactions on Medical Imaging, 2015

[8] A. v. Gladiss et al. *Compressed Sensing of the System Matrix and Sparse Reconstruction of the Particle Concentration in Magnetic Particle Imaging*. in: IEEE Transactions on Magnetics, vol. 51, no. 2, 2015

Comparing Different Compressed Sensing Reconstruction Techniques for the System Matrix in Single-Sided MPI

P. Weissert [1], A. von Gladiss [2], and T. M. Buzug [2]

[1] Medizinische Ingenieurwissenschaft, Universität zu Lübeck, patrick.weissert@student.uni-luebeck.de
[2] Institute of Medical Engineering, Universität zu Lübeck, {gladiss, buzug}@imt.uni-luebeck.de

Abstract

Magnetic Particle Imaging is a promising new medical imaging method. One of many topologies is single-sided which enables a free access to the FOV. To reconstruct images, a system matrix is acquired priorly in a time-consuming process. This time can be reduced by acquiring a randomized part of the FOV and reconstructing the full system matrix by methods of Compressed Sensing. This work applies different reconstruction algorithms on undersampled system matrices and compares the results. Furthermore, the reconstruction is modified in order to minimize the reconstruction error.

1 Introduction

In 2005, Gleich and Weizenecker presented *Magnetic Particle Imaging (MPI)* as a method for imaging the spatial distribution of superparamagnetic iron oxide nanoparticles (SPIONs) based on spatial- and time-varying magnetic fields [1]. The time-varying magnetic field excites the particles and a voltage can be measured that is induced into a receiving coil. The Fourier coefficients $(\hat{u}_k)_{k=0}^{K-1}$ of the induced voltage $u(t)$ at frequency f_k and a system matrix $\hat{\mathbf{S}}$ are used to reconstruct the spatial particle distribution $c(\mathbf{r})$. The system matrix $\hat{\mathbf{S}} \in \mathbb{C}^{K \times N}$ represents the system response to a delta sample in each of the N voxels. The linear equation is denoted in the discrete matrix-vector form as

$$\hat{\mathbf{u}} = \hat{\mathbf{S}} \cdot \mathbf{c}. \tag{1}$$

The process of acquiring the system matrix is very time-consuming. A brief overview on the basic physical and technical principles can be found in [2].

Most presented scanners have got a tube-like measurement field, which limits both the size of the FOV and the size of the imaged objects. Sattel et al. presented in 2009 [3] that a single-sided scanner overcomes the object size limit and is only constrained by the penetration depth. This single-sided MPI scanner has been extended and further developed by Graefe et al. [4].

Because of the absence of ionizing radiation, the dispensation with iodine-containing contrast agent and the high temporal resolution [5], medical applications for (single-sided)-MPI were taken into account – like the localization of the sentinel lymph node for the examination of breast cancer-cell dissemination [6].

To reduce the long acquisition time of the system matrix, either the resolution of the FOV-grid could be reduced or the system matrix could be randomly undersampled. This procedure can be subsumed under the theory of Compressed

Sensing (CS) [7]. Several CS-algorithms exist for the recovery of the complete system matrix out of undersampled data: FISTA has recently been applied to single-sided MPI [8], [9]. In this work, FISTA is compared to another algorithm: the IHT, which is firstly applied on single-sided MPI and is proposed to be superior to FISTA for non-single-sided MPI data [10].

Moreover, it is evaluated, which influence the input of the reconstruction algorithm has, for instance if the magnitude and angle of the system matrix should be reconstructed separately or in one.

2 Material and Methods

In this section, different methods will be presented, which are used for acquiring the single-sided measured data and for reconstructing the system matrix.

2.1 Compressed Sensing

The theory of Compressed Sensing for solving under-determined systems of linear equations states that if a signal has got a sparse representation in another domain, which is incoherent to the acquisition domain, the signal can be undersampled and reconstructed without loss despite of the violation of the Nyquist-Shannon sampling theorem. The kth row of the system matrix $\hat{\mathbf{S}}$ (the *system function* $\hat{\mathbf{s}}_{\mathbf{k}}$) can be expanded into the basis $\mathbf{B}$ by solving the following linear systems of equations

$$\hat{\mathbf{s}}_{\mathbf{k}}^{T} = \mathbf{B} \cdot \mathbf{v}_{\mathbf{k}}, \tag{2}$$

where $\mathbf{v}_k$ describes the basis coefficients that represent the kth system matrix row in the considered basis. To shorten the acquisition time of $\hat{\mathbf{S}}$, the number M of random sampling positions $\tilde{\mathbf{r}}_m, m = 0, ..., M$ are chosen that $M < N$,

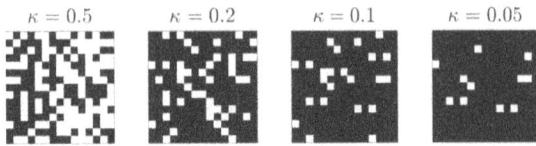

Figure 1: Random sampling grids for the system matrix. The full grid of size 15 x 15 is undersampled with different reduction rates κ. The selected voxels are indicated as white pixels.

i.e.: only M of N columns of the system matrix are recorded: $\tilde{\mathbf{s}}_k = \hat{\mathbf{s}}_k(\tilde{\mathbf{r}}_m)$. This leads to an under-determined system of equations in (2)

$$\tilde{\mathbf{s}}_{\mathbf{k}}^T = \mathbf{B_r} \cdot \mathbf{v_k}. \tag{3}$$

$\mathbf{B}_r$ is the reduced basis transformation matrix that transforms the system function into sparse domain. The aim is to reconstruct the correct solution $\mathbf{v}_k$, which can be performed by a *reconstruction in sparse domain*:

$$||\mathbf{v_k}||_p \xrightarrow{\mathbf{v_k}} \min \quad \text{subject to} \quad ||\tilde{\mathbf{s}}_{\mathbf{k}}^T - \mathbf{B_r} \cdot \mathbf{v_k}||_2 < \epsilon, \tag{4}$$

where $||\mathbf{x}||_p$ is the p-norm. The resulting vector $\mathbf{v}_k$ is inserted into (2) to calculate the complete system matrix $\hat{\mathbf{S}}$. The basis transform in (2) is the discrete cosine transform (DCT), which had the best sparsification results in [11]. The input $\tilde{\mathbf{s}}_{\mathbf{k}}^T$ in (4) can either be real or complex – if the transformation matrices $\mathbf{B_r}$ and $\mathbf{B}$ are capable of handling complex numbers, the reconstructed output $\hat{\mathbf{S}}$ will be from the same vector space as the input. This behavior makes it possible to reconstruct the magnitude and the angle of the system function separately or in one.

2.2 Data

In this work, measurement data from the newly developed 2D single-sided MPI scanner from Graefe et al. [4] is used and compared to simulated data of the same properties. The Lissajous-trajectory of the field free point (FFP) through the Field of View (FOV) is created by the excitation frequencies $f_x = 2.5$ MHz$/99 \approx 25253$ kHz and $f_y = 2.5$ MHz$/96 \approx 26042$ kHz. The receive signal is sampled with 17.5 MHz. The FOV has a size of 30 x 30 mm^2 and is discretized by non-overlapping voxels with an edge length of 2 mm, which gives 15 x 15 $=$ 225 system matrix columns. The N columns of the system matrix are the Fourier-transformed measurements of the system response to a delta sample in each of the N voxels. The full grid can be randomly undersampled with a specific reduction rate κ, where a high undersampling is equivalent to a low reduction rate. In this work, reduction rates between 50 % and 5 % (see Fig. 1) have been used.

2.3 System matrix reconstruction algorithms

For reconstructing the complete system matrix, several algorithms, which exploit the sparse structure of the transformed system matrix ($p = 1$), have been proposed: the

fast iterative shrinkage thresholding algorithm (FISTA) [9] is the current gold standard in MPI for solving equations like (4).

Moreover, the restoration of the system matrix can be regarded as an inpainting problem – Kleine et al. [10] modified an existing algorithm for MPI: the term "$\mathbf{B_r} \cdot \mathbf{v_k}$" in (4) is left-handed multiplied with a binary diagonal mask $\mathbf{M}$ indicating existing pixels with 1 and missing pixels with 0. During the iterative reconstruction process, sufficiently small values in the current $\mathbf{v}_k$ are set hard to zero (*iterative hard thresholding* (IHT)), the respective threshold δ is decreasing exponentially over the iteration steps. At the end, the missing pixels of the system function are interpolated due to the overlapping support of the cosine basis functions [10]:

$$\hat{\mathbf{s}}_{\mathbf{k}}^{T,\text{recon.}} = \tilde{\mathbf{s}}_{\mathbf{k}}^T + (1 - \mathbf{M}) \cdot \hat{\mathbf{s}}_{\mathbf{k}}^{T,(\text{itermax})}. \tag{5}$$

In this work, a comparison between FISTA and the IHT-algorithm is done. The FISTA algorithm is parametrized with 20 iterations, 3 inner iterations, a starting regularization parameter λ of 10^{-3} and a λ-decrease-factor of $3/5$ per iteration. The maximum iteration number of the IHT algorithm is set to 200 and the stopping threshold σ is chosen as the maximum absolute value of the undersampled system function divided by 64. Furthermore, a total variation (TV) penalty is applied on the current reconstructed system function during the iterations: its parameter is $\sigma/8$. This TV-correction requires *even* discretizations of the FOV in the 2 space-dimensions. The FOV has been cut to 14 x 14 pixels to match this requirement.

2.4 System matrix reconstruction quality

To compare a reconstructed to an original complete system function, the *normalized root mean square error* (NRMSE) is used:

$$\text{NRMSE}(\hat{\mathbf{s}}_{\mathbf{k}}, \hat{\mathbf{s}}_{\mathbf{k}}^{\text{recon.}}) = \frac{\sqrt{\frac{\sum_{i=1}^{N}\left(\hat{\mathbf{s}}_{\mathbf{k},i} - \hat{\mathbf{s}}_{\mathbf{k},i}^{\text{recon.}}\right)^2}{N}}}{\max\text{Value} - \min\text{Value}}, \tag{6}$$

with the maximum value in reconstructed and original system function maxValue $=$ $\max(\max_i \hat{\mathbf{s}}_{\mathbf{k},i}, \max_i \hat{\mathbf{s}}_{\mathbf{k},i}^{\text{recon.}})$ and the minimum value "minValue" defined analogously. In this work, the NRMSE is normalized to 2π for angles to be compared. Moreover, the π-periodicity of the angle has been taken into account and (6) has been adapted appropriately.

3 Results and Discussion

The performance concerning the reconstruction of the system functions of both the FISTA- and the IHT-algorithm can be inferred from Fig. 2. IHT performs slightly better than FISTA for the simulated as well as for the measured data and for all reduction rates κ.

The improvements of IHT towards FISTA at conventional MPI [10] could not be validated for single-sided MPI in

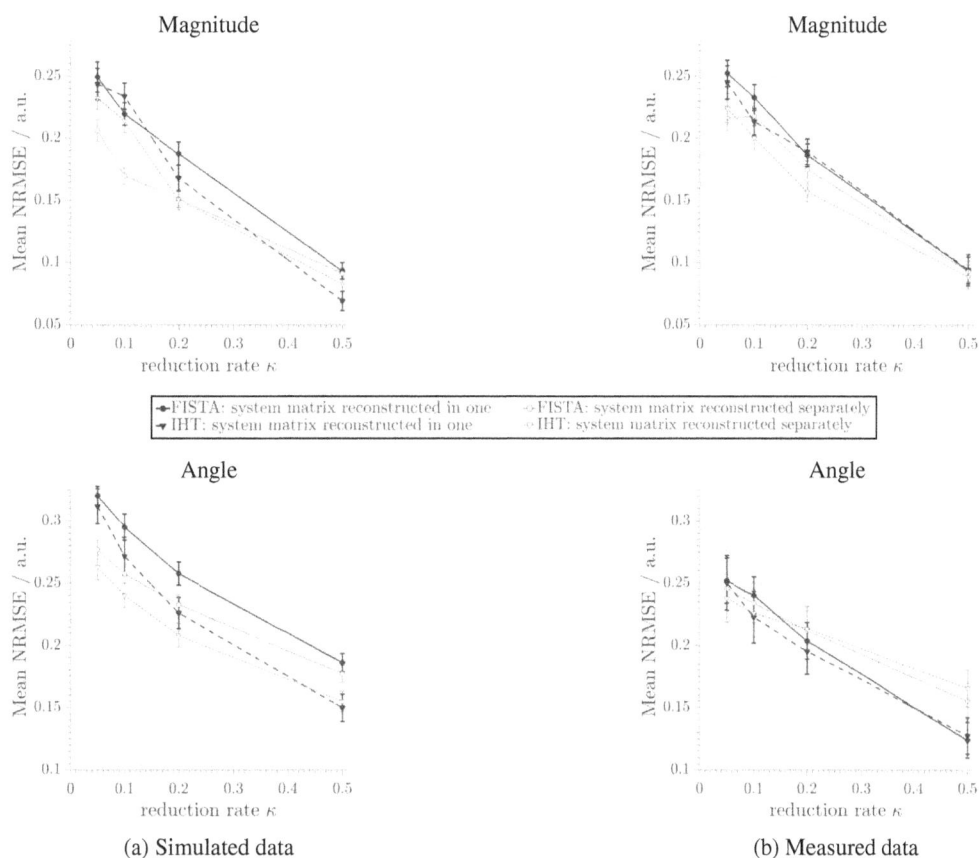

Magnitude — Magnitude — Angle — Angle

Legend:
- FISTA: system matrix reconstructed in one
- IHT: system matrix reconstructed in one
- FISTA: system matrix reconstructed separately
- IHT: system matrix reconstructed separately

(a) Simulated data

(b) Measured data

Figure 2: Comparison of the reconstruction error using simulated data (column (a)) and measured data (column (b)). NRMSE between the fully-sampled system function and the respective reconstructed system function averaged above all system functions with a sufficiently high SNR of the x-receive-channel. Length of the error bar: $2\sigma \cdot 1/4$.

our setup. For low reduction rates, this behavior can be explained by the low reconstruction quality and the large differences between the reconstructed and the fully sampled system function (see Fig. 3 and 4), which are probably due to the highly inhomogeneous magnetic field in the single-sided MPI scanner [4], [8]. Moreover, the low discretization of the FOV could be problematic, but the simulations with higher discretizations show similar results as in Fig. 2a.

For low reduction rates, the averaged NRMSE from Fig. 2 shows that the **magnitude of the system functions** should be reconstructed separately instead of together with the angle. This holds both for simulated and measured data and for both FISTA and IHT. This cannot directly be explained by watching an example system function $k = 62$ in Fig. 3 and 4: for $\kappa = 0.5$ the results look similar for in-one and separate reconstruction. For lower reduction rates, noise makes the images difficult to interpret and the proposed superiority to separate over in-one reconstruction from Fig. 2 could be a low discretization and high noise phenomenon.

The results for the reconstructions of the **angle of the system functions** show that this measurand should be reconstructed separately for the simulated data and in-one for the measured data. It is remarkable that the curves diverge in direction of the low reduction rates for the simulated data whereas the curves of the measured data approximate. This behavior of the (practically relevant) measurements is

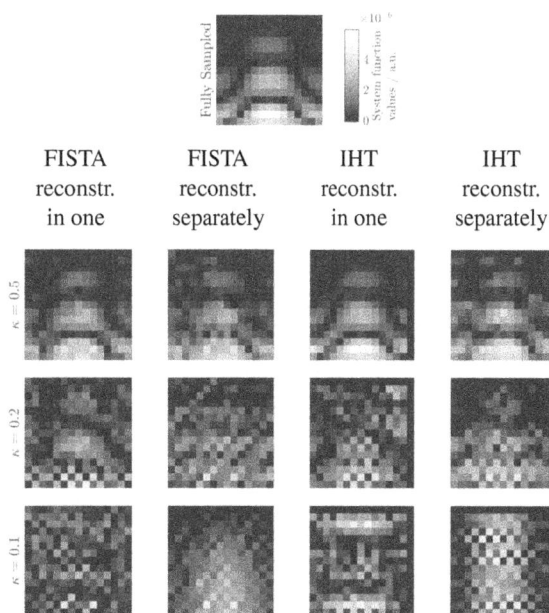

Figure 3: Reconstructed system function using simulated data. Magnitude of the CS-reconstructed system function $k = 62$ of the x-receive-channel for different reduction rates κ and reconstruction algorithms.

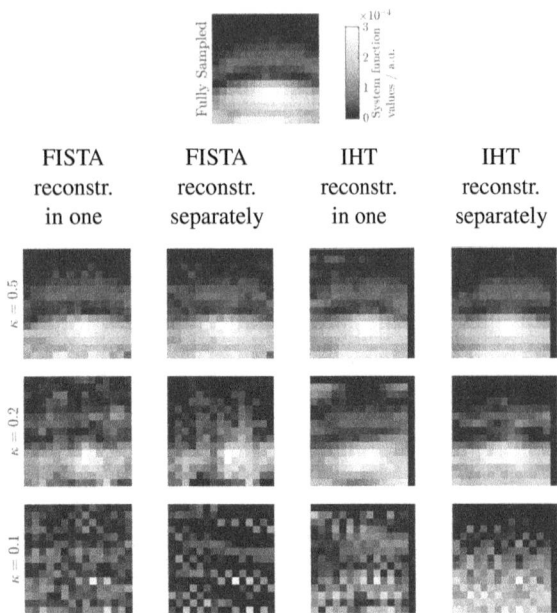

Figure 4: Reconstructed system function using measured data. Magnitude of the CS-reconstructed system function $k = 62$ of the x-receive-channel for different reduction rates κ and reconstruction algorithms.

beneficial considering the bad reconstruction results at low reduction rates (so that high reduction rates should be used in practice).

Moreover, the over magnitude and angle averaged reconstruction errors are similar for the measured and the simulated data.

4   Conclusion

The IHT-algorithm should be used to reconstruct system functions of the single-sided MPI scanner. The magnitude should be reconstructed separately and the angle together with the magnitude. This procedure would be a computational extra effort and it is clearly the question whether it is worth it. The advantage of this overhead can be estimated by calculating the reconstruction error of real phantoms. Generally, the results propose to use high reduction rates about 0.5 (i.e. low undersampling) in single-sided MPI – although high undersampling seems to be possible in conventional MPI (see e.g. [11]). This implicates that the possible time saving during the acquisition of the system matrix is not as high as desired.

To evaluate the applicability of the reconstructed system functions, those have to be applied to the voltage data of measured phantoms and a comparison of their imaging quality has to be made.

Generally, the application of Compressed Sensing seems to be more difficult in single-sided-MPI. Principles that allow higher undersampling rates have to be developed – weighted undersampling patterns considering the special magnetic field configuration in single-sided MPI could be an area of research.

Acknowledgement

The work has been carried out at the Institute of Medical Engineering, Universität zu Lübeck.

5   References

[1] B. Gleich and J. Weizenecker, "Tomographic imaging using the nonlinear response of magnetic particles," *Nature*, vol. 435, no. 7046, pp. 1214–1217, 2005.

[2] T. Knopp and T. M. Buzug, *Magnetic Particle Imaging - An Introduction to Imaging Principles and Scanner Instrumentation.* Berlin/Heidelberg: Springer, 2013.

[3] T. F. Sattel et al., "Single-sided device for magnetic particle imaging," *J. Phys. D: Appl. Phys.*, vol. 42, pp. 1–5, 2009.

[4] K. Graefe, A. v. Gladiss, G. Bringout, M. Ahlborg, and T. M. Buzug, "2D images recorded with a single-sided magnetic particle imaging scanner," *IEEE Trans. Med. Imag.*, 2015, admitted, DOI 10.1109/TMI.2015.2507187.

[5] B. Gleich, J. Weizenecker, and J. Borgert, "Experimental results on fast 2D-encoded magnetic particle imaging," *Phys. Med. Biol.*, vol. 53, no. 6, pp. N81–N84, 2008.

[6] K. Graefe et al., "Magnetic-particle-imaging for sentinel lymph node biopsy in breast cancer," in *Magnetic Particle Imaging* (T. M. Buzug and J. Borgert, eds.), vol. 140 of *Springer Proc. in Phys.*, pp. 237–241, Springer Berlin Heidelberg, 2012.

[7] E. J. Candès, J. Romberg, and T. Tao, "Robust uncertainty principles: exact signal reconstruction from highly incomplete frequency information," *IEEE Trans. Inf. Theory*, vol. 52, no. 2, pp. 489–509, 2006.

[8] A. von Gladiss, K. Graefe, M. Ahlborg, and T. M. Buzug, "Undersampling the system matrix of a single sided MPI-scanner," in *Book of Abstracts 5th IWMPI*, p. 36, 2015, DOI 10.1109/IWMPI.2015.7107021.

[9] A. Beck and M. Teboulle, "A fast iterative shrinkage-thresholding algorithm for linear inverse problems," *SIAM J. Imaging Sci.*, vol. 2, no. 1, pp. 183–202, 2009.

[10] M. Kleine, A. v. Gladiss, and T. M. Buzug, "Iterative hard thresholding for the reconstruction of an undersampled system matrix in magnetic particle imaging," in *Proc. BMT 2014*, vol. 59, pp. 628–631, 48. DGBMT Jahrestagung, Hannover, DeGruyter, 2014.

[11] T. Knopp and A. Weber, "Sparse reconstruction of the magnetic particle imaging system matrix," *IEEE Trans. Med. Imag.*, vol. 32, no. 8, pp. 1473–1480, 2013.

Reconstruction of Connected Structures in Magnetic Particle Imaging using Markov Random Field based Regularization

C. Kowalski [1], M. Maaß [2], and A. Mertins [2]

[1] Medical Engineering Science, Universität zu Lübeck, christian.kowalski@student.uni-luebeck.de
[2] Institute for Signal Processing, Universität zu Lübeck, {maass, mertins}@isip.uni-luebeck.de

Abstract

Magnetic particle imaging uses superparamagnetic nanoparticles as a tracer material for the determination of their spatial distribution. It is possible to draw inferences from the induced voltage signals and the beforehand determined system matrix about the unknown magnetic particle concentration. Because of the fact that the nanoparticles mainly spread out through the blood vessels, it might be reasonable to put an emphasis on the connected structures during the reconstruction process. In this paper, the presented problem is solved by using iterative reconstruction algorithms on the basis of Markov random fields (MRFs). While one approach employed the Gibbs prior as a penalty term for regularization, the proposed algorithm makes use of the Gibbs energy. Both algorithms are compared to the algebraic reconstruction technique (ART) during a test on an image containing blood vessels. It is shown that with the proposed method, a fast reconstruction of high quality is possible.

1 Introduction

Magnetic particle imaging (MPI) is a new method for pre-clinical imaging by determining the spatial distribution of superparamagnetic nanoparticles *in vivo* given the measured voltages in the receive coils of the MPI scanner [6]. To acquire a signal over the whole field of view, a fieldfree point (FFP) is moved along a specific path. Only tracer material in or close to this FFP is influenced and responsible for the signal over time generation. One typical movement path for the FFP is the Lissajous trajectory. The determination of the concentration is called reconstruction and usually represents an inverse and ill-posed problem. Under special circumstances, a certain preknowledge in the form of a prior distribution can be used to improve the reconstruction outcome by using regularization. In general, image reconstruction ultimately aims for the best approximation of the true image. However, because of the fact that the nanoparticles are injected and predominantly spread out through blood vessels, so that, for example, the cardiovascular system of a living mouse can be observed non-invasively [11], it makes sense to put an emphasis on connected structures during the reconstruction process of the image. Additionally, the MPI hardware is able to scan volumes with more than 40 frames per second [5], which generates a high amount of data, thus a fast yet accurate reconstruction is of uttermost importance. One approach for achieving this effect might be the usage of Markov random field (MRF) priors for regularization. After describing the MRF model and the image reconstruction problem in the following section, the results of the algorithms are presented and discussed afterwards with a conclusion.

2 Material and Methods

In order to put an emphasis on connected structures during the image reconstruction process, MRFs are used for regularization.

2.1 Markov Random Fields

The theory of MRFs is a statistical model, which can be used to analyze spatial or contextual dependencies [8]. More precisely, it is possible to estimate a true image signal from a noisy version by embedding the MRFs into a Bayesian framework [3]:

$$P(c|\mathbf{r}) = \frac{p(\mathbf{r}|c)P(c)}{p(\mathbf{r})} \propto p(\mathbf{r}|c)P(c). \qquad (1)$$

Let $\mathbf{d} = \{1, ..., m\}$ be a set of sites that indicates image pixels in a two dimensional plane. Furthermore, the set $\mathbf{D}$ consists of discrete labels that are assigned to the pixels and $\mathbf{r}$ represents the observation. In this case, the configuration $c = \{c_i \in \mathbf{D}|i \in \mathbf{d}\}$ represents a Markov random intensity field [8]. The *a priori* probability $P(C = c)$ is difficult to compute, but taking advantage of the Hammersley-Clifford theorem leads to a solution. The theorem implies that C can be described as a MRF on $\mathbf{d}$ with respect to the neighborhood system $N = N_i|\forall i \in \mathbf{d}$ if the probability distribution of the configuration follows a Gibbs distribution [9]:

$$P(c) = Z^{-1} \cdot e^{-\frac{1}{T}U(c)}. \qquad (2)$$

Z is called the partition function and is used for normalization of the term, the temperature T is a global control parameter, and $U(c)$ represents the prior energy:

$$U(c) = \sum_C V_C(c). \tag{3}$$

This energy function consists of a sum over different potential functions $V_C(c)$, whereby the cliques C are used to resolve the associations within the neighborhood. Further on, the choice of the potential function (see Table 1) determines the *a priori* knowledge of interactions between the labels of the neighborhood system [8].

Figure 1 shows the Gibbs distribution for the most common potential functions in a four-neighborhood system. Besides, the neighborhood system N needs to fulfill two conditions:
(1) $c \notin N_c$, which means that c must not belong to its own neighborhood and
(2) $c_i \in N_{c_j} \leftrightarrow f_j \in N_{c_i}$, which means that c_j belongs to the neighborhood of c_i and vice versa [9].

Table 1: List of common potential functions for pairwise cliques where Δ is the difference between c and a pixel neighbor while p and δ are free parameters [9].

Name	Potential function
Gauss	Δ^2
Besag	$|\Delta|$
Green	$\log \cosh(\frac{\Delta}{p})$
Geman and McClure	$\frac{\Delta^2}{\delta^2 + \Delta^2}$
Bouman and Sauer	$|\Delta|^p$
Hebert and Leahy	$\log(\delta^2 + \Delta^2)$

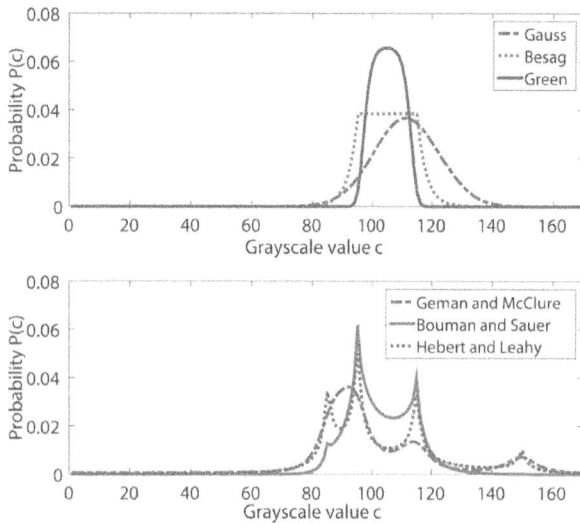

Figure 1: Example of *a priori* densities generated by different potential functions. The four neighbors of the neighborhood contained the pixel values 85, 95, 115 and 150.

2.2 Image Reconstruction

After data acquisition, it is necessary to reconstruct the image using the receive coil voltages. The underlying problem of determining the particle distribution is an inverse problem and can be described using a linear integral equation, which can be discretized in time and space so that a linear system of equations is obtained [1]:

$$\mathbf{Sc} = \hat{\mathbf{u}}. \tag{4}$$

In this case, $\mathbf{c} \in \mathbb{R}^N_+$ represents the unknown particle distribution, $\mathbf{S} \in \mathbb{R}^{M \times N}$ is the system matrix and $\hat{\mathbf{u}} \in \mathbb{R}^M$ is the measurement vector. Usually, the vector contains a measurement error and is therefore perturbed by noise. The equation can be rewritten as the following approximation with the noise $\hat{\boldsymbol{\eta}}$:

$$\mathbf{Sc} \approx \tilde{\mathbf{u}} = \hat{\mathbf{u}} + \hat{\boldsymbol{\eta}}. \tag{5}$$

Given the system matrix $\mathbf{S}$ and the measurement $\tilde{\mathbf{u}}$, the particle distribution $\mathbf{c}$ can be reconstructed. Trying to solve the linear system of equations in order to get one exact solution cannot be guaranteed, hence the problem is ill-posed and regularization is necessary for the least squares problem

$$\underset{\mathbf{c}}{\mathrm{argmin}}\{||\mathbf{Sc} - \tilde{\mathbf{u}}||_2^2\}. \tag{6}$$

In general, the system is regularized using the Tikhonov regularization with a parameter $\beta > 0$ and the Euclidean norm [4], [7]:

$$\underset{\mathbf{c}}{\mathrm{argmin}}\{||\mathbf{Sc} - \tilde{\mathbf{u}}||_2^2 + \beta||\mathbf{c}||_2^2\}. \tag{7}$$

The reconstruction can be performed using direct or iterative methods. Direct methods like the singular value decomposition have the disadvantage that the whole system matrix has to be loaded into the memory. This can be a problem depending on the computational system, especially in the case of three-dimensional MPI imaging. Therefore, iterative methods are the better choice and are less memory intensive [1], but the classical iterative methods tend to lead to slow convergence because of the large condition number of the system matrix [4]. Popular iterative methods are the conjugate gradient method and the algebraic reconstruction technique (ART), which is based on the Kaczmarz method. The ART is part of the splitting methods and works on a single matrix row at a time. To put it straight, the algorithm works by interpreting each equation of the linear system as a hyperplane in the N-dimensional solution space, whereby one sub-iteration represents the orthogonal projection of the current estimate of the solution onto the hyperplane [2]. One sub-iteration l of the j-th entry of $\tilde{\mathbf{u}}$, started with a zero vector, can be described the following way:

$$\mathbf{c}^{l+1} = \mathbf{c}^l + \frac{\tilde{\mathbf{u}}_j - \tilde{\mathbf{s}}_{\mathbf{j}}^* \cdot \mathbf{c}^l}{||\mathbf{s}_j||_2^2}\mathbf{s}_j^*. \tag{8}$$

One sweep through all rows of the matrix is considered as one full iteration. In addition, choosing the rows in a specific order also has an impact on the convergence speed of the algorithm. As an alternative to running the algorithm until convergence, stopping it after a specific number of iterations is also viewed as a tool of regularization [1].

An expansion to the ART algorithm is the reguralized Kaczmarz extended with relaxation parameters algorithm (RK-ERP) [10]. In this case, the second part of the minimization problem in (7) should not only serve as a measure for the roughness of the image but should also contain prior information about the missing image parts. To achieve this, the energy function in (3) defined by the MRF can be used as a discrete smoothing norm (see [10] for detailed derivation):

$$\underset{\mathbf{c}}{\operatorname{argmin}}\{||\mathbf{Sc} - \tilde{\mathbf{u}}||_2^2 + 2\beta U(\mathbf{c})\}. \qquad (9)$$

Algorithm 1 RKERP algorithm

1: Let $\mathbf{c}^0 \in \mathbb{R}^N, \mathbf{y}^0 = \mathbf{u}$; for $k = 0, 1, \dots$ do:
2: $\mathbf{y}^{k+1} = (1 - \alpha)\mathbf{y} + \alpha(\mathbf{y} - \frac{<\mathbf{y}, \mathbf{s}_j>}{||\mathbf{s}_j||_2^2} \cdot \mathbf{s}_j$
3: $\mathbf{u}^{k+1} = \mathbf{u} - \mathbf{y}^{k+1}$
4: $\mathbf{c}^{k+1} = (1 - \omega)\mathbf{y} + \omega(\mathbf{y} - \frac{<\mathbf{y}, \mathbf{s}_i> - \mathbf{u}_i}{||\mathbf{s}_i||_2^2}) - 2\beta\nabla U(\mathbf{c}^k)$

Equation (9) ultimately leads to the former algorithm where α, ω and β denote optimizable relaxation parameters. The total energy function $U(\mathbf{c}^k)$ takes the form of the first-derivate of the Green's function. Using the RKERP algorithm as a basis, we present the Markov regularized algebraic reconstruction technique (MRART). The MRART uses the *maximum a posteriori* probability (MAP) estimate of the Gibbs distribution (2) instead of the total energy function (9) as a prior for regularization and thus puts an emphasis on connected structures during the reconstruction process. Additionally, the Green potential function is applied as shown in Table 1 and all algorithms run in a specific order defined by the two-norm in order to start with the most important rows first in a descending sequence. At last, every negative value computed during the reconstruction is set to zero because of the fact that negative concentrations cannot occur.

Algorithm 2 MRART algorithm

1: Let $\mathbf{c}^0 \in \mathbb{R}^N$, for $k = 0, 1, \dots$ do:
2: $\mathbf{c}^{k+1} = (1 - \omega)\mathbf{y} + \omega(\mathbf{y} - \frac{<\mathbf{y}, \mathbf{s}_i> - \mathbf{u}_i}{||\mathbf{s}_i||_2^2}) + 2\beta P(\mathbf{c}^k)$

For comparison purposes, the algorithms ART, RKERP and MRART were implemented in Matlab and tested on an image with the size of 50×50 pixels (see Fig. 2) and a system matrix of the size 2381×2500. A Lissajous trajectory was used for generating the simulated two-dimensional data. The test was carried out on noiseless data in the first case and on noise perturbed data with an SNR of 40 dB in the second case. Furthermore, the used image represents a vessel in which the nanoparticles could accumulate (see Fig. 3). The parameters for the reconstruction results are described in Table 2.

In the course of testing the algorithms, the parameters were optimized manually and the Green potential function was used for evaluation as it generated the best reconstruction results based on experiments with the different potential functions defined in Table 1. The pixel values of the test

Table 2: List of parameters for the image reconstruction using the RKERP and the MRART algorithm (see Fig. 3).

	α	β	p	ω
Noiseless image	0.1	$5 \cdot 10^{-4}$	$2 \cdot 10^{-3}$	0.9
Noisy image	0.1	$5 \cdot 10^{-4}$	$2 \cdot 10^{-3}$	0.5

image ranged from 0 to 255 and a maximum of 300 iterations were made. Additionally, the peak signal-to-noise ratio (PSNR) was computed as a measure of image quality on each iteration and plotted against the particular iteration number. The result of each algorithm with the iteration number of its best PSNR was plotted (see Fig. 3).

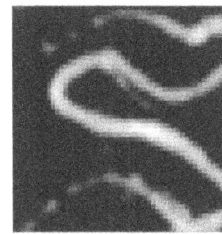

Figure 2: Test image of 50×50 pixels with vessels to simulate the accumulation of nanoparticles.

3 Results and Discussion

As it can be seen in Fig. 3, all three algorithms were successful in reconstructing the 50×50 image given the system matrix and the simulated and perturbed measurement vector. Nevertheless, all three algorithms worked differently in the noiseless (a–d) and the noisy (e–h) case. In the first case, convergence was achieved after approximately 260 iterations, and the image quality increased constantly. Furthermore, the RKERP algorithm had the lowest PSNR with a maximum of 32 dB while the MRART and the ART algorithms performed almost equally, having a higher PSNR of 37 dB. All three algorithms showed nearly the same course of the PSNR curve. The lower PSNR of the RKERP algorithm can also be seen in the reconstructed image (a), which is reflected in the lower image quality in comparison to the other two reconstruction results (b) and (c).

In the second case, the perturbed measurement vector led to another reconstruction solution. Here, all three algorithms reached a peak on a different amount of iterations (h). The ART algorithm reached the peak of 25 dB first after 9 iterations while the MRART algorithm needed 16 iterations for its peak at approximately 29 dB PSNR. In contrast, the RK-ERP algorithm reached its peak at also 29 dB not until 75 iterations. Stopping the algorithms at their PSNR peak led to the reconstruction results in Fig. 3 (e–g). These results show that, despite having the same PSNR, the RKERP reconstruction (e) needed more time but came to a solution inferior to the MRART reconstruction (f) and the ART results appeared almost equally qualitative (g). The MRART algorithm resembled a fine tradeoff regarding reconstruction speed and quality. Stopping it even earlier would generate a smoother image.

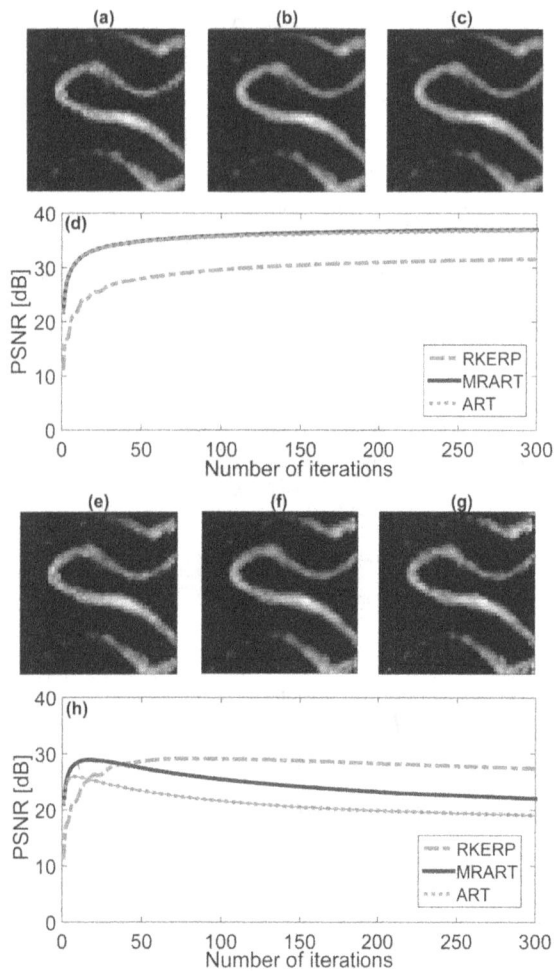

Figure 3: Reconstruction results for best PSNR of the ART, RKERP and MRART algorithms and corresponding PSNR for noiseless (a–d) and noisy case with 40 dB (e–h). (a) and (e): Image reconstruction using RKERP; (b) and (f): Image reconstruction using MRART; (c) and (g): Image reconstruction using standard ART; (d) and (h): PSNR plotted against the number of iterations for all three algorithms for both cases.

To sum up, the MRART succeeds in reconstructing a rather good approximation. Depending on the desired reconstruction result, the parameters have to be adjusted.

4 Conclusion

The proposed MRART algorithm is capable of reconstructing an approximation of the true image using the Gibbs distribution as a prior for regularization. Further on, it succeeds in putting an emphasis on connected structures while the iteration number of the algorithm is low by stopping the reconstruction process early on. This procedure also works as a regularization technique and makes it possible to reconstruct the desired image in a small amount of time. On the other hand, the three parameters β, ω and p have to be set before starting the reconstruction, which is a disadvantage compared to standard procedures. The MRART can be seen as an expansion to the RKERP algorithm and is able to work as a regularization algorithm for connected structures like vessels. This scenario is especially important in the case of MPI. In the future, the algorithm needs to be tested on more images with different sizes and higher noise levels in order to create a more realistic scenario.

Acknowledgement

The work has been carried out at the Institute for Signal Processing, University of Lübeck. In addition, I want to thank Max Schmiedel and Richard Schäfer for their support during the writing process of this paper.

5 References

[1] T. M. Buzug and T. Knopp, *Magnetic Particle Imaging*. Springer, Berlin/Heidelberg, 2008.

[2] T. M. Buzug, *Computed Tomography*. Springer, Berlin/Heidelberg, 2008.

[3] X. Descombes and E. Zhizhina, *The Gibbs fields approach and related dynamics in image processing*. In: Condensed Matter Physics, vol. 11, p. 293, 2008.

[4] A. A. Ivanov and A. I. Zhdanov, *Kaczmarz algorithm for Tikhonov regularization problem*. In: Applied Mathematics E-Notes, vol. 13, pp. 270—276, 2013.

[5] T. Knopp and A. Weber, *Local System Matrix Compression for Efficient Reconstruction in Magnetic Particle Imaging*. In: Advances in Mathematical Physics, 2015.

[6] T. Knopp and A. Weber, *Sparse Reconstruction of the Magnetic Particle Imaging System Matrix*. In: IEEE Transactions on Medical Imaging, vol. 32, no. 8, pp. 1473–1480, 2013.

[7] J. Lampe et al., *Fast reconstruction in magnetic particle imaging*. In: Physics in Medicine and Biology, vol. 57, pp. 1113–1134, 2012.

[8] S. Z. Li, *Markov random field models in computer vision*. In: Lecture Notes in Computer Science, Springer, Berlin/Heidelberg, pp. 361–370, 1994.

[9] A. Mertins, *Image recovery from noisy transmission using soft bits and Markov random field models*. In: Optical Engineering, vol. 42, pp. 2893–2899, 2003.

[10] C. Popa and R. Zdunek, *Penalized least-squares image reconstruction for borehole tomography*. In: Proceedings of ALGORITMY 2005 Conference, Vysoke Tatry–Podbanske, pp. 260–269, 2005.

[11] J. Weizenecker, B. Gleich, J. Rahmer, H. Dahnke, and J. Borgert, *Three-dimensional real-time in vivo magnetic particle imaging*. In: Physics in Medicine and Biology, vol. 54, no. 5, pp. L1–L10, 2009.

Tikhonov regularization scheme for image reconstruction in the presence of model errors

J. Mrongowius [1], J. Frikel [2], and C. Kaethner [3]

[1] Medizinische Ingenieurwissenschaft , Universität zu Lübeck, julia.mrongowius@student.uni-luebeck.de

[2] Department of Applied Mathematics and Computer Science, Technical University of Denmark, jyfr@dtu.dk

[3] Institute of Medical Engineering, Universität zu Lübeck, kaethner@imt.uni-luebeck.de

Abstract

Biomedical imaging techniques are a key part of modern diagnostics. The underlying data is usually obtained in a measurement process and is then always afflicted with errors. The data is used to reconstruct the image, often in combination with a model as a description of the imaging system. In this work, we consider reconstruction problems in the presence of data- and model-errors at the same time. Such problems arise in novel biomedical imaging modalities, such as in Magnetic Particle Imaging, where no proper model descriptions are found yet. We formulate a Tikhonov type regularization method that takes data- and model-errors into account. In numerical experiments, we show that our approach is able to outperform a classical Tikhonov regularization when the system matrix is corrupted by noise.

1 Introduction

The goal of biomedical imaging techniques, such as computed tomography (CT), magnetic resonance imaging, etc., is to provide images of the interior structure of an object. In order to achieve that, a large number of indirect measurements of the imaged object is collected and processed by appropriate algorithms. The calculation of images from measured data (reconstruction) is based on a mathematical model for the measurement process, which describes the relation between the sought object $x \in \mathbb{R}^N$ and the data $y \in \mathbb{R}^M$. In many situations this relation is linear and, therefore, the reconstruction problem can be modeled by a linear system of equations

$$y = Ax, \tag{1}$$

where $A \in \mathbb{R}^{M \times N}$ is a *system matrix* and $y \in \mathbb{R}^M$ a measurement vector. Thus, reconstructing x from y is equivalent to solving the system of equations given by (1).

If the system matrix A is invertible, the solution can formally be calculated by $x = A^{-1}y$. However, for many biomedical imaging situations this approach is not feasible or leads to unsatisfactory results because the reconstruction problem (1) is ill-posed, i.e., the system matrix is ill-conditioned [3], [4]. As a result, the reconstruction process is unstable and, hence, even small measurement errors can lead to huge reconstruction errors. Since the measured data y is corrupted by noise $n \in \mathbb{R}^M$, this can be a serious issue in practice. The realistic reconstruction problem consists in finding an approximation to x from noisy data

$$y = Ax + n, \tag{2}$$

where the noise vector n is usually not known exactly. The presence of noise in the data results in a loss of information

and makes the reconstruction problem much more challenging.

In order to stabilize the reconstruction, regularization procedures can be used [3], [4]. A classical approach is the *Tikhonov regularization*, where the solution is obtained by solving the minimization problem

$$\min_{x} \|Ax - y\|_2^2 + \rho \|x\|_2^2. \tag{3}$$

The first term in (3) is the data fidelity term, which ensures that the solution fits to the data. The second term is a regularization term that imposes a-priori assumptions on the solution by enforcing the minimizer to have a rather small Euclidean norm. A Tikhonov regularization mitigates the amplification of noise during the reconstruction by enforcing smoothness of the solutions. The extent of the regularization is thereby controlled by the *regularization parameter* $\rho \geq 0$ [3].

Besides the Tikhonov regularization, there are other techniques that can be used to stabilize the reconstruction. For example, it is possible to use a truncated singular value decomposition (TSVD) or iterative regularization techniques, such as the Landweber method, the conjugate gradient method, or the Kaczmarz method [3], [4]. All of these methods are designed to reduce the influence of data-noise to the reconstruction.

However, in practice there are further error sources besides data-noise that might generate instabilities of the reconstruction. For example, a system matrix describes the real measurement process only approximately and, hence, introduces inaccuracies to the reconstruction. For some novel imaging modalities, such as Magnetic Particle Imaging (MPI), there are no proper model descriptions found yet.

In such cases, the system matrix is recorded in a calibration procedure and all entries of the system matrix are corrupted by noise [2]. A Tikhonov regularization and other regularization methods do not take that noise into account and lead to errors in the reconstruction.

In the case of MPI, different approaches have been proposed to account for such model uncertainties. A commonly used strategy is an increase of measurements per matrix entry followed by an averaging. In [5], a subtraction of background noise from each matrix entry was used. Moreover, denoising the system matrix by use of frequency domain filters has been introduced and evaluated in [6].

The idea of these approaches is to modify the system matrix in a preprocessing step followed by some standard regularization techniques. In this work, we propose a one-step regularization strategy that does not require any preprocessing of the system matrix. For that, a solution based on Total Least Squares could be a possible access, nevertheless in our approach, we generalize the formulation of Tikhonov regularization (3) for the case of reconstruction problems with data- and model-errors. In numerical experiments, we consider the case of CT and show that our approach is able to outperform a classical Tikhonov regularization when the system matrix is corrupted by noise.

2 Theory

Our goal is to approximately reconstruct x from the noisy data y in (2), given a perturbed system matrix A. We consider a situation where the data y as well as the system matrix A are not known exactly. A classical Tikhonov regularization (3) only addresses the case of noisy data by minimizing an energy functional with respect to the sought variable x. In order to address data- and model-errors at the same time, we formulate an extended Tikhonov functional with respect to both variables, the object x and the system matrix A, as follows:

$$\min_{A,x} \|Ax - y\|_2^2 + \delta\|A - S\|_F^2 + \rho\|x\|_2^2. \quad (4)$$

Here, S denotes the given erroneous system matrix and $\|A-S\|_F^2 = \sum_{m,n} |a_{m,n} - s_{m,n}|^2$ measures the deviation of matrix A from S with respect to the Frobenius norm. The approach given by (4) is more flexible than (3) as it allows the matrix A to vary and so to optimize the reconstruction also with respect to model errors. The parameter δ controls how close A is to the given system matrix S. Note that $\delta = 0$ leads to the classical Tikhonov regularization (3). In order to calculate a solution for (4), we use alternating minimization, i.e., we initialize $A^1 = S$ and, for $n \geq 1$, iterate the procedure

$$x^n = \arg\min_x \|A^n x - y\|_2^2 + \rho\|x\|_2^2 , \quad (5)$$

$$A^{n+1} = \arg\min_A \|Ax^n - y\|_2^2 + \delta\|A - S\|_F^2 . \quad (6)$$

The first step is a classical Tikhonov regularization for x, whereas the second step determines a matrix A (given x) that minimizes the data error and the matrix error at the same time.

3 Material and Methods

We implemented the alternating minimization in MATLAB and tested it for the reconstruction problem of CT, in which case the measurement process is modeled by the Radon transform, [1]. To that end, we generated a system matrix for a given image size and a given measurement setup (as described below) by applying MATLAB's radon function to a delta-sample located at each of the pixel positions. In order to simulate model errors, we added a Gaussian white noise with noise level of 3% (i.e., highest added noise value is 3% of the largest matrix element) to the non-zero matrix elements.

For the evaluation of our approach, we use two different simulated data sets (corresponding to two different objects). First, we consider a circle phantom of size 60×60 and calculate the measurements for 180 equispaced angles distributed in the range of 0 to 179 degrees and 145 equispaced detector positions. In order to avoid an inverse crime, we evaluated the Radon transform of the circle image analytically. In addition to that, we added Gaussian white noise to the measurements with a noise level of 4%. Second, we consider the Shepp-Logan phantom of size 100×100 and calculate the measurements for 100 equispaced angles distributed in the range of 0 to 179 degrees and 145 equispaced detector positions. This time, we use MATLAB's radon function for the computation of measurement data. As in the first case, we added Gaussian white noise to the data with a noise level of 4% to simulate measurement errors.

We calculated reconstructions for both data sets by using a standard Tikhonov regularization (3) and by using the proposed method (4). In order to obtain the Tikhonov regularization, we calculated a solution to the equation $(A^T A + \rho I)x = A^T y$ by using MATLAB's backslash operator which we also use for the first step of our alternating minimization procedure (5). For the second step (6), we calculated the minimizer analytically.

In order to choose the parameters in (4), we performed a series of reconstructions for different combinations of δ and ρ by using 10 iterations of the procedure given by (5) and (6). To that end, we first set $\delta = 0$ and calculated solutions with respect to nine logarithmically spaced values of ρ, namely 10^{-4}, 10^{-3}, ..., 10^4, and choose the "best" parameter ρ^*. In the next step, we used that ρ^* to calculate solutions for logarithmically spaced values of δ, precisely 10^{-4}, 10^{-3}, ..., 10^4, and choose the "best" parameter δ^*. The strategies for choosing the "best" parameters will be discussed in the next section.

4 Results and Discussion

The result section is separated into three parts. In the first part, the results for the PSNR based choice of regularization parameters are presented, whereas in the second part results for the manually determined regularization parameters are shown. In the third part, a consideration on the number of iterations takes place.

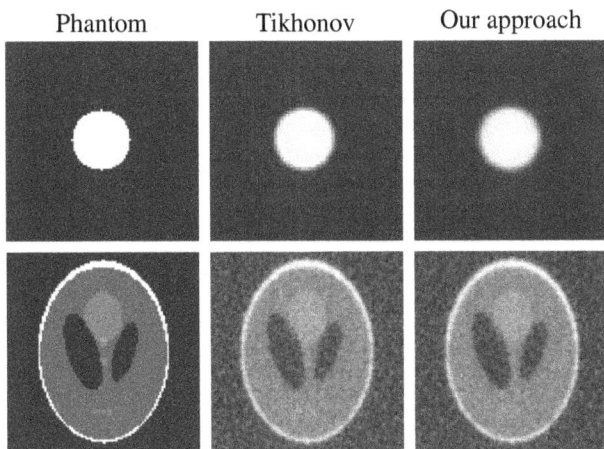

Figure 1: Reconstruction results for the PSNR based choice of regularization parameters for the circle and Shepp-Logan phantom. Our approach has been computed for ten iterations.

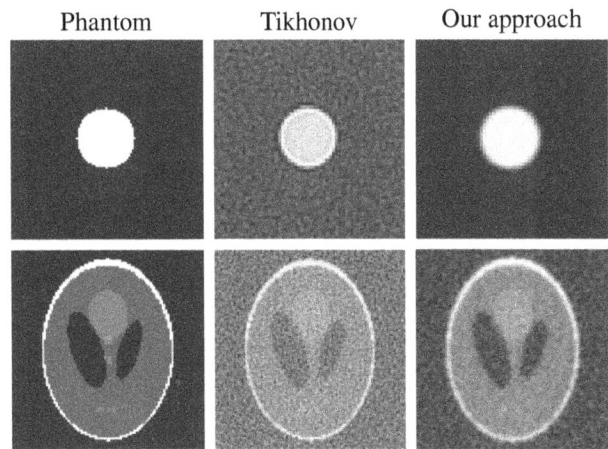

Figure 2: Reconstruction results for the manually chosen regularization parameters for circle and Shepp-Logan phantom. Our approach has been computed for ten iterations.

PSNR based choice of regularization parameters.
Here, we choose the regularization parameters in a classical way, where the regularization parameter ρ^* of the Tikhonov solution with the highest PSNR value is selected as basis for the calculation of our approach. In the second step, we choose δ^* that leads to the image with the highest PSNR value.

The results are shown for both data sets in Fig. 1. In the first row, the results for the circle phantom, while in the second row, the results for the Shepp-Logan phantom are visualized. In Table 1, the corresponding PSNR values and regularization parameters for both test setups are listed.

Considering the results of the circle phantom, the PSNR value of the result of our approach and the PSNR value of the Tikhonov solution are approximately equal. In contrast to this, the visual impression of the result of our approach shows a reduction of the background noise.

In case of the Shepp-Logan phantom, the solution of our approach results in a higher PSNR value than the Tikhonov solution. The visual impression is comparable.

Manually determined regularization parameters.
Since the choice of two regularization parameters is a task with many options, the aforementioned combination must not necessarily result in the best reconstruction results with respect to the PSNR value. In a manually chosen parameter selection, we found combinations of regularization parameters that lead to a higher PSNR value than the results that are presented above.

In this section, the best solution of our approach in terms of PSNR, independent from the best Tikhonov solution in terms of PSNR, is searched. That means, that the regularization parameter ρ^* that results in the Tikhonov solution with the highest PSNR value must not necessarily be the most suitable basis for our approach. In Fig. 2, the Tikhonov solution that serves as a basis for our approach as well as the result of our approach are visualized. It can be seen in Fig. 2 likewise in Table 1, that in both test setups, there is a parameter set for our approach that results in a

better solution, in terms of PSNR, than the best Tikhonov solution.

The visual improvements from the corresponding Tikhonov reconstruction to the solution of our approach lay evidently in the contrast and the reduction of the background noise.

Table 1: PSNR values and regularization parameters ρ and δ for the results presented in Fig. 1 and Fig. 2.

	Phantom	Image	PSNR	Reg. param. ρ	δ
PSNR	Circle	Tikhonov	22.64	10^{-2}	–
	Circle	Our appr.	22.52	10^{-2}	10^4
	Sh.-Logan	Tikhonov	16.04	10^{-2}	–
	Sh.-Logan	Our appr.	16.26	10^{-2}	10^4
Manually	Circle	Tikhonov	15.10	10^{-1}	–
	Circle	Our appr.	23.03	10^{-1}	10^2
	Sh.-Logan	Tikhonov	12.80	10^{-1}	–
	Sh.-Logan	Our appr.	16.37	10^{-1}	10^{-3}

Influence of the number of iterations. In order to evaluate the influence of the number of iterations on the reconstruction results, it is varied for our approach. In this part, only the Shepp-Logan setup is considered. The results are presented in Fig. 3, while the corresponding PSNR values can be found in Table 2. In the first and second column, the phantom and the Tikhonov solution without considering the system matrix update are shown as reference images. In the following columns, the solutions for our approach with 10, 20, 50, and 100 iterations are shown. For the results in the first row, the regularization parameters of the manually selection have been used ($\rho = 10^{-1}$, $\delta = 10^{-3}$). With the increasing number of iterations, the background noise is reduced, but also an increase of the smoothness is observable. The second row in Fig. 3 visualizes the results of manually chosen regularization parameters, where the PSNR value is increased until the 50th iteration and is higher than the best result in the first row with respect to the PSNR value. In this

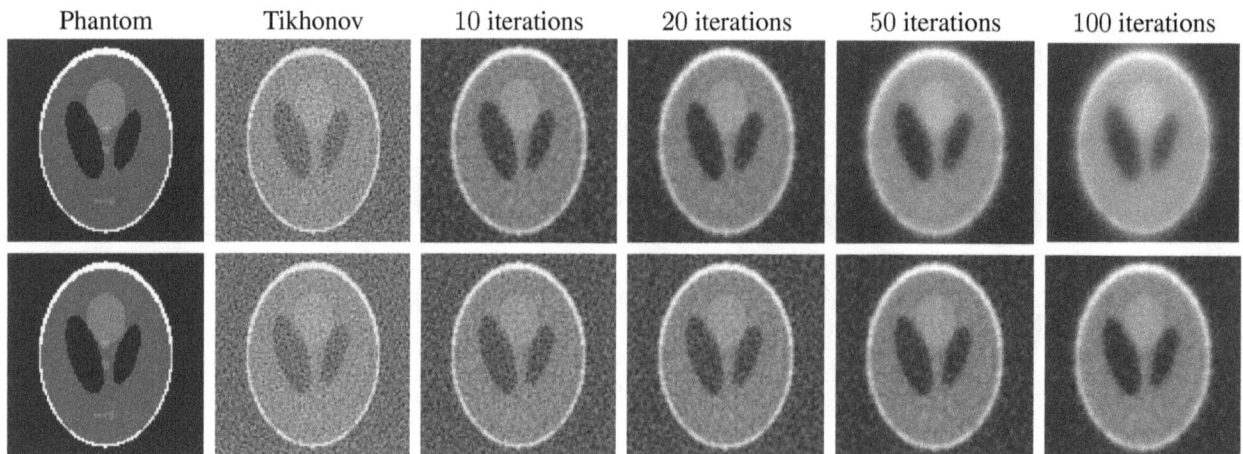

| Phantom | Tikhonov | 10 iterations | 20 iterations | 50 iterations | 100 iterations |

Figure 3: Reconstruction results for manually chosen regularization parameters $\rho = 10^{-1}$ and $\delta = 10^{-3}$ in the first and $\rho = 10^{-1}$ and $\delta = 10^3$ in the second row. The last four columns show the results for our approach with different numbers of iterations.

example, one can see the complexity of choosing the right regularization parameters. Taking only the reconstruction after ten iterations into account, the parameters in the first row seem to result in the best solution, while considering a higher number of iterations the solutions in the second row lead to an even better solution in terms of the PSNR.

In the second row, the stepwise improvement of the image quality is also observable in the contrast and the noise reduction of the image.

Table 2: PSNR values and regularization parameters δ with $\rho = 10^{-1}$ for the results presented in Fig. 3.

| | | Our approach / Iterations | | | |
δ	Tikhonov	10	20	50	100
10^{-3}	12.80	16.37	16.72	15.03	13.29
10^3	12.80	14.77	15.80	16.79	16.34

5 Conclusion

In this paper, a reconstruction problem is considered, where both data- and model-errors are present. In oder to take both error types into account, we formulated an extended variant of the Tikhonov regularization, which we processed by an alternating minimization scheme. The algorithm is tested on simulated data with synthetically produced errors and compared to the classical Tikhonov solution that originally only takes the data-errors into account.

In this work, we observed that the proposed approach is working well in a simulated environment and may lead to an improved reconstruction compared to the Tikhonov solution.

Overall, our approach shows to have potential for imaging methods with both data- and model-errors. Since it contains a high number of regularization parameters a deeper consideration of the right choice of those has to be done.

In the future work, we will test the proposed approach on measured data in order to analyze the functionality for an actual application.

Acknowledgement

The work has been carried out at the Department of Applied Mathematics and Computer Science, Technical University of Denmark, within the HD-Tomo group of Prof. Per Christian Hansen. Special thanks to Martin Skovgaard Andersen and Yiqiu Dong for their help and advises during the investigation. Thanks for the support to Prof. Thorsten M. Buzug from the Institute of Medical Engineering, Universität zu Lübeck. The internship was financially supported by the ERASMUS+ grant.

6 References

[1] T. M. Buzug, *Computed Tomography: From Photon Statistics to Modern Cone-Beam CT*, Springer Verlag, Berlin-Heidelberg, 2008.

[2] M. Grüttner, et al., *On the formulation of the image reconstruction problem in magnetic particle imaging*, Biomedical Engineering, vol. 58, no. 6, pp. 583-591, 2013.

[3] P. C. Hansen, *Discrete Inverse Problems-Insight and Algorithms*. SIAM, Philadelphia, 2010.

[4] A. Rieder, *Keine Probleme mit inversen Problemen*. Vieweg, Wiesbaden, 2003.

[5] K. Them, et al., *Sensitivity enhancement in Magnetic Particle Imaging by Background Subtraction*, IEEE Transactions on Medical, vol. PP, no. 99, 2015.

[6] A. Weber, J. Weizenecker, U. Heinen, M. Heidenreich and T. M. Buzug, *Reconstruction Enhancement by Denoising the Magnetic Particle Imaging System Matrix Using Frequency Domain Filter*, IEEE Transactions on Magnetics, vol. 51, no. 2, 2015.

11

Image Processing III

Model-based Respiratory Motion Compensation in MRI-guided Radiotherapy

I.Y. Ha [1], M. Wilms [2] and M. P. Heinrich [2]

[1] Medizinische Informatik, Universität zu Lübeck, in.ha@student.uni-luebeck.de

[2] Institut für Medizinische Informatik, Universität zu Lübeck, { wilms, heinrich }@imi.uni-luebeck.de

Abstract

Motion tracking is a computational approach used in radiotherapy. While currently used methods find good estimations for specific locations, it is still difficult to track the organ motion of the entire image domain. Especially in *gated treatment*, where the beam is on while the target is located in a pre-defined location, it could raise a problem, as there is no information on the movement of other organs at risk. To address this problem without increasing computation time, we propose a new motion estimation method for MRI-guided radiation therapy, which incorporates a pre-learned motion model and enables the approximation of a dense motion field from the sparse motion information. A sparse motion field is computed using self-similarity context descriptor and block-matching in 2D MRI slices. The method is evaluated for different numbers of sparsely sampled interest points and is able to plausibly approximate dense motion fields using sparse correspondences.

1   Introduction

Recently, magnetic resonance imaging (MRI) has been incorporated for the first time into image-guided radiation therapy (IGRT) [1]. Because of its non-invasiveness, absence of ionizing radiation and excellent soft tissue contrast, it has often been used in the planning phase for treatment settings. Currently, the first generation low-field MRI-guided radiotherapy systems are in clinical use and the first generation high-field MRI-guided radiotherapy system is planned to be released in 2017 [1]. Although the availability of MR scans during the treatment may help to alleviate the problem of multi-modal image registration, the current systems cannot provide real-time 3D volumetric images of the target area and the problem of inaccuracies caused by respiratory organ motion still remains. One of the commonly used approaches to solve this problem is *motion tracking*, which aims to estimate organ motion and provide the exact location of the tumor to enable the *gating* of the radiation beam. One variation of direct motion tracking, which does not require the implementation of fiducial markers is motion tracking based on a motion model. A patient-specific model can be generated in the planning phase of the therapy using 3D MR images taken over time (i.e. 4D) under free-breathing. This can provide the information of plausible physiological deformation. There have been significant improvements in the development of respiratory motion models. However, due to the prohibitive computational demand [2] and the use of template matching [3], which only focuses on estimating the tumor location, it may fail to track other organs at risk correctly.

By incorporating prior knowledge of plausible physiologi-cal deformation into the motion estimating step, the accuracy of intra-fraction organ motion prediction can be improved reliably, not only for a specific template location such as the tumor, but also for the entire image domain, i.e. the thorax. Moreover, this method can be realized without implanting fiducial markers and preserves the computational speed.

The aim of this work is to generate a patient-specific respiratory motion model based on 4D MR images and estimation of respiratory motion by means of sparse feature points extracted from 2D MR images acquired during the treatment in real-time. In order to do this, the following steps will be taken:

- computation of a motion model from 3D dense motion fields using the *principal component analysis* (PCA)
- feature points extraction from a 2D reference MR image using the *Harris corner detector* [6]
- computation of a sparse motion field for selected feature points using the *self-similarity context* (SSC) descriptor [9] and block-matching
- estimation of a 3D dense motion field for the entire image domain from sparse motion field based on the model using regression analysis.

2   Material and Methods

This section provides the information on the MRI data used for testing, the 3D registration and the feature point detection algorithms. Furthermore, a brief description on how the motion model was built using PCA and how the motion vectors for the feature points are computed, are given.

2.1 MRI data and image registration

Full 4D MRI data of the thoracic region from two volunteers are taken from [4]. Both datasets consist of 200 timesteps and 50 and 52 slices (5 mm) each with the 2D image dimension of 224×224 (1.2050 mm and 1.2949 mm). The dataset of *patient 1* has a total of 13 cycles of respiratory motion, whereas the dataset of *patient 2* contains 26 respiration cycles with smaller displacements inside the lung, which indicates shallower breath. In the absence of real data from a MRI-guided radiotherapy system, both datasets are divided into 2 sets containing half of the breathing cycles each. The first set of the data represents 4D MR images obtained in the planning phase and the second set is used as online MR images for testing purposes.

The computation of displacement fields (or motion fields) for each timestep is performed using the 3D registration approach from [5] by registering the timesteps 2–200 to the first timestep (the reference). The motion fields of the training data acquired from the registration are used to compute a model and are considered as the ground truth for the test data, which had to be estimated using our method.

2.2 Motion model using PCA

Three matrices for a motion field are obtained for every timestep from the 3D registration program. These are first vectorized and then concatenated to build a motion vector $\mathbf{v}_i$ for timestep i

$$\mathbf{v}_i = \left[x_1, ..., x_N, y_1, ..., y_N, z_1, ..., z_N \right]^T \in \mathbb{R}^{3N} \quad (1)$$

where x, y and z denote the displacement of each point in corresponding direction and N is the number of voxels in the image. A 3D dense motion model is generated using the PCA, which finds the principal components (PC), where the first component accounts for the largest variance of the data and the next component the second largest and so on. A data matrix $X \in \mathbb{R}^{3N \times m}$ is generated using all m motion vectors from the training data. The mean motion vector $\bar{\mathbf{v}}$, which contains the mean displacement values at each point, is subtracted from each motion vector $\mathbf{v}_i$ in the data matrix X. The PCs (the basis of a model) can be found by diagonalizing the covariance matrix C of data matrix X:

$$C = \frac{1}{m-1} X X^T = U \operatorname{diag}(\Sigma^2) U^T, \quad (2)$$

where U is the matrix of eigenvectors and $\operatorname{diag}(\Sigma^2)$ a diagonal matrix of the squared eigenvalues Σ^2 of C. Each column of the eigenvector matrix U is an eigenvector $\mathbf{u}_i$ and the corresponding eigenvalue describes the amount of variation in the dataset that can be described by that eigenvector. The eigenvectors form a basis i.e. a model and a 3D motion field at a certain timestep can be approximated as a weighted sum of the basis:

$$\mathbf{v} \approx \bar{\mathbf{v}} + \sum_{i=1}^{k} w_i \mathbf{u}_i \quad (3)$$

where $\mathbf{v}$ denotes the estimated dense motion field, k is the number of the basis used, $w_i \in \mathbb{R}$ are corresponding weights and $\mathbf{u}_i$ are the basis displacement fields. Since respiratory motion is rather simple compared to other natural motions, it can be approximated with a few PCs, i.e. we need a small number of eigenvectors [7], [8].

2.3 Feature points and Block-Matching

To estimate the motion during treatment, motion vectors are computed at feature points extracted from the reference 2D MR image slices using the Harris detector function of [6]. In Fig. 1, the response of the Harris detector and the feature points selected by a threshold value of 3 are shown. The threshold refers to the response value of the Harris detector, i.e. by setting it as 3, the points with response value smaller than 3 are not selected. Non-maximum suppression is used to obtain uniform distribution of the feature points.

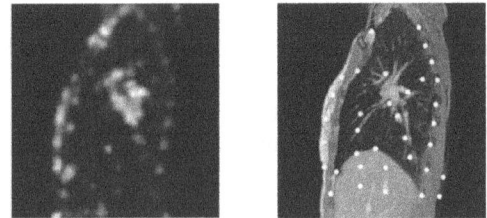

Figure 1: Response of Harris detector (left), feature points found by Harris detector are marked with dots (right).

The computation of displacement vectors for feature points is performed using block-matching. A small patch of image data is selected for a feature point in the reference image and the best match for this patch is searched for in the selected region of the target image. An example of the block-matching process for a feature point is shown in Fig. 2, where the box with the solid line describes the reference patch, the box with the dashed line is the corresponding patch in the target image and the thick solid line indicates the search region. The feature point position in the reference image is marked with a cross and the corresponding point with a circle.

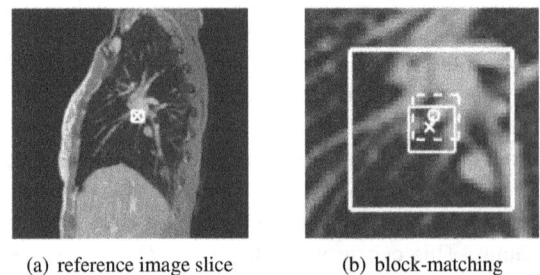

(a) reference image slice   (b) block-matching

Figure 2: Block-matching example for a feature point. (solid line: reference patch, dash line: target patch, thick line: block-matching region in the target image)

Although an MR image is shown in Fig. 2, to be able to find a correct corresponding point regardless of the difference in image qualities, the algorithm for computing a displacement vector should not be influenced by local changes in image contrast. The self-similarity context (SSC) descriptor proposed in [9] is used for this reason, to determine the correspondence of points between reference and test images. SSC is based on the context of a point in its neighborhood (patch), and is therefore insensitive to image noise and invariant to local image contrast changes. The self-similarity descriptor $\mathcal{S}(I, \mathbf{x}, \mathbf{y})$ is defined as:

$$\mathcal{S}(I, \mathbf{x}, \mathbf{y}) = \exp\left(-\frac{\mathcal{SSD}(\mathbf{x}, \mathbf{y})}{\sigma^2}\right) \qquad \mathbf{x}, \mathbf{y} \in \mathcal{N} \quad (4)$$

where $\mathcal{N}$ denotes a certain neighborhood and $\mathbf{x}$ and $\mathbf{y}$ define the center of reference and target patches in the image I. σ^2 is the noise estimate and $\mathcal{SSD}(I, \mathbf{x}, \mathbf{y})$ is the sum of squared distances between two points.

Based on the SSC descriptors, the displacement vectors for the selected feature points are computed using the block-matching algorithm, which results in a sparse motion field. The use of sparse feature points reduces computation time significantly, but it only provides a fraction of information of the full motion field that we need to estimate. However, the dense displacement fields can be reconstructed based on this information and by including prior knowledge from our motion model using regression analysis.

2.4 Reconstruction of dense motion fields

A motion field can be approximated by the linear combination of the basis with appropriate weights. We can find optimal weights $\mathbf{w}^*$ by least squares fitting as:

$$\mathbf{w}^* = \underset{\mathbf{w}}{\mathrm{argmin}} \, \|U_s\mathbf{w} - (\mathbf{v}_s - \overline{\mathbf{v}}_s)\|_2^2 \in \mathbb{R}^k. \quad (5)$$

The index s indicate that the vector or matrix is sparse, consisting of only the values at the feature points, i.e. the vectors and matrices with this index have the length M, where $M < N$ is the number of feature points. The dense motion field for the whole image can be reconstructed using $\mathbf{w}^*$ and original basis $U \in \mathbb{R}^{3N \times k}$, where k denotes the number of eigenvectors used for the estimation. A similar approach has been applied for optical flow computation in [3], [8].

3 Results and Discussion

We have evaluated our method by comparing the motion fields estimated by the sparse displacement fields. For a quantitative evaluation, the mean Euclidean distances (the mean estimation errors) between the ground truth motion field and the estimated dense motion fields are computed:

$$d(\mathbf{p}, \mathbf{q}) = \frac{1}{N} \sum_{i=1}^{N} \|\mathbf{p}_i - \mathbf{q}_i\|_2 \qquad \mathbf{p}_i, \mathbf{q}_i \in \mathbb{R}^3 \quad (6)$$

where $\mathbf{p} \in \mathbb{R}^{3 \times N}$ denotes the 3D motion field from the registration algorithm, $\mathbf{p}$ the approximated dense motion field and N the number of points in $\mathbf{p}$ and $\mathbf{q}$. To evaluate our method, we compare $d(\mathbf{p}, \mathbf{q})$ values for the results of following input data:

1. The ground truth dense motion field approximated by the PCA model
2. The ground truth sparse motion field approximated by the PCA model
3. The sparse motion field computed using the block-matching
4. The sparse motion field without large outliers ($\Delta < 3$ mm, $\Delta < 7$ mm)

For the input data 2, 3 and 4, the estimation of the dense field is made using different numbers of feature points, i.e. for different sparsities of the motion field. For the input data 4, the Euclidean distances between displacement vectors of ground truth and block-matching results for each feature point are calculated and defined as Δ. The feature points with $\Delta > 3$ mm (about 40% of the points) and $\Delta > 7$ mm (about 10%) are eliminated from the sparse fields to evaluate the influence of the accuracy and the sparsity of the input data.

Prior to the evaluation, parameters for the feature point detection, the SSC descriptor and the block-matching as well as the number of eigenvectors k for the model are determined. The mean estimation errors for a selected test data with different numbers of eigenvectors are compared to determine the model. Due to outliers in sparse motion field computed by the block-matching algorithm, the estimation of the dense motion field became worse, when too many eigenvectors were used. The best motion estimation was made when 84% of variation ($k = 5$) can be described for *patient 1* and 74% ($k = 7$) for *patient 2*. Heuristic methods are used to determine parameters for the SSC descriptor ($\sigma = 0.2$ and the distance between patch centers $\delta = 3$) and the block-matching (*patient 1*: block size 5, window size [5, 8] and *patient 2*: block size 3, window size [4, 7]). Since result of the estimation became worse with a large window, the values are selected according to the maximum displacement values for each direction.

The evaluation result is shown in Table 1. The errors were computed using masks to account just for the motion inside the patient's body. It is clear from (c) and (f) that a dense 3D motion field for the entire image domain can be adequately approximated with a small number of displacement vectors. The approximation with more feature points (500 and 1000

Table 1: Mean estimation errors from different input fields (300 feature points for (c)–(f)). (GT: ground truth, BM: block-matching)

Input field	patient 1 (mm)	patient 2 (mm)
(a) no compensation	2.57 ± 0.99	1.38 ± 0.57
(b) dense field (GT)	0.80 ± 0.17	0.59 ± 0.13
(c) sparse field (GT)	0.79 ± 0.17	0.59 ± 0.13
(d) BM	1.89 ± 0.70	1.10 ± 0.39
(e) BM ($\Delta < 7$ mm)	1.17 ± 0.42	1.06 ± 0.39
(f) BM ($\Delta < 3$ mm)	0.89 ± 0.26	0.68 ± 0.18

points) resulted in similar estimation errors, with an average difference less than 0.1 mm (data not shown). Comparison using different parameters for block-matching in the lung region to enable a better consideration of lung motion also made no improvement in estimations (data not shown).

Although the number of feature points used for the sparse motion field does not affect the estimation of the dense motion field using our model, the accuracy of displacement vectors had an influence on the estimation. As shown in Table 1, the dense motion field obtained via block-matching made almost no compensation of motion. By analyzing the

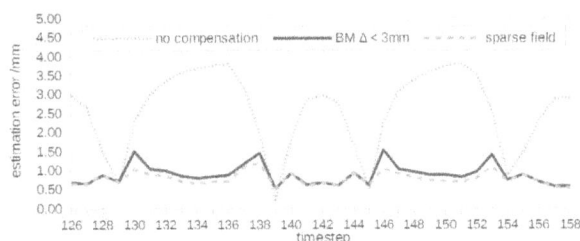

Figure 3: Mean estimation errors of dense motion field estimated using 300 feature points.

block-matching result qualitatively, we discovered that this is caused by the discrepancies between displacement vectors obtained by block-matching and the ground truth. To analyze the robustness of the model, we have omitted the feature points with large differences from the sparse motion field (Table 1 (e) and (f)). Although the sparsity is increased by doing this, the estimation errors have decreased. Fig. 3 shows the estimation errors made by sparse motion field computed by block-matching with such feature points.

4   Conclusion

In this paper, the feasibility of estimating 3D organ motion with sparse input data from 2D MR slices using the patient-specific motion model is studied. A motion model is constructed from planning 4D MRI data using PCA and tested with sparse displacement fields obtained using block-matching of SSC descriptors. It is shown that dense 3D motion fields can be approximated adequately from sparse displacement fields using a least-squares fitting based on the respiratory motion model. However, outliers in the sparse displacement fields can have a considerable influence on the estimation of the dense field and should be dealt with a proper method. The number of feature points used, i.e. the sparsity of the input field seems to have less influence on the estimation as far as the displacement vectors approximate the motion accurately.

For further improvement of this approach, an iterative method can be implemented to deal with the outliers in sparse motion field. Estimating motion for individual feature point locations can lead to less accurate result in the case of discontinuous sliding motion. This problem can be solved by an additional coupling term [10], which attempts to keep the initially estimated motion vector and the sparse motion field regularized by the model. In the future, we plan to integrate this coupling term and further refine our approach by using a faster implementation e.g. for the SSC

descriptors and the block-matching.

Acknowledgement

The work has been carried out at Institut für medizinische Informatik, Universität zu Lübeck.

5   References

[1] U. Oelfke, *Magnetic Resonance Imaging-guided Radiation Therapy: Technological Innovation Provides a New Vision of Radiation Oncology Practice*. Clinical Oncology, vol. 27, no. 9, pp. 495–497, 2015.

[2] C. Buerger, T. Schaeffter and A. P. King, *Hierarchical adaptive local affine registration for fast and robust respiratory motion estimation*. Medical Image Analysis, vol. 14, no. 4, pp. 551–564, 2011.

[3] F. Preiswerk et al., *Model-guided respiratory organ motion prediction of the liver from 2D ultrasound*. Medical Image Analysis, vol. 18, no. 5, pp. 740–751, 2014.

[4] D. Boye, G. Samei, J. Schmidt, G. Szekely and C. Tanner, *Population based modeling of respiratory lung motion and prediction from partial information*. SPIE Medical Imaging 2013 : Image Processing, vol. 8669, pp.86690U-1–86690U-7, 2013.

[5] M. P. Heinrich, M. Jenkinson, M. Brady and J. A. Schnabel, *MRF-based deformable registration and ventilation estimation of lung CT*. IEEE Transaction on Medical Imaging, vol. 32, no. 7, pp.1239–1248, 2013.

[6] P. D. Kovesi, *Matlab and Octave functions for computer vision and image processing*. Available: http://www.peterkovesi.com/matlabfns [last accessed on 18.01.2016].

[7] R. Li et al., *On a PCA-based lung motion model*. Physics in Medicine and Biology, vol. 56, no. 18, pp. 6009–6030, 2011.

[8] J. Wulff and M. J. Black, *Efficient sparse-to-dense optical flow estimation using a learned basis and layers*, in: IEEE Conference on Computer Vision and Pattern Recognition (CVPR) 2015, IEEE, Boston, MA, pp. 120–130, 2015.

[9] M. P. Heinrich, M. Jenkinson, B. W. Papiez, M. Brady and J. A. Schnabel, *Towards realtime multimodal fusion for image-guided interventions using self-similarities*, Medical Image Computing and Computer-Assisted Intervention (MICCAI) 2013, LNCS 8149, pp. 187–194, 2013.

[10] M. P. Heinrich, B. W. Papiez, J. A. Schnabel and H. Handels, *Non-parametric discrete registration with convex optimisation*, in: Biomedical Image Registration, Springer International Publishing, pp. 51–61, 2014.

Phase-Based Motion Processing

K. Kreft [1] and A. Mertins [2]

[1] Medizinische Ingenieurwissenschaft, Universität zu Lübeck, karl.kreft@student.uni-luebeck.de
[2] Institute for Signal Processing, Universität zu Lübeck, mertins@isip.uni-luebeck.de

Abstract

The goal of this work is to use a motion processing scheme to magnify and attenuate motions in videos and to subsequently extend it to reduce motion artifacts in MRI. It is based on a method to reveal temporal variations in videos that are difficult or impossible to see with the naked eye and display them in an indicative manner. This paper implements this scheme of video decomposition (Laplace-, complex steerable- and Riesz Pyramids), temporal filtering, amplification and reconstruction, and extends it by an additional decomposition method, the 2D short-time Fourier transform. Several videos are amplified using this method and are being compared to the original implementation. The resulting amplified videos achieve similar magnification, but are noisier than the previous implementation.

1 Introduction

A vast amount of phenomena exhibit motions, that are too small to see with the naked eye and attempts have been made to make these motions visible in videos. In [8], Lagrangian motion processing is described, which tracks motion over time. Motion vectors are calculated and amplified, followed by imprinting an estimation of the amplified motion into the frame. This method is computationally costly, but is capable of amplifying fine point motion by a large amount. It is also susceptible to noise. Lagrangian motion processing was the initial approach to the motion amplification issue. Contrary to the Lagrangian approach, the Eulerian approach ([1]-[2]) does not track motions in videos. It averages motions over time periods and amplifies them directly. It is better than Lagrangian motion processing for small motions and smooth structures and does not assume particular types of motion. It is also susceptible to noise at larger amplifications in it's first implementation [3], but very resistant to noise in the phase-based implementation [1]. The Eulerian method is used in this paper.

1.1 Phase in Signals and Images

While [3] amplifies real valued motion information in videos, [1] changed this approach and manipulates the phase information of videos. According to [5], phase information is more important than amplitude information for image- and audio reconstruction. Oppenheim [4] has shown that by Fourier transforming two images of the same size and swapping the phase, the reconstructed images are more similar to the image whose phase was used. In [5], a similar experiment that uses audio data is described. Two audio files, one with a male and the second one with a female voice, were phase-swapped and the reconstructed record-

ings were – when played to a human listener – recognizable as the ones whose phase was used, both in gender and meaning of the sentence.

The importance of phase information in images has inspired its implementation in various tasks such as edge and corner detection, image segmentation and motion processing [1]. Phase is highly resistant to noise and contrast distortions - features desirable in image processing.

2 Materials and Methods

The publicly accessible MATLAB code from [3] was used to reverse engineer a phase-based motion magnifying approach proposed by [1] and was later improved by implementing different decompositions. 9 videos from [3]'s online repository - varying from small 280 x 280 pixel videos to 1776 x 904 pixel greyscale videos - were motion magnified using the same methods as ([1]-[3]) and a 2D-short-time Fourier transformation (STFT) implementation. The length was 6 to 33 seconds and the filmed objects were amongst other things humans, trees, guitar strings and cranes. All videos showed subtle motions, almost unperceivable to the human eye.

2.1 Amplitude and Phase

The frequency spectrum X of a signal x can be separated into amplitude and phase. Amplitude mag is defined as

$$mag = \sqrt{\Re(X)^2 + \Im(X)^2} \qquad (1)$$

where $\Re(X)$ is the real part and $\Im(X)$ the imaginary part of the frequency spectrum, while the phase ϕ is denoted as

$$\phi = Arg\left(\frac{\Im(X)}{\Re(X)}\right). \qquad (2)$$

Here, Arg is the argument or the angle between the imaginary and real part of the frequency spectrum, corrected by π for negative and $\frac{\pi}{2}$ for zero inputs.

2.1.1 Global Phase Translation

A global circular shift of a signal is represented by a change in phase without affecting the amplitude. This is equally true for both 1D and 2D signals.

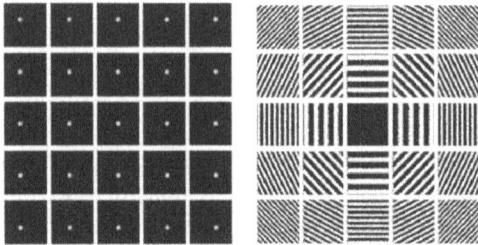

Figure 1: Correlation between global translation and phase difference. The left image shows translation of a smooth circle by steps of 4 pixels. The right figure shows the corresponding phase difference. Amplitude does not change.

The phase difference between a globally shifted object in an image and its unshifted counterpart is predictable and identical for diverse simple phantoms if the shift direction is known, so that one can shift objects reliably by manipulating the phase. Figs.1 and 2 show some properties of globally shifted images.

Figure 2: Different phantoms (upper row) were shifted by $[-1-2]$ pixels. Phase difference (lower row) between different unshifted and shifted phantom is almost identical.

2.1.2 Local Phase Translation

While the global phase manipulation performs well for global translations in an image or video, there are adjustments to be made when analyzing complex or multiple translations. When analyzing real-life videos, one does seldom deal with global motion, but more often local motion of different length and direction. Phase-extracting methods with limited spatial support like wavelet transform or 2D-STFT are being utilized to analyze these motions [1].

2.2 Image Decomposition

The first step in magnifying videos is to extract their local phase information. This Paper utilizes decompositions already used by ([1]-[3]). These uncompressed image representations require more memory than uncompressed original images. This is called *overcomplete* representation. The

steerable pyramid used in [1] requires 12 to 56 times more memory than the original image and is thus up to 56 times overcomplete.

2.2.1 2D Short Time Fourier Transform

Using the STFT to analyze complex motions in an image is comparable to subdividing the image into smaller overlapping windows. This limits the impact a local motion can have on the whole image, but also limits the length of motions and thus the possible magnification or compensation. Fig. 3 shows the logarithmic frequency spectrums of a 2D STFT representation.

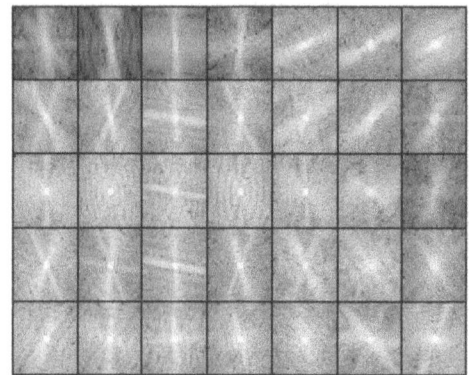

Figure 3: Visual representation of a 32 pixel window length 2-dimensional short-time Fourier transform. Decomposed image shown in Fig.6.

2.2.2 Laplace Pyramid

In [1], the standard Laplace pyramid by [7] was used for image decomposition into different spatial frequency bands by low-pass filtering and downscaling. The image is blurred and subsampled multiple times, resulting in increasingly smaller, more blurred images. This blurring produces non-aliased sub-bands, a desirable feature for image- or motion processing. Fig.4 implies the pyramid structure of this representation. It was used in the first motion magnifying approach of this kind in [3]. It has since then been replaced by other pyramid structures, which are subsequently introduced.

2.2.3 Complex Steerable Pyramids

The steerable pyramid [6] is an overcomplete multiresolution decomposition with oriented, non-aliased sub-bands and steerable filters. The pyramid (representation shown in Fig.5) is built by applying its transfer functions to the discrete Fourier transform of an image to decompose it into different spatial frequency bands. In the frequency domain, the process to build the pyramid is given by

$$\tilde{I}_R = \sum \tilde{S}_{\omega,\theta} \Psi_{\omega,\theta} \tag{3}$$

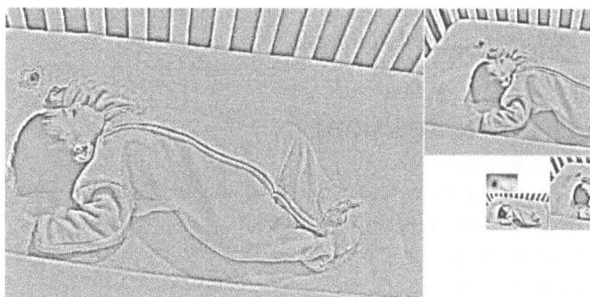

Figure 4: Visual representation of a Laplace Pyramid. Big images are high-pass residue. Small images are band-pass and low-pass.

where $\tilde{I}_R$ is the image representation, $\tilde{S}_{\omega,\theta}$ are spatial frequency (ω) bands with orientation θ and $\Psi_{\omega,\theta}$ are oriented steerable pyramid bands. Reference [1] uses 8 orientations and half-octave band filters, which produces 12 to 56 times overcomplete image representations and is computationally intensive. The complex steerable pyramid was initially used in this paper, but later replaced by the faster, less overcomplete Riesz pyramid.

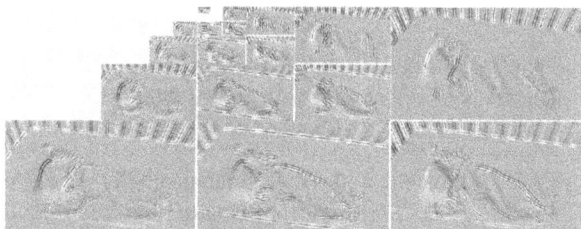

Figure 5: Visual Representation of a real-valued 4 orientation, 4 scales steerable Pyramid.

2.2.4 Riesz Pyramid

The Riesz transform is a multidimensional extension based on the Hilbert transform introduced by [9]. It was used by [1] to improve their previous method by lowering the computation time. The implementation using Riesz pyramid shifts motion in its dominant orientation rather than every direction like the steerable pyramid, but conserves the quality of the previous, slower implementation.

2.3 Temporal Filtering

The second step in magnifying videos is to average movement information according to their frequency. To amplify only medium duration motions (e.g. 15 frames), one can average 15 frames (or image phase information) and obtain the averaged motion from these frame-sets. Short and long frequency motions will not be amplified in this case.

2.4 Phase-Based Motion Processing

The authors of [1] improved the linear Eularian magnification by manipulating the phase of the image and not the spatial image itself. Using the steerable Pyramid or Riesz decomposition, the video is divided into frequency bands $S_w(x,t)$ at space x and time t. The amplified frequency band $\hat{S}_w(x,t)$ is calculated by

$$\hat{S}_w(x,t) := S_w(x,t)e^{(i\alpha B_w)} \qquad (4)$$

where α is the amplification factor and B_w is the temporal bandpassed phase. It improved [3] by increasing the maximum amplification amount and remove the striking noise susceptibility of the previous approaches.

3 Results

The goal of this paper was to rework the "Phase-Based Video Motion Processing" method to fit the requirements of motion artifact reduction in magnetic resonance imaging.

3.1 Phase-Based Motion Processing

The MATLAB application was able to magnify all the videos found in [1]'s repository except for 2 very big videos due to memory limitations. This could be improved by rewriting the code, which for now loads all frames into memory at the same time. The target frequency is controllable via an input parameter and makes it possible to selectively amplify slow or fast motions. The amplification amount is also controllable, but the optimal amount has to be found empirically for every video and target frequency. Strong magnification introduces noise- and ghost artifacts. Since no computational method to quantify a motion magnified video was available, frames of an original and motion magnified video were compared by overlaying the start and the end of a breathing motion (Fig. 7 and 8). The zip fastener was colored using Photoshop for better visibility. As one can see, the unmagnified image has both zipper positions on the same spot since it is a sub-pixel motion, while the magnified image has a distance of about 5 pixels between both zipper positions.

The resulting single core MATLAB code application ran on a Xeon E3-1200 with 12GB of memory and was able to process a 272 x 480 black-and-white video with 300 frames (10 seconds of video footage) in 20 seconds, which is not yet real-time, but easily achievable by porting the code to fast languages as C++. The code is excellent for parallel computing on graphic cards since every frame must be decomposed and reconstructed independently. Even the temporal filtering step is suited for parallel computing. In this implementation, the Parallel Computation Toolbox for multi-CPU was used to speed the process up, halving the computation time with 4 cores. Computing using the slower steerable pyramid implementation with 4 orientations and 4 scales took about 3 minutes on the same machine, but yielded slightly more artifact-free results. Using a half-octave bandwidth with 8-orientations approach as in [1] should increase the computation time by a factor of 4. The implementation using the 2D STFT took about 40 seconds with window size of 32 pixel and step size of 16 pixel. The image needed to be padded to avoid border artifacts, which increased the computation time further. The resulting magnification was not as expressive as using the pyramid implementation, but less noisy. Slight ringing artifacts degraded the image, because a rectangular STFT window was used. Hann or Hemming windows could not be used, because they introduced obvious grid artifacts.

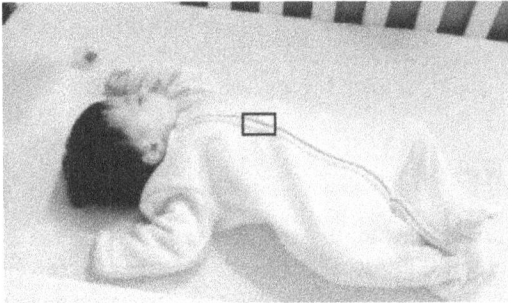

Figure 6: One video frame of a sleeping baby. Breathing is hardly visible in original video.

Figure 7: Overlay of original breathing in and breathing out frames. Local motion averages one pixel.

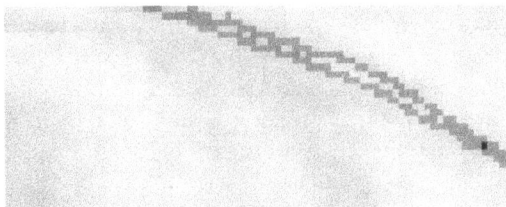

Figure 8: Overlay of magnified breathing in and breathing out frames. Local motion averages 5 pixels.

4 Outlook

For the purpose of using this motion processing scheme for magnetic resonance imaging one has to rethink the temporal filtering aspect. Video motion processing operates by temporal filtering between frames, but MRI imaging only yields one single spatial frame, so that the modification of the method is not trivial. The MRI records k-space rows in constant chronological orders and the pyramid decomposition has qualities, which can separate these frequency band.

So if one can separate these frequency bands in the corrupted MRI k-space representation and analyze the motion between frequency frames, it should be possible to correct motion corrupted MRI images.

Acknowledgement

The work has been carried out at the Institute for Signal Processing, Universität zu Lübeck.

5 References

[1] N. Wadha, M. Rubinstein, F. Durand and W. T. Freeman, *Phase-Based Video Motion Processing*. in: ACM Transactions on Graphics, vol. 32, No. 4, 2013.

[2] N. Wadha, M. Rubinstein, F. Durand and W. T. Freeman, *Riesz Pyramids for Fast Phase-Based Video Magnification*. in: Computational Photography (ICCP), 2014.

[3] H. Wu, M. Rubinstein, E. Shih, J. Guttag, F. Durand and W. T. Freeman, *Eulerian Video Magnification for Revealing Subtle Changes in the World*. in: ACM Transactions on Graphics, vol. 31, No. 4, 2012.

[4] A. V. Oppenheim and J. S. Lim, *The Importance of Phase in Signals*. in: Proceedings of the IEEE, vol. 69, No. 5, pp. 529-541, 1981.

[5] Nikolay Skarbnik, Yehoshua Y. Zeevi and Chen Sagiv, *The Importance of Phase in Image Processing*. CCIT Report, vol. 773, 2010.

[6] E. P. Simoncelli and W. T. Freeman, *The Steerable Pyramid: A Flexible Architecture For Multi-Scale Derivative Computation*. in: Proceedings of the 1995 International Conference on Image Processing, vol. 3, 1995.

[7] P. Burt and E. Adelson, *The Laplacian Pyramid as a Compact Image Code*. in: IEEE Trans. Comm., vol. 31, No. 4, pp. 532–540, 1983.

[8] C. Liu, A. Torralba, W. T. Freeman, F. Durand and E.H. Adelson, *Motion Magnification*. in: ACM Trans. Graph., vol. 24, pp. 519–526, 2005.

[9] D. Garbor, *Theory of Communication. Part 1: The Analysis of Information*. in: Journal of the Institution of Electrical Engineers - Part III: Radio and Communication Engineering, vol. 93, No. 26, pp. 429-441, 1946.

Parallelization of Motion Correction Algorithms for Multi-Band fMRI Data

E. Franke [1], N. Scheel [2], and A. Madany Mamlouk [2]

[1] Medizinische Informatik, Universität zu Lübeck, eric.franke@student.uni-luebeck.de
[2] Institute for Neuro- and Bioinformatics, Universität zu Lübeck, {nscheel,madany}@inb.uni-luebeck.de

Abstract

For resting-state functional MRI data, preprocessing is a fundamental and urgently required step to obtain meaningful data for investigation. It also is a cpu-intensive step and subject to optimization, both in quality and in runtime. The runtime is a restricting factor which will slow down research. This paper discusses the runtime optimization implemented at the Institute for Neuro- and Bioinformatics. The optimization consists of the conversion from local and serial computing to parallel networkcomputing, the efficient use of network resources and the deployment of a parallelization framework in an easy-to-use manner. We will show, that the computing time for a dataset consisting of 208 subjects was reduced from 336 hours to 42 hours, which allows a more efficient and flexible workflow when working with resting-state functional MRI data. Furthermore, this optimization enables the option to alter the preprocessing algorithms to be more complex and cpu-intensive if needed.

1 Introduction

In the last few decades, task-based functional MRI was, beside other methods, commonly used to investigate the human brain and detect its functional regions. This method was particularly succesfull in identifying brain regions which react to a specific stimulus. Researchers were able to identify and map brain regions which correspond to different stimuli (light, shapes, faces, ...) and tasks. Besides its advantages, this method is very specific and provides no information about the general functionality of the human brain. This limitation led to the development of an alternative fMRI-technique: the resting-state fMRI. Resting-state refers to a state, where the subject is not engaged in any task, but also not meant to fall asleep [1]. Measuring resting-state fluctuations in the blood oxygenation level-dependent (BOLD) signal, as measured by resting-state fMRI, may provide valuable information about the neural functional architecture of the human brain [2]. Unfortunately, there are some factors that can mask the BOLD-signal, such as temperature fluctuations, head motion, respiration and the heart rate [1]. Therefore, preprocessing to minimize the effects of unwanted variability is needed and a multitude of different approaches exists [2]. The methods are usually independent of the experimental design and prepare the data for statistical analysis [1]. This paper discusses a runtime-optimization of the cpu-intensive preprocessing algorithm ICA-AROMA [3] (see section 2) as an example. Because ICA-AROMA multiplies the size of the resting-state fMRI data during its execution by a factor greater then seven (from 628MB to 4900MB), the available network storage at the Institute for Neuro- and Bioinformatics is not sufficient to support the calculation and storage of many subjects at once. This leads to the current workflow to calculate and store the data in a serial manner on the user's local workstation. The disadvantage of this approach can be seen in Table 1.

execution	duration	execution	duration
1	96m 32s	16	95m 12s
2	96m 52s	17	98m 21s
3	96m 47s	18	98m 25s
4	95m 52s	19	97m 12s
5	97m 55s	20	98m 32s
6	97m 10s	21	97m 57s
7	96m 39s	22	97m 11s
8	97m 26s	23	98m 08s
9	97m 22s	24	96m 07s
10	97m 20s	25	97m 00s
11	97m 53s	26	97m 28s
12	96m 55s	27	98m 31s
13	97m 05s	28	95m 11s
14	97m 10s	29	97m 23s
15	97m 30s	30	96m 52s
ø		97m 12s	

Table 1: Runtime data for 30 executions of ICA-AROMA on the local workstation "pc28.inb.uni-luebeck.de".

On average, the execution of ICA-AROMA takes 97 minutes and 12 seconds to finish. Depending on the local workstation, one may not have the opportunity for parallel execution. The resting-state fMRI dataset used at the Institue for Neuro- and Bioinformatics was recorded at ultra-high resolution using multiband imaging and contains 208 subjects. This leads to a theoretical runtime of 336 hours for serial computing. One should keep in mind, that ICA-AROMA is just one part of a much bigger preprocessing-pipeline. Since statistical analysis always depends on a sufficiently big sample, the runtime optimization is an important step to be able to work with high-resolution resting-state fMRI data.

2   Material and Methods

To examine and optimize the runtime of the preprocessing algorithms, we used the Nathan Kline Institute-Rockland Sample (NKI-RS) [4]. This sample is a fully-fledged, ultra-high resolution multiband dataset consisting of 208 resting-state fMRI recordings freely available for research. It was created with the endeavor to improve research in neuro-science and psychiatry. On this dataset, the preprocessing algorithm ICA-AROMA [3] was applied. This algorithm is the newest approach to automatically remove motion artifacts caused by headmovement during the data acquisition in MRI. Before applying this algorithm, motion correction needs to be applied. Therefore, the FMRIB Software Library (FSL) from the Analysis Group of Oxford was used [5]. FSL is a "comprehensive library for analysis tools for functional, structural and diffusion MRI brain imaging data" [5]. The used functions are McFLIRT [5], BET [6] and SUSAN [7]. McFLIRT is a tool for intra- and inter-modal brain image registration. It is fully automated and considered robust and accurate. BET deletes tissue in the image, which does not belong to the brain. SUSAN is a tool for noise reduction, using nonlinear filtering. It is similiar to a smoothing algorithm, with the advantage to preserve the underlying structure of the data. The runtime data of Table 1 includes the computation time of these FSL methods. The working environment consists of a workstation, a network drive with a capacity of 250 GB per user and nine accessible servers. The specifications of the servers can be seen in Table 2.

name	cpu	ram
sfb01	48@2.2 GHz	264 GB
sfb02	48@2.2 GHz	132 GB
grid20	4@2.5 GHz	8 GB
grid22	4@2.5 GHz	8 GB
grid23	4@2.5 GHz	8 GB
grid24	4@2.67 GHz	8 GB
grid25	4@2.67 GHz	8 GB
grid26	4@2.67 GHz	8 GB
grid27	4@2.67 GHz	8 GB

Table 2: Specifications of the available servers.

sfb01 and sfb02 are dedicated to compute cpu-intensive cal-

culations and are frequently used. 208 high-quality recordings and the multiplication of the data by a factor greater than seven, leads to the fact that the network capacity is not sufficient to do all the calculation on the network drive. Theoretically, a minimum of 915 GB would be sufficient to compute and store the results of ICA-AROMA on the network drive. For that reason, a shell-framework was implemented, which allows to easily integrate an external hd drive in the process. Fig. 1 shows the workflow of the framework.

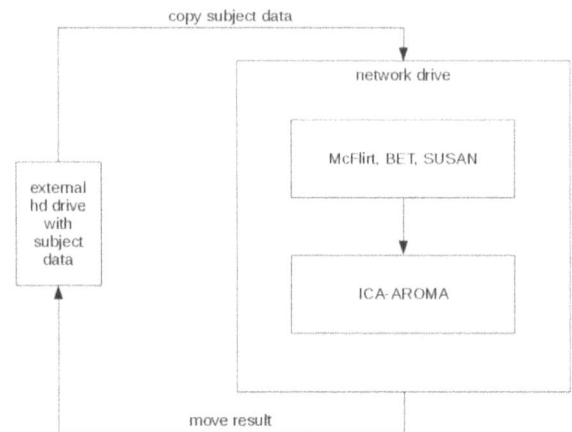

Figure 1: Conceptual visualization of one iteration with the integration of an external hd drive in the preprocessing step.

In this setup, the external hd drive contains the NKI-RS dataset with a total of 208 subjects. The framework copies one resting-state fMRI sample to a specified location on the network drive. Then, the algorithm will be applied. After the computation is done, the result will be moved back to the external hd drive and the copied data will be deleted. At this point, the network drive contains no data related to the resting-state fMRI subjects. This will be repeated, until the dataset is preprocessed. This setup solves the problem with network drive capacity. To optimize the runtime, the data needs to be computed simultaneously. To achieve that on the in Table 2 specified servers, we used the Torque Queue System for parallel computing. The Torque Queue System provides control over batch jobs and distributed computing resources. Since the subjects are independent of each other in terms of the preprocessing algorithms, our framework was easily adjustable to parallel computing. Furthermore, the I/O on the external hd drive needs to be managed. Since every process copies data from the external hd drive at the start and writes the results back at the end, this is a possible bottleneck of the system. The framework was configured, so that a maximum of one instance at a time is eligible to access the external hd drive. Since the initial copying of the data just takes a few seconds and the processes finish at different points in time, the additional idle time is negligible. With these adjustments, the computing was distributed on the servers and therefore done simultaneously. Every in Fig. 2 specified server computed a sample of 20 subjects to determine the mean runtime needed by every server to estimate the new total runtime on the full NKI-RS dataset.

3 Results and Discussion

The results received from the shell-framework are seperated into two groups: FSL-methods and ICA-AROMA. The reason is to investigate, if either of the tasks performs significantly better when computed on a specific server-group. While the sfb-servers have a great quantity of RAM, the grid-server's CPUs are clocked higher by 0.3 to 0.47 GHz. Table 3 displays the combined runtime for McFLIRT, BET and SUSAN on every machine.

host	McFLIRT, BET and SUSAN		
	min	max	∅
sfb01	56m 03s	64m 04s	59m 20s
sfb02	57m 20s	63m 25s	59m 54s
grid20	27m 49s	28m 07s	27m 56s
grid22	28m 22s	36m 15s	32m 19s
grid23	28m 10s	35m 32s	31m 38s
grid24	27m 04s	33m 57s	30m 02s
grid25	27m 15s	33m 07s	29m 15s
grid26	27m 04s	34m 25s	29m 29s
grid27	27m 00s	33m 41s	29m 30s
sfb	56m 03s	64m 04s	59m 36s
grid	27m 04s	36m 15s	30m 01s
all	27m 04s	64m 04s	36m 06s

Table 3: Runtime data for the FSL-functions of the available servers with a sample size of n = 20.

The average computation time to finish the task was 36 minutes and 6 seconds. For nine parallel executions, this results in approximately 4 minutes computation time per subject. While, compared to the grid-servers, the sfb-servers took on average double the amount of time to finish the task. One may conclude, that the CPU is the determining factor in this observation. Despite comparable specifications, the local workstation is, with 20 minutes computation time, significantly faster than any of the grid-servers. Within the different server-groups, the results were homogeneous and the span of the computation times was relatively small. This is to be expected from the grid-servers, since they all share similar specifications. Although it is a little surprising, that the lower-tier grid20-server was the fastest among them. The sfb-servers however have a large difference in their RAM-specification: 256 GB for sfb01 compared to 128 GB for sfb02. This leads to the conclusion, that the requirements for these methods are met by every server and the difference in computation time has another reason. One possible explanation is the usage of the servers. While the sfb-servers are on average occupied with a minimum of 8 to 10 jobs, the grid-servers have rarely any jobs running on them besides their standard network-tasks. The local workstation is the least occupied. As mentioned in Section 1, the size of the data increases during the execution by a factor greater then seven, which results in a lot of I/O-processes. If the server is already occupied with other I/O-tasks, this may lead to a bottleneck. This assumption is supported by

Table 4. It contains the runtime data for the ICA-AROMA step, which shows similiar results.

host	ICA-AROMA		
	min	max	∅
sfb01	191m 07s	238m 09s	211m 25s
sfb02	183m 37s	215m 20s	200m 49s
grid20	115m 19s	126m 33s	121m 00s
grid22	131m 32s	148m 47s	139m 11s
grid23	129m 00s	155m 47s	138m 19s
grid24	124m 43s	148m 55s	134m 36s
grid25	119m 29s	143m 22s	130m 41s
grid26	123m 35s	148m 35s	134m 11s
grid27	122m 55s	147m 49s	132m 35s
sfb	183m 37s	238m 09s	206m 07s
grid	115m 19s	155m 47s	140m 05s
all	115m 19s	238m 09s	149m 12s

Table 4: Runtime data for ICA-AROMA of the available servers with a sample size of n = 20.

Again, the computation time seems to be mainly dependend on the existing I/O traffic. sfb02 performed on average better than sfb01, despite having only half as much RAM. The same observation was made for the grid20-server. This server performed the best among the grid-cluster despite having less cpu power than grid24 and upwards. Additionally, every server performs significantly worse than the local workstation, which shares nearly identical specifications with the grid-servers. For nine parallel executions with an average computation time of 149 minutes and 12 seconds, one instance of ICA-AROMA is calculated in approximately 16 minutes and 35 seconds. Assumed that every server preprocesses one subject of the NKI-RS dataset at a time, one subject will take 20 minutes and 35 seconds to finish. Hence, the new theoretical computation time for the full dataset is 71 hours, 21 minutes and 20 seconds. This can be further improved by computing two instances of ICA-AROMA per server simultaneously. Tests with two instances per server showed, that the computation time for the sfbs stayed the same, while the grid-servers were slowed down by approximately 20%. This observation confirms the assumption again, that the computation time mainly depends on the existing I/O. The new per-subject-time is 12 minutes and 6 seconds, which is sufficient to compute all subjects within 42 hours, which is 12.5% of the time needed, if the subjects were computed consecutively on the local workstation. The comparison of the per-subject execution times can be seen in Fig. 2.

4 Conclusion

Exorbitantly large computation times are both expensive and restricting. They can slow down the research process significantly and limit the possibility to implement new approaches. The estimated 336 hours of computation time do not even include other preprocessing algorithms, which

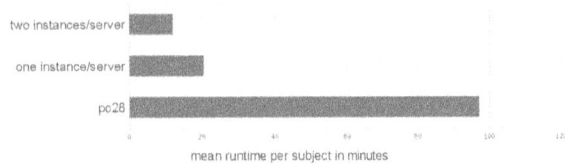

Figure 2: Comparison of the per-subject mean execution times between consecutive computation on the local workstation pc28 and parallel execution with one/two subject(s) per server.

are also crucial to create meaningful resting-state fMRI data and are applied after ICA-AROMA. Reducing the runtime of the ICA-AROMA algorithm to 12.5% of its native time was a crucial step to overcome these problems. The shift from local to network computing has even more advantages. The local workstation was strongly occupied while the calculations where executed. Using cpu-intensive tasks on the local workstation while it executed ICA-AROMA even further slows down the process. Using our framework, this resource is completely available for other tasks and does not interfere with the preprocessing anymore. Since the shell-framework is programmed in a general-purpose-way, it can be used for different tasks on all kinds of data. This feature makes it valuable for future projects with similar problems. With this framework, a long-term, general-purpose, cost- and resource-efficient solution was implemented which allows researchers to compute their data independently of their local resources in an efficient manner.

Acknowledgement

The work has been carried out at the Institute for Neuro- and Bioinformatics, Universität zu Lübeck.

5   References

[1] Huettel, S. A., Song, A. W., McCarthy, G. (2008). *Functional Magnetic Resonance Imaging, Second Edition*, Sunderland, Mass: Sinauer Associates.

[2] Cole, D. M., Smith, S. M., Beckmann, C. F. (2010). Advances and pitfalls in the analysis and interpretation of resting-state FMRI data. *Front. Syst. Neuroscience* 4:8. doi:10.3389/fnsys.2010.00008

[3] Pruim, R. H. R., Mennnes, M., van Rooji, D., Llera Arenas, A., Buitelaar, J. K., Beckmann, C. F. (2014). ICA-AROMA: A robust ICA-based strategy for removing motion artifact from fMRI data. *Submitted for Publication.* doi:10.1016/j.neuroimage.2015.02.064

[4] Nooner, K.B., Colcombe, S.J., Tobe, R.H., Mennes, M., Benedict, M. M., Moreno, A.L., ... Milham, M.P. (2012). The NKI-Rockland Sample: A Model for Accelerating the Pace of Discovery Science in Psychiatry. *Frontiers in Neuroscience*, 6 (October), 152. doi:10.3389/fnins.2012.00152

[5] Jenkinson, M., Bannister, P., Brady, M., Smith, S. (2002). Improved optimization for the robust and accurate linear registration and motion correction of brain images. *Neuroscience, 17(2)*, 825-841. doi:10.1016/S1053-8119(02)91132-8

[6] Smith, S. M. (2002). Fast robust automated brain extraction. *Human Brain Mapping*, 17(3):143-155, November 2002.

[7] Smith, S. M., Brady, J. M. (1997). SUSAN - a new approach to low level image processing. *International Journal of Computer Vision*, 23(1):45-78, May 1997.

Design of a Motion Phantom for Magnetic Resonance Imaging

M. Schmiedel [1], A. Möller [2], M. A. Koch [3], and A. Mertins [2]

[1] Medizinische Ingenieurwissenschaft, Universität zu Lübeck, max.schmiedel@student.uni-luebeck.de
[2] Institute for Signal Processing, Universität zu Lübeck, {moeller,mertins}@isip.uni-luebeck.de
[3] Institute of Medical Engineering, Universität zu Lübeck, koch@imt.uni-luebeck.de

Abstract

Even today, dealing with motion artifacts in magnetic resonance imaging (MRI) is a challenging task. Image corruption due to spontaneous body motion complicates diagnosis. In this work, an MRI phantom for rigid motion is presented. It is used to generate motion-corrupted data, which can serve for evaluation of blind motion compensation algorithms. In contrast to commercially available MRI motion phantoms, the presented setup works on small animal MRI systems. Furthermore, retrospective gating is performed on the data, which can be used as a reference for novel motion compensation approaches. It is shown that retrospective gating based on motor trigger signals can be performed and motion-corrected images can be reconstructed using the proposed setup. Moreover, the importance of preprocessing the MRI raw data, e.g. phase-drift correction, is demonstrated. The gained knowledge can be used to design an MRI phantom for elastic motion.

1 Introduction

Magnetic resonance imaging (MRI) is one of the most important imaging modalities in clinical routine today. Especially, the superior tissue contrast and the absence of ionizing radiation make MRI a vital tool for diagnosis. These upsides come with a longer time for image acquisition compared to computed tomography, which makes MRI particularly susceptible to motion artifacts [1]. However, in particular in thorax imaging, there is body motion due to respiration and cardiac motion.

Breath holding is one way to avoid respiratory motion artifacts [2] but it is inconvenient and not suitable for all patients and imaging schemes. This and cardiac motion, which is inevitable, make it necessary to deal with the problem of motion artifacts on the level of data acquisition or data processing. Ways to do so are prospective and retrospective gating. In both cases additional information to the complex k-space data, is needed. It is desirable that motion corruption could be estimated just with the complex MRI raw data without additional measurements or trigger signals. One approach for such a blind motion compensation algorithm is presented in [3]. To test this blind motion compensation algorithm on real reproducible measurement data, a motion phantom needs to be utilized. Even though, there are commercially available MRI motion phantoms, none of them can be used in a small animal MRI system.

In this work, an MRI phantom for rigid motion is presented, which can be used in a small animal MRI system. Furthermore, results of motion corrupted imaging and retrospective gating are displayed.

2 Material and Methods

For all experiments, an 1 T ICON™ MRI system (Bruker, Ettlingen, Germany) was used. It is designed for imaging small rodents and is based on a permanent magnet. The gradient system provides gradients up to 360-390 mT/m. All data were acquired with a mouse body transmit-receive coil [4].

2.1 Imaging Sequence

For all images, data were acquired using a fast low angle shot sequence (FLASH) [5]. FLASH is a gradient echo sequence and uses an equilibrium state of the longitudinal magnetization for fast data acquisition [5]. Therefore, the repetition times are chosen much shorter than the T1 relaxation times of the imaged object. The remaining transversal magnetization is spoiled by RF spoiling. For excitation, a flip angle smaller than $90°$ is used, so that after each excitation, sufficient longitudinal magnetization remains for excitation by the following RF pulse. After a few excitations an equilibrium state of the longitudinal magnetization is reached and a constant signal strength is available. For a good signal-to-noise ratio (SNR) a strong signal with a high echo amplitude is needed. The echo amplitude for basic gradient echo sequences like FLASH can be described as

$$A_E = M_z^0 \frac{1 - e^{-T_R/T_1}}{1 - \cos(\alpha) \cdot e^{-T_R/T_1}} \sin(\alpha) e^{-T_E/T_2^*} \quad (1)$$

with M_z^0 the longitudinal magnetization in thermal equilibrium, T_R the repetition time, T_E the echo time, α the flip angle, T_1 the longitudinal relaxation time, and T_2^* the transverse relaxation time as shown in [6].

In this work, repetition times of 50 ms, echo times of 6 ms and flip angles of 15° were used. The direction of object motion was chosen as phase-encoding direction. For periodic motion, this may cause motion artifacts if

$$T_{acq} > \frac{1}{2mf_0} \tag{2}$$

with T_{acq} the acquisition time for all phase-encoding steps, the mth harmonic of the periodic movement and f_0 the frequency of the object motion [6]. For the first harmonic and the chosen 0.43 Hz motion frequency, (2) limits the acquisition time to 1.16 s if motion artifacts are to be avoided. With 3.2 s acquisition time was chosen in a way that motion artifacts will appear.

2.2 Phase Drift Correction

In case of retrospective gating, one have to deal with longer times for imaging, since multiple repetitions are needed. For this work, a data acquisition time of 10 min for 200 repetitions was needed. In these dimensions, phase-drifts become relevant. Over time the total transverse magnetization drifts in phase but stays constant in magnitude for the same time after excitation. The oscillating signal here in complex notation can be described as

$$S(t) = S_0 \cdot e^{i(\omega_{rf}t + \phi)} \tag{3}$$

with S_0 being the magnitude of the signal, ω_{rf} the precession frequency and ϕ the initial phase. For short imaging times, ϕ can be assumed as constant. However, over longer times, ϕ is not constant and results in a translation in image space. For combining data which is spread over several minutes in time, as needed for retrospective gating, a phase correction has to be performed as a phase subtraction so that

$$\phi_{in} = \phi_{ac}(t) - \phi_{co}(t) \tag{4}$$

with ϕ_{in} the phase for the initial repetition, $\phi_{ac}(t)$ the actual phase for a chosen repetition and $\phi_{co}(t)$ the phase-correction value. Therefore, the sampled data was corrected by multiplication with a time dependent term

$$e^{i(-\phi_{co}(t))}. \tag{5}$$

The k-space for every repetition was corrected using one specific $\phi_{co}(t)$. Therefore, (4) becomes discrete and dependent on the number of repetition. The repetition-dependent $\phi_{ac}(t)$ is determined by plotting the phase for all phase-encoding steps with the same number and the same motor position. The result is fitted with a 4th degree polynomial. The fitting is shown in Fig. 1.

If the phase-drift remains uncorrected, wave-like artifacts appear due to not corresponding phase values of the combined data of different repetitions. This effect is shown subsequently in the results section.

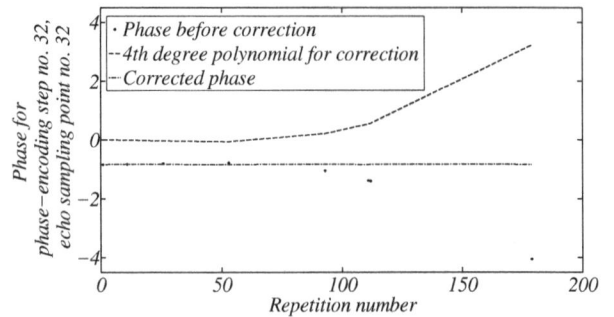

Figure 1: Drift corrupted phase was fitted with a 4th degree polynomial and the result was used to get a constant phase for all repetitions

2.3 Motion Phantom

The motion phantom is designed in a way that rigid object motion can be imaged with an MRI system.

It consists of an inner part, placed inside the MRI system, and an outer part, placed right next to the MRI system. The outer part consist of an elctro motor RE50 (maxon motor GmbH, München, Germany) and a translation stage. It is used to generate a force to deform the inner part of the phantom. Both parts are connected via a polyethylene line. The electro motor of the outer part is controlled using a laptop and a C++ script. It rotates periodically and the rotation is transferred to a translation stage. This results in the periodical movement shown in Fig. 3. The movement is transferred to the inner phantom part via the line and causes a translational relative movement between a small diameter test tube (length/inner diameter/outer diameter: 100/20/22mm) and a fixed large diameter test tube (length/inner diameter/outer diameter: 90/13/18mm). To reverse the translation of the inner phantom part, an elastic band is used. The imaged object, an agarose cylinder, is placed in the small diameter test tube. The agarose has a concentration of 0.75% and is prepared under laboratory conditions, cooking agarose powder (Natura, Hannover, Germany) in water for five minutes. A schematic overview of the phantom is shown in Fig. 2.

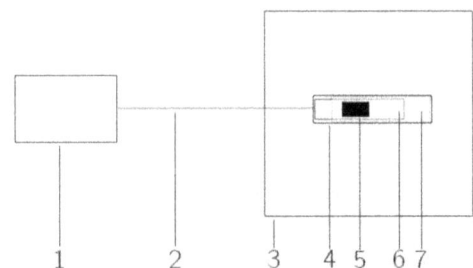

Figure 2: Overview of the motion phantom: (1) translation stage/motor, (2) line, (3) MRI system, (4) elastic band, (5) agarose cylinder, (6) moved test tube, (7) fixed test tube.

2.4 Triggering for Retrospective Gating

For retrospective gating, trigger signals of the phantom and the MRI system are needed to assign the measured data to the corresponding phantom motion status. The trigger signals are rectangular voltage changes and denoted as pulses in the following.

The motor control sends a trigger signal with a pulse width of 100 μs every 27° of motor rotation which corresponds to a linear movement of the motion phantom of 75 μm. With these pulses the motion of the phantom can be approximated. Therefore, the movement between two trigger signals is assumed to be linear. The result is shown in Fig. 3.

The MRI system sends a trigger signal with a pulse width of 1 ms before every phase-coding step. The time for frequency coding is approximated as two times the echo time. With the used motor rotation speed of 100 rpm and an echo time of 6 ms this results in an object motion of 10 μm during frequency coding. Compared to a voxel size of 781.25 μm in the direction of movement, the motion during frequency encoding is considered negligible.

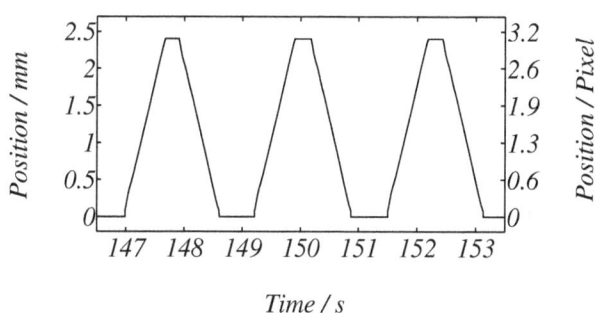

Figure 3: Section of the assumed movement of the translation stage. The movement between two trigger signals was approximated as linear.

The trigger pulses are recorded using a data-acquisition card UR22 (Steinberg,Hamburg, Germany) and the audio recording software Audacity (version 2.1.2) with a sample rate of 44.1 kHz.

2.5 Image Reconstruction and Retrospective Gating

During data acquisition, the k-space is fully Cartesian-sampled but the contained data is phase-drift- and motion-corrupted. Therefore, the phase-drift correction has to be carried out prior to reconstruction.

To show the influence of object motion on the reconstructed images, a 2D inverse fast Fourier transform is applied to the k-space data.In case of the retrospectively gated images, before applying a 2D inverse fast Fourier transform, data binning was executed as follows.

The motor position was discretized in intervals with an interval width of 0.2 mm. With a total moving distance of 2.5 mm this makes 13 bins for the gating process. Now the data

for individual k-space lines are sorted into bins according to the motor position at the time of their echo middle. All bins contain data for all rows of k-space. The difference between data of different repetitions of a specified phase-encoding step in one bin can be considered negligible, so the repetition to be used can be chosen randomly within one bin.

3 Results and Discussion

As it can be seen in Fig. 4a, image reconstruction for an static agarose cylinder was performed successfully. With the chosen imaging parameters, an SNR of 32 could be achieved. It was calculated manually by the ratio of mean object values and mean background values.

In Fig. 4b, one out of a series of images which were corrupted due to object motion can be seen. The mean square error (MSE) between the motion-corrupted image and the image without motion amounts to 140. Compared to a mean signal value in the region of interest of 150, the MSE indicates artifacts in the motion-corrupted image along the phase-encoding direction. Visual examination reveals that there are noticeable artifacts which can be recognized in the direction of motion. The object appears to be shorter in the motion-corrupted image compared to the image without motion.

3.1 Gating

The results of the gating process can be seen in Fig. 4d-f. The movement was split in 13 bins for gating and the positions 0 mm, 1 mm and 2 mm were chosen for visualization. The position 0 mm corresponds to a pause in the periodic movement which can be seen in Fig. 3. Therefore the gated image for position 0 mm, should be the same as the image with no object motion for a perfect gating process and as a consequence, the MSE between these two images should be zero. In fact the MSE amounts to 39, despite that, visually there can not be identified much difference. The reason for that is, that the mean value in the region of interest in the gated image is 12 less than in the motionless one. That is caused by the not entirely perfect equilibrium state of the longitudinal magnetization. The motionless image was acquired with one repetition, whereas the gated image consists of the first seven repetitions. An observation of the raw data revealed that the signal at the beginning of a data acquisition is slightly higher than in the following repetitions. Therefore, the gating for this position can be considered successful despite the MSE of 39.

For perfect gating, the image at position 1 mm should be a shifted version of Fig. 4a and 4c. Because of that the image (pos. 1 mm) was manipulated applying a shift in the opposite direction and than the MSE to the image with no motion was calculated. The result is an MSE of 62, which is much less than that between the motion-corrupted image and the image in rest, but more than the MSE between the gated image (0 mm) and the image in rest. The fact of the smaller error reveals that a correction has been performed,

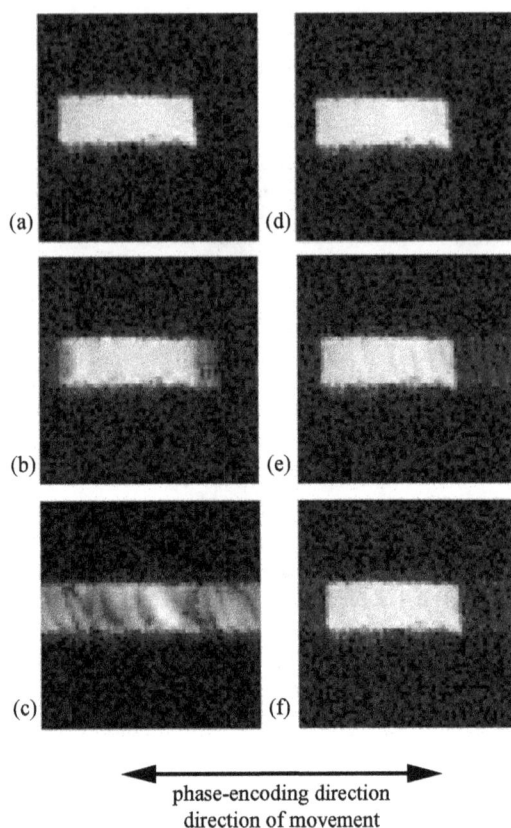

Figure 4: Shown are images (64×64) of the agarose cylinder: (a) no motion (b) periodical movement (c) gated with periodical movement and without phase correction (d) gated with periodical movement and phase correction (pos. 0mm) (e) gated with periodical movement and phase correction (pos. 1mm) (f) gated with periodical movement and phase correction (pos. 2mm).

but it seems as it has been less successful than in the case for the position 0 mm. This is probably caused by the phase-drift. Even though a phase-drift correction was performed, which had a big impact that can be recognized by comparing Fig. 4d-f with Fig. 4c, it seems that there is still data corruption left. Experiments with the data that is chosen for the gating process revealed that the gating result becomes better if the chosen data belongs to successive repetitions (data not shown).

For the gated image at position 2 mm, the results are similar to the position 1 mm. The image was also shifted and then an MSE of 72 was calculated. The MSE is slightly higher than in the 1 mm case. The reason for that is probably that the low-frequency components, which have a big impact on the gray-value components in the reconstructed image, could be less well combined due to phase-drift corruption in the 2 mm case than in the 1 mm case.

3.2 Motion Phantom

Analysis of the gated images revealed that there was an object motion of about 3 pixels. With a voxel width in the direction of movement of 0.78 mm this corresponds to 2.3

mm of object motion. Therefore, it can be assumed that almost the full movement of the translation stage could be transferred to the inner phantom. This validates the chosen phantom design.

The agarose cylinder was successfully used as a signal source, but its edges appeared not to be smooth. This was caused by the chosen way of extracting a cylinder out of an agarose block.

4   Conclusion

It was shown that retrospective gating based on motor trigger signals can be performed. The preprocessing of the raw data, e.g. phase-drift correction, needs to be improved. This processing step is key to the quality of the gating results. In this work, rigid object motion was implemented. In further examinations, the gating process as implemented needs to be examined with elastic motion and therefore the motion phantom has to be modified. If successful, the data can be used as reference for blind motion compensation algorithms as presented in [3].

Acknowledgement

The work has been carried out at the Institute for Signal Processing, Universität zu Lübeck. All MRI measurements has been carried out at the Institute of Medical Engineering, Universität zu Lübeck. I would particularly like to thank Christian Clauß, Christian Kowalski and Richard Schäfer for the contentual input.

5   References

[1] M. A. Tavallaei, P. M. Johnson, J. Liu, and M. Drangova, *Design and evaluation of an MRI-compatible linear motion stage*. Medical Physics 43, American Association of Physicists in Medicine, Alexandria, pp. 62-71, 2016

[2] M. Zaitsev, P. Mclaren, and M. Herbst, *Motion Artifacts in MRI: A Complex Problem With Many Partial Solutions*. Journal of Magnetic Resonance Imaging 42, International Society for Magnetic Resonance in Medicine, Concord, pp. 887-901, 2015

[3] A. Möller, M. Maaß, and A. Mertins, *Blind Sparse Motion MRI with Linear Subpixel Interpolation*. Bildverarbeitung für die Medizin 2015, pp. 510-515, 2015

[4] Bruker BioSpin MRI GmbH, *Icon:System Manual*. Document No: T02985_02_02, 2012.

[5] W. R. Nitz, V. R. Runge, S. H. Schmeets, W. Faulkner, and N. K. Desai, *Praxiskurs MRT*. Georg Thieme Verlag KG, Stuttgart, pp. 58-59, 2007.

[6] Z.-P Liang, and P. C. Lauterbur, *Principles of Magnetic Resonance Imaging*. IEEE Press, New York, pp. 226-267, 2000.

SliceTracker: An open-source platform for prostate motion tracking during in-bore targeted MRI-guided biopsy

P. A. Behringer [1], C. Herz [2], T. Penzkofer [3], K. Tuncali [2], C. M. Tempany [2] and A. Fedorov [2]

[1] Medizinische Ingenieurwissenschaft, Universität zu Lübeck, peter.behringer@student.uni-luebeck.de
[2] Radiology, Brigham and Women's Hospital, Harvard Medical School, {cherz,ktuncali,ctempany,fedorov}@bwh.harvard.edu
[3] Institute of Radiology, Charité Universitätsmedizin Berlin, tobias.penzkofer@charite.de

Abstract

In-bore targeted Magnetic Resonance (MR)-guided Prostate Biopsy is a promising approach that could overcome high false-negative occurring in Transrectal Ultrasound (TRUS) biopsy. Tracking of the biopsy targets throughout the biopsy procedure remains challenging for clinical staff, but is crucial for biopsy accuracy. We present an open-source software to support intra-procedural tracking of the prostate and biopsy targets using B-Spline deformable image registration. The software is implemented in the widely used 3D Slicer. Our evaluation include validation of registration accuracy, computational time and sensitivity to inter-user variability. We retrospectively evaluated data from 25 clinical in-bore MR-guided prostate biopsy cases with a total of 343 registrations. We observed improvement of Dice similarity coefficient (DSC) by 0.17 ($p<0.0001$, range 0.02-0.48) on average. Computation time was 17.83 ± 6.98 sec (range 5.22-43.92). Our approach is not sensitive towards different operators. In conclusion, we present a functional software tool that is ready for prospective evaluation.

1 Introduction

Prostate Cancer (PCa) remains a leading cause of cancer mortality in the USA and worldwide [1]. In Prostate Cancer Diagnoses, characterization of tumor aggressivness relies on the analysis of histopathological tissue samples, which are taken from the patient during a prostate biopsy [2]. Traditionally, prostate biospy is performed using the non-targeted transrectal ultrasonography (TRUS)-guided biopsy. During the TRUS biopsy, the biopsy targets are randomly taken by cutting out tissue with a pre-defined needle pattern. Several studies showed that the classic TRUS biopsy found only 77-80% of cancers after the first TRUS exam [4]. This high false-negative ratio leads to repeated biopsy procedures, frustrating both patients and clinicians [5].

In the recent years, targeted sampling of suspicious regions has emerged as an effective alternative to the traditional sextant biopsy [3]. Such targeted approaches require multiparametric MRI (mpMRI) exam for localizing suspicious regions, which are then re-identified by means of image registration in the intra-procedural imaging. However, prostate motion during the course of biopsy, which can last for over an hour, can complicate localization of the suspected lesion. Therefore, tracking of the pre-defined target location is required to enable accurate targeting.

In this paper, we present an open-source platform to facilitate re-identification of the cancer targets and their tracking throughout the course of the procedure. In order to spatially correlate biopsy targets and the intra-procedural scan, we use deformable image registration to compensate for the high deformation of the prostate that occurs when an endorectal coil is used during the pre-procedural imaging. Applying deformable registration and visually evaluate the results is a time-consuming and complex task, requiring specialized expertise and remains challenging for clinical staff. Our contribution is the development and integration of a software solution to support in-bore MR-guided prostate biopsy procedures, developed for clinical operator. The software can be used as a module (plugin) for the widely used 3D Slicer (version 4).

2 Material and Methods

We first present the overall setup and clinical workflow of the targeted in-bore MRI-guided prostate biopsy to establish the requirements for the software development. We follow with the description of our approach to the development of the platform. Finally, we present our evaluation approach, which is concerned with the accuracy, consistency and computational performance of the image registration.

2.1 In-bore Targeted MRI-guided Prostate Biopsy

The In-bore transperineal targeted MRI-guided biopsy procedure can be devided into two stages. First, mpMRI is acquired prior to the procedure and the cancer suspicious targets are localized using 3D Slicer. Segmentation of the prostate gland and peripheral zone is created on the T2-weighted image. During the procedure, the patient is immobilized and sedated on the movable table top. Imaging involved two types of imaging sequences: (1) axial T2w MRI planning scan (voxel size $0.5 \times 0.5 \times 3$ mm, imaging time 4 min) obtained in the beginning of the biopsy procedure for the purposes of target identification and (2) a series of lower resolution T2w MRI needle confirmation images (voxel size $0.75 \times 0.75 \times 3$ mm, imaging time ~ 1 min) collected after needle placement to visually assess targeting accuracy.

2.2 Image Registration

The purpose of image registration is to track the tumor suspicious regions throughout the procedure, so that the interventionalist can re-identify the cancer targets in the intra-procedural images. Therefore, image registration is the most critical component of the workflow, since its result may have direct effect on the accuracy of biopsy sampling. Our custom deformable image registration strategy requires limited user interaction and is based on earlier developed methodology [6],[9]. Segmentation of the prostate gland is created manually in the T2w scan (planning scan). During the course of the procedure, subsequent needle confirmation images are registered automatically. Registration step is implemented in the BRAINSFit [10] module of 3D Slicer, that uses a hierarchical approach which includes 6, 9, and 12 degrees of freedom transformations, followed by B-spline deformable transformation, with mutual information as the similarity metric and is based on ITK[1][7]. In the case of registration failure due to large prostate motion, software workflow allows for manual segmentation of the prostate in the current needle confirmation image.

2.3 Description of the Software

Fig. 2 shows the clinical workflow implemented in our software. As a pre-procedural step, the patient is selected from the database and the diagnostic data can be reviewed in order to confirm target positions. Data connection between the research workstation and the clinical workstation is established in order to receive intra-procedural DICOM data. After the first planning scan is received, a coarse manual segmentation of the prostate gland is prepared using a semi-automated procedure (this segmentation is used for the initialization of the registration algorithm and does not need to be very accurate [6]). Upon completion of the registration, registered images and pre-procedural biopsy targets are examined side-by-side with the capability to switch between different registration stages. Intra-procedural registration and evaluation are applied every time a needle confirmation scan is received.

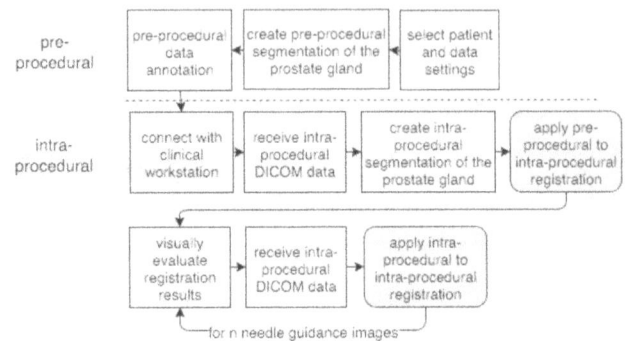

Figure 1: Steps of the procedural workflow. The developed software platform provides support for the intra-procedural phase of the workflow.

2.4 Evaluation

Our evaluation consists of two components. First, we evaluated registration accuracy and computational performance. Registration quality was first evaluated using visual assessment for each image pair, since annotation of images for quantitative assessment would be very time-consuming due to the large number of images. Quantitative evaluation was done using Dice Similarity Coefficient (DSC) between the manual segmentation of the high-resolution T2w image propagated to the last needle confirmation image through the chain of registration transformations, and the manual segmentation of the last needle image. The segmentations used for DSC assessment were prepared by an expert radiologist with specialization in abdominal imaging, and were not used in the registration process. Second, we evaluated sensitivity of registration to the variability in segmentation of the prostate gland. This was done by comparing registration results performed by two readers with different level of training. Neither of the readers had medical training. First reader had multi-year hands-on experience in prostate gland contouring, while the second reader had a brief training and no prior experience. Evaluation was done retrospectively using datasets collected during clinical MR-guided prostate biopsy procedures.

3 Results and Discussion

The software was implemented as a module within the 3D Slicer extension SlicerProstate[2], which provides a collection of modules to facilitate (1) processing and management of prostate image data and (2) utilizing prostate images in image-guided interventions. Functionality and user interface are not discussed here due to the lack of space, and demonstrated in the videos available online[3]. A total number of 343 needle confirmation images from N=25 clinical cases with a median of 15 needle confirmation images (range 2-26) were used in the evaluation of the registration functionality.

---

[1]https://www.itk.org

[2]https://github.com/SlicerProstate
[3]https://vimeo.com/131041774

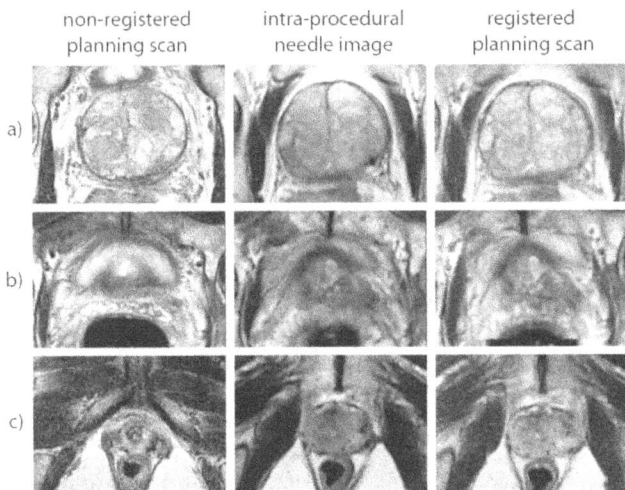

Figure 2: Example cases for demonstrating registration quality. Left: planning scan without registration. Middle: Intra-procedural needle image. Right: Result of our registration approach. Rows a), b) and c) show three different biopsy cases as representative examples for excellent, moderate, and poor registration quality.

In 8 clinical cases large prostate motion for at least one of the needle confirmation images caused registration failure and manual segmentation of the needle confirmation image was required (i.e., in 15 out of 343 needle confirmation images). Improved alignment of the prostate gland between the registered planning scan and the needle confirmation images was confirmed visually for all 343 registrations. We provide an interactive website[4] that can be used to visually assess the registration quality for each of the image pairs. Registration quality was characterized as excellent in 226 images, moderate in 108 images, and poor in 9 images by criteria such as alignment of inner landmarks, alignment of the outer prostate contour and amplitude of irregularities. Fig. 2 demonstrates examples for each registration quality in row a), b) and c) respectively. Fig. 3a shows assessment of the gland segmentation overlap before and after registration for the final needle confirmation image. We observed improved DSC in all 25 cases, with an average improvement by 0.17 (p<0.0001) and a range from 0.02 (Case 22) to 0.48 (Case 28). Fig. 3b shows the summary of gland segmentation overlap (DSC) before and after registration for the final needle confirmation image using two different sets of non-expert segmentations with different levels of training (average difference in DSC of 0.01 and maximum difference of 0.06 (p>0.05)). Mean $\pm$ standard deviation (SD) of the computation time for registration of one needle confirmation image was 17.83 ± 6.98 sec (range 5.22-43.92), which is compatible with the clinical constraints of the workflow. Fig. 3c shows the distribution of computation times across 25 cases. DSC improved from 0.68 ± 0.13 (range 0.31-0.89) before registration to 0.84 ± 0.06 (range 0.67-0.93) for ITKv4.

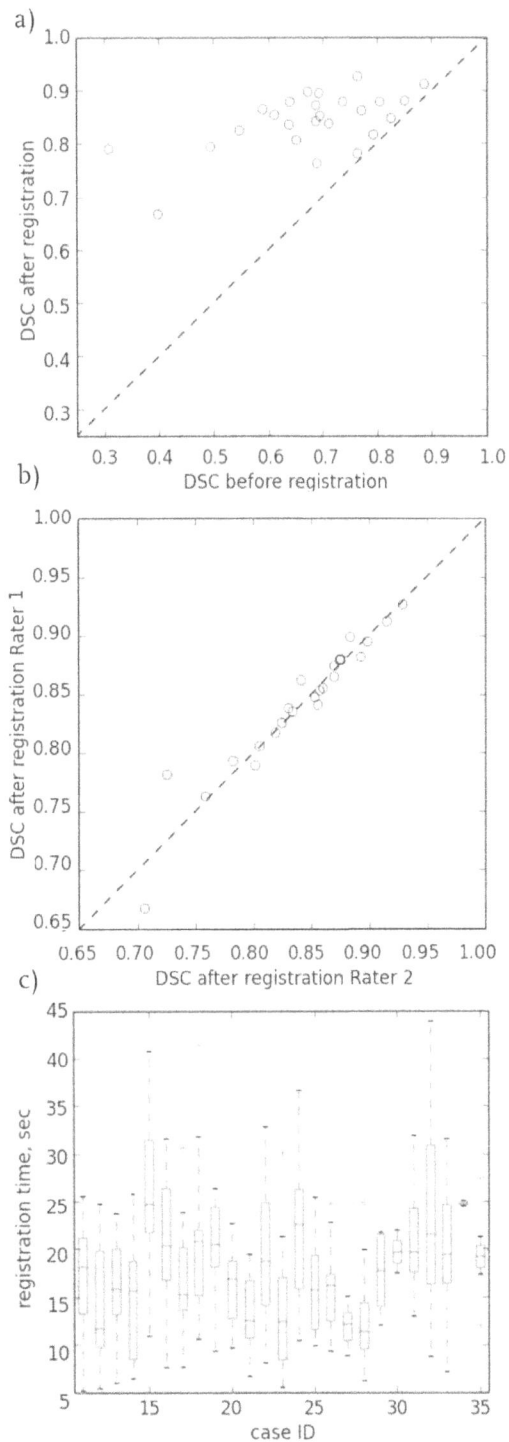

Figure 3: a) Summary of gland segmentation overlap (DSC) before and after registration for the final needle confirmation image. b) Gland segmentation overlap (DSC) after registration for the final needle confirmation image comparing two sets of non-expert segmentations with different levels of training. c) Summary of the computational time for registration of all needle confirmation images across all 25 cases using ITKv4. Median, lower and upper quartiles (bottom and top 25% of the data) and the extreme values within the $1.5 \times$ interquartile range)

---

[4] http://slicerprostate.github.io/ProstateMotionStudy

4   Conclusion

The study was motivated by the practical need for a tool to support intra-procedural biopsy workflow and enable motion compensation for cancer suspicious target tracking, addressing clinical users. We presented an open-source end-user solution, implemented as an extension to the widely used 3D Slicer software, for intra-procedural tracking of the prostate gland and the biopsy targets throughout the procedure. We evaluated the underlying registration approach and demonstrated an improvement of gland alignment in all 25 cases. Our evaluation showed that the registration approach is not sensitive to the differences in initialization due to the variability in segmentation of the prostate gland by different readers. We note however, that registration results with ITK version 4 showed moderate to high irregularities in a subset of approximately 28 of 343 images (see Fig.2, right column, c): upper right side of prostate gland), that did not appear in the later version ITK version 3. We could reduce those irregularities by using a dilated version of the propagated mask for deformable registration phase within our ITKv4 implementation. Evaluation of registration results is always a challenging problem. The two commonly used approaches rely on evaluation of the overlap of the segmented structures, captured by DSC or a similar metric, and on the Landmark Registration Error (LRE). It has been recognized that the use of structure overlap may not characterize the performance of a registration method well [8]. Manual annotation of the images with anatomical landmarks is time-consuming (especially when hundreds of images need to be annotated), and is particularly difficult in the prostate that has a limited number of salient points. Neither DSC nor LRE can capture unrealistic or inaccurate deformations within the outlined regions or in the areas with no landmarks. These observations motivated us to develop an online resource that enables visualization of the registration results for each of the 343 registrations. Improvement in DSC of the total gland segmentation is large: from 0.49 to 0.79. Our study has several limitations. Although design of the software was performed in coordination with the target clinical users, we have not evaluated it prospectively during biopsy procedures. Our evaluation was limited to the intra-procedural motion compensation step. As with any open-source software, the functionality will be refined in the course of its applications in clinical trials. In conclusion, we presented a functional software tool to support intra-procedural prostate biopsy intended for clinical users. Prospective evaluation during clinical research MRI-guided prostate biopsy procedures under the guidance of a clinical operator is warranted. Although the motivating application for this development was prostate biopsy, we aim to investigate other use cases to make the software more generic for other procedures that require intra-procedural target tracking and motion compensation.

Acknowledgement

The work has been carried out at the Surgical Planning Laboratory (SPL) at the Brigham Women's Hospital in Boston. This work was supported in part by the National Institutes of Health through grants U24 CA180918, R01 CA111288 and P41 EB015898.

5   References

[1] P. Boyle, B. Levin et al., *World Cancer Report 2008.* in: IARC Press, International Agency for Research on Cancer, 2008.

[2] O. Andrén, K. Fall, L. Franzén, S.-O. Andersson, J.-E. Johansson and M. A. Rubin *How well does the Gleason Score predict Prostate Cancer Death? A 20-year Follow up of a Population Based Cohort in Sweden.* in: J. Urol, vol. 175, pp. 1337-1340, 2006.

[3] N. B. Delongchamps, M. Peyromaure, A. Schull, F. Beuvon, N. Bouazza et al. *Prebiopsy Magnetic Resonance Imaging and Prostate Cancer Detection: Comparison of Random and Targeted Biopsies.* in: J. Urol, vol. 189, pp. 493-499, 2013.

[4] K. A. Roehl, J. A. V. Antenor and W. J. Catalona, *Serial Biopsy Results in Prostate Cancer Screening Study.* in: J. Urol, vol. 167, pp. 2435-2439, 2002.

[5] J. Tokuda, K. Tuncali, I. Iordachita, S.-E. Song, A. Fedorov, S. Oguro, A. Lasso et al., *In-bore Setup and Software for 3T MRI-guided Transperineal Prostate Biopsy.* in: Phys. Med. Biol., vol 57, pp. 5823-5840, 2012.

[6] A. Fedorov, K. Tuncali, F. M. Fennessy, J. Tokuda, N. Hata, W. M. Wells, R. Kikinis and C. M. Tempany, *Image Registration for Targeted MRI-guided Transperineal Prostate Biopsy.* in: J. Magn. Reson. Imaging, vol. 36, pp. 987-992, 2012.

[7] L. Ibanze, W. Schroeder, L. Ng et al., *The ITK software guide: the insight segmentation and registration toolkit.* in: Kitware, Inc., vol. 5, 2003.

[8] T. Rohlfing, *Image Similarity and Tissue Overlaps as Surrogates for Image Registration Accuracy: idely used but unreliable.* in: IEEE Trans. Med. Imaging, vol. 31, pp. 153-163, 2012.

[9] A. Fedorov, K. Tuncali, T. Penzkofer, J. Tokuda, S. Song, N. Hata and C. M. Tempany, *Quantification of Intra-procedural Gland Motion during Transperineal MRI-guided Prostate Biopsy.* in: International Society for Magnetic Resonance in Medicine (ISMRM), 21st annual meeting, Salt-Lake City, USA, 2013.

[10] H. Johnson, G. Harris, and K. Williams, *BRAINSFit: Mutual Information Rigid Registrations of Whole-Brain 3D Images, Using the Insight Toolkit.* in: Inside J., 2007

Infinite Science Publishing is a University Press and Academic Printing Incorporation. It provides a publication platform for excellent theses as well as scientific monographies and conference proceedings for reasonable costs.

These publications enable scientists and research organizations to reach the maximum attention for their results.

The service of Infinite Science Publishing comprises the entire range from the publication of print-ready documents up to cover design as well as copy-editing of single articles.

Infinite Science Publishing is an imprint of the Infinite Science GmbH, a University of Lübeck spin-off and service partner of the BioMedTec Science Campus.

www.infinite-science.de/publishing

Infinite Science GmbH
MFC 1 | BioMedTec Wissenschaftscampus
Maria-Goeppert-Str. 1, 23562 Lübeck
book@infinite-science.de

Infinite Science
Publishing

www.ingramcontent.com/pod-product-compliance
Lightning Source LLC
Chambersburg PA
CBHW081149250326
R18032300002B/R180323PG41598CBX00006B/3

9 783945 954188